Environmental Science and Engineering
Subseries: Environmental Engineering

Series Editors: U. Förstner • Robert J. Murphy • W.H. Rulkens

Jorge Marx Gómez, Michael Sonnenschein,
Martin Müller, Heinz Welsch,
Claus Rautenstrauch (Eds.)

Information Technologies in Environmental Engineering

ITEE 2007 - Third International ICSC Symposium

With 181 Figures

Springer

EDITORS:

PROF. DR. JORGE MARX GÓMEZ
PROF. DR. MICHAEL SONNENSCHEIN
PD DR. MARTIN MÜLLER
PROF. DR. HEINZ WELSCH
CARL VON OSSIETZKY UNIVERSITÄT
OLDENBURG
FAKULTÄT II -INFORMATIK, WIRTSCHAFTS-
UND RECHTSWISSENSCHAFTEN
26111 OLDENBURG,
Germany

PROF. DR. CLAUS RAUTENSTRAUCH
OTTO-VON GUERICKE-UNIVERSITÄT
MAGDEBURG
INSTITUT FÜR TECHNISCHE UND
BETRIEBLICHE INFORMATIONSSYSTEME
39106 Magdeburg
Germany

ISSN 1863-5520
ISBN 10 3-540-71334-4 **Springer Berlin Heidelberg New York**
ISBN 13 978-3-540-71334-0 **Springer Berlin Heidelberg New York**

Library of Congress Control Number: 2007922328

Springer is a part of Springer Science+Business Media
springeronline.com

Cover design: deblik, Berlin
Production: A. Oelschläger
Typesetting: Camera-ready by the Editors

Printed on acid-free paper 30/2132/AO 543210

Preface

Potentially dangerous environmental changes are happening in the atmosphere, oceans, animal habitats and places where hazardous materials are used, or have been discarded without adequate environmental protections. These increasing problems that also affect human health demand for interdisciplinary approaches where engineers, natural scientists, economists and computer scientists work together. Information technology has become significant to all scientific groups and fields involved in environmental engineering: Model based systems which enable the study of environmental changes have been developed and are being extended to manage those environments. New paradigms for designing objects to enable easy disassembly and recovery of components contribute to reuse. Web-based information systems enhance public awareness to environmental changes and allow participation in decision making. Developments in exploiting alternative energy sources are reducing dependence on non-renewable resources. Numerical economy-environment models contribute to cost-benefit analysis of environmental policy, and environmental monitoring and accounting systems facilitate market-based environmental regulation. Further advance is only possible if scientific teams have adequate experience, methods and tools for investigation of the changes in the environment. Success requires a high level of organization related to technical as well as scientific and human aspects of information handling. This book publishes the results of the ITEE 2007 conference where information about the topics above has been presented and discussed among environmental engineers, computer scientists and economists.

March 2007

Jorge Marx Gómez
Michael Sonnenschein
Martin Müller
Heinz Welsch
Claus Rautenstrauch

Editors

Prof. Dr. Jorge Marx Gómez

Carl von Ossietzky Universität Oldenburg
Fakultät für Informatik, Wirtschafts- und Rechtswissenschaften
Department für Informatik
Ammerländer Heerstr. 114-118
26129 Oldenburg
Germany

jorge.marx.gomez@uni-oldenburg.de

Prof. Dr. Michael Sonnenschein

Carl von Ossietzky Universität Oldenburg
Fakultät für Informatik, Wirtschafts- und Rechtswissenschaften
Department für Informatik
Ammerländer Heerstr. 114-118
26129 Oldenburg
Germany

michael.sonnenschein@informatik.uni-oldenburg.de

PD Dr. Martin Müller

Carl von Ossietzky Universität Oldenburg
Fakultät für Informatik, Wirtschafts- und Rechtswissenschaften
Institut für Betriebswirtschaftslehre und Wirtschaftspädagogik
Ammerländer Heerstr. 114-118
26129 Oldenburg
Germany

martin.mueller@uni-oldenburg.de

Prof. Dr. Heinz Welsch

Carl von Ossietzky Universität Oldenburg
Fakultät für Informatik, Wirtschafts- und Rechtswissenschaften
Institut für Volkswirtschaftslehre und Statistik
Ammerländer Heerstr. 114-118
26129 Oldenburg
Germany

welsch@uni-oldenburg.de

Prof. Dr. Claus Rautenstrauch

Otto-von-Guericke-Universität Magdeburg
Institut für Technische und Betriebliche Informationssysteme
Universitätsplatz 2
39106 Magdeburg
Germany

claus.rautenstrauch@iti.cs.uni-magdeburg.de

Organization

Organizing Committee

Nico Brehm
Elke Daniels
Jorge Marx Gómez
Moritz Grohmann
Nils Heyer
Martin Müller
Thomas Path
Dirk Peters
Claus Rautenstrauch
Boris Richter
Jürgen Sauer
Dirk Schlehf
Andreas Solsbach
Michael Sonnenschein
Ute Vogel
Margret Warns
Heinz Welsch
Manuela Wüstefeld

Program Committee

Ralf Antes, Uni Halle-Wittenberg, Germany
Hans-Knud Arndt, Otto-von-Guericke-Universität Magdeburg, Germany
Ioannis Athanasiadis, IDSIA - Istituto Dalle Molle di Studi sull' Intelligenza Artificiale, Switzerland
Christian Becker, Universität Heidelberg, Germany
Christoph Bey, Université de La Rochelle, France
Rainer Bitsch, BTU Cottbus, Germany
Simme Douwe Flapper, Eindhoven University of Technology, Netherlands
Albrecht Gnauck, TU-Cottbus, Germany
Stefan Gößling-Reisemann, Universität Bremen, Germany
Oliver Günther, Humboldt-Universität Berlin, Germany
Rüdiger Hohmann, Otto-von-Guericke-Universität Magdeburg, Germany
Ralf Isenmann, University of Bremen, Germany
Horst Junker, FHTW - University of Applied Science Berlin, Germany
Kostas Karatzas, Aristotle University of Thessaloniki, Greece

Sponsors

Universitäts-Gesellschaft
Oldenburg e.V. (UGO)
Postfach 49 01
26039 Oldenburg
Germany

EWE STIFTUNG
Stiftung des bürgerlichen Rechts
Tirpitzstraße 39
26122 Oldenburg
Germany

OFFIS e.V.
Escherweg 2
26121 Oldenburg
Germany

ifu Hamburg GmbH
Große Bergstraße 219
22767 Hamburg
Germany

Stadt Oldenburg
Wirtschaftsförderung
26105 Oldenburg
Germany

Carl von Ossietzky Universität
Oldenburg
Fakultät für Informatik, Wirtschafts- und Rechtswissenschaften
Ammerländer Heerstr. 114-118
26129 Oldenburg
Germany

ICSC Interdisciplinary Research
(Head Office)
P.O. Box 279
Millet, Alberta T0C 1Z0
Canada

Keynotes

Air Pollution

Corporate Sustainability Reporting

Environmental Policy and Management

Environmental Risk Assessment

Life Cycle Assessment

Spatial Applications

Water Pollution

Workshop: Corporate Environmental Management Information Systems

Our Energy Supply Future - 10 BULLENSEE ASSUMPTIONS

Werner Brinker

CEO, EWE AG, Oldenburg, Germany

Keynote

The global energy sector is currently undergoing a period of fundamental change. Increasing demand is threatening to outstrip the world's limited energy resources, resulting in high and very volatile commodity prices. At the same time, industrialized nations are becoming increasingly dependent on energy imports from just a handful of countries, and the world is waking up to the negative and irreversible consequences of climate change. These developments have a major bearing on the future energy supply situation and require new and sustainable solutions. The challenges are great: government and business must work together to meet them if end consumers are to have a reliable, affordable supply of energy into the future.

EWE is committed to contributing to this process, and has sought to push the debate forward at the political and scientific levels with its ten "BULLENSEE ASSUMPTIONS". These assumptions about our energy future, and the recommendations derived from them, were developed jointly by EWE and some of the leading experts from Germany's research and scientific communities. The name "Bullensee" is taken from the place where the authors met to formulate the assumptions and recommendations Lake Bullensee, a small lake near Bremen in North Germany.

The Bullensee strategy is based on three core principles: the world must achieve greater energy savings, it must significantly improve its energy efficiency, and it must make greater use of renewable. The measures developed by government and business to address these principles must be economically efficient, they must be environmentally sustainable, and they must not jeopardize security of supply.

Since its inception, the BULLENSEE think-tank has grown, allowing more in-depth discussion. The most recent sessions have been dedicated to BULLENSEE ASSUMPTION No.8: the role of distributed micro-CHP, that is, micro-cogeneration plants with an electric output of less than 5 kilowatts. Micro-CHP plants will have an important part to play in the energy supply mix of the future because they are very efficient, simultaneously producing useable heat and power close to the locations where the energy is used.

Using the insights gained from both the BULLENSEE and the Micro-CHP discussions as the starting point, EWE will continue to forge ahead with forward-looking projects, develop innovative products and thus play a long-term role in shaping our energy supply future.

Sustainability Impact Assessment – The Role of Numerical E3-Models

Christoph Böhringer

Oldenburg University , Ammerländer Heerstr. 114-118,
26129 Oldenburg, Germany, e-mail: boehringer@uni-oldenburg.de

Keynote

Sustainability Impact Assessment (*SIA*) of economic, environmental, and societal effects triggered by governmental policies has become a central requirement for policy design. The three dimensions of *SIA* are inherently intertwined and subject to trade-offs. Accomplishing one objective frequently means back-pedalling on another. Therefore, research activities on *SIA* increasingly aim at developing quantitative tools for trade-off-analysis along the *SD* dimensions. Since economics is the study of trade-offs, there is plenty for economists to contribute in order to make the concept of *SD* operational.

The quantification of trade-offs calls for the use of numerical model techniques in order to assess systematically and rigorously the interference of the many forces that interact in the economy thereby affecting potential *SD* indicators. Compared to stylized analytical models, the numerical computer-based approach facilitates the analysis of complex non-linear system interactions and the impact assessment of structural policy changes. In the end, the decisions how to resolve potential trade-offs must be taken on the basis of societal values and political decisions. However, model-based analysis puts decision making on an informed basis concerning sustainable development rather than on fuzzy or contradictory hunches.

A major challenge in building quantitative *SIA* tools is the policy makers' demand for comprehensive coverage of potentially important policy impacts. *SIA* tools must identify "the chain of significant cause-effect links from the … [policy] measures … through to any sustainability impact" and produce "comparable indicators of the magnitude and dimensions of each sustainability impact" as an input into policy formulation and implementation. There is a wide range of quantitative models for assessing the causal chains between a proposed policy change and its potential economic, environmental, and social impacts. In general, there is no specific model, which fits all requirements for comprehensive *SIA*, but rather a package of models or methods depending on the policy measure or issue to be assessed and the availability of data.

Regarding the appropriateness of quantitative *SIA* we investigate the use of so-.called energy-economy-environment (E3) models that are concerned with linkages between economic activities, energy transformation, and associated environmental impacts. We find that operational versions of E3-models have a good coverage of central economic indicators.

Environmental indicators such as energy-related emissions with simple direct links to economic activities are widely covered, whereas indicators with complex natural science background such as water stress or biodiversity loss are hardly represented. Societal indicators stand out for very weak coverage, not at last because they are vaguely defined or incommensurable. Our analysis identifies prospects for future modeling in the field of integrated assessment that link standard E3-models to theme-specific complementary models with environmental and societal focus.

The Material Side of Virtualization

Lorenz M. Hilty

Technology and Society Lab, Empa Swiss Federal Laboratories for Materials Testing and Research, Lerchenfeldstr. 5, 9014 St.Gallen, Switzerland

Keynote

The Environmental Informatics community could recently celebrate its 20th anniversary [Tochtermann et al. 2006; Hilty et al. 2006a]. The application of Information and Communication Technology (ICT) to problems of environmental research and management has made considerable progress and contributes to sustainable development [Hilty et al. 2004a]. Moreover, ICT has the potential to virtualize processes that would otherwise consume considerable amounts of material and energy; virtual meetings, for instance, could avoid 97-98% of the CO_2 emissions of physical meetings [Hischier and Hilty 2002].

The time, space, material and energy needed to provide a unit of ICT service have roughly decreased by a factor of 1,000 since the first PC was sold. It seems therefore natural that researchers and industries using ICT in the environmental field ignore the environmental impacts caused by ICT hardware – they are just negligible compared to the environmental benefits that can be realized with the applications.

Paradoxically, it is the progress in ICT hardware efficiency that has made ICT a part of the problem, too. The global mass and energy flows caused throughout the hardware life cycle are increasing due to the widespread use of ICT products and their decreasing useful lives. The environmental problems caused by the production, use and disposal of ICT hardware are solvable in principle; they are not as hard as the discrepancy between – e.g. – growing mobility and CO_2 reduction goals. But problems can only be solved if they are not neglected.

The Technology and Society Lab at Empa has explored some environmental and social aspects of ICT production, use and disposal at a global scale under the umbrella of the "Sustainability in the Information Society" research program funded by the ETH Board. This research was based on Life Cycle Assessment (LCA) studies, prospective Technology Assessment (TA) studies and technology cooperation projects with partners in emerging economies (China, India and South Africa). These projects have identified the following issues:

- *Production*: The scarcity of some chemical elements (e.g. Indium used for LCDs) may constrain some development paths in the near future.
- *Use*: The overall energy demand of ICT infrastructure tends to increase, but it is possible to go against this trend.
- *Disposal*: There is a huge informal electronics recycling sector in Asia with relevant health and environmental impacts [Hilty 2005]. Recycling systems as they have been implemented in Switzerland and now in the EU (WEEE directive), are indispensable because electronic waste contains toxic as well as valuable substances. However, miniaturization of devices and the trend towards pervasive computing [Hilty et al. 2004b] create new challenges for electronic waste recycling. In the long run, ICT industry will have to find ways to avoid toxic and scarce materials.

The vision of a sustainable information society, i.e. a society which virtualizes those processes that are no longer acceptable for ecological reasons and keeps the material side-effects of ICT small at the same time – such a vision is not completely unrealistic. However, using more ICT does not automatically "create" a sustainable information society, as our observations and models clearly show [Hilty et al. 2006b]. Political strategies are needed if ICT is to serve sustainability.

References

Hilty LM, Seifert EK, Treibert R (eds) (2004a) Information Systems for Sustainable Development. Idea Group Publishing, Hershey

Hilty LM, Som C, Köhler A (2004b) Assessing the Human, Social, and Environmental Risks of Pervasive Computing. Hum Ecol Risk Assess, vol 10, edn 5, pp 853-874

Hilty LM (ed) (2005) Environmental and Social Impacts of Electronic Waste Recycling. Special issue. Environmental Impact Assessment Review, vol 25, edn 5

Hilty LM, Page B, Hrebicek J (eds) (2006a) Environmental Informatics. Special issue. Environmental Modelling & Software, vol 21, edn 11

Hilty LM, Arnfalk P, Erdmann L, Goodman J, Lehmann M, Wäger P (2006b) The Relevance of Information and Communication Technologies for Environmental Sustainability – a Prospective Simulation Study. Environmental Modelling & Software, vol 21, edn 11, pp 1618-1629

Hischier R, Hilty LM (2002) Environmental impacts of an international conference. Environmental Impact Assessment Review, vol 22, edn 5, pp 543-557

Tochtermann K, Scharl A (eds) (2006) EnviroInfo 2006. Managing Environmental Knowledge. Shaker, Aachen

Electric Power Management and a Clean Environment – a Discrepancy and an Algorithmic Challenge

Horst F. Wedde

School of Computer Science, University of Dortmund, Germany

Keynote

For maintaining an efficient and reliable supply of energy throughout the next decades under a growing shortage in fossil fuel, renewable energies will be further utilized and integrated into the current electric infrastructures. Such sources are based on solar or wind power, on renewable fuel like linseed oil, or on hydrogen technology. They are used for electric power generation in typically highly distributed small or mid-size facilities. The sources are inexhaustible, and through coupling electric and heat energy (e.g. in block heat and power plants) the technical efficiency is well over 90%. Furthermore these energy sources are ecologically clean.

Establishing clean or renewable energy sources involves the key problem of adequate management for the networked power sources:

- Typically, producers are at the same time consumers, and vice versa;
- Production plant or facilities are of small or medium size. They are widely distributed.
- They are operated on quite different supply and voltage levels which, in the current architectures, are opposed to trading, or purchasing, energy within the power grids.
- For the same reason they are excluded from serving as local (or regional) nodes of reserve capacities (in case of power failures), or as providing balancing energy in case of excess production or unexpected excess consumption that will not be requested for consumption. Even worse, due to the unpredictability from a global perspective, there will be traditional facilities (e.g. coal-based power plants) established that run continually (and mostly quite predictably) side by side with renewable sources just to make up for failures or unpredictable power declines. In case of excess production the renewable sources are disconnected from the grid in order to keep the balance, in terms of voltage

level and cycle frequency. At any rate this causes an enormous waste of energy.

The major drawback for the traditional approach is a conceptually centralized management approach for producing and trading electric energy as well as for safety and security. While such approaches are even quite error-prone as recent experience shows more and more frequently they rely at the same time on forecast mechanisms, within prediction periods ranging from 15 minutes to 1 year. Only under these circumstances the variation of capabilities as normal for sources based sun or wind power appears to be non-manageable.

In this lecture we describe several first phases of the joint R&D funded project DEZENT between the School of Computer Science and the College of Electrical Engineering at the University of Dortmund, devoted to decentralize and adaptive electric power management through a distributed real-time multi-agent architecture. Unpredictable consumer requests or producer problems, under distributed control or local autonomy, will be handled through distributed (P2P) real-time negotiation algorithms involving agents on different levels of negotiation, on behalf of producers and consumers of electric energy. *The key idea is that these negotiations take place (locally and regionally) on such a short-term basis that negotiation can be considered to be constant during these time intervals (ca. 40 msec) while the electric distribution processes can be finalized within the same time frame*. Despite the lack of global overview we are able to prove that in our model no coalition of malicious users could take advantage of extreme situations like arising from an abundance as much as from any (artificial) shortage of electric power that are typical problems in "free" or deregulated markets. Our multi-agent system exhibits a very high robustness against power failures compared to centrally controlled architectures. In extensive experiments we have demonstrated how, in realistic settings of the German power system structure, the novel algorithms can cope with unforeseen needs and production specifics in a very flexible and adaptive way, taking care of most of the potentially hard deadlines already on the local group level (corresponding to a small subdivision). We further demonstrate that under our decentralized approach customers pay less than under any conventional (global) management policy or structure.

We will report on our current progress which includes preparations for extensive real field studies in Southern Germany.

References

Handschin E, Krause O, Wedde HF, Lehnhoff S (2006) The Emerging Communication Architecture in Electrical Energy Supply and Its Implications. In: Proceedings of the CRIS Workshop 2006 – Influence of the Distributed and Renewable Generation of Power System Security, Magdeburg, Germany, Dec. 6-8, 2006

Krause O, Lehnhoff, S (2006) Die Rolle dezentraler Energieversorgung im Netz der Zukunft. In: 4. Fachkongress für die Energiewirtschaft, Essen, Deutschland, Sept. 20, 2006, IT-Trends Energie. http://www.it-trends-energie.de, access: 2007-02-13

Wedde HF, Lehnhoff S, Handschin E, König D, Krause O (2006) Real-Time Multi-Agent Support for Decentralized Management of Electric Power. In: Proceedings of the ECRTS'2006, July 5-7, 2006. Dresden/Germany; IEEE. Computer Society Press

Wedde HF, Breuer FT, Freund W, Handschin E, König D, Neumann H (2004) Decentralized Real-Time Management of Largely Unpredictable Power Needs and Supply. In: Proceedings of the 28th IFAC/IFIP Workshop on Real-Time Programming (WRTP2004), Istanbul

A Spatial Assessment of Air Pollution Impact of a Highway Project Using GIS

M.L. Agrawal[1], B. Maitra[2], M.K. Ghose[3]

[1] Reader, Civil Engineering Department, National Institute of Technology, Raipur (C.G.), India 492010, e-mail: mohanlalagrawal@yahoo.com

[2] Assistant Professor, Civil Eng. Dept., Indian Institute of Technology, Kharagpur (W.B.), India 721302, e-mail: bhargab@civil.iitkgp.ernet.in

[3] Senior Scientist, Regional Remote Sensing Service Center (ISRO), Kharagpur (W.B.), India 721302, e-mail: mkghose2000@yahoo.com

Abstract: In the recent years Government of India as well as different provincial governments have taken up several ambitious high-way developmental activities to accelerate the economic growth of India. The increased vehicular traffic caused due to highway development is a major source of air pollution. Carrying out environ-mental impact assessment (EIA) is an essential part of major road development projects in India. Also, the assessment of impacts on air pollution is an important aspect of EIA. In the present paper, a methodology has been presented for incorporating spatial data in GIS environment for assessing air pollution impact of a highway project.
Keywords: environmental impact assessment (EIA), highway development project, Geographic Information System (GIS).

1 Introduction

Infrastructure is the basic need for accelerating the economic growth and development of a country. Highway development is one of the important infrastructure development activities. In the recent years, the need for developing adequate road transportation infrastructure has been realized by Government of India and various state Governments. Major road development projects have been formulated at national level under National Highway Development Project (NHDP), which will strengthen and widen 14,000 kilometers of National Highways by 2007 [Maitra et al. 2002]. Various State Governments have also started projects for upgrading some of the state highways. All these efforts are intended to provide improved road transportation facilities for passenger and goods traffic.

Vehicular emission contributes significantly towards the total air pollution load. Development of roads encourages the growth of traffic, which eventually increases the air pollution level of the surroundings. Vehicular emissions are of particular concern since these are ground level sources and therefore, have the maximum impact on the population of the surroundings. For sustainable road development, it is necessary to understand the impact of the road development on the air quality. A suitable methodology is also required for assessment of impact on air quality due to the development of high-ways.

In the process of assessing impacts on air quality, it is necessary to consider the concentration levels of different pollutants and their spatial distributions. Both of these aspects are important as the detrimental effect of a pollutant depends on its concentration and the concentration level varies with the distance from the source of emission. For a realistic assessment of impact, it is necessary to relate the concentration levels with the area affected. Altogether it is necessary (i) to model the level of pollutants (ii) understand their spatial distribution, (iii) define suitable impact categories for each pollutant and (iv) express the impacts in relation to both impact categories and the area affected under each category.

The line source model (IITLS) developed by Indian Institute of Technology (IIT) Delhi, is based on Gaussian dispersion approach with suitable modifications in dispersion parameters [Goyal and Ramakrishna 1999]. The modifications in dispersion parameters included in the IITLS model makes it more suitable for Indian conditions as compared to other models normally used for the same purpose. In the process of air pollution mode-ling and mapping vast data is used, which is spatial in nature. Therefore, a suitable tool for spatial database management, analysis and display of results is of immense use for air pollution impact assessment due to vehicular traffic. The other aspect, which demands the spatial analysis of air pollution impacts, is to determine the extent of dispersion of pollutants and their concentration at different places surrounding to the highway. Since early 90's, attempts have been made by researchers to use GIS in transportation related air quality modeling and management [Moragues and Alcaide 1996; Andersons et al.1996; Briggs et al.1997; Li et al. 1999; Sharma et al. 2002].The above discussions supports the utility of GIS for analyzing spatial data in connection with air pollution modeling and impact assessment.

For presenting the spatial distribution of impacts, it is necessary to express the pollutant's concentration in different ranges instead of discrete values. Therefore, for assessment of air pollution impacts, a quantification technique considering the area affected under a particular impact category, is useful. The main objective of the present work is to improve the rationality of air pollution impact assessment process by integrating IITLS model with a GIS database by considering the spatial distribution of the impacts with reference to a case study of highway project in India. Carbon Monoxide (CO) is considered here for demonstrating the methodology adopted in this study. The impact is assessed for the base year (1999) as well as for horizon years (2010 and 2020).

2 Study Area

In order to demonstrate the methodology for modeling and mapping air pollution in a GIS framework, a stretch of National Highway in India and its surroundings is considered. The study stretch considered in the present work is a part of National Highway (NH-60) between Jaleshwar to Kharagpur in India. This stretch of highway is in the process of being upgraded from two lanes to six lanes. To generate air pollution impact map, about 15 km on either side of the highway is considered so that the maximum possible impacts can be captured.

3 Methodology

The major steps followed for the integration of GIS database with IIT line source model (IITLS) include (i) estimation of air pollution emission rate (q) due to highway traffic; (ii) modeling and mapping of dispersion of pollutants in atmosphere; and (iii) quantification of impacts. The detail process adopted for each step is discussed below.

4 Estimation of Air Pollution Emission Rate (q) due to Highway Traffic

To estimate the emission rate of different pollutants caused by highway traffic, it is required to estimate the classified traffic volume and use suitable emission factor for each vehicle category.

- Traffic volume data- The traffic volume data for the base year (1999) is taken form secondary source [NHAI 2000] and for predicting the traffic volume for horizon years (2010 and 2020), the growth factor is taken equal to 8 percent per annum as per NHAI [2000]. Traffic volumes are categorized in three classes namely heavy vehicle (heavy commercial vehicles and buses), medium vehicles (light commercial vehicles, diesel driven cars, jeeps and pick up vans), and light vehicles (petrol driven cars and all two wheelers). The classified traffic volume data used for the work are presented in Table 1.

Table 1. Maximum hourly traffic volume (source: [NHAI 2000]*)

Category of Vehicles	Base Year* (1999)	Horizon Year-I (2010)	Horizon Year-II (2020)
Light vehicles	145	338	729
Medium vehicles	31	973	157
Heavy vehicles	236	550	1187

– Emission factors- It is a very tedious and complex task to obtain suitable emission factors for different vehicle types as the emission factor depends on type of fuel, condition of engine, driving cycle, age of the vehicle, speed of vehicle, driving mode etc. Goyal and Ramakrishna [1998] have worked out emission factors for different categories of vehicle for Indian conditions and the same are adopted for the present study. The emission factors for different category of vehicles, as used in the present work, are mentioned in Table 2.

Table 2. Emission factors for different category of vehicles

Pollutant Emitted	Emission Factor (gm/vehicle/km)		
	Heavy vehicles	Medium vehicles	Light vehicles
Carbon Monoxide	11.2	12.4	4.5

Emission rate of a pollutant caused by each category of vehicle is calculated by multiplying the emission factor (gm/vehicle/km) of the vehicle category with the number of vehicles under the same category passing the road stretch in one hour. The emission rate thus obtained in gm/km/hr is converted to gm/m/sec for use as input to IITLS model for the prediction of pollutant concentration.

4.1 Modeling and Mapping of Dispersion of Pollutants in Atmosphere

In order to predict the dispersion of pollutants in atmosphere due to highway traffic, the IIT line source (IITLS) model [Goyal and Ramakrishna 1999], as given in Equation (1), is adopted. The value of H is taken as 0 for ground level concentration of pollutants and therefore, the model in Equation (1) reduces to Equation (2).

$$C_{(x,0)} = \frac{2q}{(2\pi)^{1/2}\sigma_z u} \exp\left[-\frac{1}{2}\left(\frac{H^2}{\sigma_z}\right)\right] \quad (1)$$

Where,

- q is the pollution emission rate per unit length of road (gm/m/sec),
- u is the mean wind velocity (m/sec),
- H is the height of source (m),
- σ_z is the vertical dispersion parameter (m) and
- C is the concentration of air pollutants at downwind distance x when wind is normal to the line source

$$C_{(x,0)} = \frac{2q}{(2\pi)^{1/2}\sigma_z u} \quad (2)$$

For mapping the air pollution impacts due to highway traffic, IITLS model is integrated with GIS database using following major steps.

- The value of emission rate (q) of the Carbon Monoxide generated is estimated separately for each category of vehicles by multiplying emission factor to the number of vehicles passing in unit time and then added together for all three categories to get the total emission rate (q).
- The buffers for radial distances from centerline of road are generated for 50m intervals up to a distance of 15km on either side of the road using ERDAS IMAGINE 8.3 [1997] GIS software.
- The dispersion of the air pollutants from centerline of the road is modeled with the help of IITLS model in GIS environment using ERDAS IMAGINE 8.3 in "model-maker" option by taking the dispersion of air pollutant as a function of horizontal distance across the road (x) from the road centerline and the vertical dispersion coefficient (σ_z). Vertical dispersion coefficient (σ_z) is taken as per Pasquill-Gifford's chart [Gifford 1961].
- The concentration of air pollutant for the base year and horizon years for each pixel is obtained and air pollution maps are generated for base year (1999) as well as horizon years (2010 and 2020).

In order to validate the IITLS model, the estimated values are compared with the observed values for the base year (1999). The measured concentration of air pollutants in the base year (1999) at four locations [NHAI 2000], are used for the validation purpose. A comparison of modeled and measured values of Carbon Monoxide is shown in Table 3. It may be observed from Table 3 that at Jaleswar and also at Dantan, the modeled values are good in agreement with the observed ones. However, at Belda and Kharagpur the observed values are marginally on higher side than the modeled values, which is due to the poor road surface condition between Belda and Kharagpur in the base year. As the road surface condition is expected to improve significantly after the road improvement works, no changes are made in the model for prediction of Carbon Monoxide in horizon years.

Table 3. Comparison of observed and modeled values of air pollutants

Pollutant	Location wise pollutant conc. (mg/m^3)							
	Jaleswar		Dantan		Belda		Kharagpur	
	O^a	M^b	O^a	M^b	O^a	M^b	O^a	M^b
CO	1.95	1.93	1.91	1.93	2.08	1.93	2.04	1.93

O^a-Observed Value,
M^b- Modeled value

4.2 Quantification of Air Pollution Impacts

In order to quantify the air pollution impacts, a new approach for quantification of impacts is adopted following the procedure mentioned below:

- The concentration of pollutant varies with respect to the horizontal distance normal to the road centerline. Therefore, the concentrations of pollutant at different points are expected to be different. For obtaining the spatial distribution of impacts it was necessary to present the CO concentration in different ranges instead of discrete values. The range of pollutant levels due to CO is classified in four categories based on ambient air standards prescribed by CPCB [2000] (refer to Table 4). Air pollution impact maps are generated for the base year as well as horizon years.

Table 4. Pollutants level range and corresponding impact category

Air Pollutant	Range of pollutant conc. in mg/m^3			
	No Impact (NI)	Slight Adverse Impact (SAI)	Moderate Adverse Impact (MAI)	High Adverse Impact (HAI)
CO	<1.0	1.0-3.0	3.0-5.0	>5.0

– Area under each class of impact is determined for the base year as well as horizon years and the change in area affected in each class are observed.

5 Results and discussion

Air pollution impact map is generated for Carbon Monoxide using ERDAS IMAGINE 8.3, GIS software for the base year [1999] as well as horizon years (2010 and 2020). Impact map generated for Carbon Monoxide is shown in Fig. 1 to demonstrate the spatial distribution of impact. The impact map shows that the extent of impact due to vehicular traffic increases rapidly with the increase in traffic volume over time. The area surrounding to the highway having a narrow strip on either side is mostly affected due to growth of traffic in the horizon years.

The percentage areas under each category of impact in base year as well as horizon years are shown in Fig. 2. The percentage areas under different impact categories are found to change over time due to the increase in traffic volume. In the base year (1999), most of the area is under no impact category and negligible area lies under slight adverse impact category. However, in the horizon years (2010 and 2020), the area under no impact category is found to decrease and areas under slight adverse impact category as well as moderate adverse impact category are found to increase over time due to growth of traffic volume. No area is found under high adverse impact category in base year as well as horizon years.

6 Conclusion

A methodology is used for the estimation of air pollutants concentration and its dispersion in surroundings due to highway traffic by integrating IITLS model with geographic information system (GIS). Impact map generated in GIS environment could add to the perception of visual effect of the impacts and therefore, are more capable in pursuing the extent of pollution, which increases rapidly over time due to increase in traffic volume. For the case study considered in the paper, it is found that in the base year (1999) most of the area is subjected to no impact category. But, in the future (i.e. horizon years 2010 and 2020) substantial areas are expected to be under slight adverse impact and moderate adverse impact categories. No area is found under high adverse impacts category in base year as well as horizon years. Altogether, the integration of GIS database with IITLS model for air pollution impact assessment due to highway development project and the advantages of using spatial data and analysis are demonstrated. Similar approaches may be followed for modeling or assessing the impacts on other pollutants due to the development of highway projects.

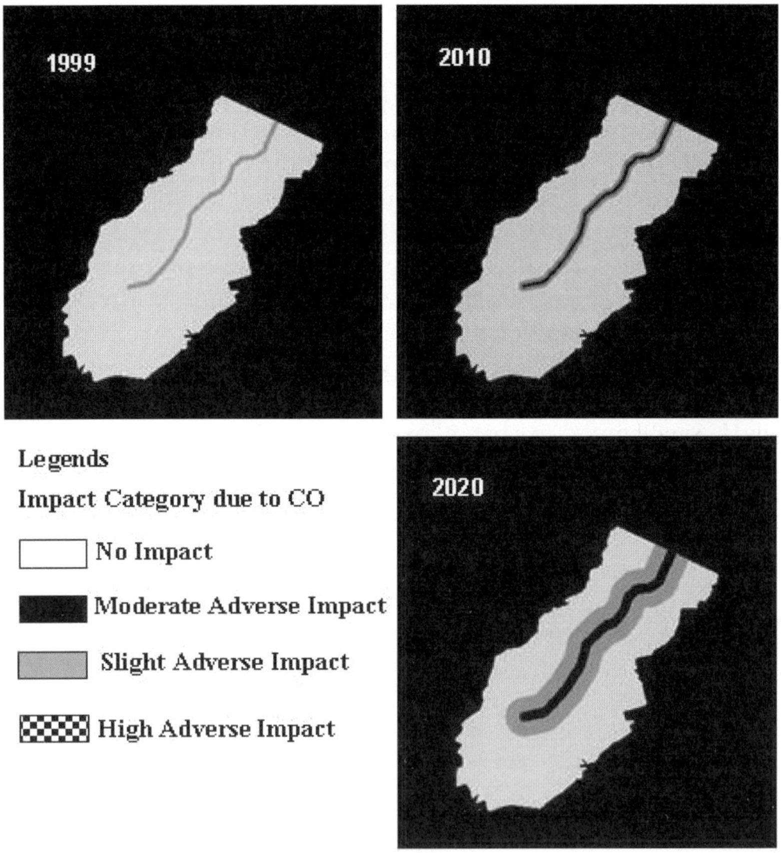

Fig. 1. Impact map due to Carbon Monoxide (CO)

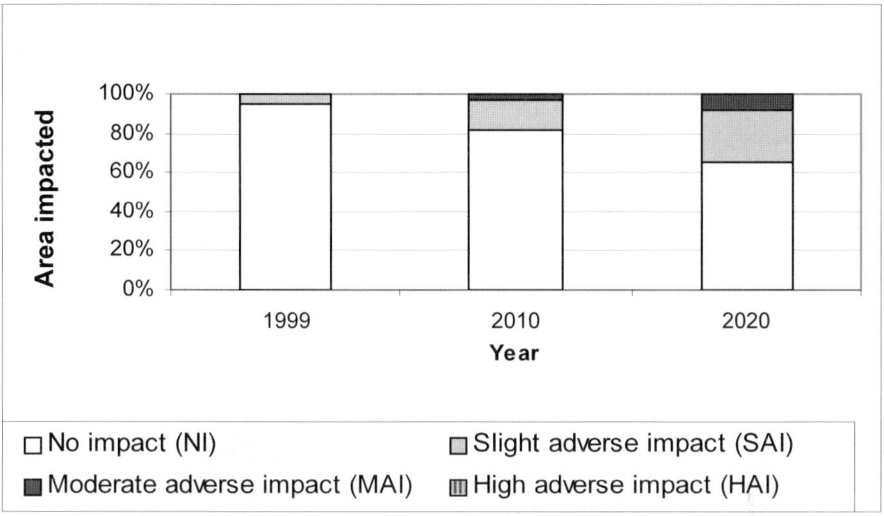

Fig. 2. Percentage area impacted under different impact categories

References

Anderson WP, Kanaroglou PS, Miller EJ, Buliung EL (1996) Simulating automobile emissions in an integrated urban model. In: Transportation Research Record 1550. Transportration Research Board, Washington, pp 71-80

Briggs DJ, Collins S, Elliott P, Fischer P, Kingham S, Lebert E, Pryl K, Van Reeuwijk H, Smallbone K, van der Veen A (1997) Mapping urban air pollution using GIS: a regression based approach. In: International Journal of Geographic Information Sciences, vol 11, edn 7, pp 699-718

Central Pollution Control Board (CPCB) (2000) Air Quality Status and Trends in India. In: National Ambient Air Quality Monitoring Series (NAAQMS) vol 14, 1999-2000

ERDAS IMAGINE Ver. 8.3 (1997) ERDAS IMAGINE TOUR Guide. ERDAS Inc. Atlanta, Georgia

Gifford FA (1961) Use of Routine Meteorological Observations for Estimating Atmospheric Dispersion. In: Nuclear Safety, vol 2, edn 4, pp 47

Goyal P, Rama K (TVBPS) (1998) Various Methods of Emission Estimation of Vehicular Traffic in Delhi. In: Transport Research Part D, vol 3, edn 5, pp 309-317

Goyal P, Rama K (TVBPS) (1999) A Line Source Model for Delhi. In: Transportation Research Part D, vol 4, pp 241-249

Li X, Wang W, Li F, Deng X (1999) GIS Based Map Overlay Method for Comprehensive Assessment of Road Environmental Impact. In: Transportation Research Part D, vol 4, pp 147-158

Maitra B, Sarkar JR, Sikdar PK (2002) Road Transportation in India: Emerging Issues and Challenges. In: Proceedings of National Seminar on Road Transportation in India: Emerging Trends and Techniques, pp 4.87- 4.94

Moragues A, Alcaide T (1996) The Use of a Geographical Information System to Assess the Effect of Traffic Pollution. In: The Science of The Total Environment, vol 189/190, pp 267-273

National Highway Authority of India (NHAI) (2000) Environmental Assessment Report Including Resettlement Action Plan. Rehabilitation and Updating, 4/6 Lanes Widening of Baleshwar-Kharagpur Section of NH-60, Detailed Project Report, vol 4

Sharma N, Bhandari K, Rao P, Shukla, A (2002) GIS Applications in Air Pollution Modeling. http://www.gisdevelopment.net, access: 2007-02-13

The Air Quality Management System AirQUIS

Leiv Håvard Slørdal, Harold McInnes, Terje Krognes

Norwegian Institute for Air Research, P.O. Box 100, NO-2027 Kjeller, Norway

1 Introduction

AirQUIS [AirQUIS 2006] is a computer software system for air quality management, developed by NILU (Norwegian Institute for Air Research). It consists of an Oracle database, various models, and a graphical user interface.

The system offers functionality for collection and statistical evaluation of meteorological- and air quality data from monitoring stations, a module that provides a user friendly treatment of emission data, and models for calculating emissions, wind fields, pollutant dispersion and exposure. A detailed description of AirQUIS is given in section 2 below.

In this paper we will also present results from a recent research project where AirQUIS was applied to study the air quality in Oslo, Norway. The scope of this research project was to investigate to what degree the inhabitants were exposed to exceedances of the National air quality standard for NO_2, PM_{10}, and benzene. The model evaluation performed as part of this study is presented in section 3, and some of the project results are presented in section 4. Finally, some concluding remarks are offered in section 5.

2 Description of the Air Quality Information and Management System (AirQUIS)

2.1 System Architecture

The typical AirQUIS user will create a complex collection of input data (Megabytes to Gigabytes), and create model scenarios and program configurations by selecting or specifying a few tens to a few hundreds of parameters. The user will launch complex model calculations that run for some minutes to several days on a good PC. The results will amount to Megabytes, sometimes Gigabytes of additional data to store, and the user may spend substantial time on interpretation and presentation of results.

AirQUIS is built as a single-user program that runs on a PC. The program can use a local single-user database, or a central (networked) multi-user database. The database engine is Oracle (currently version 9.2). The user interface is written in Visual Basic 6 (VB6) and is run under Microsoft Windows XP or 2000. Model computations are performed in additional program modules (some written in FORTRAN) that also run on the user's PC. Only the database function can currently be served by a networked node. In a distributed environment, each user only has access to the computational power of the local PC, but many PC's may be served simultaneously by one database server.

A set of input data, ranging from topography and population data to meteorological data, air quality monitoring data and the large and complex emission inventory, is collected in the database as a *user project*. The project will also collect model computation results. For scenario setup and data evaluation the user may have several projects open on one PC. During model computations one AirQUIS project will consume all resources on one CPU. Standard Oracle tools are used to create dumps of the entire database, or of single user projects (which are *Oracle users* from the database point of view).

2.2 System Functionality

Since AirQUIS contains a wide variety of tools, it has been organized as separate modules, served by a user friendly graphical user interface (GUI). AirQUIS contains the following modules: *Measurement*, *Emission Inventory*, *Models* and *Geographical Information System (GIS)*

2.2.1 The Measurement Module

Measurements of air quality and meteorological parameters are stored in the database with metadata for each measurement station. Typical metadata for a station are station name, geographical coordinates, altitude and station type. The measurements are stored as time series. Each station may have several time series with different compounds, units and measurement positions. Data for stations and time series are presented in spreadsheets, and graphs, where the user may view and edit data.

AirQUIS provides tools for statistical calculations such as descriptive statistics, percentile calculations, counting observations above bounds, wind roses and Breuer Diagrams. The wind rose combines time series for wind direction and wind speed and gives the distribution of wind from different directions in specified classes of wind speed. The Breuer Diagram combines time series for both wind direction and air quality, and is an efficient tool for tracking major sources of air pollution. AirQUIS provides graphical presentation of data from the statistical calculations. An example of the AirQUIS graphical user interface is shown in Figure 2.1

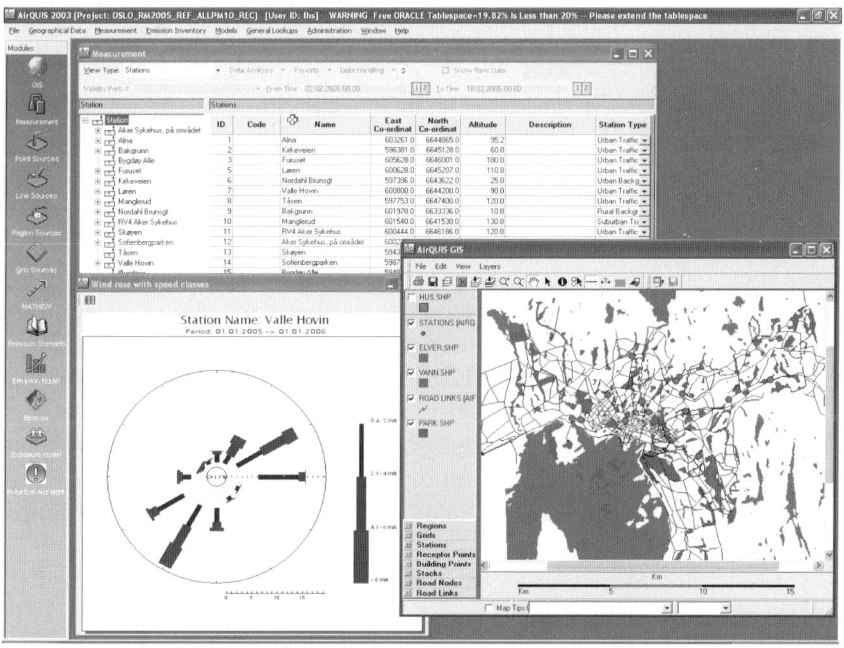

Fig. 2.1. Example of the AirQUIS graphical user interface when applying the measurement module. Shown are the measurement spread sheet, the wind rose representation and an open GIS window.

Measurements may be imported into time series in the database by drag and drop from an Excel sheet. It is also possible to import data automatically from specified files at scheduled times. Automatic import should be used together with systems for retrieving data from stations.

2.2.2 The Emission Inventory Module

The AirQUIS Emission Inventory provides a database containing all data relevant for handling sources of air pollution and emission calculations, together with tools for assessing and updating the data.

Air pollution from industrial plants and refineries is mainly emitted through stacks. These stacks are treated as point sources in AirQUIS. Together with emission or consumption data, information such as location, stack height, stack diameter, gas temperature and gas velocity are necessary parameters for point sources. This information can be viewed and edited in AirQUIS.

Emissions from traffic are mainly handled as so-called line sources. Line sources are represented as road links. A road link is a road segment between two road nodes where data such as road slope, traffic amount, vehicle distribution and speed, are assumed homogenous. Each road link is connected to a time variation, which distributes the emissions within days, weeks and years. Calculations of emissions from road traffic are based on a set of emission factors for different compounds for each vehicle class and fuel type.

More diffuse sources are distributed on an area basis either in administrative areas, such as counties, municipalities etc, or in regular grids. These are treated as area sources by AirQUIS. Typical area sources are emissions from domestic heating, harbor activity, and small and medium size enterprises. Area emission data may either be stored as emission or as fuel consumption per time unit. Consumption data are converted to emissions simply by multiplying with their associated emission factors. Area distributed emission and consumption data can be connected to time variations.

2.2.3 The Models module

AirQUIS provides models for the calculation of emissions, wind fields, pollutant dispersion and exposure. All definitions necessary for running the models are stored in the database and may be updated by the user. The results from model calculations may be viewed in spreadsheets, plotted in graphs or presented on maps.

Emissions are calculated based on the line-, point-, and area source information defined in the Emission Inventory. Before running the emission model, the user must define an emission scenario, which specifies the sources to be included in the calculation. For line sources the user must select road links to be included, and for point sources the stacks must be selected.

A diagnostic wind field model is used to calculate three dimensional wind fields. As input this model requires wind and temperature data from one or more measurement stations within the model domain. These data must be specified in a meteorological setup before the model is run.

Based on data from the emission and wind field models, the dispersion model calculates the concentration of a selected air quality compound. The air dispersion model implemented in AirQUIS is a combined 3D Eulerian grid model, with embedded sub-grid line- and point- source models, for urban and local-to-regional scale applications.

Dispersion calculations may be started either manually or automatically at scheduled times. An automatic version of AirQUIS has for several years been used operationally to produce 48 hours forecasts of ambient concentrations of NO_2, PM_{10} and $PM_{2.5}$ in Oslo (and 5 other Norwegian cities). Since these air quality forecasts require prognostic wind fields as input, an automatic import of wind fields from an operationally run weather forecast model (MM5) is built into this forecast version of AirQUIS [Ødegaard et al. 2006].

Ambient air concentrations can be calculated at the individual positions of buildings, (hereafter referred to as building points), and within the domain-covering, three-dimensional model grid (hereafter simply referred to as grid cell values). When performing exposure calculations the inhabitants of the considered buildings are assigned to building point concentrations while the rest of the population are assigned to concentration values computed in the grid cells containing the location of their home address. The exposure module within AirQUIS calculates the number of people exposed for concentrations above user defined threshold values in all grid cells and in the building points.

2.2.4 The Geographical Information System (GIS) Module

The Geographical Information System is used as a platform for integrating the presentation and selection of measurements, emissions and model results. The GIS is directly linked to the database from which data may be presented together with map themes from shape files. This allows results from the models, measurement stations, road links, point sources and area sources to be presented together with basic map data such as topography, land/sea, rivers, park areas and buildings.

3 System Evaluation

An evaluation of the AirQUIS performance has recently been performed as part of a research project that was carried out for the Norwegian city of Oslo [Slørdal et al. 2006a]. In this evaluation study, observed values of NO_2 and PM_{10} at two sites, Kirkeveien and RV4, were compared with the corresponding concentration values calculated by AirQUIS for a 7 month period in 2005.

Statistical comparisons between measured and calculated NO_2 values at the two sites are shown in Table 3.1. On average the NO_2 levels are overestimated at RV4, while there is a somewhat stronger under-prediction at Kirkeveien. At both stations high values of the correlation coefficient are found, but the rather high values of the standard error of the regression line indicate a considerable spread around this line. A direct comparison of the measured and calculated hourly NO_2 concentrations for a one-month period (February 2005) is presented in Figure 3.1 for RV4 and Kirkeveien, respectively. These plots clearly reveal the high degree of correlation between the measured and calculated values.

Table 3.1. Statistical comparison between calculated and observed hourly values of NO_2 in Kirkeveien and RV4 for the 7 months simulation period.

NO_2	Mean value (μg/m³)		St. deviation (μg/m³)		Max. value (μg/m³)	
	Measured	Calculated	Measured	Calculated	Measured	Calculated
RV4	38.1	42.4	25.8	34.0	132.6	142.8
Kirkeveien	44.5	32.5	29.8	27.1	178.3	126.8

	Correlation coefficient	Slope of linear regression line	Intercept point of regres. line	Standard error of regres. line
RV4	0.68	0.91	8.1	25.0
Kirkeveien	0.64	0.60	6.2	20.8

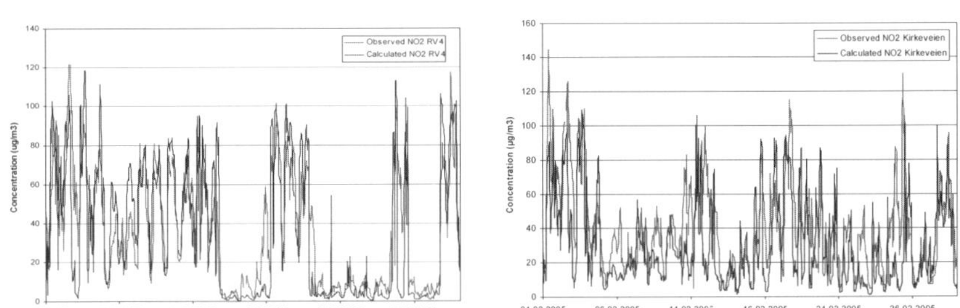

Fig. 3.1. Hourly values of NO_2 at RV4 and Kirkeveien in February 2005.

The statistical comparison between the measured and calculated PM_{10} values at RV4 and Kirkeveien are shown in Table 3.2. When compared with

Table 3.1 it is seen that the deviations between predicted and observed values are somewhat larger for PM_{10} than for NO_2. This is also to be expected, since the uncertainties associated with the emission estimates are larger for PM_{10} than for NO_2. PM_{10} concentrations are clearly overestimated at RV4 and slightly underestimated at Kirkeveien. The correlation coefficients are lower than for NO_2, but still larger than 0.5. As for NO_2 the values of the standard error indicate a substantial spread around the regression line. A direct comparison of the hourly measured and calculated PM_{10} concentrations at the two measurement sites are shown in Figure 3.2 for the month of February 2005. Again, the overall impression is that the model reproduces the observations rather well.

Table 3.2. Statistical comparison between calculated and observed hourly values of PM_{10} in Kirkeveien and RV4 for the 7 months simulation period.

PM_{10}	Mean value (μg/m^3)		St. deviation (μg/m^3)		Max. value (μg/m^3)	
	Measured	Calculated	Measured	Calculated	Measured	Calculated
RV4	31.3	43.2	30.3	53.4	260.1	416.9
Kirkeveien	29.9	25.9	25.0	30.2	220.4	273.0
	Correlation coefficient		Slope of linear regression line		Intercept point of regres. line	Standard error of regres. line
RV4	0.54		0.95		13.3	44.9
Kirkeveien	0.52		0.63		7.2	26.0

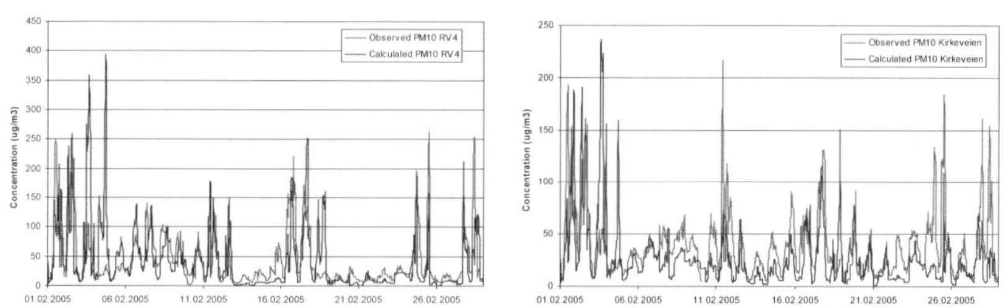

Fig. 3.2. Hourly values of PM_{10} at RV4 and Kirkeveien in February 2005

4 Application of AirQUIS for air quality assessment

4.1 Estimation of exposure levels and source contributions for the year 2005

As part of the research study referred to in section 3, AirQUIS was used to calculate outdoor concentration levels of NO_2, PM_{10} and benzene, and to estimate the number of inhabitants in Oslo that are living in areas where exceedances of the Norwegian ”National Air Quality Target” are computed. This target specifies that, during a year, no more than 8 hours are allowed with (hourly mean) NO_2 concentration levels above 150 μg/m^3, no more than 7 days with (daily mean) PM_{10} concentrations above 50 μg/m^3, and, finally, the yearly averaged benzene concentration is not allowed to exceed 2 μg/m^3 [Report No. 21 2005].

In this study the “present” situation was analyzed by peforming calculations for the year 2005. Since our focus was on the higher concentration levels, and experience has shown that these levels are encountered during the winter/spring season, no calculations were made for the summer period, May 1. – September 30. The reason why the highest concentrations are found in wintertime is the combination of frequently occurring stable atmospheric conditions (poor dispersion conditions) and enhanced emissions of PM_{10}, emanating from the use of studded tyres in the car fleet and from domestic wood burning. In order to estimate the yearly mean average for benzene, available benzene measurements were used to scale the calculated long term (7 months) average benzene values.

Ambient air concentrations and population exposure were calculated at both the building points and at the model grid cells. Since the air quality target for NO_2 allows 8 hours above 150 μg/m^3, the 9th highest hourly NO_2 concentration computed during the simulation period was stored in each grid cell and building point. If this value was above 150 μg/m^3, the inhabitants of the particular building, or grid cell, were defined as exposed to an exceedance of the target. A similar procedure was employed for PM_{10}, in which exceedances were encountered for the persons living in areas where the 8th highest daily averaged value was higher than 50 μg/m^3. For benzene, exceedances were simply related to the estimated yearly averaged value.

The resulting gridded concentration field for the 9th highest hourly NO_2 values for Oslo is presented in Figure 4.1 below.

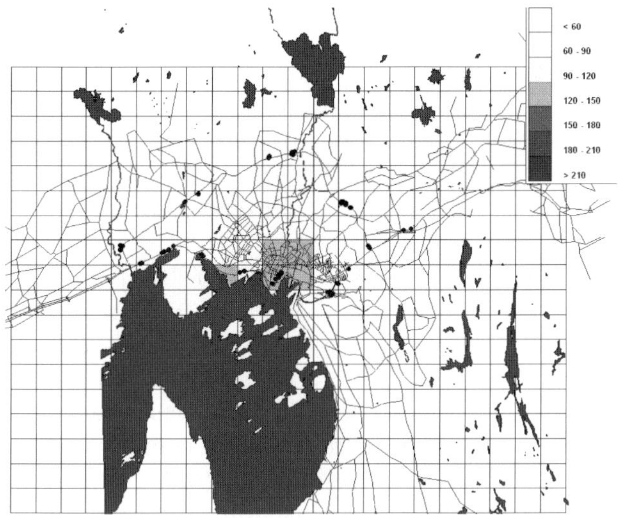

Fig. 4.1. The 9th highest hourly grid value of NO_2 (μg/m^3) for Oslo in 2005. The black dots are illustrating the building points where the 9th highest hourly NO_2 value is above the national target of 150 μg/m^3.

As seen from this figure no exceedances, i.e., no values above 150 μg/m^3, are computed in the model grid. Exceedances are only estimated at a few near-road building points. These buildings are shown as black dots along the main road system in Figure 4.1. The exposure results show that only 652 inhabitants, i.e., 0.12 % of the total population within the model domain, are exposed to exceedances.

The gridded concentration field for the 8th highest daily PM_{10} values is presented below in Figure 4.2. As seen from this figure large areas of Oslo experience exceedances on the grid, or urban background, level, i.e., grid cell concentration values above 50 μg/m^3 are estimated. As expected the model also predicts exceedances at many building points, as illustrated by the black dots along the main road system in Figure 4.2 . In total, it is estimated that 235,849 inhabitants, i.e., 44.8 % of the total population, are exposed to exceedances.

The corresponding model results for benzene (not shown) reveal that just a few grid cells in the central city area experience exceedances with yearly average concentration levels above 2 μg/m^3. However, the model again predicts exceedances in a rather large number of buildings along the main road system. In total it is estimated that 56,547 inhabitants, i.e., 10.7 % of the population, are exposed to exceedances of the benzene target.

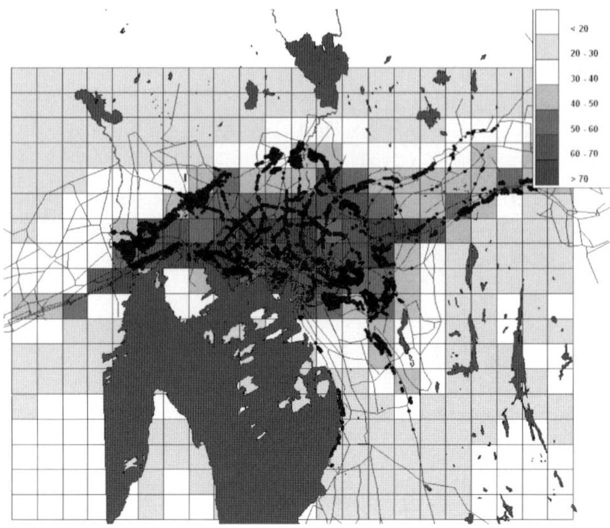

Fig. 4.2. The 8th highest daily grid value of PM_{10} ($\mu g/m^3$) for Oslo in 2005. The black dots are illustrating the building points where the 8th highest daily PM_{10} value is above the national target of 50 $\mu g/m^3$.

4.2 Projections towards 2010

By projecting the emissions towards their expected future values, the above exposure calculations for PM_{10} were also recalculated for the year 2010 [Slørdal et al. 2006b]. This exercise was restricted to PM_{10} since the largest exceedances in 2005 were estimated for this compound. In the calculation for 2010, identical meteorological and regional background data were applied as in the 2005 simulation. With regard to the emissions, however, expected changes in emissions from both wood burning and road traffic were incorporated in the emission inventory applied for 2010. The emission changes from wood burning incorporated the effects of expected replacements of old stoves with new clean burning ovens. The changes in traffic emissions included the combined effects of an expected increase in traffic, and reduced emissions due to improved motor technology and better fuels. The overall effect is a clear reduction in the expected particle emissions from road traffic. In addition, a slight decrease in the use of studded tyres was also assumed in 2010, which contributes to a further reduction in the traffic emissions. Since the influence of the other local sources are of little importance, these sources were kept unaltered.

The resulting gridded concentration field for the 8th highest daily PM_{10} values in 2010 is presented in figure 4.3. As seen from this figure concentration values above 50 μg/m^3 are still estimated in some grid cells in the central city area. As expected the model also predicts exceedances in a large number of building points, as illustrated by the black dots along the main road system in Figure 4.3. In total it is estimated that 66,071 inhabitants, i.e., 12.6 % of the total population, will be exposed to exceedances. This is a marked reduction compared to the 2005 situation.

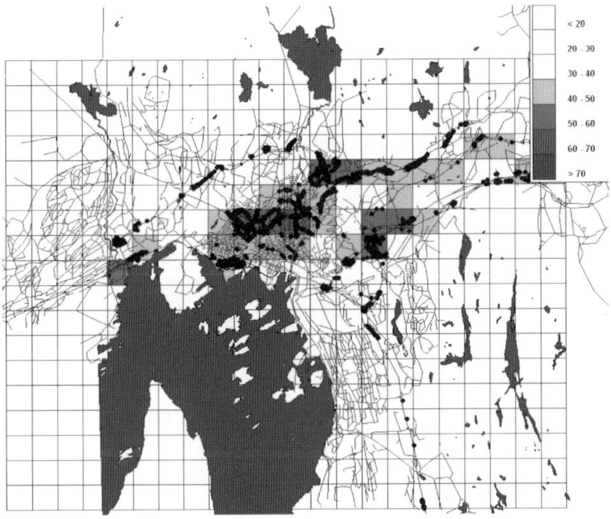

Fig. 4.3. The 8th highest daily grid value of PM_{10} (μg/m^3) for Oslo when projecting the emission inventory towards 2010. The black dots are illustrating the building points where the 8th highest daily PM_{10} value is above the national target of 50 μg/m^3.

As part of the PM_{10} projection work for Oslo, several abatement scenarios were also investigated. One of these scenarios included the combined effect of 1) decreasing the percentage of studded tyres in the car fleet, from 15 % in the 2010 projection to 5 % in the abatement scenario, 2) allowing only use of new clean burning wood stoves, and 3) reducing all speed limits on roads with existing limits above 60 km/h down to 60 km/h (a measure implemented to lower the suspension of particles from the road surface). When applying this emission scenario, the 8th highest daily PM_{10} concentration field presented in Figure 4.4 is calculated. As seen from this figure the abatement measures lead to a strong reduction in the concentration levels compared with the levels predicted in the 2010 simulation (Fig. 4.3). The grid averaged concentration levels are now well below the limit value of 50 µg/m^3, e.g., below 40 µg/m^3, and there is a marked drop in the number of buildings that are experiencing concentration levels above 50 µg/m^3. The inhabitants of these buildings now add up to a total of just 872 persons.

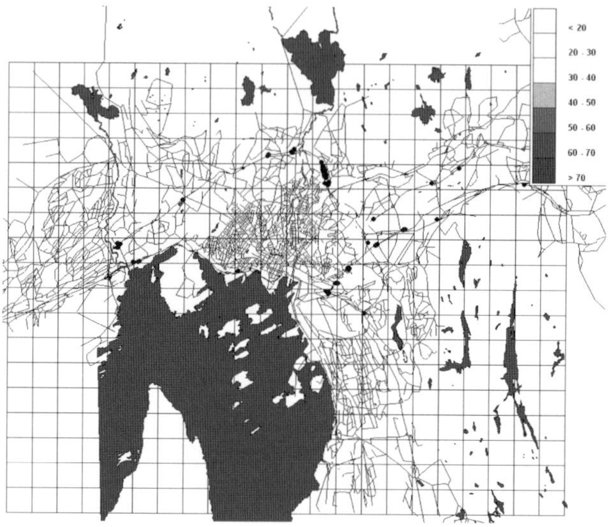

Fig. 4.4. The 8th highest daily grid value of PM_{10} (µg/m^3) for Oslo when performing additional abatement measures on the projected 2010 emission inventory. The black dots are illustrating the building points where the 8th highest daily PM_{10} value is above the national target of 50 µg/m^3.

5 Concluding remarks

Air quality assessment requires a variety of tools for model calculations and data handling. The air quality mangement system AirQUIS has been developed in order to provide easy access to such tools. The combination of a graphical user interface and a flexible database makes the system user friendly and efficient in spite of its complexity.

The evaluation presented in section 3 shows that AirQUIS calculates ambient concentrations of NO_2 and PM_{10} that generally agree well with the observations. The estimated correlation coefficients are rather high for both species, even though better agreement is found for NO_2 than for PM_{10}. The output of the evaluation work, therefore, clearly helps in justifying the use of AirQUIS in air quality applications.

Furthermore, this type of model system represents an important tool in assessing present and future urban air quality. Recalculations with prescribed changes in emissions can give valuable information on the qualitative effect of various types of abatement measures. Moreover, model simulations of this type can be helpful in detecting problem areas, as far as air quality is concerned, at an early stage, and thereby provide local authorities with important information with respect to city planning.

Acknowledgement

The Norwegian Pollution Control Authority (SFT) has supported this work financially. In addition the authors gratefully acknowledge the active involvement from colleagues at NILU.

References

AirQUIS (2006) AirQUIS. http://www.airquis.com, access: 2007-02-16

Report No. 21 [2004–2005] to the Storting. (2005) The Government's Environmental Policy and the State of the Environment in Norway. (Recommendation from the Ministry of Environment of 18 March 2005, approved in the Council of State on the same date. (White paper from the Bondevik II Government)

Slørdal LH, McInnes H, Laupsa H, Walker, SE (2006a) Dispersion and exposure calculations of PM_{10}, NO_2, and Benzene for Oslo and Trondheim for the year 2005. Kjeller (NILU OR 90/2006)

Slørdal LH, McInnes H, Laupsa H (2006b) Evaluation of abatement measures for PM_{10} in Oslo and Trondheim for the year 2010. Kjeller (NILU OR 82/2006)

Ødegaard V, Gjerstad KI, Bjergene N (2006) Bedre byluft. Evaluering av prognosemodell for meteorologi og luftkvalitet vinteren 2005/2006. Norwegian Meteorological Institute (met.no report no. 8/2006), Oslo

Danger Assessment of Potential Resources of Air Pollution once Natural Disasters Occur

Tabassom Hashemi Farzad[1], Ebrahim Omidi Garekani[2]

[1] Islamic Azad University, Science and Research Campus, Tehran, Iran, e-mail: tabbassom@yahoo.com

[2] Faculty of Geodesy & Geomatics Eng, K. N. Toosi, University of Technology, Tehran, Iran, e-mail: e_omidi@sina.kntu.ac.ir

Abstract: Tehran with an area of 780 square kilometers and a population over 10 millions is one of the most polluted cities in the world. Concentration and accumulation of the industrial areas and their inappropriate location are the main reasons of pollution. While, concerning Tehran's topographical and geological status and existence of many active faults, these areas can potentially be considered the main pollutant resources of air pollution once natural disasters occur. Therefore creating a suitable headstock for both optimized management to control air pollution and a management of crisis seems to be necessary.

An Optimum crises management needs identification of risky places before the crises happen. As forecasting and site selection are of the most important functions of Geospatial Information Systems (GIS), in this research GIS is used for modeling and estimation of risk rate. Also for assessment of collected data and classification of urban areas regard to risk rate, we used GIS technology which is able to analyze spatial and attribute data simultaneously.

The first step for this, is defining the risk parameters. In this research, three parameters are defined as main ones: nearness to faults, nearness to potential resources of air pollution and pollutant volume in these resources.

So at first, digital maps of Tehran faults was provided by GIS and saved as an input layer. But inventory of second and third parameters (nearness to potential resources of air pollution and pollutant volume in these resources) would need a time-consuming and costly field work.

For identifying the potential resources of air pollution, we had to gather statistics from industrial centers of Tehran such as chemical stores, fuel storages and position of these resources and volume of pollutants would help us in determining the remained two parameters. The position was de-

termined by GPS and the volume of pollutants was recorded in designed forms.
While defining the combination model in GIS, the gain of emission factors of Inflammable materials was also used for more exact estimation of risk rate. Also for determination of distribution radius (determination of risk radius) and distance which is under effect of pollution, we used estimation of plum rise and meteorology. Using the Gaussian model is the basic of mathematics calculations. After determining the combination model, the model was introduced to software by using overlay functions in GIS environment. Hence, buffering analysis and overlaying of different risk layers were of the most important functions which we used in GIS section of this project.
Therefore, the risky areas around pollutant resources were identified concerning the pollutants' volume. Then through a weighted overlay, different risk layers and buffers of Tehran's faults, together with an estimation of probable pollution rates at the time of the incident were found.
On the basis of the conclusion of this research, not only a database for distribution status of potential resources of pollution was provided, but also it was possible for managers and plan designers to make and propose a monitoring plan and security decisions for identified resources.
Keywords: Crises management, GIS, Fault, Potential resources of air pollution

1 Introduction

One of the most important problems in Tehran (capital of Iran) is the air pollution phenomena. Geographical position of Tehran which is surrounded by high mountains from north, east and south east, has changed situation in a way that air pollution is high in most days of year. Centralization of most pollutant industries in the wind way of dominant west wind which transfer the pollution to city, low amount of annual raining, incomplete combustion of fuel as a result of shortage in oxygen (because Tehran is located in the height of 1,200 meter from ocean level), inversion phenomena in most days of year and extra production and enter of vehicles to city public transportation are the most important factors in air pollution of Tehran. In addition to all of the above, existence of Mash-Fasham fault, north of Tehran fault, Ray fault, which surround Tehran and other fault, which are parallel to mail ones in direction and also Damavand extinct volcano has caused Tehran and it's industries to be a potential source of air pollution in case of natural disasters.

2 Classification of main potential sources of air pollution in Tehran

Main foci relating to maintenance and use of chemical materials

- Huge chemical industries units
- Centers and units for maintenance and storage of chemical materials
- Centers and units for offering and distribution of chemical materials

Main foci relating to combustible materials

- Stations which offer fuels (like service or petrol stations)
- Reservoirs for storage of petroleum materials (petrol or benzene reservoirs)
- Stations which offer gas (gas stations)
- Fuel transferring lines

Maintenance centers and goods storage

- Storage of combustible goods

3 Study and assessment of amount and manner of pollution diffusion from main foci

One of the good samples for these kinds of foci is the refineries and units which purify petroleum materials. We have introduced number, transmission manner, location and other features of them before.

Sulfurous hydrogen which converts to SO_2 rapidly attendant to volatile hydrocarbons (some of them have sulfur contents) and others like these, are the most important pollutant which comes from refineries and purifying units. We should mention Tehran refinery as the most important unit which is at the first place of pollutant releasing and being hazardous compare to other units of this kind.

Transmission radius of pollutants produces by this refinery is between 100 meter (minimal) and 1,500 (maximal) which depends on the height and diameter of chimney (flue pipe) and the type of output parameters and also environmental factors and situations, hence settlement of maximal density pollutant in these distances from the funnels is expected.

Whereas, in fires more than 98 percent of diffused pollutant from refineries and purifying units are enumerated between the parameters similar to those caused by combustion, its expected to have diffusion of large amount of un-combusted hydrocarbons and aerosols from refinery's output in addition to carbon oxides, sulfur oxides, azotic oxides and photo-chemical oxides.

In case of chemical factories, we can point to acid and alkali producing factories, poison and pesticide producing units and also tiner and color material, rubber and adhesive, inks and coatings producing factories; which are of the most important industries that directly effect and cause air pollution. Also, they are enumerated between the potential sources of air pollution.

In assessing the amount of diffused pollutant from fires (potential sources and materials becoming actual), we used coefficient factor of combustible material diffusion. Also, in determining the transmission radius (risky and hazardous radius) and the distance which is under the effect of pollution, we used factors like: assessment of the plum rise from chimney (effective height of raised plum), mean speed of wind in Tehran and some other environmental factors.

However, statistical mathematics, based on famous method called "Gaussian distribution" was the basic of our mathematical calculation and assessments.

4 Final results of (estimation) assessments (calculation done on assumptive data)

Tiner and color material producing units

- Mean distance of effective transmission radius (distance which covers the most part of pollution): 400 meter
- Mean capacity considered for each center: 10,000 kg/day
- Mean amount of main diffused pollutant
- CO = 900kg, NO_x = 480kg, C_xH_y = 1,300kg, TSPM = 350kg, Mists & Fumes = 630kg

Rubber and adhesive producing units

- Mean distance of effective transmission radius (distance which covers the most part of pollution): 300 meter
- Mean capacity considered for each center: 5,000kg/day
- Mean amount of main diffused pollutant

- $CO = 500kg$, $NO_x = 350kg$, $C_xH_y = 750kg$, $TSPM = 450kg$, Mists & Fumes = 350kg, Cl & Halids = 80kg

Inks and coating producing units

- Mean distance of effective transmission radius (distance which covers the most part of pollution): 150 meter
- Mean capacity considered for each center: 3,000kg/day
- Mean amount of main diffused pollutant
- $CO = 300kg$, $NO_x = 150kg$, $C_xH_y = 450kg$, $TSPM = 270kg$, Mists & Fumes = 210kg

Acid and alkali producing units

- Mean distance of effective transmission radius (distance which covers the most part of pollution): 200 meter
- Mean capacity considered for each center: 2,000kg/day
- $Co = 150kg$, $No_x = 100kg$, $C_xH_y = 250kg$, $TSPM = 180kg$, Acid mists = 140kg, $So_x = 90kg$

Poison and pesticide producing units

- Mean distance of effective transmission radius (distance which covers the most part of pollution): 200 meter
- Mean capacity considered for each center: 10,000kg/day
- Mean amount of main diffused pollutant
- $CO = 600kg$, $NO_x = 500kg$, $C_xH_y = 900kg$, $TSPM = 850kg$, Mists & Fumes = 450kg, Cl & Halids = 200kg

Petroleum materials purifying units and refineries (production and storage of petroleum materials)

Tehran refinery

- Mean distance of effective transmission radius (distance which covers the most part of pollution): 1,000 meter
- Mean amount of main diffused pollutant
- $CO = 900kg$, $NO_x = 480kg$, $C_xH_y = 1,300kg$, $TSPM = 350kg$, Mists & Fumes = 630kg, $SO_x = 1,500kg$

Other refinery (Nafte pars, Nafte Tehran, Iranol)

- Mean distance of effective transmission radius (distance which covers the most part of pollution): 750 meter

- Mean amount of main diffused pollutant
- CO = 4,500kg, NO_x = 2,400kg, C_xH_y = 6,700kg, TSPM = 1,800kg, SO_x = 1,000kg

Other units which produce chemical materials

- Mean distance of effective transmission radius (distance which covers the most part of pollution): 300 meter
- Mean capacity considered for each center: 5,000 kg/day
- Mean amount of main diffused pollutant
- CO = 375kg, NO_x = 90kg, C_xH_y = 550kg, TSPM = 340kg, Mists & Fumes = 200kg; SO_x = 185kg

All of the markets and stations which offer and sell chemical materials

- Mean distance of effective transmission radius (distance which covers the most part of pollution): 400 meter
- Mean capacity considered for each center: 10,000 kg/day
- Mean amount of main diffused pollutant
- CO = 3,000kg, NO_x = 2,100kg, C_xH_y = 4,500kg, TSPM = 2,700kg, SO_x = 1,800kg, Mists & Fumes = 2,000kg, Cl & Halides = 480kg

5 Data gathering

Main information layers created in the software regardless of importance and applicable sources are listed below:

- the type of potential source
- the type of materials in each potential source
- working capacity of each potential source
- geographical coordination of these potential sources (latitude and longitude of each source)
- the type of probable pollutants diffused from potential sources becoming actual in case of unpredictable disasters (like fires)
- density of each pollutant parameter diffused, on the basis of calculated assessments (determined functions and equations)
- Vastity of each classified pollution zones on the basis of extremity of probable pollution diffusion (vastity of risky zones caused by pollution diffusion).

Information about the location of storages was achieved by earth assessment (field study) which was accompanied by sampling with GPS.

6 Data collection in data base

As some information had alternative and diverse structure, we had to convert all the formats to a same one which was "Shape file". On the other hand, all data related to each information layer were classified in specific files (each layer has its own file).

Information bank (descriptive data) were also complemented in main layer (data layer), so information related to all of the data were classified in specific files in Coverage or Shape file format as said above.

7 How we analyze data

In this stage we specified buffer zones for assessing and determination of risky and most hazardous regions by creating relative model and also an effective region for fuel offering places. By selecting the relative layer and giving appropriate instructions to software, the effective region for each layer was specified and then classified in a Shape file format.

The fuel offering places are divided into four different classes on the basis of their capacity:

- less than 50,000 liters
- between 50,000 and 100,000 liters
- between 100,000 and 150,000 liters
- more than 150,000 liters

For each class, we considered an effective layer (buffer).

For the first class with the capacity less than 50,000 liters a 100 meter region was considered which is divided into two other 50 meter regions itself.

For the places with capacity between 50,000 and 100,000 liters, we specified a 200 meter (radius) region which divides into two 100 meter regions.

For the places of third and forth class we considered a 300 meter (radius) region - which divide into three other regions (100 meter each)- and 400 meter (radius) region – which is divided into four 100 meter regions- respectively. We have classified the layers in attention to these subdivided regions which determine the zones with the lowest dangers to highest ones.

Also for main fuel reservoirs which contains five huge reservoirs and ray refinery, we specified such a region.

On this basis, information related to amount of So_2, TM, HC, CO, NO_2 was also specified for each fuel offering place and hence effective radius was determined for each.

The described data are also available in information bank (DBE) in each layer.

Four effective layers (1,100 meter radius) was determined for each layer (buffer) on the basis of reservoir's capacity and other data from each of the five important reservoirs, Kan, Besat, Ghoochak, Mehrabad and Ghods; also for the Ray refinery, we specified four effective region which are five kilometer in radius for each.

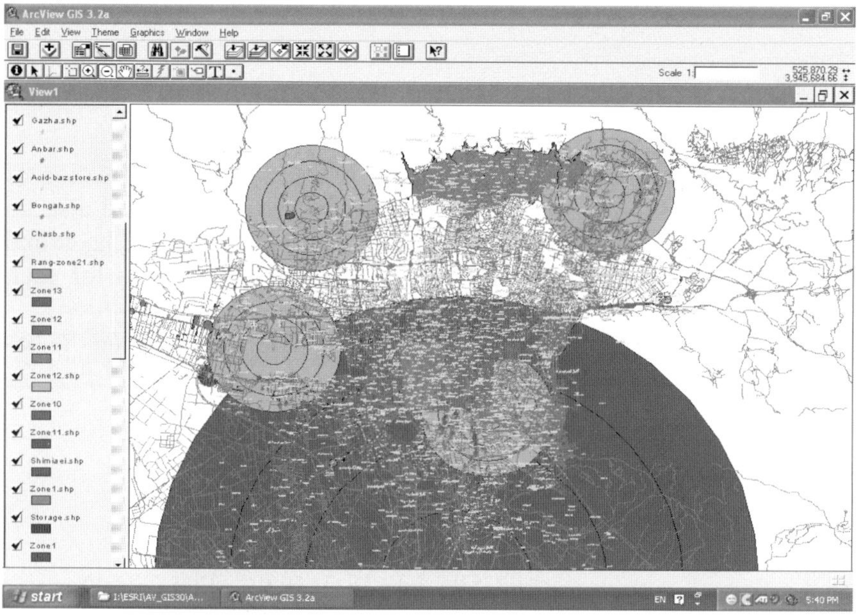

Fig. 1. Overview of layers

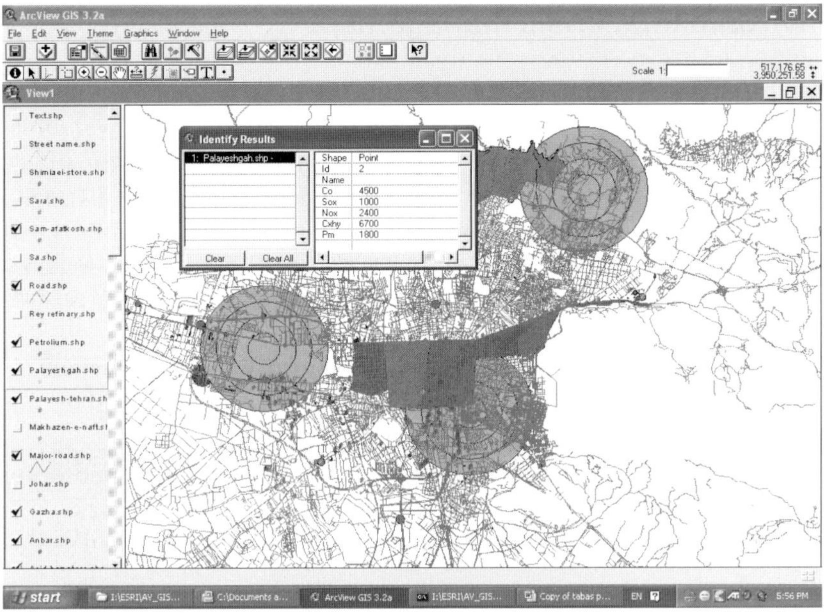

Fig. 2. Ability to work with quality tables like identify

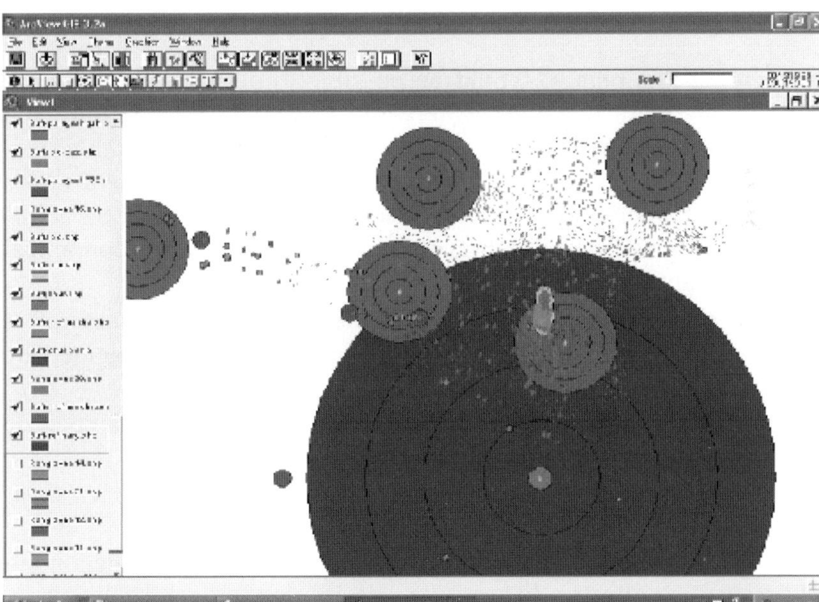

Fig. 3. Risk radius preview

8 Environmental crisis management

In attention to determination of extremity and vastity of effective area for each center in case of natural disasters and also specification of susceptible spots and priority of different zones (on basis of vulnerability situation), we can use it in crisis management in case of occurring natural disasters. In this way, proper knowledge about amount, transmission and number of these sources, can offer the opportunity to reduce the pollution load by modifying legislations and provisions for qualitive and quantitive control and continuous monitoring before the incidents occur.

- Transmission of industries with similar application to out of city zone
- Providing and declaring the detection, inspection and control plans for potential sources of pollution on basis of determined priority.
- Providing and declaring the measuring and environmental monitoring plans for potential sources.
- Providing and declaring the security plans for potential sources of pollution.

9 Conclusion

By implementing this plan, it would be possible to derive and access the main information about potential centers and sources of air pollution and also extremity and effective domain of them in case of unpredictable incidents. Also determination of risky points and priority of different zones would be possible inattention to their vulnerability. So, it would be possible to use the achieved data for preventing and reducing the economical and social disadvantages.

- showing the location of all or a part of potential sources in Tehran
- showing the location of each potential source in Tehran (separated by the type of source activity)
- showing the location of all or a part of potential sources in a specific zone in Tehran (separated by different Tehran zones)
- showing the location of each potential source in a specific zone (separated by the type of source activity)
- showing the vastity of pollution caused by potential source becoming actual (vastity of pollution diffused by each source)
- showing the polluted regions in each 22 zones of Tehran by separation of potential source exist in them

- showing the polluted regions distribution in all part of the city or in each zone by separation of the type of pollutant (chemical, particles, pollutant gas, etc.)
- showing different regions caused by extremity of pollution in each source (showing the diffused pollution distribution in each polluted region)
- getting report from data in each information layer (separated by the type of information)
- getting report from several information layers simultaneously (separated by each Tehran zones)
- getting report from data available in bank by separation of the type of pollutant (CO, NO_2, SO_2, etc.)
- getting report from data available in bank by separation of amount and extremity of diffused pollution
- getting report from data by separation of risky regions for probability of pollutant diffusion from each of the potential sources.
- Determination of all potential sources hazardous probability.
- determining the 12th zone as the most hazardous zone in Tehran because of probable diffusion of hazardous materials and also because of having the highest potential for polluting the air in case of unpredictable disasters.
- determining the Tehran refinery as the highest potential industry unit because of having the most reservoirs for storage of combustible materials.

  Specification of each unit risk level in attention to the following factors:

1. Centers and foci which store or sell chemical materials
2. Centers and foci which store fuels
3. Centers and foci which offers fuels
4. Centers and foci for storage

References

Noroozi V (2002) Monitoring system management, forcasting and informing in critycal condition of air pollution in Tehran. MSc thesis Science and Research Campus, Tehran, Iran

Poorkermani M, Aryan M (1998) Potentiality for earthquakes in Iran. Shahid Beheshti University

Using Topic Maps for Sustainability Reporting

Hans-Knud Arndt[1], Henner Graubitz[1], René Klesinski[2]

[1] Otto-von-Guericke Universität Magdeburg, Universitätsplatz 2, 39106 Magdeburg, Germany, e-mail: {hans-knud.arndt, graubitz}@iti.cs.uni-magdeburg.de

[2] Otto-von-Guericke Universität Magdeburg, Universitätsplatz 2, 39106 Magdeburg, Germany, e-mail: klesinski@web.de

Abstract: In this paper we propose a solution of how to publish and visualize Inter-net-based sustainability reports. After the Global Reporting Initiative (GRI) published the final version of the third generation of GRI-guidelines more and more organizations start to publish sustainability reports on their Internet sites. But these reports are split by environmental topics. Readers are confronted by an in-formation flood without understanding the relationships between all topics in an organization. With the usage of the eXtensible Business Reporting Language (XBRL) we propose a solution of how to transform sustainability reports into Topic Maps (XTM).
Keywords: Sustainability Reporting, Topic Maps, eXtensible Markup Language (XML), XML Topic Maps (XTM), Global Reporting Initiative (GRI), G3-Guidelines, eXtensible Business Reporting Language (XBRL), Financial Reporting Taxonomies Architecture (FRTA)

1 The eXtensible Business Reporting Language (XBRL)

The eXtensible Business Reporting Language (XBRL) is a standard primarily pushed by the American Institute of Certified Public Accountants (AICPA), the largest professional organization of Certified Public Accountants (CPAs) in the United Stated of America. Normally financial information will be stored different in companies. E.g. much of the same financial information in headquarters and its subsidiaries will be stored. But the granularity and the formats of all financial information are different. That's why it makes it hard to exchange this information between these organizations.

While using standards of the approved eXtensible Markup Language (XML) the XBRL standard gives shareholders, banks, investors, or other interested parties the possibility to exchange financial reporting in all its different procedures. These reporting can easily delivered via networks and analyzed with common software products.

In contrast to similar exchange formats which follows the same goal (Research Information Exchange Markup Language (RIXML), Markup Language for the Financial Information eXchange (FIX) Application Message Layer (FIXML), Market Data Definition Language (MDDL), or Open Financial Exchange (OFX)) the XBRL standard can be described as the state of the art standard to exchange financial information and is mainly by international companies. Additionally XBRL is the most widespread format in the North America.

XBRL specifies the syntax of a report and can be defined as a number of report concepts as well as its respective contents [XBRL 2006a]. Both will be separated into instances and taxonomies [XBRL 2006b]. An instance file uses xlinks (eXtended links, [W3C Xlink 2001]) to corresponding taxonomy files which consists of a W3C XML Schema modeling descriptions. Depending on which taxonomy is used these files have the following content:

- *Labels*: A label gives the element in an instance a name. Additionally, while using attributes to corresponding foreign languages and labels, XBRL is able to display instances polyglot.
- *References*: A reference linkbase gives XBRL users the possibility to add information to a label. For example a label can be explained better via a linkbase to its corresponding paragraph or to external descriptions.
- *Definition*: This linkbase is used to bring all different labels into a hierarchical order so that a father-child or child-father relationship can be build up. E.g. this definition linkbase offers the possibility to display sub-instances or subtotals.
- *Presentation*: The presentation linkbase orders child elements in a better granularity. While using the definition linkbase for father-child relationship this linkbase specifies brother-sister relationship.
- *Calculation*: It is the heart of the taxonomies linkbases and contains information about indicator relationships between its corresponding elements. This expression of semantic meaning can also be used to model semantic connections [Binstock et al. 2005].

All taxonomy files in sum give instance files the possibility to display its content in the right format at the right place [XBRL 2006c]. A set of specific taxonomy files is also called Discoverable Taxonomy Set (DTS).

Compared to the XBRL specification taxonomies can be separated into three levels: *recommended taxonomies* must have the same official status of the XBRL specification itself. *Approved* ones are taxonomies which are compliant the XBRL specification and have been developed by the XBRL interest group. In contrast *acknowledged* taxonomies are taxonomies which are used compliant to the XBRL specification but are developed by other interest groups.

2 The Standardization in Sustainability Reporting

Sustainability reports can be described as a development of environmental reports. They contain information about a sustain development of the specific organization. After financial reports sustainability reports are one of the most important possibilities to distribute information about an organization. Special global acting organizations use this possibility to perform social respective duties. It is also a possibility to establish an additional marketing tool for an organization.

Sustainability reports are based on the three pillars environmental, social and economic performance and describe each pillar and existing interrelations between effected pillars. Sustainability reports have been most influenced by the Global Reporting Initiative (GRI), a non-governmental international organization launched in 1997 as a joint initiative of the Coalition for Environmentally Responsible Economics (CERES) and the United Nations Environment Programme (UNEP). The GRI describes its vision "that reporting on economic, environmental, and social performance by all organizations becomes as routine and comparable as financial reporting." Practically this will be done by distributing global applicable guidelines. [Morhardt 2002] expects that "its guideline will become the de facto standard for sustainability reporting worldwide".

With the help of an international network from all kind of business and civil society the third generation (G3) of guidelines has been released and changed after the GRI G3 conference from a former "draft version" to a "ready for use" one [GRI 2006]. The G3-Guidelines can be divided into three main parts: Part 1 defines report content, boundary, and quality, while Part 2 describes standard disclosures. Part 3 contains general reporting notes.

2.1 G3-Guidelines Part 1: Defining Report Content, Quality, and Boundary

Part 1 contains ten basic topics which should be included into a sustainability report. These topics form the basic of the report and can be described the essentials of a report:

1. *Stakeholder inclusiveness:* All stakeholders inside and outside of an organization should be identified and mentioned in the report. The organization has to concentrate on these stakeholders and their feedbacks.
2. *Materiality*: All topics and indicators which effect economic, environmental, and social impacts or stakeholders decisions should be reported. Additionally materiality describes at which threshold topics and indicators should be reported.
3. *Sustainability context:* To get a better overview of an organization and its sustainability the report should describe topics and indicators in wider context of sustainability. Interrelations between sustainability topics and non-topics become clearer.
4. *Completeness:* The organization must report all material which influences the economic, environmental, and social performance. With completeness all its boundaries, scopes, and time periods should be declared.
5. *Balance:* Reports should include all, positive and negative aspects of the organization. Only this guarantees a non-biased view of the organization.
6. *Comparability:* The reporter has to maintain consistency in boundary and scope of its reports, disclose any changes, and re-state previously reported information.
7. *Accuracy:* Following the accuracy principle the reporter reduces errors to a minimum while concentrating on a high degree of exactness. That means stakeholder will make prompt decision because they trust these information.
8. *Timeliness:* Reports should be published frequently in a fixed period. Normally both, the stakeholders "timetable" and the schedule of the information should be respected and inline.
9. *Clarity:* To reach the highest level of credibility the published sustainable report describes all processes, procedures, and assumptions in an understandable and accessible way.
10. *Reliability:* Reported information should be gathered, recorded, compiled, analyzed, and disclosed in a way that would resist every examination, internally and externally.

2.2 G3-Guidelines Part 2: Standard Disclosures

Part 2 mentions basic content which should appear in a sustainability report:

1. *Strategy and Analysis:* Sustainability reports must include a statement of the representatives of an organization e.g. the chief executive officer (CEO). The representative should describe the key impacts, risks and chances which are involved in the publication of the report.
2. *Organizational Profile:* Various information about an organization must be submitted to the report. On the one hand regional, national and international roles should not been missed, on the other hand meta information like the name of the organization, the brands, products and ser-vices should been added.
3. *Reported Parameters:* This guideline is split into 4 parts: the report profile requires information about reporting dates, the cycles of the report, or contacts for questions or feedbacks. The "Report Scope and Boundary" part clarifies which information is reported and to which persons these information are addressed. While describing e.g. data measurement techniques in this part the report is able to show its content and its boundaries. The third part, named "GRI Content Index", describes how a certain report refers to the G3-guidelines. The last part "Assurance" describes internal and external assurances for the sustainability report.
4. *Governance, Commitments, & Engagement:* "Governance", the first part of this section, formalizes the policy, responsibilities, and competencies of the organization. The "Commitments to External Initiatives" presents an overview about the non-organizational/external social, economic and environmental initiatives of the reporters. The last part "Stakeholder Engagement" enumerates all stakeholder engagement beside during the reporting period.
5. *Management Approach and Performance Indicators:* The last part of the standard disclosures includes economic, environmental and social indicators, its management approach and a description of the used performance indicators. The social indicators can be categorized e.g. by Human Rights or Society, the economic indicators impact on the economic conditions, while the environmental indicators impacts e.g. on living and non-living natural systems, including all ecosystems.

2.2 G3-Guidelines Part 3: General Reporting Notes

The G3-Guidelines finalize with general reporting notes. It includes advices of how to gather data, the report form and its frequency, and advices of how to choose assurances.

3 Topic Maps

Organizations in these days concentrate on publishing their sustainability on Internet sites, via ad-hoc messages, news-ticker or print adds like magazines, brochures or newspapers. Typically sustainability is ordered by its topics e.g. strategy, employees, or environment. But this kind of publication results in one big drawback: interested readers loose the big picture while reading a sustainability of only one topic.

Normally all topics influence each other, or one topic is a result of another one. To prevent these drawbacks in computer science the standard ISO/IEC 13250:2000, also known as Topic Maps (TMs), exists. TMs, represented as an idea of the Davenport Group [Pepper 2000], give you an idea of how to visualize knowledge via information technology.

While reading an explanation about a topic and visualizing the associations between other topics TMs sustainability interested user get the chance to understand the big picture of one topic. Additionally TMs can be seen as a new way of e-learning: while using digital publications (TMs) users are able to visualize topics and they can interact with the knowledge. Knowledge in a TM is represented by topics; each topic has an association to another topic which fulfils its information. Additional information can be achieved by so called occurrences: they describe which other sources (e.g. pictures of a topic) can be used to fill up the information of a topic.

TMs can be described as a derivate of the semantic web. The roots of TMs can be found in the structures of how to process knowledge in information technology like classification, building up thesauri or creating indexes for quicker search. While modifying the standard ISO/IEC 13250: 2000 in 2002 the TopicMaps.org author group lift up TMs to ISO 13250: 2003. One of the big reasons for creating a new standard was the limit possibility of using TMs in the Internet [Park and Huntington 2002].

The reliable standard eXtensible Markup Language (XML) was used to create a new TM XML standard (XTM). By using a XML schema TMs got the chance not only to be visualized. Also different datasets and information sets respectively could be merged or used as input for new topic maps. E.g. specific words can be explained by other XTMs from other Internet sources.

One of the big features is the specification of the element scope: XTMs can be represented for different interest groups without recalculating the whole topic map. People interested in an environmental sustainability report of organizations get a distinct view of environmental actions and will not be flooded by financial information. Only by interacting with the scopes of a XTM these people get the chance to request for financial information and receive the association between environmental and financial information.

All these XTM items make a topic map a perfect tool to display and represent sustainability XBRL. Different interest groups will be provided with specific information they are searching for without loosing the big picture in the background [Arndt and Günther 2004]. Additionally while using existing external XTMs these information can be explained and fulfilled without rewriting these information.

4 Transformation Process: Generating an XBRL Topic Map (XBRLTM)

In this section we give an idea of how to transform XBRL sustainability reports into topic maps. All these files are based on the Financial Reporting Taxonomy Architecture (FRTA) 1.0. This section is split into three parts: part 1 describes a proposal of how to transform an XBRL taxonomy to a XTM automatically. To reduce duplicate elements and unnecessary complexity of the produced XTM files part 2 describes processes of how to aggregate parts of elements. In the last part we describe of how to specialize different scope for special user groups.

All transformations are based on a created XBRL taxonomy using sustainability reports. These reports are based on Thiessen [2006] and Arndt et al. [2006].

The process of how to create a Topic Map out of XBRL files can be done in three steps: the first step is the generation of a general valid XTM out of the box, while the second fulfils this XTM with information step by step using the XBRL taxonomy. Part 3 uses external or internal files to visualize different scopes of the created XTMs (not discussed in this paper).

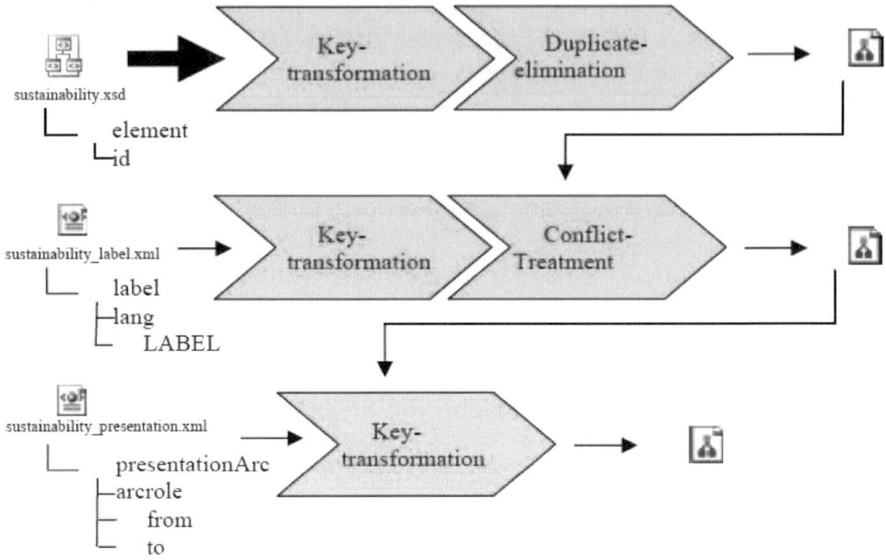

Fig. 1. Transformation Process

4.1 Creation of the XTM

Following [XBRL 2006b] as suggested for XBRL taxonomies elements the specification of schema files can be separated into syntactically and semantic layer. We focus on syntactically ones. Every linkbase contains – for every element – locators. This locator references with the support of its arc elements the equivalent of the schema file. References can be done individually or in pairs. E.g. referencing a father-child node in pairs means also to reference the child-father node. Because the schema file contains XBRL taxonomy root elements this file can be used as the starting point to transform XBRL into XTM. Attributes of the elements will be used as distinct topic identifiers (id) in the XTM. Because the attribute concatenate different father-child nodes our method concentrates on the last part fragment of the attribute name. Another step is the de-duplication of the topics: duplicate attributes will be eliminated by merging its additional information with previous information. Additional XBRL information like data type will be eliminated.

```
< element
  name =" sustainReport "
  id =" sustainReport "
  type =" xbrli:tupleType "
  substitutionGroup =" xbrli:tuple "/>
< element
  name =" sustainReport
      . strategyAndAnalysis "
  id =" sustainReport
      . strategyAndAnalysis "
  type =" xbrli:tupleType "
  substitutionGroup =" xbrli:tuple "/>
```

→

```
< topic id =" sustainReport "/>
< topic id =" strategyAndAnalysis "/>
```

Fig. 2. Extracting the Ids

4.2 Fulfill Information and Extraction of the Base Name

Through the label linkbase and the corresponding previous extracted ids we are now able to extract readable names polyglot. The locator of each element in the label link base corresponds to an element in the schema file. Because we already extracted the corresponding id we can add the base names to the corresponding topic. Element names can be displayed polyglot; additionally these extracted attributes can be used as scopes.

```
< loc
  xlink:type ="locator"
  xlink:href ="sustainability.xsd
        # sustainReport"
  xlink:label ="t_sustainReport"
  xlink:title ="sustainReport"/>
< label
  xlink:type ="resource"
  xlink:label ="t_sustainReport_de"
  xlink:title ="t_sustainReport_de"
  xlink:role ="../ label / standard"
  xml:lang ="de">
  Nachhaltigkeitsbericht          </ label >
< label
  xlink:type ="resource"
  xlink:label ="t_sustainReport_en"
  xlink:title ="t_sustainReport_en"
  xlink:role ="../ label / standard"
  xml:lang ="en">
  Sustainability Report           </ label >
```

```
< topic id ="de">
  < baseName >
    < baseNameString > Deutsch    </ baseNameString >
  </ baseName >
</ topic >
< topic id ="en">
  < baseName >
    < baseNameString > Englisch   </ baseNameString >
  </ baseName >
</ topic >

<! -- Hinzunahme der Sprache im
jeweiligen Scope-->
< topic id ="sustainReport">
  < baseName >
    < scope >
      < topicRef xlink:href ="#de"/>
    </ scope >
    < baseNameString >
    Nachhaltigkeitsbericht
    </ baseNameString >
  </ baseName >
  < baseName >
    < scope >
      < topicRef xlink:href ="#en"/>
    </ scope >
    < baseNameString >
    Sustainability Report
    </ baseNameString >
  </ baseName >
</ topic >
```

Fig. 3. Extracting Polyglot Descriptions

To receive child-father association two linkbases can be used: the definition linkbase and the presentation linkbase. We concentrate on Thiessen [2006] and use the presentation linkbase. The first containing presentation Arc element will be used to represent the XBRLTM hierarchical structure. The additional meta information “from” and “to” will be used to create brother-sister associations.

```
< loc
  xlink:type =" locator "
  xlink:href =" sustainability.xsd
        # sustainReport
        . strategyAndAnalysis "
  xlink:label =" t_sustainReport
        . strategyAndAnalysis "
  xlink:title =" sustainReport
        . strategyAndAnalysis "/>
< presentationArc
  xlink:type =" arc"
  xlink:from =" t_sustainReport
        . strategyAndAnalysis "
  xlink:to =" t_sustainReport "
  order =" 1 "
  xlink:show =" replace "
  xlink:arcrole =" .../ arc/ child-parent "
  xlink:actuate =" onRequest "
  xlink:title =" ... "/>
< presentationArc
  xlink:type =" arc"
  xlink:from =" t_sustainReport "
  xlink:to =" t_sustainReport
        . strategyAndAnalysis "
  order =" 1 "
  xlink:show =" replace "
  xlink:arcrole =" .../ arc/ parent-child "
  xlink:actuate =" onRequest "
  xlink:title =" ... "/>
```

```
< topic id=" child-parent ">
  < baseName >
    < baseNameString > child-parent </ baseNameString >
  </ baseName >
</ topic >
< topic id=" child ">
  < baseName >
    < baseNameString > child </ baseNameString >
  </ baseName >
</ topic >
< topic id=" parent ">
  < baseName >
    < baseNameString > parent </ baseNameString >
  </ baseName >
</ topic >
<! – Etablierung der Hierarchie durch Assoziation-->
< association >
  < instanceOf >
    < topicRef xlink:href =" #child-parent "/>
  </ instanceOf >
  < scope >
    < topicRef xlink:href =" sustainReport ">
  </ scope >
  < member >
    < roleSpec >
      < topicRef xlink:href =" #child "/>
    </ roleSpec >
    < topicRef xlink:href =" #strategyAndAnalysis "/>
  </ member >
  < member >
    < roleSpec >
      < topicRef xlink:href =" #parent "/>
    </ roleSpec >
    < topicRef xlink:href =" #sustainReport "/>
  </ member >
</ association >
```

Fig. 4. Creating Associations

4.3 Specialization of Different Scopes for Special User Groups

The final step of the XBRL to XTM transformation will be the step of adding different member groups to the corresponding scopes. This can be done by using external files or by using available DTS containing member groups (see Fig. 5).

```
< association >
  < scope >
    < topicRef xlink:href =" #insurance "/>
    < topicRef xlink:href =" #aktionaer "/>
  </ scope >
  < member >
    < topicRef xlink:href =" #sustainReport "> </ topicRef >
    < topicRef xlink:href =" #ist-in-statementCEO "> </ topicRef >
  </ member >
</ association >
```

Fig. 5. Adding Different Members to the Scopes

5 Conclusion

In the last couple of years the environmental awareness of organizations changed. Especially global acting organizations started to publish sustainability reports on their Internet sites. With guidelines [GRI 2006] the GRI provide detailed proposals of how to build up these reports.

In this paper, we suggested a method of how to transform and visualize sustainability reports in a better way. Normally these reports are published separated by environmental topics. By using the format of XTM organizations can provide the complete report with the possibility of a scoped Internet-based visualization. Topics can be displayed individually for special user groups. While using the XTM specification sustainability reports can be fulfilled with external information like paragraphs, or dictionary descriptions.

As input we used sustainability reports which have been transformed to an XBRL Financial Reporting Taxonomy Architecture as suggested in Arndt et al [Arndt et al. 2006]. This input has been transformed to XTMs. On the one hand an organization possesses an ordered exchangeable format which gives this organization the possibility to transform or store these reports software independent. On the other site this format can be used to transform and create XTM and visualize a wide-spread information flood in a compact view.

References

Arndt H-K, Günther O (2004) Environmental Reporting of Organizations – Extending the XML-based Standardization Using Topic Maps. In: Minier P, Susini A (eds) Sh@ring (18th International Conference Informatics for Environmental Protection, EnvioInfo CERN Geneva 2004, 21.10.-23.10.2004), Part 2. Geneva, pp 172-183

Arndt H-K, Isenmann R, Brosowski J, Thiessen I, Marx-Gomez J (2006) Sustainability Reporting Using the eXtensible Business Reporting Language (XBRL). In: Tochtermann K, Scharl A (eds) Managing Environmental Knowledge 20th International Conference on Informatics for Environmental Protection, Graz, 06.09.-08.09.2006). Shaker, Aachen, pp 75-82

Binstock C, Hoffman C, v Egmond R, Walenga, W (2005): Comparing XML and XBRL. http://www.ubmatrix.com/Documents/XBRLComparedToXML-2005-07-06%20(4).pdf, access: 2006-01-17

eXtensible Business Reporting Language (XBRL) (2006a) XBRL FAQ. http://www.xbrl.org/Faq.htm, access: 2006-01-17

eXtensible Business Reporting Language (XBRL) (2006b) Extensible Business Reporting Language 2.1, Recommendation -2003-12-31 and Corrected Errata -2005-11-07 http://www.xbrl.org/Specification/XBRL-RECOMMENDATION-2003-12-31+Corrected-Errata-2005-11-07.htm, access: 2006-01-17

eXtensible Business Reporting Language (XBRL) (2006c) Guidance on XBRL Instance Documents and Reports. http://www.xbrl.org/InstanceGuidance/, access: 2006-01-05

Financial Reporting Taxonomies Architecture (FRTA) 1.0 (2005) Recommendation dated 2005-04-25, Corresponding to XBRL 2.1 Recommendation

Global Reporting Initiative (GRI) (ed) (2006) Sustainability Reporting Guidelines, Version 3.0, 24 November 2006, Amsterdam

International Organization for Standardization ISO, International Electrotechnical Commission IEC (eds) ISO/IEC 13250 Topic Maps - Information Technology Document Description and Processing Languages (ISO/IEC 13250:2002, edn 2). Geneva

Morhardt E J (2002) Clean, green, and read all over. Ten rules for effective corporate environmental and sustainability reporting. ASQ Quality Press, Milwaukee

Park J, Huntington S (2002) XML Topic Maps: Creating and Using Topic Maps for the Web. Addison Wesley Longman, Amsterdam

Pepper S (2000) The TAO of Topic Maps - Finding the Way in the Age of Infoglut. In: XML Europe 2000 (ed) conference proceedings, 12-16 June 2000, Le Palais des Congrès de Paris, Paris, France. GCA, 2000

Thiessen I (2006) Konzept und Implementierung eines XBRL Schema zur computergestützten Unternehmensberichterstattung, Diploma Thesis. Technische Universität Kaiserslautern

World Wide Web Consortium (W3C) (ed) (2001) XML Linking Language (Xlink) Version 1.0, http://www.w3.org/TR/xlink, access: 2006-01-17

Conception of System Supported Generation of Sustainability Reports in a Large Scale Enterprise

Christian Grünwald[1], Jorge Marx Gómez[2]

[1] Volkswagen AG, P. O. Box 1837, 38436 Wolfsburg, Germany
e-mail: christian.gruenwald1@volkswagen.de

[2] Oldenburg University, Department of Informatics,
Ammerländer Heerstr. 114-118, 26129 Oldenburg, Germany
e-mail: jorge.marx.gomez@uni-oldenburg.de

Abstract: For a presentation of activities and performance concerning a sustainable development, specific sustainability reports are generated for companies. Characteristics represent an essential part and furthermore a basis of sustainability reports. These indicators collect facts related to economy, ecology and human resources in a concentrated form.
In connection with a case study in cooperation with Volkswagen AG a concept for a system supported generation of sustainability reports is being developed. A result-oriented approach includes actual performance indicators of sustainability reporting in the concept.
The system concept is presented in form of an integrated concept, as in the case study existing source systems for basis information are connected to this module.

1 Motivation

Specific sustainability reports are generated in companies with the target of presenting activities and performance concerning a sustainable development.

The term of sustainable development is, among other things, described by the definition of Brundtland-Commission [1987]: A development that satisfies the requirements of the present without running the risk that future generations are not able to satisfy their own requirements [Loew and Hjalmarsdottír 1996].

As an essential part and even a basis of sustainability reports, characteristics allow a quantitative collection of facts in a concentrated form. Business management characteristics are applied for an illustration of business structures and processes, oriented to decisions. In this case the representation of the individual indicators can be increased by performing comparisons of characteristics in form of business-, time- or target-performance comparisons. The group of environmental characteristics in this case is identified by the connection of facts related to business and environment. The task of collecting required data for a system for environmental characteristics represents a big challenge. However, if basic data for a system for characteristics already exist in a company in a large part, the generation of environmental characteristic numbers is possible, requiring little time and costs [Seidel et al. 1998].

In a more detailed view, the term of sustainability can be seen as a model of three columns. Besides the group of environmental characteristics, a sustainability report contains elements of ecologic and economic characteristics.

This viewpoint is also shared by numerous companies. The result of an analysis of the general principles of companies, which was performed by the Institute of German Economy Cologne in 2004, shows the following facts: Nearly 80 percent of the principles correspond with the general understanding of sustainability, containing ecologic as well as human resources-related targets besides economic aspects [Biebeler 2004].

Several approaches have been developed in the task of automating the generation of the corresponding reports, the sustainability reports.

Considering the trend of moving from print-media to an interned-based online-reporting [Isenmann et al. 2006b], a first result recommends using an XML-based schema for sustainability reporting [Breid 2004]. In further development, an XBRL-taxonomy based on the G3 reporting guidelines of GRI as well as specific XSL-stylesheets have been defined [Isenmann et al. 2006a]. The Global Reporting Initiative (GRI) was founded in 1997 and aims the target of development and distribution of a guideline for sustainability reports that is adaptable worldwide [Brauweiler et al. 2003]. Not to mention last, a concept and a prototype including an engine for workflow-based generation and administration as well as a module for an automated distribution and target-group-oriented presentation of the reports were designed [Isenmann et al. 2006a]. Moreover, approaches of connecting standard source systems to this software have been developed [Isenmann et al. 2006b].

Especially for small and medium scale companies concepts concerning automation as well as receiver-orientation seem to be suitable, as these companies usually show low resources for personnel, budget and information-/ communication-infrastructure [Isenmann et al. 2005].

In large scale enterprises, not rarely a complex system infrastructure exists containing several modules of heterogeneous platforms [Grünwald 2006]. Additional information systems already support subtasks of financial or environmental reporting and have to be considered for an integration.

Based on a scenario of a large scale enterprise, the procedure of the development as well as an architecture for a system supporting the generation of sustainability reports will be described in the following paragraphs.

2 Process Analysis

In connection with a case study in cooperation with the German car manufacturer Volkswagen AG, a concept for a system supported generation of sustainability reports is being developed. The applied approach is target-oriented: In the centre of view is the compliance with existing guidelines for sustainability reports, for example the Sustainability Reporting Guidelines V 3.0 of the Global Reporting Initiative (GRI). The presented work describes the actual state of the project.

In this connection, as a first step, the initial state of the case study was analyzed. The basis for the reports primary consists of characteristics of three fields: Environmental protection, human resources and economics. In form of interviews with the respective departments as well as by the use of existing documentations, the actual procedure of the determination of characteristics was analyzed. The results were illustrated with the help of ARIS toolset in form of process models. Figure 1 shows the entire model.

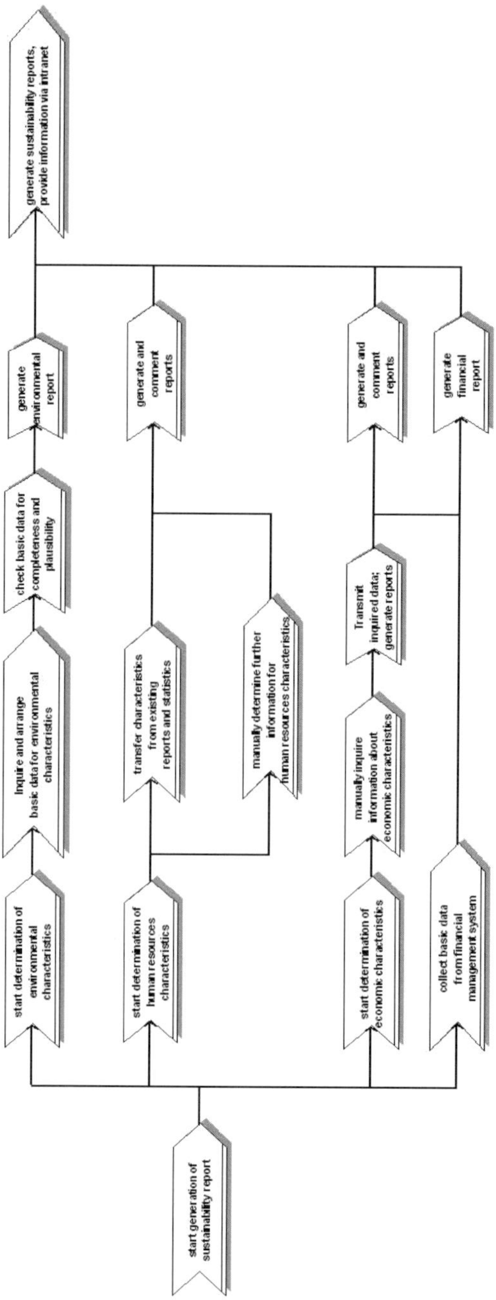

Fig. 1. Process model of generating sustainability reports

Analogous to the dimensions of sustainability, the entire presented process can be segmented into three process parts. The determination and generation of environmental characteristics proceeds manually in the initial state: After the inquiry, the required basic data are transferred to a central responsible employee. At this position, the information is checked manually in terms of data completeness and plausibility. If all basic data were accepted, the generation of environmental characteristics follows.

Different data formats as well as non-uniform communication lines in the transfer of basic data cause additional processing costs for data aggregation to generate the characteristics.

In the future, a special tool for environmental characteristics will support this process for the most part [Griese 2006].

In the field of human resources, existing statistics and reports are used to determine the characteristics. Analogous to the field of environmental characteristics, basic data is additionally entered manually in this process part. Also the generation of the characteristics occurs manually, using standard office tools.

Against that, source systems exist in the field of economic characteristics for a part of the basic data. One basis information in this case are financial statements of the reporting systems. Further, additional required basis information is determined manually.

3 Conception

3.1 Procedure in the project

Target of this project in connection with the case study is a concept for system support for the most part of the described process of generating sustainability reports.

As a first step, a comparison of the requirements determined by existing guidelines for sustainability reports and the available information in the case study is carried out. One example for the guidelines in this case is the Sustainability Reporting Guidelines of GRI. Based on this information, the procedure for the determination of basic data that are required for the generation of the characteristics is defined. Depending on the availability of information, there are three alternatives in this case:

1. Connection of source systems, if basic data for sustainability-characteristics are already managed in data processing systems.
2. Import of existing reports that contain information about sustainability characteristics into a system for the support of sustainability reports. For this alternative, reports have to be in a uniform structure.

3. Web-based inquiry of basis information for sustainability characteristics that are neither available in existing systems nor in statistics or reports that can be automatically read in. In this case suitable forms for data input have to be provided.

3.2 Development of the concept

Based on the analysis results of existing processes as well as available data sources and software, a system architecture for the support of generating sustainability reports was derived. The architecture is shown in Figure 2.

As already described in the section before, three variants are possible for the determination of basic data. Information, that is available in data processing systems of environmental protection, economics or human resources, is imported via specific interfaces into a module for collecting basic data. Additional, an interface of this system component exists that allows the automated import of existing reports and statistics (MS Excel-, MS Word files, etc.), that have been generated in connection with tasks of several departments. The third variant represents the manual input of basic data by the responsible user. Therefore suitable forms are provided, that can be accessed web-based by the user.

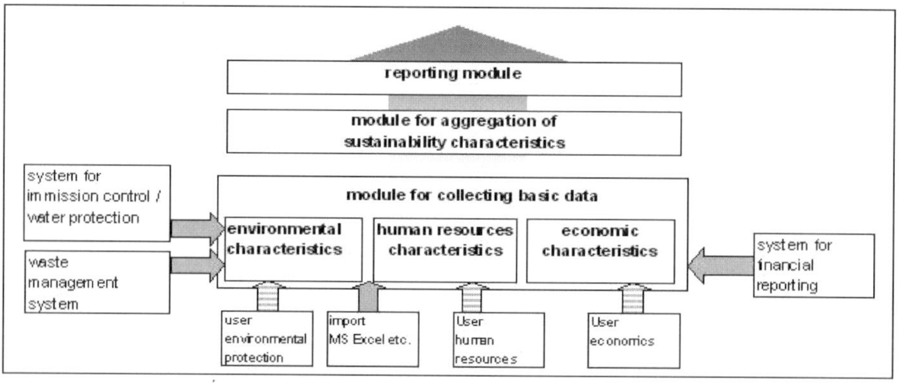

Fig. 2. System architecture for the support of the generation of sustainability reports

A realization based on java-techniques is recommended in this case, combined with XML-based data processing and -transfer [Riekert and Tochtermann 1999; Arndt et al. 1999]. The component for data acquisition is realized by a separated module. Using an identical system module for the three different tasks of environmental, human resources- and economics-related characteristics, results in less costs of implementation, compared to the use of separated tools. Further, rules for process controlling and plausibility checks of data can be included in the system module.

To ensure a flexible further processing of the basic data in form of different kinds of reports, the components for determination, aggregation and reporting are modular separated. Examples for reports in this case are: Parts of the financial statements, environmental reports or even the sustainability report.

For further processing the characteristics are aggregated, according to the determined requirements for sustainability characteristics and visualized by a standard reporting tool.

4 Conclusions and outlook

The development of a concept for a system supported generation of sustainability reports represents the setting of a task in connection with the described project.

In form of a case study, an initial, practice-related scenario was specified. In this connection existing performance indicators for sustainability reports were determined after a comprehensive process analysis. On this basis existing data sources were identified.

The requirements for data output and the information of existing data sources can be compared with the help of these results.

Based on the information about processes and data sources, a system architecture for a technical support of generating sustainability reports was derived.

Future steps will contain a definition of the structure of the used forms for web-based data inquiry, a description of the interfaces to source systems and for data import as well as a detailed specification of functionality and techniques of the application.

References

Arndt H-K, Fugmann T, Günther O (1999) Umweltmanagementinformationssysteme: Der Arbeitsplatz. In: Rautenstrauch C, Schenk M (eds) Umweltinformatik '99: Umweltinformatik zwischen Theorie und Industrieanwendung (13. Internationales Symposium, Magdeburg 1999). Metropolis, Marburg, pp 244-255

Biebeler H (2004) Ergebnisse der Unternehmensbefragung zum nachhaltigen Wirtschaften. Institut der deutschen Wirtschaft Köln, Köln

Brauweiler J, Heling K, Kramer M (2003) Internationales Umweltmanagement - Umweltmanagementinstrumente und -systeme. In: Kramer M, Brauweiler J, Helling K (eds) Internationales Umweltmanagement. Gabler, Wiesbaden

Breid, Joachim (2004) Konzeption und Implementierung eines XML-Schema zur zielgruppenspezifischen internetbasierten Unternehmensberichterstattung. Umgesetzt am Beispiel der Nachhaltigkeitsberichterstattung gemäß der Guidelines der Global Reporting Initiative. Diploma thesis, TU Kaiserslautern

Griese S, Grünwald C, Meyer C, Schrul G (2006) Konzeption und Einsatz umweltrelevanter Systeme in der Volkswagen AG. In: Rey U, Lang-Koetz C, Tuma A (eds) Betriebliche Umweltinformationssysteme – System- und Unternehmensübergreifender Daten- und Informationsaustausch. Shaker, Aachen, pp 157-182

Grünwald C, Marx Gómez J (2006) Technical Concept of an Integrated Environmental Information System – Case Study Volkswagen AG. In: Tochtermann K, Scharl A (eds) EnviroInfo 2006 – Managing Environmental Knowledge (Proceedings of the 20th International Conference "Informatics for Environmental Protection" Graz, Austria). Shaker, Aachen, pp 245-252

Isenmann R, Arndt H K, Brosowski J, Marx Gómez J, Schappert M, Thiessen I (2006a) Software-Werkzeug zum system- und organisationsübergreifenden Austausch betrieblicher Nachhaltigkeits-Information – neuere Entwicklungen. In: Rey U, Lang-Koetz C, Tuma A (eds) Betriebliche Umweltinformationssysteme – System- und unternehmensübergreifender Daten- und Informationsaustausch (13. Tagung der Fachgruppe Betriebliche Umweltinformationssysteme), pp 39-60

Isenmann R, Brosowski J, Schappert M, Marx Gómez J (2005) Software-Werkzeug für Single Source Cross Media Multiple Requirement Sustainability Reporting. In: Lang C V, Rey U (eds) Betriebliche Umweltinformationssysteme – Best Practice und neue Konzepte (12. Tagung der Fachgruppe Betriebliche Umweltinformationssysteme der Gesellschaft für Informatik e.V.). Shaker, Aachen, pp 219-239

Isenmann R, Jordan T, Marx Gómez J (2006b) Workflow-supported Creation and Administration of XML-based Sustainability Reports. In: Tochtermann K Scharl A (eds) EnviroInfo 2006 – Managing Environmental Knowledge (Proceedings of the 20th International Conference "Informatics for Environmental Protection" Graz, Austria). Shaker, Aachen, pp 95-98

Loew T, Hjalmarsdottir H (1996) Umweltkennzahlen für das betriebliche Umweltmanagement. IÖW, Berlin

Riekert WF, Tochtermann K (1999) Hypermedia im Umweltschutz – Erfahrungen aus dem Arbeitskreis Tagungsband 13. Symposium Umweltinformatik, Magdeburg. Metropolis, Marburg, pp 195-205

Seidel E, Clausen J Seifert E K (1998) Umweltkennzahlen. Vahlen, München

Information Management for Sophisticated Environmental Reporting

Ralf Isenmann

University of Bremen, Institute for Project Management and Innovation (IPMI), and Research Center for Sustainability Studies (artec), Wilhelm-Herbst-Strasse 12, 28359 Bremen, Germany, phone: +49 421 218 8229, fax: +49 421 218 8222, e-mail: ralf.isenmann@innovation.uni-bremen.de, web: http://www.innovation.uni-bremen.de

Abstract: The contribution describes an information management approach that elevates the orthodox "one size fits all" disclosure practice of environmental reports to a sophisticated digital stage, using environmental statements according to the European Union Eco-Management and Audit Scheme (EMAS) as example. The information management approach is illustrated along three basic elements: stakeholder analysis and information requirement analysis (representing information demand), XML-based document engineering (modeling information supply), and IT supported reporting system (cross matching information supply and demand). As a result, environmental statements could be developed from universal documents on print media and thus being a mere EMAS requirement to valuable environmental communication vehicles which provide substantial and reliable information in a tailored fashion and are available on various media – due to an underlying single source cross media principle.

1 Introduction to corporate environmental reporting

According to a recent contribution to corporate environmental reporting [Marshall and Brown 2005], nowadays it is merely a question of how to report on environmental issues, and no longer whether to report at all. Marshall and Brown [2005] argue that environmental reporting is becoming part of companies' daily affairs, even entering the business mainstream. Regardless of nationality and differences in country results, this is not just true for organizations with environmental management systems in place, environmental pioneers, and sector leaders, but also for many global players and multinationals [SustainAbility and UNEP 2002; Raar 2002; KPMG

2005], and an increasing number of medium-sized [Remmert 2001; Clausen et al. 2001] and even small sized companies [European Commission 2002] whose activities either result in high environmental impacts or are suspected of causing them. Examples abound in the pharmaceuticals, chemicals, mining, transport, electronics and automotive sectors [FEA 2001; Kolk 2004; Reddy 2005]. Consequently, environmental reporting has become of competitive relevance [Fichter 1998] and strategic importance [Larsen 2000], at least in certain industries, and also for SMEs.

Among the number of vehicles companies are using for the communication of environmental issues [Brophy and Starkey 1996], reports could be seen as the primary and leading instruments. Reports are playing a pivotal role because of their unique credibility and reliability stakeholders ascribe to them, perhaps as they are usually combining qualitative data providing descriptions as well as quantitative data offering facts and figures.

Due to the voluntary status in most countries and a lack of generally accepted standards in terms of contents and structure, there are different approaches of environmental reports currently emerging. For example, the ACCA [2001] identified seven major types like

- compliance based environmental reports,
- toxic release inventory based reports,
- eco balance reports,
- performance based environmental reports,
- product focused environmental reports,
- environmental and social reports, and
- sustainability reports as an integrated fashion (triple bottom line).

The differences between these approaches are depending in parts on nationality and the degree of supplementing environmental issues with social and financial ones.

Probably one the most applied approaches in Europe are environmental statements according to the European Union Eco-Management and Audit Scheme (EMAS). Currently more than 5.000 sites are EMAS registered [European Commission 2007]. Environmental statements could be understood as environmental reports that fulfill the EMAS requirements. Environmental statements include a minimum of required contents. Their overall structure is defined and standardized and usually based on underlying environmental management accounting systems.

These features make environmental statements an excellent source and thus they are regarded here also as a good starting point moving towards integrated approaches like sustainability reports as they assess a company's integrated performance [Schaltegger and Burritt 2000]. For example, they address environmental aspects in monetary terms, and measure a company's impact on nature in physical terms [Burritt et al. 2002]. Reports based on environmental management accounting systems serve as a solid basis for reliable information. These underlying systems are particularly needed to provide integrated performance indicators like eco-efficiency.

2 Reporting requirements according to EMAS

The European Union Eco-Management and Audit Scheme (EMAS) is a voluntary policy instrument and management tool which acknowledges organizations that improve their environmental performance on a continuous basis. EMAS registered organizations are legally compliant, run an environment management system, and evaluate and report on their environmental performance through the publication of an independently verified environmental statement. These publications are recognized by the EMAS logo which guarantees the reliability of the information provided [European Communities 2001].

EMAS was first implemented in 1993 and then revised in 2001. Since its early applications more than 13 years ago, EMAS has rapidly grown to a field of research with increasing relevance to companies, general public, and administration, even through the eyes of non-participants [Clausen et al. 2002]. Through an environmental statement organizations communicate their environmental performance to interested parties, target groups, and other stakeholders. An environmental statement must include a number of detailed requirements specified in EMAS II Annex III, point 3.2. An environmental statement must include at least:

- "a clear and unambiguous description of the organization registering under EMAS and a summary of its activities, products and services and its relationship to any parent organizations as appropriate";
- "the environmental policy and a brief description of the environmental management system of the organization";
- "a description of the environmental objectives and targets in relation to the significant environmental aspects and impacts";
- "a summary of the data available on the performance of the organization against its environmental objectives and targets with respect to its significant environmental impacts. The summary may include figures on pollutant emissions, waste generation, consumption of raw material, energy and water, noise as well as other aspects indicated in Annex VI. The data should allow for year-by-year comparison to assess the development of the environmental performance of the organization";

- "a description of all the significant direct and indirect environmental aspects which result in significant environmental impacts of the organization and an explanation of the nature of the impacts as related to these aspects (Annex VI)";
- "other factors regarding environmental performance including performance against legal provisions with respect to their significant impacts";
- "the name and accreditation number of the environmental verifier and the date of validation."

3 Environmental reporting challenges from an information management perspective

Environmental reporting is a multifaceted, rapidly developing field, influencing a company's communication strategy and image profile as well as its organization, staff, accounting systems, and particularly its underlying information management, and IT capabilities [Isenmann and Marx Gómez 2004a]. Despite certain difficulties companies are struggling at present, there are three trends of strategic relevance for information management facing EMAS registered organizations and environmental reporters today or at least in the near future [Isenmann 2004]:

- integration of financial and social issues into environmental reports,
- provision of reports on various media, and
- fine tuning reports according to users' needs and preferences and exactly meeting numerous standards, guidelines, and recommendations.

Today, an orthodox disclosure practice merely providing isolated environmental statements and standalone environmental reporting instruments on printed media do not seem to be sufficient any longer. A substantial amount of information, matters of communication style, and the provision of tailored reporting instruments and customized communication vehicles on various media are required [Braun et al. 2001; Isenmann and Kim 2006]. Further, environmental reporting is only successful if the underlying information management and accounting systems are appropriate.

4 Information management approach for sophisticated environmental reporting

From a business information systems perspective, an information management approach for sophisticated environmental reporting consists of at least three elements [Isenmann and Marx Gómez 2004b] (Fig. 1):

- Information demand: stakeholder analysis and information requirement analysis,
- Information supply: XML-based document engineering, and
- Cross matching: IT supported reporting system.

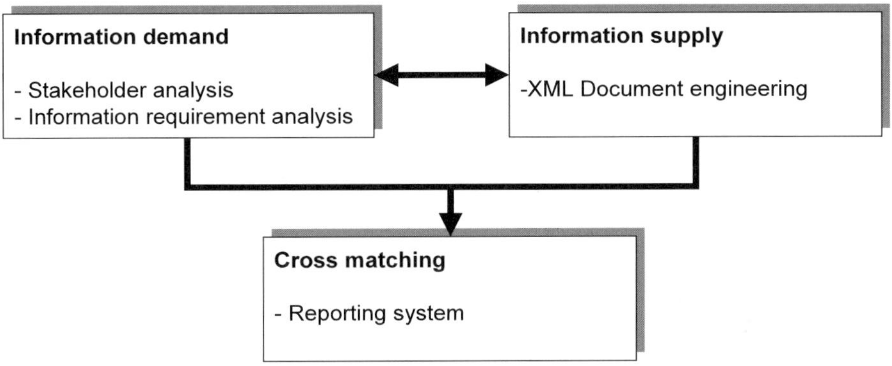

Fig. 1. Information management for sophisticated reporting, illustrated

4.1 Information demand: Stakeholder analysis and information requirement analysis

The starting point of any information management approach is a stakeholder analysis identifying the primary users and typically asking: Who are the relevant stakeholders, including the critical ones? Which key target groups inside and outside the company require information via environmental reporting? According to Lenz [2003] the primary key target groups interested in environmental reports can be arranged in a stakeholder map with four clusters:

- Financial community, including investors, insurance agents, and financial analysts;
- business partners, including employees, customers, and suppliers;
- diffuse groups, including media representatives, neighbors, and consultants;
- normative groups, including local authorities, respective legislators, pressure groups, and standard setting institutions.

To some extent, the users within a certain cluster are indicated through fairly homogeneous information needs. Following stakeholder analysis and identification of primary users, the next step is to study the needs and other preferences expected to be met in report form and content. Such an analysis of stakeholder information requirements determine relevant contents that target groups expect and the preferences they require regarding form, layout, design, media and distribution channel. There is consensus that meeting users' needs is needed for successful environmental reporting (e.g. EMAS II, Annex III, point 3.6).

At present, however, little work has been done to conceptualize users' information needs, especially concerning distribution channels, presentation styles, and media preferences [Azzone et al. 1997; van Dalen 1997]. A considerable analysis on stakeholder information requirements may help answering this need [Lenz 2003] (Fig. 2):

- Employees are interested in environmental performance of their employers and companies.
- Suppliers and business partners in supply chains and manufacturing networks need environmental information regarding resource efficiency, regulatory compliance, new product and service opportunities.
- Investors, financial analysts, and investment consultants are increasingly interested in environmental issues and their financial interrelations. They notice that environmental reports are a valuable source when assessing a company's overall performance.

The stakeholder information requirement analysis clearly demonstrates that target groups usually have heterogeneous information needs. These different needs cannot be fully satisfied or easily met just by reporting "as usual" through an orthodox disclosure practice. The results of the two analyses lend themselves to the creation of specific user profiles. For each of the key target groups, a profile of their information needs will be established that comprises content requirements, preferences as to the reporting form, and secondary requirements such as the distribution channel.

4.2 Information supply: XML-based document engineering

The results of stakeholder analysis and deeper insights of stakeholder information requirements are used for document engineering, indicating the IT-heavy area where contents, structures, procedures, and design of reporting instruments and other communication vehicles are defined. This leads to the questions: How should an advanced environmental report look like? What contents should be included? Who should be addressed? On what devices should the report be available? Which standards or guidelines need to be adhered to? Here, certain aspects of report structure, contents, and layout are explicitly considered.

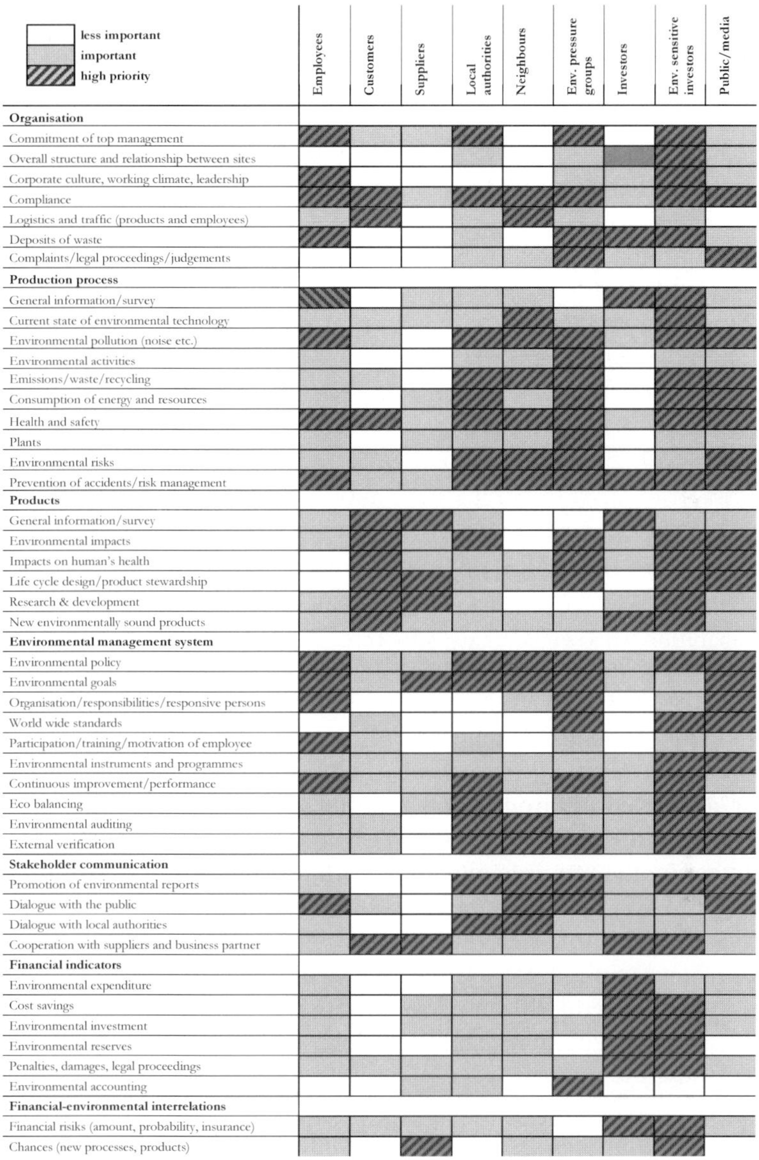

Fig. 2. Stakeholders' information needs for environmental reporting [Lenz 2003, p 232]

Computer scientists [Arndt and Günther 2000], IT experts [Glushko and McGrath 2005] and other reporting professionals [DiPiazza and Eccles 2002] propose the eXtensible Markup Language (XML) as the preferred data format for any advanced reporting approach. The suitability of XML is particularly based on the separation between contents (semantics), report structure (logical order), and representation (layout and style) (Fig. 3).

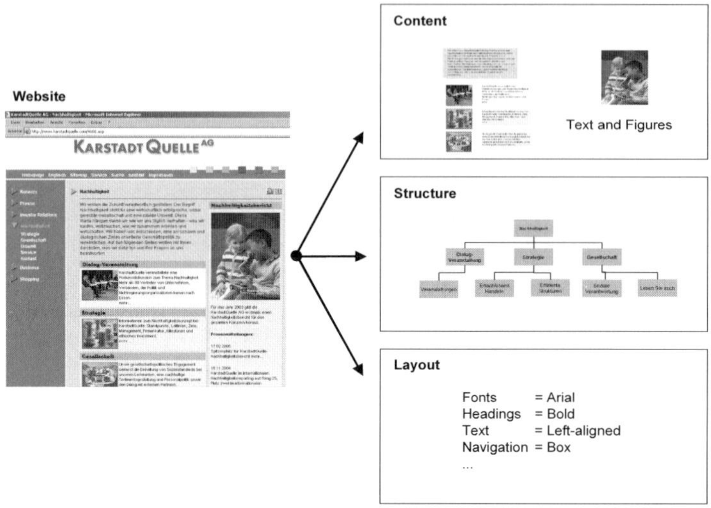

Fig. 3. Separation between contents, structure, and layout of a report using XML

The core of XML-based document engineering is the development of a so-called XML schema. A schema defines the semantics and overall pool of contents in a basic structure for a certain group of documents, in this case, for advanced environmental reports particularly meeting the needs according to EMAS. Of this pool of structured contents, tailored reports exactly meeting the requirements of certain user, user groups, or guidelines (including EMAS) can be prepared in an automated fashion.

The development of a schema is a sophisticated work that needs employing a sound methodology [Brosowski et al. 2004]. As companies usually want to fulfill several different requirements at once, various environmental reporting regulations, standards, and guidelines have been analyzed in a case study, including:

- EMAS II,
- the international standard ISO 14001 on "environmental management systems" [DIN 1996],
- the German standard DIN 33922 "environmental reports for the public" [DIN 1997],
- the early international guideline on "company environmental reporting" [UNEP and SustainAbility 1994],
- its German counterpart "environmental reports – environmental statements" [future and IÖW 1994], and

- a publicly available specification (PAS) on "data exchange between enterprise resource planning (ERP) systems and environmental information systems" [Lang et al. 2003].

From these reporting requirements, available reports, and users' needs and preferences, a catalogue of 115 semantic components specified through certain data types was identified (Fig. 4) and then arranged to a document model (Fig. 5). Using exclusively the EMAS II requirements, 48 semantic components are needed to create an environmental statement.

ID	Description	r/o	Source	Generic Identity
1	foreword	o	future 6.1, II	foreword
2	organisation	r	EMAS II, A III, 3.2	organisation
3	organisation description	o	instances	orgDescription
4	corporate culture	o	users	corporateCulture
5	relationship to parent organisation	o	EMAS II, A III, 3.2	parentOrg
6	sites	o	future 6.1, I	sites
...	... ...	...	...	...
98	economic-environmental interdependences	o	users	econEnvInterdep
99	financial risks	o	users	financialRisks
100	financial chances	o	users	financialChances
101	formalities	r	EMAS II, A III, 3.2	formalities
102	imprint	o	instances	imprint
103	publisher/author/originator	o	DIN 33922, 5.6, instances	authorOriginator
104	publicationDate	o	instances	publicationDate
105	reporting period	o	DIN 33922, 5.6	temporalCoverage
106	date of next report	o	future 6.1, X	nextReport
107	responsibility and participation in env. rep.	o	future 6.1, III	reportTeam
108	contact	o	DIN 33922, 5.6	contact
109	verification	o	UNO 5, I, 11	verification
110	verifier name	r	EMAS II, A III, 3.2	verifierName
111	verifier accreditation number	r	EMAS II, A III, 3.2	verifierAccredNo
112	verifier address	o	DIN 33922, 5.6	verifierAddress
113	verifier statement	o	future 6.1, X	verifierStatement
114	verification date	r	EMAS II, A III, 3.2	verificationDate
115	additional information	o	future 6.1, X	additionalInfo

Fig. 4. 115 relevant semantic components for environmental reports, extract

Currently, this schema is blended into an already existing XBRL Financial Reporting Taxonomies Architecture (FRTA 2005). This reference architecture for sustainability reports based on XBRL (eXtensible Business Reporting Language) will also meet the requirements of GRI's G3 (GRI 2006), the third generation of GRI guidelines [Arndt et al. 2006] (Fig. 6).

Fig. 5. Schema for advanced environmental reports, illustrated

The Global Reporting Initiative (GRI) is a non-governmental international organization launched in 1997 as a joint initiative of the Coalition for Environmentally Responsible Economics (CERES) and the United Nations Environment Programme (UNEP). The goal of GRI is to enhance quality, rigor and utility of sustainability reporting, particularly by developing globally applicable guidelines. Despite the voluntary nature applying its guidelines, the GRI is a rather catalyzing player for a standardized approach in the field and hence Morhardt [2002, p 32] expects that “its guideline will become the *de facto* standard for sustainability reporting worldwide". Further, he concludes, organizations "almost cannot avoid meeting the GRI standard in any case" [Morhardt 2002, p 38].

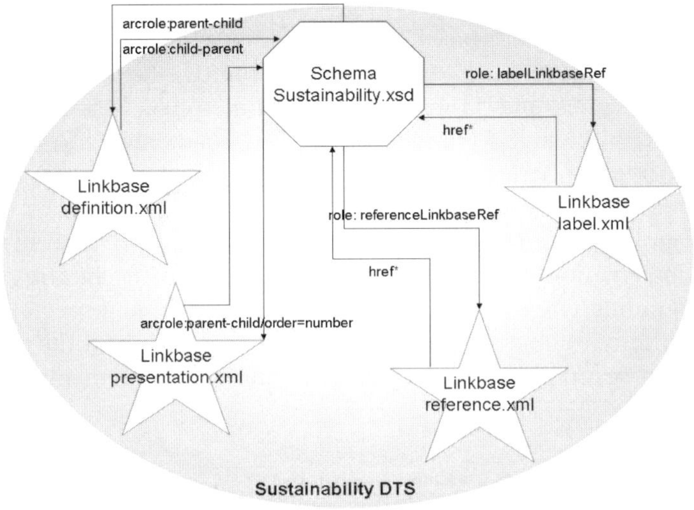

Fig. 6. Reference architecture for sustainability reports

4.3 Cross matching: IT supported reporting system

An IT supported reporting system has to carry out the cross matching between information supply and information demand. Therefore, few software tools are available, e.g. doWEB Online Reporting and Analysis (doCOUNT Gmbh), Enablon Sustainable Development (enablon), Corporate Sustainability Management Software (Proventia Solutions), or SoFi (PE Europe GmbH). Borrowed from [Jones and Walton 1999, 416, own emphasis], there is a need to define an environmental reporting system "that develops [...] disclosures in a *holistic* manner in *all* media."

In order to meet this need, a promising IT supported reporting system has been developed as a practical application [Isenmann et al. 2005]. This development is a joint project embedded in and promoted through the Environmental Informatics community. Currently, the reporting system is implemented as a software prototype using current internet technologies and services. At the heart of its IT architecture lies Cocoon, a Java-based, modular structured Open Source publishing framework, able to process XML schemas and transform and format XML documents. It is thus suitable to perform single source cross media reporting. Environmental reports à la carte are made possible, prepared by machine processing, and generated in an automated manner. The underlying IT architecture allows report contents to be stored, retrieved, edited, updated, controlled and then output cross media in a variety of ways (single source cross media principle).

Currently, a set of interfaces are being developed with the aim that report contents can be taken from accounting records or extracted from other information sources and IT systems, e.g. Enterprise Resource Planning Systems (ERP) like mySAP and SAP R3, Environmental Information Systems (EIS) like Umberto, or Web Content Management Systems (WCM) like RedDot.

When applying the multitude of new opportunities that could be taken from the information management approach above offer an array of benefits for upgrading environmental reporting on the whole, especially for the provision of environmental statements (Fig. 7). These benefits can be described along seven objectives which environmental statements may fulfill [European Communities 2001].

Objectives	**Benefits from internet support**
Documentation	Ease of updating, multi-usability, exchangeability, comfortable retrieval, powerful search facilities, hypertext document structure
Information	Increasing relevance and value for decision making, e.g. through customization, multi-usability of contents, availability in computer-based media
Communication	Opportunities for interactivity and dialogue instead of strict monologue and one-way-communication, e.g. through e-mail, chat, newsgroup, forum, online-communities etc.
Innovation	Opportunities for learning mechanisms, stakeholder input continious exchange of ideas and knowledge, e.g. through online relationships with a number of key target groups
Commitment	More transparency, e.g. through global access around the world and public availability usually without any extra costs
Accountability	Incorporation of accounting and reporting despite different data sources and without any media clashes
Public Relations	Transition to a "quasi public effort" of engaging and involving stakeholders, e.g. through feedback forms, stakeholder commentaries or online "challenger reports"

Fig. 7. Support for the provision of environmental statements

5 Conclusions

Numerous target groups are no longer satisfied solely with environmental statements and other reporting instruments on print media or mere electronic duplicates. Environmental reporting is becoming increasingly relevant for decision-making, and responding to multiple inquiries that a variety of stakeholder groups are making to companies is really time-consuming and costly. Rather than endure these procedures, companies are recognizing the value in having a readily information management for providing the data needed. Many of the questions asked are already answered in meaningful reports.

Hence, it would be helpful to have a proper software tool in place supported through an efficient information management: Users could extract the information they need from a publishing database, and create an automatically generated tailored report themselves, i.e. users could generate their own “reports à la carte”, simply selecting keywords, clicking on preferences on a menu or choosing a certain guideline – perhaps creating an environmental statement according to EMAS, or a comprehensive sustainability report at one's fingertips.

An environmental statement could be an excellent source and thus is regarded as being a core element of any integrated reporting approach like sustainability reporting [GFEM and GFEA 2006], be it available on print media, or posted on the world wide web:

- Firstly, it provides a “true and fair view” as it guarantees the reliability of the information provided. Environmental statements are independently verified, a still open question for sustainability reporting.
- Secondly, an environmental statement includes integrated performance indicators like eco-efficiency. Such quantitative data are crucial to uncover and highlight the mutual interrelations between environmental, social, and financial aspects. They are essential to make the integrated performance transparent and help to conceptualize the “triple bottom line” [Elkington 1997], i.e. the core theme for corporate sustainability.

For example, [Weleda 2003], a German company in the pharmaceutical industry, and [Heidelberger Versorgungs- und Verkehrsbetriebe GmbH 2005], a German public utility and transportation service, are adopting this strategy. In so doing, they have integrated a validated environmental statement into their sustainability report.

Use and benefits of the presented information management approach are not restricted to environmental disclosure practice and its reporting methods only. Due to its generic nature, the approach provides guidance to any document-related reporting domain, be it traditional isolated reporting like financial and social reporting, or sustainability and CSR (Corporate Social Responsibility) reporting as emerging integrated examples.

References

Arndt H-K, Günther O (eds) (2000) Environmental markup language (EML). Metropolis, Marburg
Arndt H-K et al. (2006) Sustainability reporting using the eXtensible Business Reporting Language (XBRL). In: Tochtermnann K, Scharl A (eds) Managing Environmental Knowledge. Shaker, Aachen, pp 75-82
Association of Chartered Certified Accountants (ACCA) (2001) Environmental, social and sustainability reporting on the world wide web. ACCA, London
Azzone G et al. (eds) (1997) A stakeholders' view of environmental reporting. In: Long Range Planning, vol 30, edn 5, pp 699-709
Brophy M, Starkey R (1996) Environmental Reporting. In: Welford R (ed) Corporate Environmental Management. Earthscan, London, pp 177-198
Brosowski J et al. (2004) XML schema for sustainbility reports meeting the needs of the GRI guidelines. In: Minier P, Susini A (eds) Sh@ring. Tricorne, Geneve, pp 184-193
Burritt R et al. (2002) Towards a comprehensive framework for environmental management accounting. Australian Accounting Review, vol 12, edn 2, pp 39-50
Clausen J et al. (2001) The INEM sustainability reporting guide. International Network for Environmental Management (INEM), INEM, Hamburg
Clausen J et al. (2002) The state of EMAS in the EU. Institute for Ecological Economy Research, Berlin
Dalen M van (1997) Company environmental reporting. University of Humberside (UK), Zeeland Polytechnic Flushing, Holland
DiPiazza SA, Eccles RG (2002) Building corporate trust. Wiley, New York
Elkington J (1997) Cannibals with forks. Capstone, Oxford
European Commission (2002) European SMEs and social and environmental responsibility. Observatory of European SME 2002 edn 4, Office for official publications of the European Communities
European Communities (2001) Regulation (EC) edn 761/2001 of the European Parliament and of the Council of 19 March 2001 allowing voluntary participation by organisations in a Community eco-management and audit scheme (EMAS). Official Journal of the European Communities, L 114
European Commission (2007) The Eco Management and Audit Schema (EMAS). http://ec.europa.eu/environment/emas/registration/sites_en.htm, access: 2007-01-06
Fichter K (1998) Umweltkommunikation und Wettbewerbsfähigkeit. Metropolis, Marburg
Financial Reporting Taxonomies Architecture (FRTA) 1.0 (2005) Recommendation dated 2005-04-25, Corresponding to XBRL 2.1 Recommendation
future e.V., Institute for Ecological Economy Research (IÖW) (1994) Environmental reports – environmental statements. IÖW, Osnabrück
German Federal Environmental Agency (FEA) (2002) Global voluntary corporate environmental reporting. FEA, Berlin

German Federal Environmental Ministry (GFEM), German Federal Environmental Agency (GFEA) (2006) Guidance document for voluntarily developing an EMAS environmental statement into a sustainability report. GFEM, Berlin

German Institute for Standardisation eV (DIN) (1996) Environmental management systems – specifications with guidance for use. Beuth, Berlin

German Institute for Standardisation eV (DIN) (1997) DIN 33922. In: Environmental reports for the public. Beuth, Berlin

Glushko RJ, McGrath T (2005) Document engineering. MIT, Cambridge

Heidelberger Versorgungs- und Verkehrbetriebe (2005) Verantwortung im Wettbewerb. Nachhaltigkeitsbericht und Umwelterklärung. HVV, Heidelberg

Isenmann R (2004) Internet-based sustainability reporting. In: International Journal of Environment and Sustainable Development (IJESD), vol 3, edn 2, pp 145-167

Isenmann R et al. (2005) Further developments of a software tool providing tailored sustainability reports. In: Hřebiček J, Ráček J (eds) Networking environmental information. Masaryk University, Brno, pp 684-691

Isenmann R, Kim K (2006) Interactive sustainability reporting. In: Schaltegger S et al. (eds.) Sustainability accounting and reporting. Springer, Berlin, pp 533-555

Isenmann R, Marx Gómez J (2004b) How to provide customized environmental reports properly. In: Scharl A (ed) Environmental online communication. Springer ,London, pp 173-182

Isenmann R, Marx Gómez J (2004a) Editorial. Developments in environmental reporting. In: International Journal of Environment and Sustainable Development (IJESD), vol 3, edn 1, pp 1-4

Kolk A (2004) A decade of sustainability reporting. In: International Journal of Environment and Sustainable Development, vol 3, edn 1, pp 51–64

KPMG (2005) International survey of corporate responsibility reporting. KPMG, Amsterdam

Lang C et al. (2003) Data exchange between ERP-systems and environmental information systems. In: PAS 1025. Beuth, Berlin

Larsen LB (2000) Strategic implication of environmental reporting. In: Corporate Environmental Strategy, vol 7, edn 3, pp 276-287

Lenz C (2003) Empfängerorientierte Unternehmenskommunikation. Köln

Marshall SR, Brown D (2003) Corporate environmental reporting. In: Business Strategy and the Environment, vol 12, edn 2, pp 87-106

Morhardt EJ (2002) Clean, green, and read all over. ASQ, Milwaukee

Raar J (2002) Environmental initiatives. In : Corporate Communications, vol 7, edn 3, pp 169-183

Reddy S (ed) (2005) Sustainability reporting. ICFA University Press, Hyderabad

Remmert B (2001) Ein Unternehmen auf dem Weg zur Nachhaltigkeitsberichterstattung. In: Zeitschrift für Wirtschafts- und Unternehmensethik, vol 2, edn 1, pp 87-100

Schaltegger S, Burritt R (2000) Contemporary environmental accounting issues. Greenleaf, Sheffield

SustainAbility, United Nations Environment Programme (UNEP) (2002) Trust us. Beacon, London

United Nations Environment Programme (UNEP), SustainAbility (1994) Company environmental reporting. In: Technical report, vol 24. Paris
Weleda (2003): Transparenz 3. Weleda, Schwäbisch-Gmünd

Data Quality Mining: Employing Classifiers for Assuring Consistent Datasets

Fabian Grüning

Oldenburg University, Germany,
e-mail: fabian.gruening@informatik.uni-oldenburg.de

Abstract: Independent from the concrete definition of the term "data quality" consistency always plays a major role. There are two main points when dealing with the data quality of a database: Firstly, the data quality has to be measured, and secondly, if is necessary, it must be improved. A classifier can be used for both purposes regarding consistency demands by calculating the distance of the classified value to the stored value for measuring and using the classified value for correction.
Keywords: data mining, data quality, classifiers, ontology, utilities

1 Introduction

A good introduction of the main topics of the field of "data quality" can be found in [Scannapieco et al. 2005] where a motivation is given and relevant data quality dimensions are highlighted. Having discussed an ontology that describes such a definition and the semantic integration of data quality aspects into given data schemas using an ontological approach in Grüning [2006] we now come to the appliance of data quality mining algorithms to estimate the consistency of a given data set and suggest correct values where necessary. This is one of the four identified algorithms needed for the holistic data quality management approach to be developed in a projected funded by a major German utility. One of its goals is to provide an ICT infrastructure for managing the upcoming power plant mix consisting of more decentralized, probably regenerative, and sustainable power plants, e.g. wind power and biogas plants, and combined heat and power generation together with the conventional power plants. As many decisions for controlling relevant parts of the system are made automatically, good data quality is vital for the health of the overall system, as false data leads to wrong decisions that may worsen the system's overall performance. The system is used for both day-to-day business and strategic decisions. Examples for those decisions are the regulation of conventional power plants with the wind forecast in mind to provide an optimal integration of the sustainable power plants like wind power plants into the distribution grid. A strategic decision might be the decision where another wind park is built by taking statistical series of wind measurements into account.

The system contains costumer data as well as technical data about the distribution grid and power plants. The data is critical for the company as well as the state as it contains information about vital distribution systems so that concrete information about the data cannot be given in this paper. Therefore the example given later in this paper will only contain a simple list of dates. The paper focuses more on the concepts of the approach discussed beforehand.

The paper is structured as follows: First we are going to give a short summary of the term data quality mining and the dimensions belonging to it with focus on consistency. We than are going to reasonably chose a concrete classification algorithm that fits our needs in the examined field. The process of using a classifier for checking consistency in data sets is going to be described in the following section giving an example of the algorithm's performance. We are going to touch the subject of using domain experts' knowledge through employing ontologies and eventually getting to conclusions and further work to do.

2 Excursus: About Data Quality Mining

The definition of data quality by Redman [1996] defines four different data quality dimensions: accuracy, consistency, currency as a specialization of timeliness constraints and correctness. After having discussed the semantics of those dimensions in the previous paper we now concentrate on the realization of the algorithms for data quality mining, namely for checking and improving consistency.

The term "data quality mining" is meant in the way that algorithms of the data mining domain are utilized for the purpose of data quality management [see Hipp et al. 2002]. In this paper we are discussing the consistency aspect of data quality. We will explain that classification algorithms are reasonable applicable for this purpose.

2.1 Consistency as a Data Quality Dimension

Whenever there is redundancy in a data set, inconsistencies might occur. A good example is the topology of a distribution grid that consists of power supply lines and connections. An inconsistency in a relational orientated data store leads to non realizable topologies where e.g. a power supply line only has one connection or is connected more than twice. Such a data-centric problem leads to real world problems in the sense that power flow algorithms cannot be applied to the data so that management systems and the power grid get unstable or short circuit cannot be detected or are registered all the time.

This example also shows that a consistency check can only be done by considering a real world entity, here the distribution grid, on the whole and that the verification of consistency works better all the more the semantic correlation between real world entities and data schemas is realized so that relationships between the single data properties can be utilized [see Noy and Guinness 2001; Grüning 2006]. A particular good approach for assuring this correlation is using ontologies for modeling a real world extract as they explicitly keep the relationships inside of and between the examined real world's concepts in contrast to for example normalized relational data schemas.

2.2 Employing Classifiers for Consistency Checking

Classification algorithms are used to classify one data item of a data record by using the information of the remaining data items. E.g. a triple of two power supply lines and one connection implies that those two lines are connected by the very connector. This is only true if the value of the connector is in fact a valid identifier for a connector. If the connector's name is different from a certain pattern that identifies such a connector, a different resource is addressed and an invalid topology is represented. Such dependence can be learned by a classifier.

If the classified value and the stored value differ from one another, an inconsistency in the dataset might have been identified which even can be corrected by using the classified value as a clue.

Classifiers can therefore be found basically usable for finding and correcting inconsistencies in datasets and a prototype will confirm this assumption as shown in the following sections.

To check every possible consistency violation every data item has to be classified with the rest of the data record respectively. It is therefore necessary to train n classifiers for a data record consisting of n data items.

3 Using Support Vector Machines as a concrete Classification Algorithm

There are several different algorithms for classification tasks like decision trees (C4.5), rule sets (CN2), neural networks, Bayes classifiers, evolutionary algorithms, and support vector machines. A decision has to be made which algorithm fits the needs for the classification task in the field of checking consistency in data sets.

The classifiers have in common that their implementation consists of two consecutively phases: In the first phase the algorithm learns through the usage of a representative data set the characteristics of the data. This phase is called the training phase. In the second phase the algorithm classifies not known data records utilizing the knowledge gained from phase one [see Witten and Frank 2005].

There are two main points a classification algorithm has to fulfill in this kind of application:

- The dataset for the learning task in which the algorithms adapts to the characteristics of the data is in comparison to the whole data set relatively small. This is related to the fact that the data set for the learning task has to be constructed out of error-free data so that the classifier will detect and complain about data that differs from these. The labeling, i.e. the task of deciding whether a data record is correct or not, has to be done by domain experts and therefore is a complex and expensive task.
- The classification approach has to be quite general because not much is known about the data to be classified beforehand. A well qualified classification algorithm therefore needs only few parameters to be configured to be adjusted to the classification task.

Both demands are fulfilled by support vector machines (SVM) as they scale well for even small data sets and the configuration efforts are restricted to the choice and configuration of the kernel function that is used to map the training set's samples into the high dimensional feature space and the adaptation of the coefficient weighting the costs for misclassification (see Russell and Norvig [2003] for an introduction to SVMs).

A classifier's output can also be understood as a recommendation in the case where the classified value differs from the stored value. The SVM can both be used as a classification or regression algorithm making it possible to not only give recommendations for discrete but also for continuous values. The algorithm for the regression version of SVM does not differ much from the classifier version so that it is easy to be used either way. Classification and regression can be used nearly synonymously when it comes to SVM because the learning phases do not differ much from one another.

4 Prototype for Consistency Checking

The considerations made so far have to be verified by an appliance to real-world data. For this reason a prototype was developed employing YALE [see Mierswa 2006], a learning environment that allows orchestrating processes that are necessary in the field of learning algorithms. As the whole approach for data quality mining is encouraged by a German utility real data was available for testing purposes.

We will show promising results from a prototype utilizing SVMs as a classification algorithm for checking consistency in a given data set.

4.1 Phase I: Selection

To compile the training set for the first phase of the learning algorithm, a choice out of the existing data records has to be made (see Fig. 4.1).

On the one hand all relevant equivalent classes for the classification task have to be covered which is addressed by the stratified sampling, on the other hand the cardinal number of the training set has to be restricted because of the expensive labeling task for the training set (see section 3). Therefore the absolute sampling assures that only a certain amount of data records are in the training set at most.

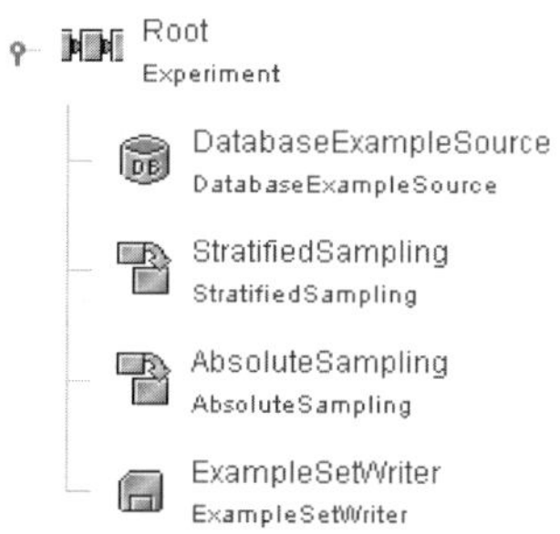

Fig. 4.1. Selection phase

The data itself is converted to interval scale [see Bortz 2005] by one of the following algorithms: If the data originally is in nominal scale the data is mapped to [0, 1] equidistantly. Ordinal data gets normalized and therefore also mapped to [0, 1] where the sequence of the data gets conserved. Strings are addressed separately: They are mapped to interval scale under a given string distance function in a way that similar strings have less distance to one another than less similar strings. The results are clusters of similar strings that get normalized to [0, 1], having obtained a certain amount of semantics.

This preprocessing phase produces data sets that only consist of interval scaled values that are therefore suitable for getting processed via the regression version of the SVM algorithm. We now can use the distance between the outcome of the regression algorithm and the mapped value as a degree of quality. The outcome of the regression algorithm can directly be used as a recommendation for the correct value.

Mentioned as a side note we do not lose any practicability by the data's preprocessing as it is still possible to establish arbitrary bounds to use the classification version of the algorithm.

4.2 Phase II: Learning

In the learning phase the classifier adapts to the characteristics of the data set. This mainly means to adjust the SVM parameter set so that it adapts optimally to the training set. As Russel and Norvig [2003] describe, this means to choose the kernel function that adapts the best to the training set and to choose the correct values for the kernel's parameters for optimal results.

The learning phase consists of several steps (see Fig. 4.2):

1. In the preprocessing phase the data sets are completed where necessary because the classification algorithm cannot handle empty data items. This is no problem as the values filled in are uniform so that they cannot be taken into account for classification because they are not characteristic for any data set.
2. The next steps are repeatedly executed to find the optimal parameter setting for the SVM: The training set is split into a learning and a classification subset as the procedure of cross validation plans. The first set is used for training the classifier and the second set is used for validating the trained classifier. Cross validation avoids a too strict adaptation to the training set so that the classifier only adapts to the characteristics of the training set and does not "mimic" it. Having done that with a defined number of combinations the overall performance of the classifier is evaluated and associated with the parameter configuration.

   The more parameter combinations of the classification algorithms are tested the better the classifier is as the result of this process. This is one of the strengths of the SVMs as only three variables are used to configure a certain SVM in the case when using the radial basis function as kernel function. The parameter space can therefore be mined quite in

great detail for finding the optimal parameter configuration so that the out coming classifier is of high quality.
3. Finally, the optimal parameter configuration is used to eventually train a classifier with the whole training set which gets stored for the last step of the process of finding inconsistent data, namely to apply the classifier to not known data records.

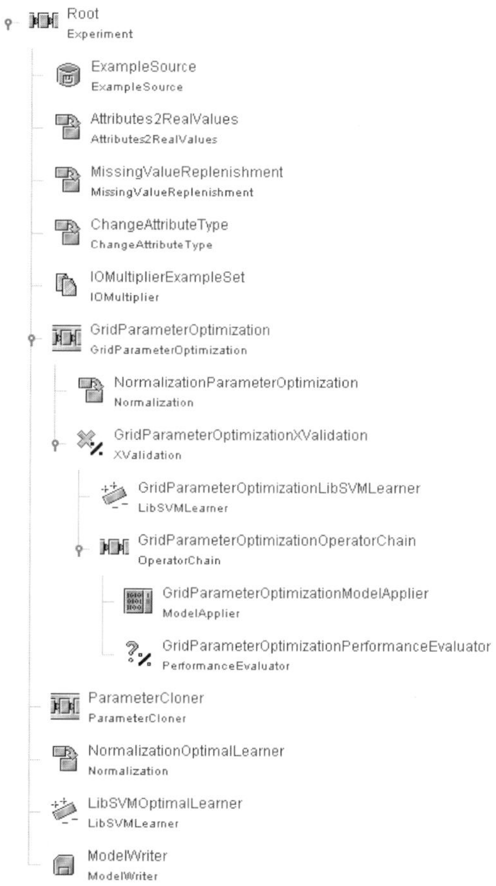

Fig. 4.2. Learning phase

4.3 Phase III: Appliance

In the last phase (see Fig. 4.3) the classifier is applied to the data records of the whole data set searching for discrepancies between classified and stored values. The more discrepancies are found the lower the data quality is regarding the consistency aspect.

As SVMs can also be used for regression, a concrete recommendation for a correct value can be made for the cases where inconsistencies occur. Such a recommendation is not only a range but a concrete value in contrast to other classification algorithms only capable of classifications, like decision trees, again showing the adequate choice of the classification algorithm.

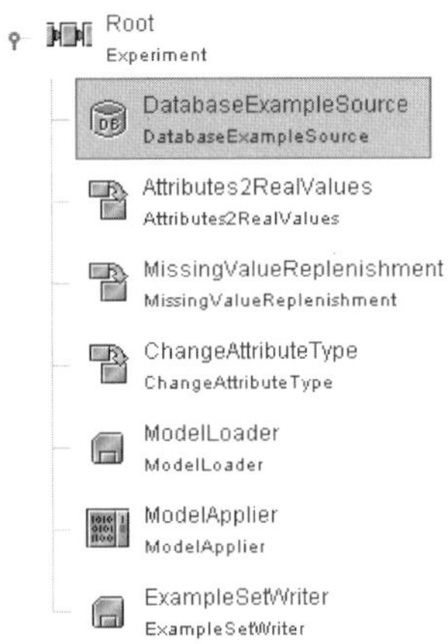

Fig. 4.3. Appliance phase

4.4 Results

A typical result is shown in Table 1. It was generated out of a training set consisting of 128 examples that were proved to be valid. The classifier was then used to find inconsistencies between the classified and the stored values.

In the examples given there are two major discrepancies between the stored and the classified value.

The first one is a result of a deliberate falsification to show the approach's functionality. The correct value had been "1,995" so that the distance relative to the remaining distances between stored and classified values is large and implies an error in the specific data set. The classified value can be used as a correction and meets the non-falsified value quite well.

The second one also shows a huge distance between the classified and the stored value although no falsification has taken place. This is an example that shows that the training set missed a relevant equivalent class so that the algorithm wrongly detects an inconsistency. The user has to mark this wrong classification. Those data sets are then included in the training set so that in the next learning phase the classifier better adapts to the data's characteristics. This procedure may be executed until the classifier has adapted well enough to the relevant data set or regularly to adapt to changes in the underlying structure of the data.

Table 1. Prototype's results sample (classified and stored values are shown)

1994.66	2000.0
1994.66	1995.0
1994.66	1995.0
1992.17	1995.0
1990.26	1990.0
1991.68	1990.0
1990.26	1990.0
1990.26	1990.0
1992.35	2003.0
[...]	[...]

5 Using Ontologies for further Improvements

As already pointed out in section 2.1 the usage of ontologies for modeling the examined real world extract is beneficial for the sake of building a classifier for the discovery of inconsistencies in data sets.

But not only the semantic coherence of the modeled concepts is useful but also further information the modeling domain expert can annotate to the identified concepts. This information is made explicit and can therefore considered to be directly usable knowledge. We gave examples in chapter 4.1 where the information about the values' scale was given by domain experts and annotated to the data scheme. These annotations can be used to configure the data quality mining's algorithms for further improvements of the achieved results by adjusting them to the needs induced by the underlying data schema and the domain expert's knowledge that would otherwise not be available or would difficulty be utilizable.

6 Conclusions and further Work

In this paper it was shown that classifiers can be employed to find inconsistencies in data sets and to give concrete recommendations for correct values. This approach was first made plausible through a discussion together with the decision to employ support vector machines as the classification algorithm and later through the results of a prototype.

For a holistic approach for data quality mining there are still the data quality dimensions accuracy, correctness, and currency open for further research. The solutions for these dimensions will be discussed in upcoming papers.

The positive influence of ontologies for the data quality mining approach in particular and checking for consistency problems in general by employing the additional semantic knowledge in contrast to other modeling techniques was highlighted.

The results presented in this paper were achieved in a project funded by EWE AG (http://www.ewe.de), which is a major German utility.

References

Bortz J (2005) Statistik. Springer Medizin Verlag, Heidelberg.

Grüning F (2006) Data Quality Mining in Ontologies for Utilities. In: Managing Environmental Knowledge, 20th International Conference of Informatics in Environmental Protection

Hipp J, Güntzer U, Nakhaeizadeh G (2002) Data Mining of Association Rules and the Process of Knowledge Discovery in Databases. In: Lecture Notes of Computer Science: Advances in Data Mining: Applications in E-Commerce, Medicine, and Knowledge Management, vol 2394/2002. Springer Berlin/Heidelberg

Noy F N, McGuinness D L (2001) Ontology Development 101: A Guide to Creating Your First Ontology. Stanford Knowledge Systems Laboratory Technical Report KSL-01-05 and Stanford Medical Informatics Technical Report SMI-2001-0880

Redman TC (1996) Data Quality for the Information Age. Artech House, Inc.

Russell S, Norvig P (2003) Artificial Intelligence: A Modern Approach. Prentice Hall

Scannapieco M, Missier P, Batini C (2005) Data Quality at a Glance. In: Datenbank-Spektrum, vol 14, pp 6-14

Witten IH, Frank E (2005) Data Mining: Practical machine learning tools and techniques, edn 2. Morgan Kaufmann, San Francisco.

Mierswa I (2007) YALE Yet Another Learning Environment. http://yale.sf.net/, access: 2007-01-31

Models for Optimization of Electrical Energy Delivery for Regional Utilities

Swantje Nikolai

University of Applied Science OOW (Wilhelmshaven)/
Technical University Clausthal

Abstract: Cost reduction is a constant problem for modern regional utilities. Due to the liberalization of the energy market new possibilities arise. Energy is ordered out of the 110 kV- network of a TSO. Costs for the use-of – system charges and costs for energy trading at the spot market on the European Energy Exchange (EEX) have to be minimized. Technical, economic and legal restrictions must be brought together to reach this aim.
There is an approach on the MV- Level chosen to investigate the possibilities of a regional utility. To have a reasonable size of control-, state- and environmental variables the MV- network is simplified by a mathematically and technically equal model witch includes the same complexity of all relevant variables as well as all types of side conditions.
After the development of the formal model the mathematical description for three time horizons ('intraday', 'day ahead', 'long term') will be developed concerning different entrepreneurial targets relating to the different time horizons. This leads to a discrete- continuous optimal control problem.
For investigation of the applicability of different optimization methods the properties of the optimization models were researched and shown by an example regarding the time horizon 'day ahead'.

Symbols

A_{DP}	max. application a day
$BZ(t)$	operating state of the dispatching potential
h	max. cumulative time steps for a time horizon
$\underline{I}_{12}(t)$	current

I_d	max. valid current
$K_{DP,BZ}$	costs for operating
$K_{DP,WW}$	costs for work
$K_{DP,Z}$	cost for activation
K_P	costs for power according to use- of – system charges
$P_{Bedarf}(t)$	power demand
$P_{DP}(t)$	generated real power of the manipulate facility
$P_{Entnahme}(t)$	demanded real power from the TSO
P_{max}	power level
$Q_{DP}(t)$	generated reactive power of the manipulate facility
$Q_W(t)$	generated reactive power of the WEA
SMP(t)	spot- market- prices
T_{max}	max. time steps of the dispatching potential
T_{mind}	min. time steps of the dispatching potential
T_{Pause}	min. interval time of the dispatching potential
ZS(t)	activation variable of the dispatching potential

1 Motivation

In order to optimize costs liberalization and regulation of the electricity market gives the motivation to energy companies to research new strategies for interventions in grid management. Probabilities for reducing the costs can be found in the power deliveries drawn by the operator of the upper voltage level (TSO). The costs for these power deliveries have to be minimized according to technical, economic and legal restrictions. The delivery processes can be affected by anticipated interventions in grid management. For these interventions decentrally available facilities are used. Therefore a set of values for available load and generation is needed, which lead to minimizing cost, considering different entrepreneurial targets and different planning intervals. To determine these set of values optimization models will be formulated, which form an adequate description of the technical and economic processes also considering side conditions. [Contreras et al. 2002; Lehtonen et al. 1995; Singh et al. 1998]

2 Modeling Approach

The object of investigation is a regional utility. For modeling the technical base, one has to consider the structure of the energy system first. The network can be separated in medium- voltage partial networks which are connected with each other and are supplied due the 110kV- power line. The 110kV- network itself is not operated by the considered company.

The shown structure allows an aggregated look on each partial network. Power flows in and out of a partial network and can be accumulated by separating in manipulable and not-manipulable power flows, which represent the control and the environmental variables respectively. The manipulable power flow is separated once more in available load and generation so it can be considered as one load and one generation respectively.

Because of its structure this energy system can be represented by a principle circuit with aggregated power flows (Fig. 1).

Concerning the economic/ legal modeling only the companies partially process 'grid operation' and 'power supply' are considered.

Given these simplifications the present problem is already very comprehensive. It is characterized by a huge number of control- and the environmental variables and by different extensive alternating effects. Because of its dimension and complexity the total process should be replaced by a reduced process in order to develop a mathematical model. This reduced process has to be comparable to the total process in its main characteristics.

The reduced process has to have the following characteristics:

1. Quality conservation in description of the economic and technical interrelations
2. Reduction of dimension:
   Smallest possible intricacy of control, environmental and state variables
3. Constant complexity:
   Consideration of all types of
   - control, environmental and state variable
   - side conditions

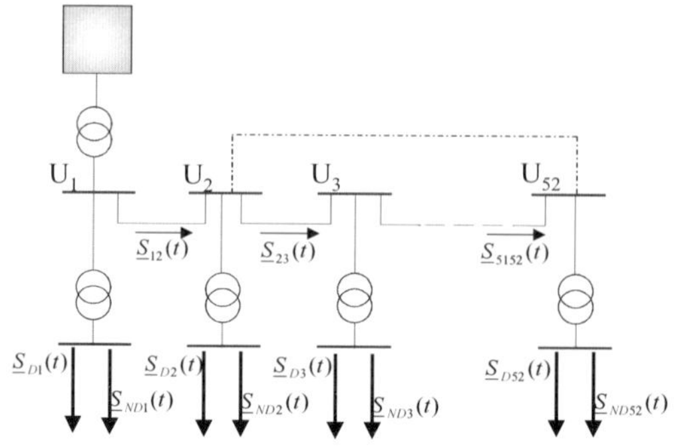

Fig. 1. principal circuit of the modeling energy system

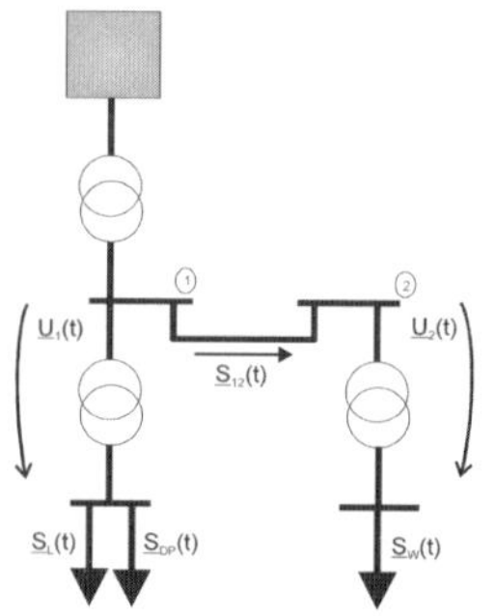

Fig. 2. principal circuit of the reduced energy system

For reaching the demanded reduction of dimension by conservation the complexity the principal circuit is aggregated by a substitute circuit with two partial networks (Fig. 2). The description of the economic and technical interrelations is nearly untouched by this aggregation.

All types of state variables will be combined in one substitute control variable which concentrates all kinds of side conditions in itself.

The vector of environmental variables $\vec{x}_u(t)$ is given by

$$\vec{x}_u(t) = \begin{pmatrix} P_L(t) \\ Q_L(t) \\ P_W(t) \\ P_E(t) \\ Q_E(t) \\ SMP(t) \\ P_{Durch}(t) \end{pmatrix}$$

$P_L(t)$: demanded real power of the load
$Q_L(t)$: demanded reactive power of the load
$P_{W,}(t)$: generated real power of WEA
$P_{E,}(t)$: generated real power
$Q_{E,}(t)$: generated reactive power
SMP(t): Spot market prices
$P_{Durch}(t)$: conveyed real power for strange costumers

and the vector of control variables $\vec{x}_s(t)$ is given by

$$\vec{x}_S(t) = \begin{pmatrix} BZ_{DP,uw}(t) \\ P_{DP,uw}(t) \\ Q_{DP,uw}(t) \\ Q_{W,uw}(t) \end{pmatrix}$$

BZ_{DP} (t): operating state of the manipulate facility
P_{DP} (t): generated real power manipulate facility
Q_{DP} (t): generated reactive power manipulate facility
Q_W (t): generated reactive power of WEA

3 Formal Model

A mathematical optimization model is always an abstraction of reality and therefore not complete. Mostly different reasonable models with different mathematical structures can be developed. The selection of a qualified mathematical description depends fundamentally on two criteria:

- Quality of approximation in description of the total process
- Solvability/ application of optimization methods/ handling

There is no precondition with respect to optimization classes (e.g. determined solver), so a most adequate description is the determinant criterion.

3.1 Mathematical Models

Mathematical models for the three time horizons

- 'intraday' (considers the present day),
- 'day ahead' (considers the planning for next day) and
- 'long term' (considers strategic planning based on the statistic year)

were developed considering different entrepreneurial targets. The targets of the two different processes 'grid operation' and 'power supply' have to be separated because of the legal claim of 'unbundling'. That means the compartments for grid operation and for trading have to be separated under company law. A combination of these targets represents the entrepreneurial target of the 'total process' which one will also consider in order to compare results. Because of the plurality of the models the time horizon 'day ahead' is displayed exemplary in the following.

Optimization model for the total process in time horizon 'day ahead'

Objective function:

$$K_{K,DA}(\vec{x}_S(t),\vec{x}_U(t)) = F_{N,DA}(\vec{x}_S(t),\vec{x}_U(t)) + F_{H,DA}(\vec{x}_S(t),\vec{x}_U(t)) - F_{D,DA}(\vec{x}_S(t)) \longrightarrow \min \tag{1}$$

considering use- of– system charges

$$F_{N,DA}(\vec{x}_S(t),\vec{x}_U(t)) = \begin{cases} \max(P_{Entnahme}(t) - P_{\max}) \cdot K_P & ,\max(P_{Entnahme}(t) - P_{\max}) > 0 \\ 0 & ,\max(P_{Entnahme}(t) - P_{\max}) \le 0 \end{cases} \tag{2}$$

fulfillment of power demand at the spot market at European energy exchange (EEX)

$$F_{H,DA}(\vec{x}_S(t),\vec{x}_U(t)) = \sum_{t=t_{start,DA}}^{t_{end,DA}} P_{Bedarf}(t)\cdot SMP(t)\cdot \Delta T \qquad (3)$$

and costs for the use of dispatching potential

$$F_{D,DA}(\overset{r}{x}_S(t)) = \sum_{t=t_{start,DA}}^{t_{end,DA}} K_{DP,WW}\cdot P_{DP}(t)\cdot \Delta T - K_{DP,B}\cdot \sum_{t=t_{start,DA}}^{t_{end,DA}} BZ_{DP}(t) - K_{DP,Z}\cdot \sum_{t=t_{start,DA}}^{t_{end,DA}} ZS_{DP}(t) \qquad (4)$$

Technical and legal restrictions for the use of dispatching potential:

$$\sum_{t=t_{start,DA}}^{t_{end,DA}} BZ_{DP}(t) \le h_{DA} \qquad (5)$$

$$AS_{DP}(t)\cdot T_{\text{mind}} - \sum_{\tau=t-T_{\text{mind}}}^{t-1} BZ_{DP}(\tau) \le 0 \qquad (6)$$

$$ZS_{DP}(t)\cdot T_{Pause} + \sum_{\tau=t-T_{Pause}}^{t-1} BZ_{DP}(\tau) \le T_{Pause} \qquad (7)$$

$$\sum_{t=t_{start,DA}}^{t_{end,DA}} ZS_{DP}(t) \le A_{DP} \qquad (8)$$

$$\begin{aligned} AS_{DP}(t)\cdot T_{\max} - \sum_{\tau=t-T_{\max}-1}^{t-1} BZ_{DP}(\tau) \ge 0 \quad &, AS_{DP}(t)=1 \\ AS_{DP}(t)\cdot T_{\max} + \sum_{\tau=t-T_{\max}}^{t} BZ_{DP}(\tau) \le T_{\max} \quad &, AS_{DP}(t)=0 \end{aligned} \qquad (9)$$

Power range of the dispatching potential:

$$BZ_{DP}(t)\cdot(-P_{DP,\max}) - P_{DP}(t) \le 0 \qquad (10)$$

$$BZ_{DP}(t)\cdot(-P_{DP,\min}) - P_{DP}(t) \ge 0 \qquad (11)$$

$$BZ_{DP}(t)\cdot(-Q_{DP,Ab,\max}) - Q_{DP}(t) \le 0 \qquad (12)$$

$$BZ_{DP}(t) \cdot Q_{DP,Auf,\max} - Q_{DP}(t) \geq 0 \tag{13}$$

Technical- legal restrictions of the grid:

$$0{,}9 \cdot U_N \leq \underline{U}_2(t) \leq 1{,}1 \cdot U_N \tag{14}$$

$$\underline{I}_{12}(t) \leq I_d \tag{15}$$

Mathematical structure of the optimization model

Formulation of the entrepreneurial target and the technical and economic side conditions on the basis of the reduced configuration leads to a discrete- continuous optimal control problem with the following structure:

- optimal control problem:
  This optimization model is related to optimization class ‚functional optimization'. That means the vector of control variables is a vector of time functions.
- discrete- continuous control variables (mixed integer):
  The model includes both discrete (binary) and continuous control variables.
- With restrictions:
  The model includes both technical and legal restrictions for state and control variables.
- No Markow- property:
  State of the process at time t is not independent of state at time t-1.

3.2 Applicability of optimization methods

Different classes of optimization methods are applicable for the problem class 'optimal control problems' including

- Analytic methods
- Enumeration methods (e.g. branch and bound, Dynamic Programming)
- Target- oriented search techniques (e.g. Newton method)
- Heuristic search techniques (e.g. Monte Carlo, Evolution Strategies)

For using analytic methods, thorough and specific knowledge of the function of environmental variables is needed. Furthermore it is a precondition that the objective function is continuously differentiable in sections. In this case both are not given so these optimization methods are not applicable.

It is the same with the precondition concerning the differentiability in the case of the target- oriented search techniques. For locating the global optimum using these search techniques it is an additional precondition that the optimization model is convex. Thus these methods are not applicable either.

Enumeration methods (not total enumeration) are based on a realization of the process as a stage process. The Markow- property is a precondition of this method. In this case these methods are not available because the function value of every step of the process depends on the future steps. It seems that optimization problems with this kind of mathematical structure can only be solved adequately by search techniques. The question if standardized heuristics (e.g. Evolution Strategies) are promising for this type of problem should be answered by analyzing properties of the solution space. [Bomze and Grossmann 1993; Kost 2003; Richter 1982; Sebastian and Sieber 1981]

4 Properties of the optimization model 'day ahead'

For analyzing the solution properties of the optimization model scenarios with different environmental values and different types of dispatching potential were researched. Basis is the optimization model for the reduced process.

All facilities have the same power range starting with P_{Min} = 0MW to P_{Max} = 160 MW with a discretisation of 1 MW.

The facilities were classified by their operating restrictions:

- *Base type*: operation is possible once a day for a one time step (1h)
- Type *generation*: operation is possible once a day for one to six time steps
- Type *demand*: operation is possible three times a day for one time step each, it has an interval time of one time step

4.1 Impact of the operation restrictions

The size of the solution space is given by the number of all possible solutions. There are $161^{24} \approx 9.2 * 10^{52}$ possible solutions. Depending on the operation restrictions valid solutions for the different dispatching facilities are given:

- *Base type* → 161*24 = 3864 valid solutions ≈ $4{,}2*10^{-48}$ %
- Type *generation* → 161*130 = 20930 valid solutions ≈ $2{,}28*10^{-47}$%

- Type *demand* → 1818*24 = 43632 valid solutions ≈ $4{,}74 \cdot 10^{-47}$ %

These calculation samples show a dominant impact of the operation restrictions on the structure of the solution space. The number of valid solutions is marginal compared to the dimension of the solution space.

4.2 Impact of the entrepreneurial targets

The following results are displayed exemplary for one scenario from of the type *base type.* The serial number of the discrete control variable is equal to the hour of operation. Qualitative characteristics, which can be identified for this type of dispatching potential, can be transferred on the other types. The described characteristics then appear in repeated.

The function differs for the different entrepreneurial target. Following targets were researched:

Derivative trade: considers the process 'power supply'
Derivative grid: considers the process 'grid operation'
Derivative total: considers the process 'total process'

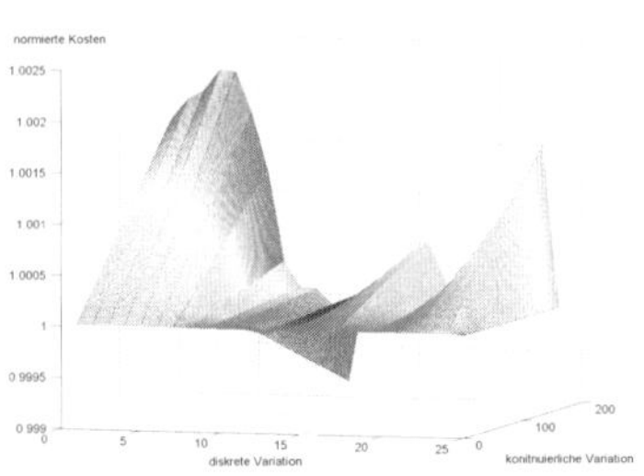

Fig. 3. objective function derivative trade

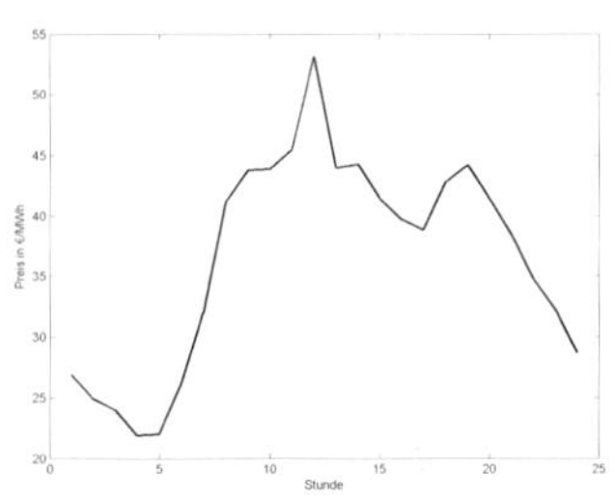

Fig. 4. spot market prices

The objective function of derivative trade shows a plurality of distinct maxima and minima as well as a few distinct minima (Fig. 3). The variation of costs depends on the function of the spot market prices (Fig. 4)

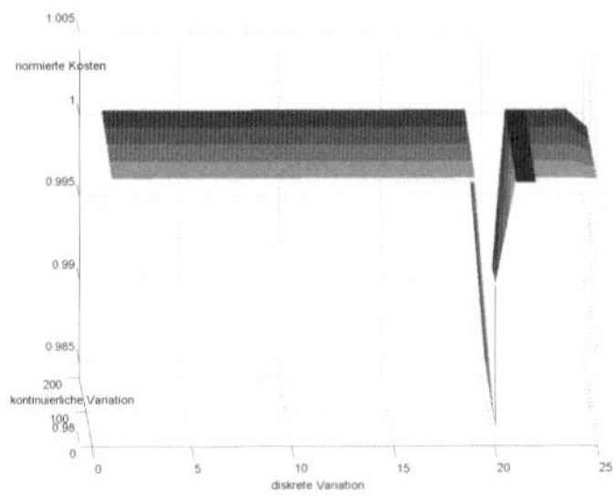

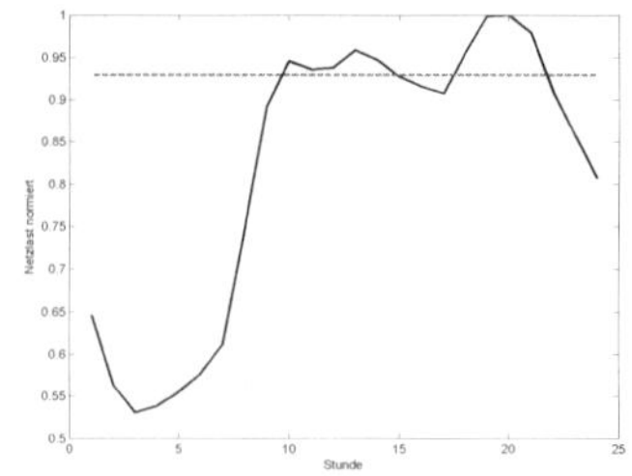

Fig. 5. objective function derivative grid

Fig. 6. grid load with power level

The graph of the costs for derivative grid shows also some low and a few distinct minima (Fig. 5). For this entrepreneurial target the variation of costs depends on the time of exceeding the power level (Fig 6).

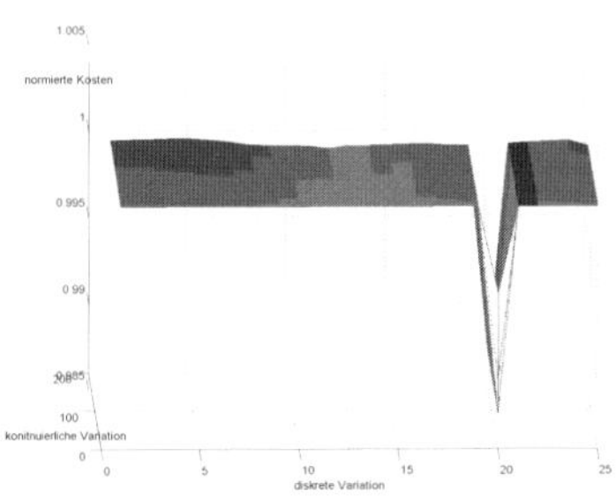

Fig. 7. objective function derivative total

The objective function of the derivative total shows a dominant impact of the entrepreneurial target of grid operation (Fig. 7).

For all entrepreneurial targets the objective function is not convex.

4.3 Valuation of the properties

The number of expected valid solutions in case the parameters have practice relevance is very low as it is shown in 'Impact of the operation restrictions'. Furthermore there are only a few distinct minima for all entrepreneurial targets as it is shown.

Standardized heuristics reach these distinct minima only at a low rate of probability. Thus it seems that these methods are not applicable. Therefore heuristics were needed which work with meta- information in order to find the less distinct minima in the small valid solution space.

The discrete control variable is the important factor for reaching the less distinct minima as it is shown in the figures of the operative functions. This is similar for all three derivatives.

There is low interconnection between the discrete and the continuous control variable so they can be considered in separate models.

5 Decomposition of the optimization model

Because of its properties the complex optimization model can be separated and formulated as two separate partial models, one discrete and one continuous model. They can be handled hierarchical. First qualified (able to give an improvement) combination of operating states with constant power will be identified. After that the power can be adapted.

Because of the decomposition among other things the outcome is a discrete optimization model. This model has a state space which dimension is reduced compared to the original model.

Objective function of the discrete model:

$$K_{K,DA}(\vec{x}_S(t),\vec{x}_U(t)) = F_{N,DA}(\vec{x}_S(t),\vec{x}_U(t)) + F_{H,DA}(\vec{x}_S(t),\vec{x}_U(t)) - F_{D,DA}(\vec{x}_S(t)) \longrightarrow \min \tag{16}$$

considering use- of– system charges

$$F_{N,DA}(\vec{x}_S(t),\vec{x}_U(t)) = \begin{cases} \max(P_{Entnahme}(t) - P_{\max}) \cdot K_P & ,\max(P_{Entnahme}(t) - P_{\max}) > 0 \\ 0 & ,\max(P_{Entnahme}(t) - P_{\max}) \le 0 \end{cases} \tag{17}$$

fulfillment of power demand at the spot market at European energy exchange (EEX)

$$F_{H,DA}(\vec{x}_S(t),\vec{x}_U(t)) = \sum_{t=t_{start,DA}}^{t_{end,DA}} P_{Bedarf}(t) \cdot SMP(t) \cdot \Delta T \tag{18}$$

and costs for the use of dispatching potential

$$F_{D,DA}(\vec{x}_S(t)) = \sum_{t=t_{start,DA}}^{t_{end,DA}} K_{DP,WW} \cdot P_{DP}(t) \cdot \Delta T - K_{DP,B} \cdot \sum_{t=t_{start,DA}}^{t_{end,DA}} BZ_{DP}(t) - K_{DP,Z} \cdot \sum_{t=t_{start,DA}}^{t_{end,DA}} ZS_{DP}(t) \tag{19}$$

Technical and legal restrictions for the use of dispatching potential:

$$\sum_{t=t_{start,DA}}^{t_{end,DA}} BZ_{DP}(t) \leq h_{DA} \tag{20}$$

$$AS_{DP}(t) \cdot T_{\text{mind}} - \sum_{\tau=t-T_{\text{mind}}}^{t-1} BZ_{DP}(\tau) \leq 0 \tag{21}$$

$$ZS_{DP}(t) \cdot T_{Pause} + \sum_{\tau=t-T_{Pause}}^{t-1} BZ_{DP}(\tau) \leq T_{Pause} \tag{22}$$

$$\sum_{t=t_{start,DA}}^{t_{end,DA}} ZS_{DP}(t) \leq A_{DP} \tag{23}$$

$$\begin{aligned} AS_{DP}(t) \cdot T_{\max} - \sum_{\tau=t-T_{\max}-1}^{t-1} BZ_{DP}(\tau) \geq 0 \quad &, AS_{DP}(t) = 1 \\ AS_{DP}(t) \cdot T_{\max} + \sum_{\tau=t-T_{\max}}^{t} BZ_{DP}(\tau) \leq T_{\max} \quad &, AS_{DP}(t) = 0 \end{aligned} \tag{24}$$

Constant power of the dispatching potential:

$$BZ_{DP}(t) \cdot (-P_{DP,const}) - P_{DP}(t) = 0 \tag{25}$$

6 Summary

In order to minimize the costs for energy delivery drawn from the upper voltage level optimization models were developed. Therefore a modeling approach was designed which represents the main technical and entrepreneurial processes. Based on this modeling approach mathematical optimization models for different time horizons and different entrepreneurial targets were formulated.

Because there were no preconditions with respect to optimization classes a most adequate description of the processes was aspired. An optimization model with a high quality of approximation in the description leads to a discrete-continuous optimal control problem with restrictions.

There are a lot of optimization methods for this kind of problem but in this case most of them are not applicable because of the mathematical structure. The only type of optimization methods which can be used without simplifications on the models are heuristic search techniques.

For a well-founded choice of a possible heuristic solution properties of the optimization model 'day ahead' was analyzed. The analysis arrives to the conclusion that standardized heuristics are not applicable for this optimization problem because they will find the distinct minima only with a very low rate of probability. Thus heuristic which apply meta-information were needed.

Another conclusion of the analysis is that the optimization model can be separated and formulated as two partial models which can be solved separately.

References

Bagemihl J (2003) Optimierung eines Portfolios mit hydro-thermischen Kraftwerkspark im börslichen Strom- und Gasterminmark. Diss. Universität Stuttgart, Fakultät Energietechnik

Bomze IM, Grossmann W (1993) Optimierung- Theorie und Algorithmen: Eine Einführung in Operations Research für Wirtschaftsinformatiker. BI- Wiss.-Verlag, Mannheim

Contreras J, Losi A; Russo M, Wu FF (2002) Simulation and evalutation of optimisation problem solutions in distributed energy management systems. In: IEEE Transactions on Power Systems, vol 17, edn 1, pp 57 – 62

Hable M (2004) Beitrag zur Energieeinsatzoptimierung mit evolutionären Algorithmen in lokalen Energiesystemen mit kombinierter Nutzung von Wärme- und Elektroenergie. Diss. Technische Universität Dresden, Fakultät für Elektrotechnik und Informationstechnik,

Kost B (2003) Optimierung mit Evolutionsstrategien, edn 1. Harri Deutsch Verlag, Frankfurt am Main

Lehtonen M et al. (1995) Distribution energy management in the environment of de- regulated electricity market. In: Energy Management and Power Delivery. Proceedings of EMPD '95, vol 2, pp 516-521

Richter K (1982) Dynamische Aufgaben der diskreten Optimierung. Akademie Verlag, Berlin

Sebastian H-J, Sieber N (1981) Diskrete Dynamische Optimierung, edn 1. Akademische Verlagsgesellschaft Geest u. Porting, Leipzig

Singh N, Kliokys E, Feldmann H, Kussel R, Chrustowski R, Joborowicz C (1998) Power system modeling and analysis in a mixed energy management and distribution management system. In: IEEE Transactions on Power Systems, vol 13, edn 3, pp 1143- 1149

Implementing and Evaluating the Common Information Model in a Relational and RDF-based Database

Stefan Schulte[1], Rainer Berbner[1], Ralf Steinmetz[1], Mathias Uslar[2]

[1] Multimedia Communications Lab (KOM), TU Darmstadt, Germany,
e-mail: {schulte, berbner, steinmetz}@kom.tu-darmstadt.de

[2] OFFIS, Business Information Management, Oldenburg, Germany,
e-mail: uslar@offis.de

Abstract: During the last decade, the Common Information Model (CIM) has evolved to an extensive ontology for the domain of energy markets. As the CIM does only offer an UML model for the implementation of its objects, an ER model or relational database schema has not been released. Therefore, it is necessary to create a CIM based database schema in order to persist CIM data in a relational database. This schema could either be constructed based on the former mentioned UML model as a relational database schema or based on an already existing RDF/XML serialization of CIM as an RDF database. This paper evaluates these two implementations of the CIM.
Keywords: Common Information Model, CIM/XML, RDF, power system data exchange, energy management systems

1 Introduction

Not only the electrical power outages in the US and Switzerland in 2003 have shown the need for fast and automatic data exchange between companies which operate in power markets. Furthermore, factors like the increasing deregulation of such markets and the accompanied distinction between network operators and companies accessing power networks resulted in a need for integrated energy management systems (EMS). Such systems are based on a common management and a capable data exchange format [Becker et al. 2000].

In this paper, different approaches for persisting CIM/XML data in a relational database are proposed and evaluated. In section 2 we provide basic information about CIM and its serializations. In section 3 we will propose a mapping from CIM/XML to standard SQL by using the Java API *HP Jena* based on a translation from the CIM UML model to a relational database schema. A performance evaluation of this approach is presented in section 4. Finally, section 5 presents a conclusion of the findings of this paper, some related work and recommendations for future work.

2 The Common Information Model and its serializations

As a data exchange format should also be suitable for the data interchange with other companies, it is beneficial to use an approved standard instead of proprietary formats. Nowadays, it seems as the Common Information Model (CIM) will be the most common data exchange format for energy markets in the future. Core elements of CIM have been adopted in International Standard IEC 61970-301, other elements are currently drafts and will be standardized as IEC 61970-302 and -303. CIM is an extensive ontology for the domain of electric utilities and is available in different formats, e.g., in Rational Rose UML, XMI and RDF Schema. The model has recently also been released in Web Ontology Language OWL [Britton and deVos 2005]. More information on the CIM can be found in [Uslar 2006; Uslar et al. 2005].

Deploying CIM leads to a lot of fundamental decisions, for instance if it is advantageous to sustain the currently used database or to implement a new database schema. As CIM does not offer any official database model and data exchange is based on the RDF representation of CIM (called *CIM/XML*), it seems consequential to establish an RDF-based database instead of a relational database schema. However, it is quite unlikely that it is possible to introduce a completely new data management in a company; instead, it is necessary to map CIM/XML data from and to the existing database.

3 Mapping RDF to SQL and vice versa

As aforementioned, currently there is no officially released implementation of CIM in a relational database schema. Hence, it is necessary to create a database schema and map CIM/XML data from RDF to SQL in order to write the information into a database. In a standard scenario, one company A would extract data from its relational database, convert the data to

CIM/XML and send it to another company B. B will reconvert the data to SQL and write it into its own database.

Converting data from RDF to SQL and backwards is the main challenge of the information exchange between the two parties. Unfortunately, this mapping between the two data types cannot be done automatically as database schema and RDF schema cannot be matched without multiple adjustments. In the following two subsections the mapping from RDF to SQL and vice versa will be briefly described.

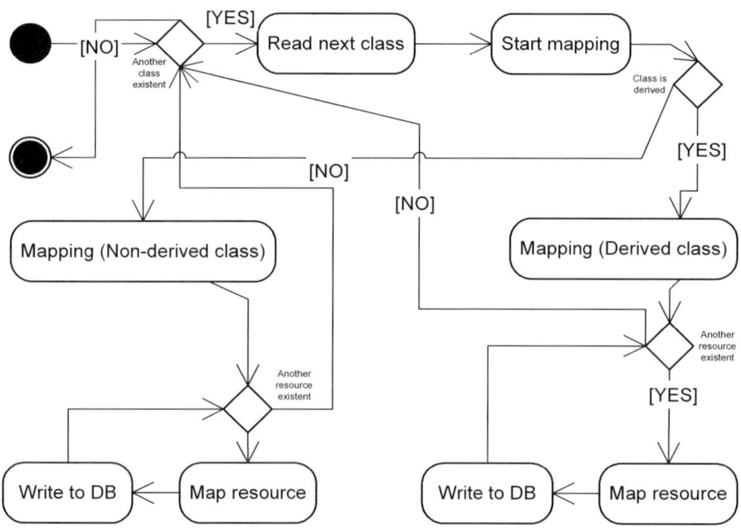

Fig. 1. Mapping data from CIM/XML to a relational database

3.1 Mapping data from CIM/XML to a relational database

The database schema is based on the approach suggested by [Podmore and Robinson 2000] – in real world examples it is highly probable that the data has to be converted from the original relational database schema to a more appropriate form first or that the mapping algorithms have to be extended. The mappings as well as other functionalities (e.g. methods for performance measurement) have been implemented in the demonstrator program *JACIM.* JACIM uses Hewlett-Packard's Open Source Java-API *Jena Semantic Web Framework* (HP Jena) to read, parse and write RDF Data and is able to map CIM/XML data to SQL statements, write these statements into a relational database and backwards.

In the database, every CIM class is implemented using a separate table; attributes and relationships are represented by the rows of this table.

The mapping from CIM/XML to SQL and insertion of the data into the relational database is based on the activity diagram outlined in Figure 1.

First, the test data is read and parsed by Jena's RDF/XML parser *ARP*. Afterwards, all RDF data is available in form of one RDF model. This model is iterated and every resource (i.e., one RDF subject) is handled on its own. Resources include several RDF statements, i.e. objects and the ID of this particular subject. After iterating the statements, all RDF data is available in form of one (if the particular CIM class is not derived from another CIM class) or more (in every other case) SQL statement(s). These SQL statements are executed and written to the database.

As primary keys of derived classes are also foreign keys which point to the primary keys of the derived class' superclasses, statements regarding superclasses have to be carried out first in order to avoid insertion errors. Relationships to other classes are realized as foreign keys, too. This implies that classes which are the "destination" of such a relationship have to be inserted in the database prior to the actual class.

Primarily, the mapping of derived and non-derived classes differs in the point that in the case of derived classes, all superclasses have to be known before the mapping starts. Both activities "Map resource" (cp. Fig. 1) include a case differentiation. These case differentiations refer to the statements of a resource:

- In the first differentiation, statements which are primary keys or part of a n..n-relationship are separated and handled separately from other statements.
- These other statements are distinguished in the second case differentiation: Statements representing relationships are handled different from other attributes.

3.2 Mapping data from a relational database to CIM/XML

The mapping from a relation database to CIM/XML is outlined in Figure 2. Data from the database is selected with a *Natural Join* which joins the actual class with all its superclasses. The result of this join is a Java ResultSet which is processed result by result. Analog to the case differentiation presented in section 3.1, a case differentiation has to be done when mapping data from the relational database to CIM/XML. However, in this case it is not complex to such an extent; basically it is necessary to verify if a row from the ResultSet is a primary key, foreign key or normal attribute.

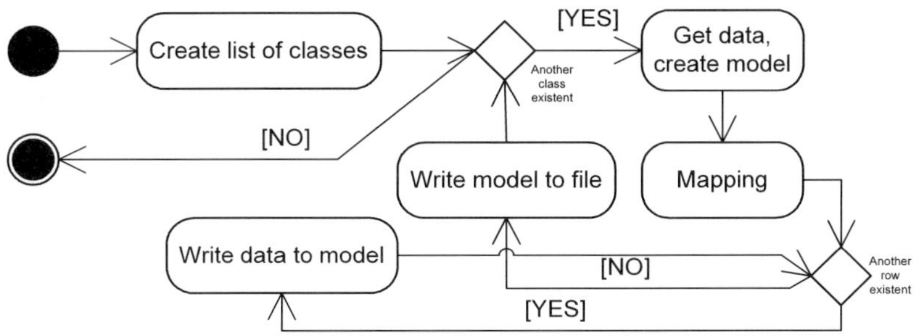

Fig. 2. Mapping data from SQL to CIM/XML

4 Evaluation

Based on the mapping introduced in section 3, the implementation of CIM/XML in a relational and RDF-based database has been evaluated. The implementation using a relational database is named *SQL-approach*; its equivalent for a RDF-based database is called *RDF-approach*.

Test data is provided by the CIM Validation Tool (CIM VT, Areva60-2006-03-17.rdf) and includes 30 CIM classes and 23352 RDF statements in 3,036 resources.

Both approaches are evaluated with the same test data. However, it is necessary to prepare the data for the SQL-approach. In the original test file, the data is listed according to the name of the actual CIM class. This presents no problem to RDF tools, as relationships between resources do not have to be existent at the moment a resource is parsed or written. Anyhow, if writing SQL statements to a database, all foreign keys have to be existent or insertion errors (cp. section 3.1) will emerge. Therefore, the sequence of resources has to be changed so that every foreign key points to an already existing resource.

All implementation and performance tests are carried out on a modern notebook with 1.5 GHz and 768 megabytes main memory. As the objective of this paper is a comparison between RDF- and SQL-approach, there is no need for a separated database server etc. Microsoft Windows XP Home SP2 is used as operating system; Microsoft SQL Server 2000 Developer Edition SP4 is employed as database. Java SDK is used in version 1.50_02, HP Jena in version 2.4 and jTDS (Open Source JDBC driver) in version 1.2. Furthermore, version 10 revision 7 of CIM is applied in our test environment.

The evaluation consists of two parts: In section 4.1, the architecture and implementation efforts of both approaches are presented; section 4.2 compares the performance of both implementations. section 4.3 summarizes the finding of the evaluation and gives a recommendation for future handling of CIM integration.

4.1 Architecture and implementation efforts

Using a relational database to store CIM data implies that data has to be mapped from CIM/XML to the relational database schema and backwards. In the case of using a RDF-based database, such a mapping is obsolete. Therefore, the implementation efforts of the SQL-approach are significantly higher than those of the RDF-approach.

A higher implementation effort leads to several disadvantages of the SQL-approach in comparison to the RDF-approach. First of all, a complex implementation increases the probability of errors, especially if special cases have to be handled.

Secondly, the used mapping algorithms have to be extended if it is intended to store CIM classes which have not already been handled in our demonstrator program. The used test data contains only a small part of CIM classes but is already handling a lot of special cases. As a result, the SQL-approach is less flexible and maintainable. The flexibility is also reduced due to the fact that all foreign keys have to be defined first in order to avoid insertion errors.

Finally, the SQL-approach needs a defined database schema while the RDF-approach creates all necessary database tables by itself.

4.2 Performance

In this subsection the results of our performance tests are presented. The performance tests are carried out on three different test data sets which are all parts of the test data provided by CIM VT. The first data set contains 11 resources and 22 statements, the second 472 resources and 3,470 statements. The last data set contains all data from CIM VT (3,036 resources, 23,352 statements). Every performance test is carried out three times in a new database in order to minimize measurement errors.

As it can be seen in Figure 3, the RDF-approach is slower than the SQL-approach when writing data from the CIM/XML-file to the database.

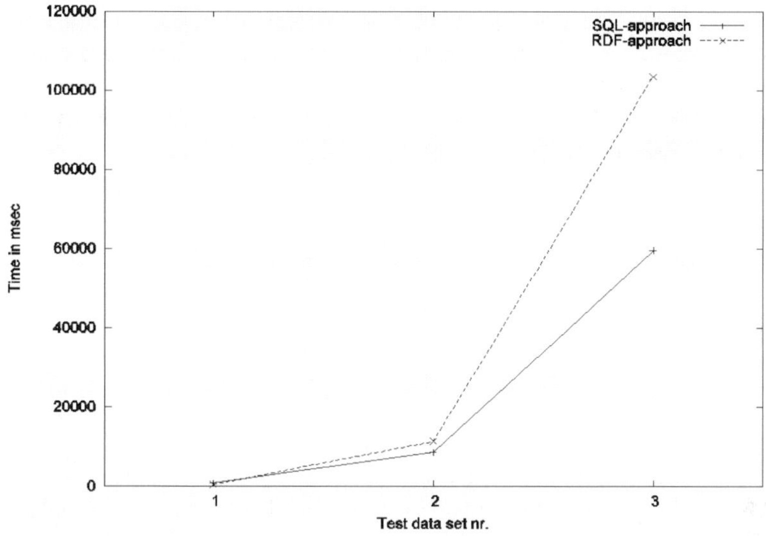

Fig. 3. Writing CIM data to database

The difference in the measured values increases disproportionately: The measured time for the SQL-approach grows roughly commensurate with the number of statements written to the database while the written statements per time unit decreases for the RDF-approach if a higher number of statements is examined.

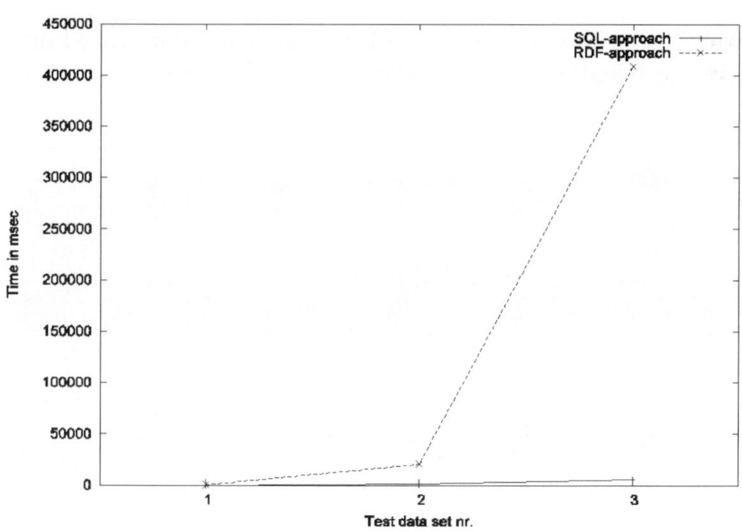

Fig. 4. Writing CIM data from database to file

As it can be seen in Figure 4, the results for writing from the database to a CIM/XML-file are quite similar, but the gap between the measured values is even bigger.

Again, the difference in the measured values increases disproportionately for the RDF-approach. In contrast, the measured values for writing the CIM/XML-file increase only slightly if using the SQL-approach.

Examining the present performance test results leads to some questions as it was not likely that the SQL-approach would come off better than the RDF-approach. The difference between the performances is especially large if writing from the particular database to the CIM/XML-file.

As both approaches invoke Jena's RDF parser and writer in the same way, the difference in the performance numbers has to result from other reasons.

Apparently, the database connection of Jena is much slower than the standard database connection provided by Java which is used in the SQL-approach. Furthermore, standard SQL statements seem to perform better than the statements Jena creates to write data to the RDF-based database.

All in all, the performance tests lead to an unexpected but unambiguous result: The SQL-approach is superior to the RDF-approach in terms of performance.

4.3 Conclusion

The evaluation leads to mixed results. While the RDF-approach is preferable due to its lower implementation efforts and better liability to errors, extensibility, maintainability and flexibility, the SQL-approach features a much better performance.

To assess which approach is easier to apply in a real-world scenario, more criteria should be taken into account: e.g., almost every company operates a SQL compatible database. Therefore, introducing a completely new technology like an RDF-based database is nearly impossible particularly with regard to other involved information systems. On the other hand, CIM/XML and similar formats gain in importance and need to be integrated in the information systems of companies involved in power markets.

The mapping between RDF-based CIM/XML and the SQL-based database has to take place somewhere: Either between the other systems and an RDF-based database or between CIM/XML and the relational database. However, from a business perspective the integration of the SQL-approach is easier as only the information system for im- and exporting CIM/XML data has to be endorsed with a mapping to the existent information systems.

It should be noted, that the scenario in this paper is quite perfect as the relational database is designed in accordance with CIM/XML (cp. section 3.1).

All things considered, the SQL-approach is preferable at the moment.

5 Summary and future work

In this paper we presented the CIM as a tool to exchange data between two companies. Even though the RDF-approach has got lower implementation efforts and better liability to errors, extensibility, maintainability and flexibility, the SQL-approach is recommended due to its better performance and easier integration into existing information systems.

To our knowledge, there are no published comparisons between the both approaches to integrate CIM as it was carried out in this paper. Nevertheless, there is some existent related work we want to mention. First of all, *D2RQ* (http://sites.wiwiss.fu-berlin.de/suhl/bizer/D2RQ/) offers a way to treat non-RDF databases as virtual RDF graphs and could provide another way to integrate CIM and databases. Furthermore, [Theoharis et al. 2005; Grosse 2003] have investigated the use of different RDF-based databases and tools.

Parallel to the research efforts presented in this paper, CIM was introduced in Web Ontology Language (OWL); as the name implies, OWL is better suited to represent ontologies than RDFS. OWL will probably be the preferred CIM format in the future and there will be only restricted support of CIM/XML. Therefore, we will examine the use of CIM/OWL in the future.

Furthermore, we will enhance the mapping from a relational database to CIM/XML data. At the moment, this mapping works but has got some shortcomings, e.g., with case sensitivity. Although the mapping provides semantically identical data (compared to the original test data), HP Jena uses a different RDF-syntax than the parser used by CIM VT. Accordingly, the resulting data is syntactically different.

As our mapping algorithms are too inflexible and have to be extended if mapping CIM classes which are not part of the test data from CIM VT, new algorithms will supersede the used ones.

Another aspect of our future work is the examination of alternative RDF-parsers and -writers. As it was outlined in section 4.3, Jena's performance could be better when writing to the database and parsing data from the database to RDF. Alternative frameworks like Sesame (http://www.openrdf.org/) could be used to improve the performance of the RDF-approach.

Last but not least, [Britton and deVos 2005] propose a scenario where CIM could be used as the foundation for a Service-oriented Architecture (SOA). In the future, we will use CIM as a way to add a semantic base to the SOA reference architecture WSQoSX introduced by [Berbner et al. 2005].

References

Becker D et al. (2000) Standards-Based Approach integrates Utility Applications. In: IEEE Computer Applications in Power, pp 13-20

Berbner R et al. (2005) An approach for the Management of Service-oriented Architecture (SoA)-based Application Systems. In: Proceedings of the Workshop Enterprise Modelling and Information Systems Architectures (EMISA 2005), pp 208-221

Britton JP, deVos A N (2005) CIM-Based Standards and CIM Evolution. In: IEEE Transactions on Power Systems, vol 20, pp 758- 764

Grosse J (2003) XML Clearinghouse Report 6: Speicherverfahren und Werkzeuge für RDF/S

Podmore R, Robinson M (2000) Introduction to CIM. PowerData Corporation. Incremental Systems Corporation

Theoharis Y et al. (2005) Benchmarking Database Representations of RDF/S Stores. In: 4th International Semantic Web Conference (ISWC 2005), pp 685-701

Uslar M (2006) The Common Information model for utilities: An Introduction and Outlook on Future Applications. In: Eckstein R, Tolksdorf R (eds) Proceedings der XML-Tage 2006 in Berlin, pp 135-148. http://www.XML-clearinghouse.de, access: 2007-02-13

Uslar M et al. (2005) Interaction of EMS related Systems by Using the CIM Standard. In: Leal Filho W, Marx Gómez J, Rautenstrauch C (eds): ITEE 2005: Second International ICSC Symposium on Information Technologies in Environmental Engineering: Proceedings, Otto-von-Guericke Universität Magdeburg. Shaker, Aachen, pp 596-610

Optimization of Adaptive Consumers to a Time-varying Electricity Supply

Ute Vogel, Michael Sonnenschein

Oldenburg University, 26111 Oldenburg, Germany,
e-mail: {vogel, sonnenschein}@informatik.uni-oldenburg.de

Abstract: A mayor challenge for the integration of renewable energy sources to the existing power supply system is to provide environment-friendly balance energy which compensates temporal fluctuations in the power supply of wind mills or photovoltaics. In this paper, we analyze a centralized approach for demand side management reducing the need for balance energy. The approach assumes a given set of energy consuming jobs that are freely movable within job-specific pre-defined time intervals. Using the meta-heuristics Tabu Search, a schedule of these jobs is searched that leads to an optimal match of energy demand with energy supply. Different initialization strategies for constructing an initial solution and different alternatives of Tabu Search are evaluated in order to enhance the optimization result. Using realistic data for wind power and synthesized but not unrealistic sets of up to 120,000 jobs, the optimization results show that a considerable reduction of balance energy could be possible by our load shifting method.
Keywords: DEMS, Renewable Energy, Load Management, Tabu Search

1 Introduction

In consideration of the limited fossil and nuclear energy resources, the efficient usage of renewable energy sources i.e. wind, solar energy, water-power, and biomass, will play an important role for covering our future energy demands. One mayor drawback of most renewable energy sources is their temporally unsteady production which is due to environmental influences and can not be controlled. So, there is no guarantee that renewable energy converters produce the amount of electrical energy that is needed at a certain time. As an efficient storage of electrical energy is yet not possible, the energy suppliers use hydraulic power plant or gas turbines, or buy and sell electricity at the energy exchange market [Crastan

2004] to guarantee the demand independently of fluctuations in supply form renewable sources. These control mechanisms usually are expensive in respect to economic as well as ecological costs.

Hence, we investigate whether an adaptation of the temporal varying energy demand to a prognosis of energy availability for a fixed time interval is possible. Such an adaptation can reduce peaks in power demand, compensate temporal losses in power supply, or generally 'shape' load curves corresponding to supply curves [Born 2001]. This approach seems to be worth to be investigated as on one hand temporally movable, energy consuming jobs exist that can be scheduled in a certain time interval (see next paragraph), and on the other hand the availability of energy from renewable sources, e.g. wind energy, can be predicted with sufficient precision [Ernst et al. 2002; Focken et al. 2002]. An adaptation of jobs to the fluctuating energy supply should be automated and should not need steady human interaction. The objective of the adaptation in the presented approach focus to minimize the overall energy deficit in the first place, as this means a decrease in the necessary balance energy that usually is produced by CO_2 emitting power plants. The second objective of reducing temporal peaks of the energy deficit relates to a reduction of the maximum net load when transferring the balance energy.

In this paper, we continue our work from Pei [2005] and Vogel et al. [2006], where we have investigated how energy suppliers could directly control the temporal energy consumption by shifting energy demanding jobs. We will show that using more refined initialization procedures and instantiations for the meta-heuristics Tabu Search [Glover and Laguna 2001] lead to a substantial increase in the manageable problem size as well as to a remarkable quality of the result.

First, we will briefly present the underlying model. We will use the meta-heuristic Tabu Search for the search for an optimal schedule. As this optimization technique was already described in [Vogel et al. 2006], we focus to the presentation of the variations which have been made in determining a promising initial solution for the Tabu Search algorithm and in the instantiation of Tabu Search.

2 Related Work

Stadler [2005] and Brauner et al. [2006] have analyzed measures for demand side management (DSM) by adapting the demand to a temporally varying energy supply. They observed a promising potential for small, heat-accumulating appliances. In experiments with temporally variable power tariffs, Quatsching and Hanitsch [1999] noticed good prospects for load adjustment of program-controlled appliances.

Generic algorithms for adapting the demand for electrical energy of appliances to the supply of energy have sparsely been addressed.

Hable [2004] examines the resource scheduling for power plants which integrates a controllable or switchable power consumer among others. The energy consumption of this consumer is adapted to the energy supply by adjusting its energy demand or by switching it on or off, resp. So, this model can integrate one large-scale consumer or an aggregation of small consumers but not differentiated sets of appliances.

Other approaches, e.g. Ygge [1998] and Penya [2006], address combinatorial auctions in order to meet the challenges of deregulated energy markets. A pragmatic problem of Ygge's agent-based Homebots system consists in the assignment of roles to the distribution system operators and to the utilities: These roles do not reflect the actual situation in our electrical power supply system. Penya [2006] also introduces an agent-based system for optimizing the allocation and scheduling of energy demand. His model is designed to integrate DSM-able consumers, but due to the necessary communication between agents the complexity of his method is $O(2^n m)$, where n is the number of agents (consumer tasks) and m the average number of alternatives for a task to be placed. Hence, the distributed algorithm has not been applied to realistic scenarios, but was evaluated for the n-queens problem with $n \leq 50$.

3 Model

The model used in our approach consists of adaptive energy consumers and a power supply function. The energy supply can be modeled as a discrete function $S{:}T \rightarrow R$ from a set T of sequential time steps to the amount of energy measured in abstract "energy units". It is described as an equidistant time series of available energy. As usual in the energy market, we use 15 minutes as time unit, and as time interval T we focused to $T = [0, 96]$ which characterizes 24h.

The energy consumption is modeled by energy demand sets J consisting of energy consuming, non-preemptive and independent jobs. Each of these jobs is characterized

- by the interval $[a(j), e(j)] \subset T$ for the feasible execution of the job,
- by the duration $d(j) \in T$ of the job execution, and
- by the total demand for energy $m(j) \in R$ which must be guaranteed during the execution of job j.

The starting times s(j) for each job $j \in J$ determine the schedule for the execution of jobs and the energy demand $D_{J,,s}$ which is necessary to fulfill the energy requirements

$$D_{J,s} : t \in T \mapsto \sum_{\substack{j \in J \\ t \in [s(j), s(j)+d(j)]}} m(j) \in R \qquad (1)$$

The search for an optimal schedule means to determine starting times $s(j) \in [a(j), e(j)-d(j)]$ for each Job $j \in J$, so that the resulting energy demand function $D_{J,s}$ matches the energy supply best. This means to minimize an objective function $O_J(s) = \sum_{t \in T} \| D_{J,s}(t) - S(t) \|$, where $\|.\|$ is an adequate measure and $s=(s_j)_{j \in J}$ the vector of starting times.

The number of jobs and their relocatability determine the dimension of the solution space. Hence, for a large number of jobs we get a high dimensional solution space. The problem in the form above can also be formulated as integer linear programming problem, but these are known to be NP-hard. So, we had chosen the Meta-heuristics Tabu Search for the search for an optimal solution. It produces nearly optimal solutions in acceptable time.

4 Optimization

The Meta-heuristic Tabu Search is a local search technique that enhances the performance of a local search method by using memory structures, the so called Tabu list. Tabu Search starts with a feasible initial solution. In each iteration loop, all solutions which are neighbored to the current solution are evaluated. The best neighbor is chosen as next solution. To overcome local minima, Tabu Search allows intermittent deteriorations of the next solution's quality. In this case, the Tabu list prevents infinite loops by recording solutions already visited.

Tabu Search was set up to search starting times $\mathbf{s}=(s(j))_{j \in J}$ resulting in an optimized match between the demand $D_{J,s}$:T→R and the energy supply function S: T → R. The general usage of Tabu Search for the job adaptation problem has already been presented in [Vogel et al. 2006]. So, here we will focus to the improvements of the algorithm which are mainly achieved by a more sensitive initialization and by enhanced Tabu Search variants enabling an efficient optimization of demand sets of up to 120,000 jobs instead of only 200 jobs in our former approach.

4.1 Computation of an initial solution

In our prior work, Tabu Search was started with an initial solution, which assigned the first possible starting time to each job j, i.e. s(j)=a(j).

Obviously, a more refined initial solution must take the actual energy supply into account: So, we experimented with different initialization strategies (P, N, S) besides the choice of the first possible starting time (F). These procedures start with the smallest possible starting time $t \in T$, choose as candidate set all jobs j with $t \in [a(j), e(j)-d(j)]$, which have not been scheduled yet. As the job's demands are successively subtracted from the energy supply function S, the auxiliary function $D'_{J,s'}$ describes the energy demand of all jobs already scheduled. The choice of jobs from the candidate set proceeds according in the following order: Job j is chosen

1. if t is the last possible starting time of job j, i.e. t = e(j)-d(j), or
2. if the remaining energy supply S(t) – D'J, s'(t) is greater or equal to the demand m(j) of job j, i.e. S(t) – D'J, s'(t) ≥ m(j), or
3. if the demand m(j) of job j is the minimal demand of all candidates and if the overdraft (S(t) – D'J, s'(t) < m(j)) at time t can probably be balanced in the next time step, i.e. S(t+1) – D'J, s'(t+1)>m(j).

The initialization strategies P, N and S schedule all jobs obeying the condition of step 1. Table 1 shows their differences in steps 2 and 3.

Table 1. Differences between the initialization strategies P, N, S

Initialization strategy	P	N	S
(A) Sorting of jobs in step 2	No	No	Sorted by energy demand m(j)
(B) Overdraft $D_{J,s'}(t) > S(t)$ in step 3	Avoided	Minimal overdraft allowed	Minimal overdraft allowed

To avoid phenomena due to an accidental presorting of jobs by the database, we additionally evaluated initialization strategies which explicitly choose jobs in random order in step 2. These initialization variants have resulted in initial solutions of nearly the same quality as the strategies P and N. So, they are not presented here.

4.2 Instantiation of the Tabu Search Algorithm

Tabu Search was implemented using the Tabu Search library OpenTS [Harder 2005]. The OpenTS framework allows the custom-designed instantiation of components and settings, in particular the choice of the objective function and of the neighborhood generator.

4.2.1 Objective Function

The objective function or goal function assesses the quality of a solution. The value of the objective function $O_J(\mathbf{s})$ depends on the choice of the starting times $\mathbf{s}=(s(j))_{j\in J}$ for all jobs. Given a set of possible job starting times, the starting times $\mathbf{s}$ in a new solution are chosen to minimize $O_J(\mathbf{s})$.

In [Vogel et al. 2006], we evaluated the Euclidean and the City Block metric to measure the distance between energy supply function S and energy demand $D_{J,s}$. As our goal is to avoid an energy demand $D_{J,s}(T) > S(t)$, we simplify the objective function in this paper and measure how much the energy demand exceeds the supply:

$$O_J(s) = \sum_{t\in T: S(t)<D_{J,s}(t)} D_{J,s}(t) - S(t) \tag{2}$$

To diminish peaks in the function $D_{J,s}(t)-S(t)$, the reduction of peaks has been chosen as second criterion: if the criterion is active and the objective function value of two candidate solutions is the same, Tabu Search chooses the candidate leading to lower peaks.

4.2.2 Neighborhood

In our former approach, the neighbors of a given solution $\mathbf{s}=(s(j))_{j\in J}$ has been defined by all solutions $\mathbf{s'}=(s'(j))_{j\in J}$ where $|s(j) - s'(j)| = 1$ for one job j and $s(j)=s'(j)$ for all other jobs. Here, we have experimented with a more general attempt: the k-neighborhood of $\mathbf{s}$ consists of all solutions $\mathbf{s'}$ with $|s(j) - s'(j)| = k$ for one job j and $s(j)=s'(j)$ for all other jobs. Imitating the idea of simulated annealing, we start with a high value of k and reduce it to k/2, if the k-neighborhood contains no solution. Besides, we change k randomly, if a solution s' is chosen which does not improve the former solution s: with a probability of 20% we replace it by k/2 (for $k > 1$), otherwise by 2*k.

The average number of k-neighbors and, hence the computational effort, decreases with an increasing value of k. On the other hand, the objective function can be very efficiently computed if $k = |s(j)-s'(j)| < d(j)$. The maximum computation effort is the same for all $k > d(j)$.

5 Results

The Tabu Search algorithm was executed with a fixed tabu list length of 200 and run for 20,000 iterations. The initialization strategies and the Tabu Search algorithm were performed for each combination of 18 different demand sets and eleven energy supply functions. The demand sets varied between 2,000 and 120,000 jobs with an average of 32,666 Jobs. The experiments were implemented in Java and run on an Intel® CPU, T2400 with 1.83 GHz using 1 GB Heap space.

5.1 Generation of test cases

A test case consists of an energy supply curve and a set of adaptive energy demand jobs. The energy demand sets were generated randomly: to reflect 'real life' energy demands of households the interval start times a(j) of all jobs j were generated according to a probability distribution with peaks in the morning (10% of all jobs), at noontime (40%) and a flatter, but broader peak in the evening (30%). These jobs have an average duration of about 60-70% of their respective time interval [a(j), e(j)]. The interval start times of the remaining 20% of the jobs are distributed randomly over the basic time interval [0, 96]; their average duration is about 8 steps (2h), the average interval length is about 42 steps (6h).

As energy supply curves we used data about the supply of wind energy from the power authority EnBW[1]. These functions were scaled to match the overall demand for energy, which is given by $D_J=\sum_{j\in J} d(j)\cdot m(j)$.

With the arbitrary combination of supply curves and energy demand, it could not be assessed how good the global optimum has been found with the initialization strategy and the Tabu Search optimization, resp. So, for benchmarking and test purposes we have additionally constructed a supply function for each energy demand set by setting a randomly generated, feasible start time for each job and interpreting the resulting energy demand curve $D_{J,s}$:T→R as supply function. We will refer to these test cases as "exact test cases".

5.2 Initialization

All initialization procedures (F, P, N, S) have been performed for each combination of supply curves and demand sets and for the exact test cases.

[1] http://www.enbw.com/content/de/netznutzer/strom/download_center

As expected, in all experiments, strategy F is always the worst strategy. The quality of all initialization strategies compared to the F-strategy is very good: On the average, the quality of the initial solution improves by 55-58% compared to the initial solution generated by strategy F. Table 2 shows the relative overdraft $O_{J,}(s_{init})$ / D_J of the initial solutions generated by the different initialization strategies.

The sorting of jobs before scheduling (strategy S) delivers the best initialization in about 50% of the experiments, the second best choices are strategies P and N leading to the best initialization in about 40% or 36% of all cases. Obviously a drawback of strategy S is the increased run time which is due to the effort of sorting all jobs before scheduling. The run times of the strategies F, P and N can be approximated by a linear regression with standard error < 0.00002 (strategies F,P) and 0.001 (strategy N) and a stability index > 0.9 (strategy F) up to 0.98 (strategies P, N).

Table 2. Relative overdraft $O_{J,}(s_{init})$ / D_J of the initial solution s_{init} compared to the total energy demand D_J and average run duration.

Initialization Strategy	Relative overdraft			Average run time	Avg. rel. overdraft in exact test cases
	Min.	Max.	Average		
F	19.6%	33.9%	26.4%	5 ms	31.9%
P	6.4%	17.4%	10.7%	129 ms	0.2%
N	8.1%	17.9%	11.7%	166 ms	0.6%
S	6.7%	16.8%	10.7%	2750 ms	0.3%

The performance of the initialization strategies for the exact test cases shows that the initialization procedures can nearly reconstruct an optimal match (Fig. 1).

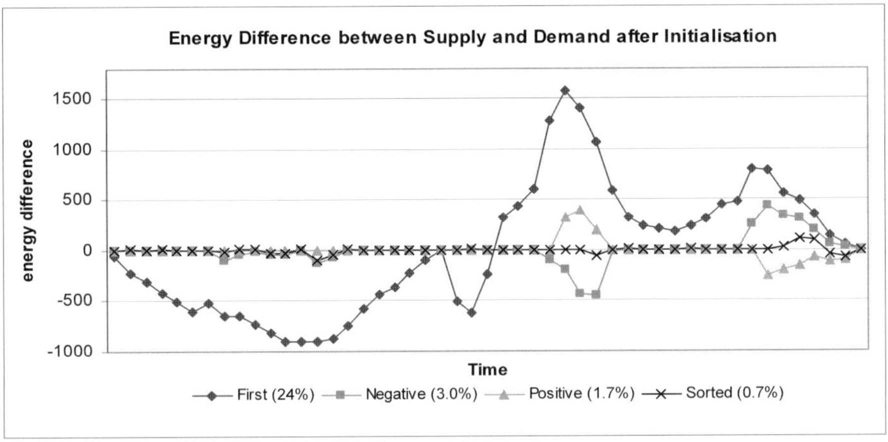

Fig. 1. Difference between supply and demand after initialization for an 'exact test case' for which an exact match exists. Strategy S (Sorted) already leads to a nearly perfect match that finally can be improved by the TS optimization. The percentage values describe the relative overdraft of the demand.

5.3 Tabu Search optimization

In each experiment, the jobs were initialized according to the best initialization method, i.e. mainly initialization strategy P or S. As the initialization delivers quite a good fit, the Tabu Search optimization does not achieve huge further improvements. The Tabu Search solutions reduce the overdraft by about 4 percentage points compared to the best initial solution for jobs sets with 40,000 jobs.

In arbitrary test cases the energy supply functions were scaled to match the overall energy demand D_J of a job set J over its time interval T. Due to the generation of the job sets causing a low energy demand in the night and peak demand around noontime, an exact match at every time step between supply and demand usually is not possible. As in these cases the optima are not known, we measured the quality of the experimental Tabu Search instantiations by means of the relative energy deficit caused be the optimised schedule compared to the overall energy demand of the job set. The improvement in the execution times of the Tabu Search instantiations was measured relative to the slowest instantiation (1) which was chosen as base scenario.

In Table 3, we compare the different instantiations of Tabu Search with the results gained by the basic instantiation – a non-dynamic 1-neighbourhood combined with the simple objective function without peak reduction (PR):

Table 3. Relative overdrafts $O_{J,}(s_{TS})$ / D_J after differently instantiated Tabu Search optimizations and average run times.

	Tabu Search instantiation		Test Cases		Exact Test Cases	
	Objective Function	Neigh-bourhood	$O_J(s_{TS})$/ D_J	Rel. run duration	$O_J(s_{TS})$/ D_J	Rel. run duration
(1)	$O_J(s)$	1	10.35%	100%	0.130%	100.0%
(2)	$O_J(s)$	dynamic	9.42%	28%	0.080%	46.5%
(3)	$O_J(s)$+PR	1	10.30%	108%	0.004%	111.0%
(4)	$O_J(s)$+PR	dynamic	9.28%	31%	0.002%	109.0%

The improvement of the solution after the optimization relative to the quality of the initial solution generated by strategy F is about 60-65%, compared to 44-50% in our former work. As strategy F means no adaptation of energy demand, this means, that the need for balance energy could be reduced by 60-65% in our scenarios.

It turns out, that in arbitrary test cases the best solutions are produced by using a dynamic neighborhood (Table 3, rows (2) and (4)). These settings lead to a faster execution than the instantiations (1) and (3) which use a 1-neighbourhood. This is due to a lesser number of candidates having to be evaluated in each Tabu Search iteration in a k-neighborhood with k>1.

For a demand set with 60,000 Jobs the run time of Tabu Search varies between 3.5 and 15.2 min. The result of the Tabu Search optimization improves the best initial solution by 3.5 percentage points.

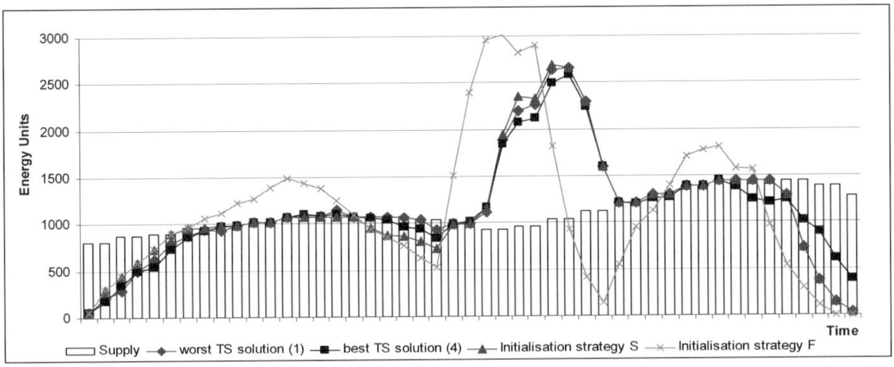

Fig. 2. Realistic supply curve (shown as bars) and load curves resulting from initial solutions and TS optimization results for a realistic test case, where an exact match does not necessarily exist. The overall energy deficit and consequently the necessary balance energy drops from 14,435 energy units (27.6%) with initialization strategy F to 8,121 energy units (15.5%) with the best TS solution (Parameter set (4)).

For test cases which allow an exact match between supply and demand, peak reduction as a secondary criterion gets an increasing importance for an optimal schedule. In these cases (3, 4) the Tabu Search algorithm was started with a nearly perfect match (see last column in Table 2) as initial solution. So choosing a candidate solution that reduces a peak in the difference between supply and demand curves seems to move ill-scheduled jobs better to their optimal starting time. But this 'fine-tuning' of the solution's quality causes increased computing time in the case of a dynamic neighborhood, because the random extension of the neighborhood distance leads to more evaluations of distant neighborhood candidates which do not improve the old solution.

The Tabu Search algorithm was generally performed for 20,000 iterations with fixed Tabu list length. Hence, the execution time can be fitted quite well (stability index > 0.978) with a linear regression function on the number of jobs per demand set. For all test cases, the optimization results seem to be very close to a local minimum, experiments setting the number of iterations to the number of jobs have shown only a marginal improvement of the solution quality (average improvement < 1%) in demand sets with more than 20,000 jobs.

6 Conclusion and Further Work

Renewable energy supply is afflicted with the problem that it can not be controlled in many cases and hence can not be adapted to the energy demand. In this paper we have shown that it is computationally feasible to fit energy demand to a time varying supply curve by means of real wind energy supply curves and sets of 2,000 to 123,000 temporally movable energy consuming jobs.

We assessed different strategies for initialization and instantiation of the optimization heuristic Tabu Search, and found that it is definitely profitable to incorporate the supply curve into the initialization of the energy consuming jobs. For instantiation, the usage of a dynamic neighborhood definition is worthwhile as it reduces the computation time and also increases the quality of the solution. For the fine tuning of solutions, the reduction of peaks in the difference between energy demand curve and energy supply curve as a second criterion led to a good convergence.

All in all, the optimization process constitutes good results in an acceptable computing time. For comparison, we will on one hand examine alternative approaches for determining an optimal solution, e.g. (integer) linear programming or other meta-heuristics.

On the other hand, we will focus to enhancements of the simplified job model to incorporate more realistic jobs which will cause the necessity for improvements in the optimization heuristic.

- The jobs are modeled as non pre-emptive: especially appliances with heat/cold accumulating characteristics (e.g. ACs or refrigerators) can be interrupted and restarted later-on.
- The assumption of independence of jobs should be dropped: the already mentioned heat/cold accumulating devices are periodically switched on and off. The next starting time depends on the length of the time interval since the last on-period and environmental conditions. Such dependencies of jobs should be integrated into the model.
- It should be possible to adapt the schedule that has been optimized in advance, to last minute changes of jobs during its execution. We need to explore and adjust online-scheduling techniques to match these requirements.
- Beside the hard constraint, given by the time interval of each job, soft constraints are significant for obtaining schedules regarding technical or social criteria. This gives rise to the need for implementing multi-criteria optimization techniques.

Last but not least, the generation of realistic test cases affords much knowledge of the energy load of real appliances. Further models should directly rely on statistical registrations of the usage of household appliances and on detailed models of equipment.

References

Born F (2001) Aiding Renewable Energy Integration through Complementary Demand-Supply Matching. PhD Thesis, Glasgow, University of Strathclyde

Brauner G et al. (2006) Verbraucher als virtuelles Kraftwerk – Potentiale für Demand Side Management in Österreich im Hinblick auf die Integration von Windenergie. Institut für Elektrische Anlagen und Energiewirtschaft. TU Vienna

Crastan V (2004) Elektrische Energieversorgung. edn 2. Springer, Berlin

Ernst B, Rohrig K, Jursa R (2002) Online-Monitoring and Prediction of Wind Power in German Transmission System Operation Centres. First Joint Action Symposium on "Wind Forecasting Techniques". Norrköping, Sweden

Focken U, Lange M, Moennich K, Waldl H-P, Beyer H-G, Luig A (2002) Short-term prediction of the aggregated power output of wind farms - a statistical analysis of the reduction of the prediction error by spatial smoothing effects. J. Wind Eng Ind Aerodyn 90, pp 231-246

Glover FW, Laguna M (2001) Tabu Search. Kluwer Acad. Publishers, Boston

Hable M (2004) Beitrag zur Energieeinsatzoptimierung mit evolutionären Algorithmen in lokalen Energiesystemen mit kombinierter Nutzung von Wärme- und Elektroenergie. PhD Thesis, TU Dresden

Harder R (2005) OpenTS – Java Tabu Search Framework. http://www.coin-or.org/OpenTS/, access: 2007-01-07

Pei J (2005) Optimierungsmethoden zur zentralisierten Laststeuerung in Stromversorgungsnetzen mit fluktuierender Einspeisung. Master Thesis, Oldenburg University

Penya Y (2006) Optimal Allocation and Scheduling of Demands in Deregulated Markets. PhD Thesis. TU Vienna

Quaschning V, Hanitsch R (1999) Lastmanagement einer zukünftigen Energieversorgung – Integration regenerativer Energien in die Elektrizitätsversorgung. BWK 10/99, pp 64-67

Stadler I (2005) Demand Response – Nichtelektrische Speicher für Elektrizitätsversorgungssysteme mit hohem Anteil erneuerbarer Energien. Habilitationsschrift, Fachbereich Elektrotechnik, University of Kassel

Vogel U, Pei J, Sonnenschein M (2006) Optimised Control of Adaptive Current Consumers - A First Approach. In: Troch I, Breitenecker F (eds) Proc 5th MathMod Vienna, vol 2, Argesim Report edn 30. Argesim-Verlag, Vienna

Ygge F (1998) Market Oriented Programming and its Application to Power Load Management. PhD Thesis, Department of Computer Science, Lund University

Distributed Power Generation: Requirements and Recommendations for an ICT Architecture

Ludger Winkels, Jan-Patrick Osterloh, Fabian Grüning, Martin Tröschel, Dave Boers, Tanja Schmedes, Mathias Uslar

OFFIS e.V., Escherweg 2, 26121 Oldenburg, Germany
e-mail: {surname}@offis.de

Abstract: Contemporary power distribution faces various new challenges. Most of those challenges have a strong impact on the ICT-structure required and on system architecture. This contribution briefly introduces changes and requirements imposed, both for trading and distribution of power. On this basis, alternatives for ICT architectures are discussed, recommendations are made on implementation choices for a meaningful sustainable solution, and communication and security challenges are addressed.
Keywords: Information and Communication Technologies (ICT), distributed generation, energy management systems (EMS), IEC standards

1 Power Generation and Distribution

1.1 New kinds of energy

The previous kind of energy generation was based upon chemical or physical conversion of resources that will eventually expire like burning oil and coal or fission. There are two main disadvantages about this proceeding: firstly, the supply of the resources will cease so that these will get more expensive in the course of time until they are not available at all. Secondly, as it is only possible to generate energy efficiently in large-scale power plants, a complex infrastructure is needed to distribute the energy to the consumers. Besides, both, the process of converting raw material, as well as the final storage of the conversion products in the nuclear power case imply severe environmental problems.

In the future of energy markets, the distributed energy production through wind and hydroelectric power plants, solar cells, and combined heat and power generation will play a major role. These energy production methods have certain characteristics. Some of these are sustainable (wind and hydroelectric power plants, solar cells), some are controllable (hydroelectric power plants, combined heat and power generation). All have in common that many small capacity power plants supply power to the grid. Also, demand side actors may be suppliers as well as consumers.

1.2 Challenges for ICT

This new situation and the emergence of energy markets [European Parliament and of the Council 2003] lead to two different challenges for ICT. On the one hand, heterogeneous power plants have to be controlled so that the supply matches the demand with regard to the stability of the energy grid. On the other hand, a platform for trading energy has to be established with requirements like high availability, currency, etc. In the following sections, various architectures and their characteristics will be discussed. Additionally, requirements of the communication infrastructure will be identified.

2 Trading

Considering electrical energy, one has to distinguish between two in principle different products: consumption power and balance power. Consumption power is considered to be the portion of the overall generated power intended for use by consumers. Its use is accounted on a quantitative basis, which means power customers, e.g. private households or industrial companies have to pay a certain lump sum per obtained kWh including production and distribution costs [Barth et al. 2006].

Unlike consumption power, balance power is not used to operate electrical equipment, but to guarantee stability of distribution grids and to ensure a constant quality of the power provided. Disturbances in the equilibrium between power generation and consumption, e.g. due to inaccurate predictions, power plant blackouts, or power line failures, manifest themselves in deviations from the mandatory voltage frequency [Nabe 2006]. These deviations can be compensated by application of balance power. To ensure undisturbed operation of the power grid, transmission providers have to hold adequate generation capacities. Contrary to the accounting mechanisms of consumption power, the possibility to supply balance power is primarily reimbursed [Theobald et al. 2003].

In order to establish an efficient competitive market it is appropriate that the operation of power grids and other business activities are separated. For example, Companies and utilities in the German liberalized energy market act in different roles (Fig. 1). With the exception of grid providers, each actor takes one or many of these roles. They have to offer the power grid to all other market participants as an independent platform. Grid providers are prohibited of otherwise engaging themselves in the market [German Federal Ministry of Economics and Technology 2005].

Consumption and balance power are provided by power plant operators. Suppliers usually buy from energy exchanges or retailers and sell consumption power to consumers. Apart from costs of purchasing the consumption power, the supplier pays a grid access fee for the connection of its customers. The consumer is charged usually for both by the supplier.

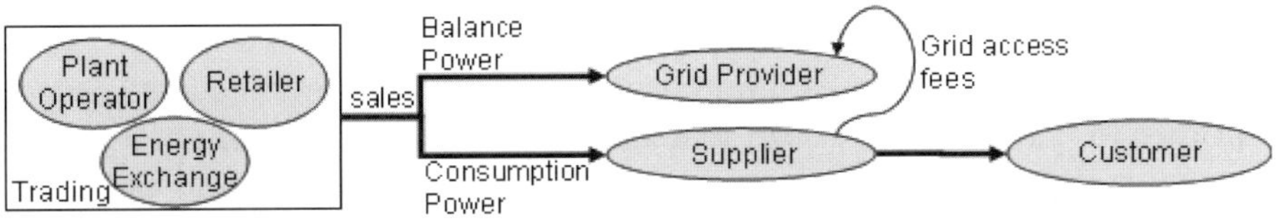

Fig. 1. Roles in the energy market

For receiving grid access fees, network carriers have to guarantee grid stability [German Federal Ministry of Economics and Technology 2005]. They obtain balancing power either directly from power station operators or from energy exchanges.

On the balancing power market, only few power station operators compete with each other. Strong restrictions such as high reliability and high minimum offer quantities prevent market access for many small companies [Verband der Netzbetreiber 2003]. Balancing power is location-dependent. In contrast to consumption power, balancing power must be fed into certain sections of the grid to ensure stability.

For proper competition it is necessary to increase the number of participants. An approach to lower the market entry barriers would be to establish co-operations between several smaller power plant operators [Theobald et al. 2003]. However, the necessary organizational and technical solutions are currently not available. ICT systems have to meet the technical and legal conditions to allow common acting of several market participants.

A similar situation exists in the market for consumption power. Private households, farmers and small industries become power producers as a result of the distributed nature of the production. The quantity of power exceeding the internal demand can be marketed, provided suitable organizational and technical solutions are established. For this target group, energy marketing is not the central activity. Therefore, suitable solutions should allow automatic trading to a large extent.

3 Power grids

In addition to trading requirements, power grids are affected by ongoing structural changes in power supplies. Grids have to deal with input from many distributed, small, and to a large extent heavily fluctuating producers. This makes scheduling, controlling and monitoring of every system from a single centralized site unattractive. Based on an anticipated, yet likely-seeming high penetration of power supplies with CHP plants, scheduling thousands of small producers from a single site would take too much time and calculating capacity, thus being inefficient. Therefore, it is reasonable to iteratively establish localized and typed clusters of plants to manage these.

Considering power grids, another requirement is guaranteed fail-safety. In order to achieve this, a grid management has to be fail-safe as well, either by means of backup plans or by avoiding a single-point-of-failure by distributing grid management functionalities. From the grid's point of view, an implementation of scheduling, controlling, and monitoring of power generators has to consider specific communication functionalities of different producers. On the one hand, there exist plants not accessible via standardized interfaces, and on the other hand the plant's autonomy poses strong restrictions on scheduling and controlling.

The power grid topologies present general conditions for grid management. Unlike traditional power plants, distributed plants feed in power at different voltage levels, including low-, medium-, and high-voltage [Barth et al. 2006]. Thus, with an increasing number of distributed power producers, the likewise growing risk [Kurrat and Schulz 2006] of current backflows damaging parts of the power infrastructure, such as voltage transformation substations or protective gear by overheating must be dealt with by means of scheduling, controlling and monitoring of power equipment at all voltage levels.

4 System architectures

The requirements mentioned above for both trading and power grid can be met using different system architectures. Figure 2 summarizes three possible architectures "centralized", "peer-to-peer" (P2P) and "super-peer" [Bischofs et al. 2006]. In the following paragraphs, the architectures with regard to power distribution and trading will be elaborated.

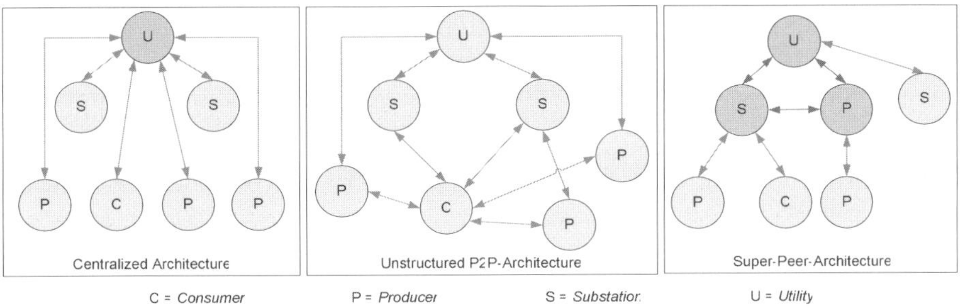

Fig. 2. System architectures

4.1 System architectures for trading

To fulfill the requirements mentioned in section 2 for both consumption power and balance power, a central trading approach or a P2P approach can be utilized.

Using a central trading platform, all information accumulates in one point for the trader in charge. A central platform provides a convenient *legal security* for all partners, as there is one responsible institution that controls and maintains the platform. Legal security can be important, when high damages – because of platform unavailability or contract violation – have to be considered. The responsible not only have to guarantee the availability of the platform, they also control the access to the platform itself, usually by means of high penalties for market entry. From the economical point of view, a better overview of the markets and hence more transparency are obtained. High costs for running systems reliably with large amounts of data can occur. Such costs are often channeled to the users in terms of high fees.

Contrary to this, there is the P2P approach, which can be fully automated (limited by market rules) for everybody to be used for trading. Also the overall amount of communication is higher than in a centralized solution, every market participant has full control over their provided data sets, and can therefore hide their internal information, with respect to legal requirements. Furthermore, real-time pricing becomes interesting because even single devices can act on the market. Low investing costs reduce the barriers to enter the markets. High availability of the overall platform services can be expected due to the sheer number of participants.

But there are also disadvantages. Every participant has to ensure all legal issues related to trading. Furthermore, due to the P2P approach, the market does not clear and no price is achieved, leading to disadvantages for some users.

Taking the requirements from section 2 and the advantages and disadvantages of both solutions into account, it appears to be useful to implement a central market place for balance power and use the P2P approach for consumption power. A trading platform should therefore support both system architectures because both types of power are utilized by large generating companies.

4.2 System architecture for the power grid

Similar to the trading platform, the power grid management can be implemented using different system architectures. In order to identify an optimal solution, one has to take requirements from section 3 such as scalability, reliability and autonomy into account.

Grid management based on a central platform makes for easy maintainability and access control for the overall system. Furthermore, a central dispatcher can supervise the global grid security and safety. Disadvantages, such as single-point-of-failures or bottlenecks of applications, can only be avoided partly by using caching-mechanisms or backup-concepts [Date 2000]. Neglected problems so far are overall system scalability and autonomy of distributed generation plants.

Using P2P solutions, the autonomy of each individual plant can be kept. Crashes of single systems (peers) can be compensated by others and do not disturb the overall system stability. One particular disadvantage of this solution is the increased amount of communication that must be handled and kept in narrow bounds by utilizing appropriate communication strategies. One further problem is a missing global entity to maintain power grid stability and to supervise decision-making processes. One solution would be a super-peer architecture where super-peers coordinate and control lesser peers. For this case, local autonomy is suppressed by global communication signals. Individual plants do not communicate directly among each other. Instead, super-peers submit orders to their subordinate peers. Crashes of super-peers can be compensated by other super-peers.

Super-peer architectures belong to the set of hierarchical architectures. Clusters of producers are organized in a hierarchical order. The communication is routed over different hierarchical levels. Within strict hierarchical architectures, each node puts its sub-nodes at risk because a failure cannot be compensated as in super-peer architectures. For the power grid and its requirements, super-peer architectures pose the best solution. So far, the problem of standardized interfaces has been omitted and will be discussed in the following sections.

4.3 Standards for communication

As mentioned before, communication with the distributed units must be laid out for different architecture scenarios. From a comparison of these scenarios, it is clear that though the mapping of the communication peers varies from architecture to architecture, the basic nature of the data and the necessity of data transmission are invariant among the architecture scenarios. In order to guarantee flexible and future-proof system architectures, both standards for communication and standardized data modeling are indispensable.

Currently, the norm families IEC 61968-13 and IEC 61970-XXX define the Common Information Model (CIM) [International Electrotechnical Commission 2003] for the semantic description of objects and attributes for utilities. This model will be discussed in depth within the next section. Additionally, IEC 61850 defines the communication standard used for medium-voltage level substations, and the norm family IEC 60870 defines communication protocols between control center and substation. Between these norms remains an urgent need for further harmonization; the integration of the distributed units in the low-voltage grid is missing. There is a pressing need for a both vertically and horizontally integrated system with consistent data models throughout the various voltage levels in the power grid. This would serve efficiency and trading. For reasons of data privacy and efficiency, one should consider data aggregation at super-peer level in super-peer architectures: super-peers need only know the overall power demand or supply needs existing in the hierarchy below.

4.4 Domain ontologies and data models

When combining several components from different vendors in order to build a distributed system, different semantics for data exchange arise. There is a need for a common vocabulary or language [Uslar et al. 2005]. One particular domain ontology for the electric utility is the Common Information Model CIM.

The CIM was originally defined the Electric Power Research Institute EPRI, a utility funded governmental institute in 1991. The CIM defines a vocabulary of relations between objects within the electric utility. Objects from different functional blocks are defined. It consists of mainly twelve packages that have a variety from contracts and legal documents to wires and transformers. Nearly all important objects and their corresponding attributes can be used to model the data exchanged between components or systems. In order to exchange CIM-compliant data, some serializations have been defined. Serializations exist mainly as XML-conform structures [International Electrotechnical Commission 2004]; the XML is used in different fashions.

XML is mainly used to define payloads utilized by EAI systems. In order to provide richer semantics for applications like net state estimation or grid model exchange [International Electrotechnical Commission 2004], RDF and OWL semantics can be used on top of the ordinary XML serializations. The OWL models provide a good possibility to use ontologies in order to achieve semantic interoperability and harmonization with other standards like IEC 61850. Therefore, the CIM can be seen as the ideal data model for the semantic electric utility.

5 Security

In distributed system architectures, apart from a standardized communication for the exchange of operational and trading data, one requires security of communication for reasons of data privacy, authenticity, and for protection against sabotage [Power system control and associated communications 2003; Stefanini 2005]. Security flaws are introduced accidentally due to the unavoidable interfaces between private and public (i.e. the Internet) network infrastructures. Thus, there is a necessity for firewalls and encryption both symmetrical and asymmetrical to guarantee tamper-proof operation of the system.

The problem with closed source security implementations is the danger of remaining security holes, often leading effectively to a "security-by-obscurity" situation, instead of real security. With open source implementations one has the advantage of auditing by a large community. Contrary, when security flaws are revealed, the infrastructure is temporarily at risk.

6 Conclusion

The need for different ICT-architectures used for trading and power grid management has been motivated. Taking the requirements mentioned into account, suitable architectures for trading and distribution were proposed. Additionally, one should not underestimate security challenges and implement a standardized and vertically integrated communications and systems infrastructure.

Acknowledgements

The authors would like to thank the EWE AG, Oldenburg, and the Ministry for Science and Culture of Lower Saxony for the partial funding of the work introduced by this contribution. Furthermore, the authors want to thank the reviewers for providing useful comments to improve the paper.

References

Barth R, Hashe B, Swider DJ (2006) Verteilte Erzeugung im deutschen Energiesystem. Institut für Energiewirtschaft und Rationelle Energieanwendung (IER). Stuttgart

Bischofs L, Warns T, Hasselbring W (2006) Peer-to-Peer Architekturen. In: Reussner R, Hasselbring H (ed) Handbuch der Software-Architektur. dpunkt verlag, Heidelberg, pp 445-462

Date CJ (2000) An Introduction to Database Systems. Addison-Wesley

European Parliament and of the Council (2003) Directive 2003/54/EC of 26 June 2003 concerning common rules for the internal market in electricity and repealing Directive 96/92/EC

German Federal Ministry of Economics and Technology (2005) Verordnung über den Zugang zu Elektrizitätsversorgungsnetzen, BGBl I, pp 2243

German Federal Ministry of Economics and Technology (2005) Zweites Gesetz zur Neuregelung des Energiewirtschaftsrechts, BGBL I, pp 1970

International Electrotechnical Commission (2003) IEC 61970-301: Energy management system application program interface (EMS-API) – Part 301: Common Information Model (CIM) Base. International Electrotechnical Commission

International Electrotechnical Commission (2004) IEC 61970-501: Energy management system application program interface (EMS-API) – Part 501: CIM RDF Schema – Revision 4. International Electrotechnical Commission

International Electrotechnical Commission (2004) Draft IEC 61970: Energy Management System Application Program Interface (EMS-API) – Part 503: CIM XML Model Exchange Format - Draft 3b. International Electrotechnical Commission

Kurrat M, Schulz C (2006) Verbundbetrieb von Mini-BHKW. Report of the VDE-Kongress 2006, Aachen

Nabe C (2006) Effiziente Integration erneuerbarer Energien in den deutschen Elektrizitätsmarkt. PhD Thesis, TU Berlin

Power system control and associated communications (2003) – Data and communication security, IEC TR 62210

Stefanini A (2005) The future of ICT for power systems: Emerging security challenges. Report of the workshop, Brussels

Theobald C et al. (2003) Regelmarkt - Gutachten zu Marktgestaltung, Beschaffungskosten und Abrechnung von Regelleistungen und Regelenergie durch die deutschen Übertragungsnetzbetreiber, Büro für Energiewirtschaft und technische Planung GmbH

Uslar M, Schmedes T, Luhmann T, Lucks A, Winkels L, Appelrath H-J (2005) Interaction of EMS related Systems by Using the CIM Standard. In: Leal Filho W, Marx Gómez J, Rautenstrauch C (eds) ITEE 2005: Second International ICSC Symposium on Information Technologies in Environmental Engineering: Proceedings, Otto-von-Guericke Universität Magdeburg. Shaker, Aachen, pp 596-610

Verband der Netzbetreiber (2003) TransmissionCode 2003: Netz- und Systemregeln der deutschen Übertragungsnetzbetreiber. Verband der Netzbetreiber - VDN - e.V. beim VDEW

Designing a Bilingual Eco-Ontology for Open and Intuitive Search

Bettina Bauer-Messmer, Rolf Grütter

Swiss Federal Institute for Forest, Snow and Landscape Research,
Zürcherstrasse 111, 8903 Birmensdorf ZH, Switzerland
e-mail: {bettina.bauer, rolf.gruetter}@wsl.ch

Abstract: In environmental databases there is a plethora of data, described in different terminologies, stored in different data structures and referring to distinct time periods and locations. Thus it is difficult for non-experts to find the data they are looking for. Adding an ontology-layer, which provides semantic enhancement, provides the underlying data structure for an open and intuitive search interface for non-experts. In this paper, we present the design of a bilingual eco-ontology using fuzzy mappings and loosely bridged ontologies which introduce a semantic layer in an existing environmental database storing data in German and French.

1 Introduction

Existing environmental databases hold a great variety of data, which are encoded in different languages, grouped into diverse inventories using their own terminologies and refer to distinct time periods. Our environmental database at the Swiss Federal Institute for Forest, Snow and Landscape Research in particular is characterized by the coexistence of diverse inventories with data in German and French. In order to help non-expert users with finding the data they are looking for, we propose to add an ontology-layer as the underlying data structure for an open and intuitive search. An *open* and *intuitive* search does not only look for exact matches of specified filter criteria but also searches for semantically similar database entries. It supports patterns of interaction which are familiar to the user. This kind of search is made possible with a natural language user interface.

Processing natural language input from users requires information systems, which "understand" up to a certain degree the input. The term "understand" refers to a mapping of conceptual structures in the users mind onto data structures. Ontologies are embodiments of shared conceptualizations [Gruber 1993] and therefore can semantically interpret and augment user input such as search terms.

Eco-ontologies are either defined as focusing on spatial "ecological niches" [Smith and Varzi 1999] or they stress the aspect of semantic heterogeneity of source communities [Fonseca et al. 2002]. In this paper the term eco-ontology refers to the later definition. Eco-ontologies represent the geospatial domain, for which no consensual ontology yet exists, and as long as the ambiguities within concepts are not clarified, it will, according to Agarwal [2005], not be possible to create one.

Building a bilingual ontology for environmental data poses challenges of different types:

- *Linguistic challenge:* The conceptualizations and vocabulary for objects in the environment change over time, vary between ethnic and cultural groups and are strongly influenced by regional dialects.
- *Bilingual challenge:* The categories of terms are built based on different criteria and do not always have the same level of detail in German and French. Direct mapping between these two languages would lead to a considerable loss in expression.
- *Heterogeneous semantics challenge:* The world is ambiguous. Datasets from different communities do not share a conceptualization. Even within a single domain, e.g., biology, taxonomies can be contradictory.
- *Toponymy challenge:* Geographic locations are often known by several names. These names depend on the scale of a map (country, canton, community). Names in rural dialects differ from those used in official maps.

The work presented here is focused on the linguistic and bilingual challenges of constructing an eco-ontology which also touch the heterogeneous semantics challenge. It builds on previous work on the design and implementation of a Web-based platform for visualizing, querying and analyzing environmental data [Baltensweiler and Brändli 2004]. The toponymy challenge will be addressed elsewhere (in preparation).

2 Linguistic Challenge: Defining a vocabulary and its classification

2.1 Local dialects and cultural differences

Depending on language, cultural background, social environment, education and many more factors the number of terms describing a given sector of the world varies enormously. Traditional Filipino Pygmies distinguish a wide variety of animals and plants: For example 15 breeds of bats, 75 avian breeds, 20 breeds of ants and 450 plants among many others [Levi-Strauss 1962]. In general, rural dialects are very rich in terms and definitions of rural everyday life. However these dialects are far less specific in other subject fields. The level of detail in describing concepts directly represents the importance for the individual. Thesaurus completeness is crucial for the success of information retrieval systems [Salton 1970]. Therefore great care must be taken to construct a thesaurus with a level of detail which corresponds to the needs of the potential users.

2.2 Historic Classifications

There are three German terms for bush, shrub and perennial herb: Busch, Strauch, Staude. Nowadays, the term Busch is a superordinate concept of Strauch and Staude. The difference between Strauch and Staude is based on the fact, that the stems of a Strauch are wooden, and the stems of Staude are soft. This hasn't always been the case. About 150 years ago there was a distinction between Strauch and Staude based on the fact, that Staude yields (edible) fruit and Strauch does not so [Grimm 1854]. Alas, there is not only a semantic shift but also the focus of attention has changed. This is reflected by the fact, that taxonomic categories are built based on different criteria, indicating that yielding fruit was an important criterion for people 150 years ago, but is no longer these days.

2.3 Orthography and spelling

The concepts of orthography and spelling emerged only recently. In 1776 Thomas Jefferson's draft of the Declaration of Independence showed many spelling inconsistencies. Even these days the German orthography reform in 1996 lead to hot tempered discussion. Language and spelling was and still is an instrument for individual expression.

3 Bilingual Challenge: Combining conceptualizations from different languages

In the last decades multilingual information retrieval has become an ever growing research area. The architectures of these systems haven't changed much over the years [Salton 1970]: thesaurus (vocabulary), translation facility (mapping functions, translation tables), hierarchical arrangement of concepts and more or less sophisticated algorithms, which represent the semantic field and context. By contrast the algorithms and data storage techniques have changed considerably since the beginnings, especially with regard to the global community participating and sharing information in the Semantic Web.

A more recent approach to multilingual information retrieval is EuroWordNet, although not being a pure ontology, it uses a top ontology to enforce uniformity and compatibility of different word-nets [Vossen 2002]. Other multilingual ontologies are being built all over the world with huge vocabularies, e.g. the Chinese-English ontology or the Arabic WordNet. Furthermore efforts are taken to combine bilingual translation services in a grid.

3.1 Taxonomic Hierarchies in German and French

In German there are three terms for bush: Busch, Strauch, Staude. In French, however, there are four words: buisson, arbuste, arbrisseau and plante vivace. The definitions of these terms (disregarding the fact that definitions vary by author) are based on different criteria. Furthermore, individual plants are categorized differently in both languages.

It is most remarkable that the classification schema is not complete and not free of overlaps. For instance plants with soft stems and a height of more than 25 cm and less than 4 m do not fit in any category. On the other hand hazelnut can be classified as both shrub and tree.

Trying to combine both German and French in a single ontology inevitably leads to inaccuracies or inconsistencies. In order to preserve the richness of detail and the historically and culturally grown categories, it is imperative to create a separate ontology per language.

In domains such as biology even in a single language several different classification schemes, so called systematics, coexist. Experts are used to reference the classification scheme whenever they report on their work.

The classification of plants shown in Figure 1 is merely one of many possibilities. Plants can be classified just as well by their life-form, for instance the height above (or below) ground of the renewal bud.

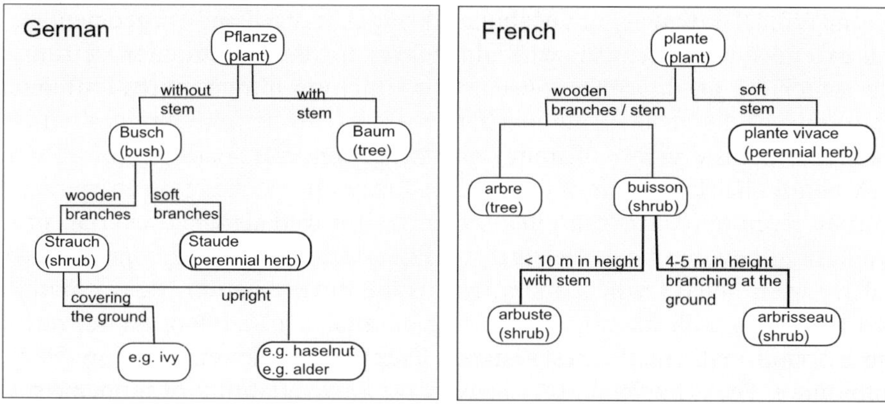

Fig. 1. Taxonomic hierarchies in German and French (Source: Meyers Lexikon, LEO Deutsch-Französisch Wörterbuch, http://dict.leo.org/frde) are built based on different classification criteria.

The classification is in many cases ambiguous, depending on the habitat and living conditions of an individual plant. Ivy for instance can grow either covering the ground or climbing at trees.

4 Design of a Bilingual Eco-ontology

The challenges discussed above lead to the following design goals for a bilingual eco-ontology:

- A bilingual eco-ontology will consist of the alignment (also known as mapping) of several ontologies, each consistent in itself. A single ontology with one root is not desirable.
- Mechanisms such as statistical or fuzzy algorithms for handling uncertainties will be introduced. One-to-one translations and alignments are not possible.
- Mechanisms for dynamic updates, maybe even self-learning or self-organization will be provided. A static ontology will soon be outdated (this design goal is not addressed here).

There are different approaches to implement ontologies, among them: Topic Maps (ISO 13250) and ontologies in OWL. While Topic Maps better support "human semantics", OWL (especially OWL DL) supports machine reasoning and automatic search engines, which is a central feature of our project.

While striving for the first two design goals we revert to the way how semantic heterogeneity is usually handled in description logics ontologies (e.g. OWL DL), that is by vocabulary bridges [Catarci and Lenzerini 1993], and extend this mechanism with algorithms for the construction of fuzzy sets which are used to map concepts for which no bridges exist. Different from state-of-the-art bridging mechanisms we consider only bridges which are based on the *equality* of source and target concepts.

A comprehensive summary of recent fuzzy approaches to web applications is given by Nikravesh [2006]. Uncertainty and semantic overlap in a single ontology can be modeled using fuzzy degrees of subsumption [Holi and Hyvönen 2006], with a fuzzy extension to description logic [Straccia 1998], or even with fuzzified OWL [Stoilos et al. 2005]. Bayesian networks use a sound mathematical (statistical) background for reasoning in fuzzy ontologies. They, however, rely heavily on the availability of probabilistic information from domain experts, which is often not available.

Unlike Bayesian networks we use *algorithms* for dealing with fuzziness. Instead of forcing a crisp one-to-one translation of German and French terms, our algorithm semantically expands both the German and French terms. A fuzzy realm of the original search term is created. We consider this combination of loosely bridged ontologies with fuzzy mappings, as shown in Figure 2, as an original contribution to research.

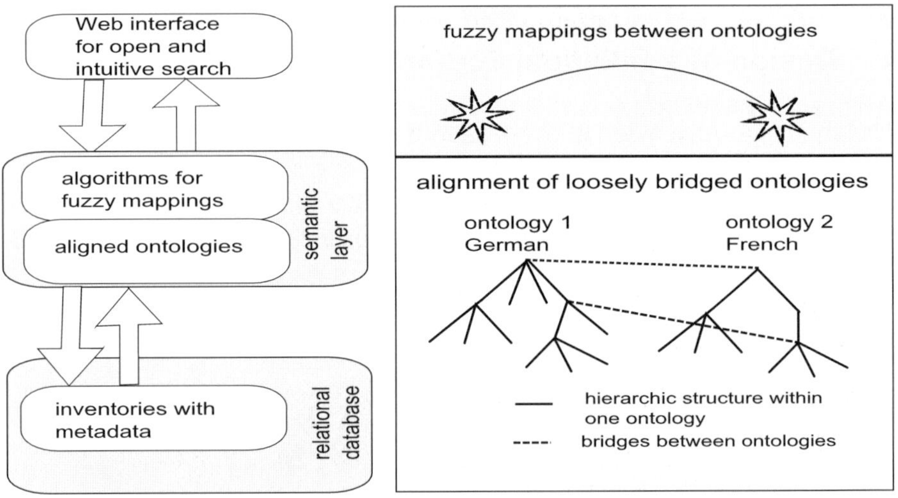

Fig. 2. Introducing a semantic layer (consisting of algorithms for fuzzy mappings and aligned ontologies) in an existing eco-database environment.

To sketch the approach in more detail, we assume that the user enters the term "Staude" into the search form and looks for relevant resources in any encoding language (i.e., German or French). Since there is no bridge between the concept denoted by the search term and a semantically equivalent concept in the French ontology, the algorithm climbs the German concept hierarchy until it finds a concept for which a bridge to the French ontology has been defined. In the example this is the concept `Pflanze` which is defined as being equal to the concept `Plante` by the bridging axiom `Pflanze` ≡ `Plante`. While proceeding, the algorithm adds the names of all concepts visited including their children's to the set of search terms, in the example the terms "Staude", "Busch" and "Strauch".

Starting from the target concept of the bridge, the algorithm then retrieves all subsumed concepts and adds their names to the set of search terms, in the example the terms "arbre", "plante vivace", "buisson", "arbuste", and "arbrisseau". In pseudo code the described algorithm can be put in the following way:

```
Declaration
item:         the given search term
targetitem:   the translated term of item
parent:       hypernym of item (generalization)
child:        hyponym of item (specialization)
query-terms:  list of search terms. At the begin-
              ning query-terms contains the search
              term item. At the end of the algo-
              rithm it contains additional, seman-
              tically related terms.

Main Program
  if bridge-exists(item) then
     /* add the search term itself */
     add-to-query-terms(item)
     /* add the via bridging translated search
        term */
     add-to-query-terms(targetitem)
  else query-expand(item)
End Program

Subroutine query-expand(item)
  if NOT bridge-exists(item) then
      /*add the specialized terms of item */
      include-children(item)
      /* back-tracking */
      query-expand(parent)
  else  /*bridge exists */
      /*add the children of the translated
        search term*/
      include-children(targetitem)
End Subroutine
```

During execution the algorithm constructs a complex concept, in our example Staude ∪ Busch ∪ Strauch ∪ Arbre ∪ Plante_vivace ∪ Buisson ∪ Arbuste ∪ Arbrisseau.

This concept represents to one part a well-defined extension of the original concept Staude, which assumes that the user is also interested in resources which are closely related to those addressed by the search term (cf. the concept of open search as introduced in the first section). To another part it represents a *fuzzy* extension of the original concept with respect to the certainty that also objects in French corresponding to those denoted by the German "Staude" are included in its (set-theoretic) interpretation: The disjunction of the subsumed concepts does not necessarily have the same extension as the subsuming concept. There might be plants which cannot be classified as either "arbre", "plante vivace", "buisson", "arbuste" or "arbrisseau".

5 Conclusion

Natural language is a highly dynamic system, with significant semantic changes over time and semantic differences between social-cultural groups. Nevertheless it is the elementary postulate of science, that there is an order in nature [Levi-Strauss 1962], which is reflected by human conceptualizations, that can be modeled as ontologies.

In combining the rigid frame of ontologies with the flexible techniques of fuzzy algorithms, a promising approach to a data structure supporting an open and intuitive search in a bilingual environmental database was designed. The implementation will make use of standard Semantic Web technology offering the possibility of future extensions to the semantic layer in a similar way as described (e.g., inclusion of zoological taxonomies). Since the algorithms are a key component of the architecture, it will be interesting to analyze how modifications thereof will affect recall and precision of predefined queries as well as system performance. Further research on the integration of the eco-ontology into the virtual data base project [Frehner and Brändli 2006] is planned for the near future.

Acknowledgements

The authors sincerely thank Jürg Schenker, Martin Hägeli and Martin Brändli for the fruitful discussions and leadership that made this research possible. This research was funded and conducted in cooperation with the Swiss Federal Office for the Environment (FOEN).

References

Agarwal P (2005) Ontological considerations. GIScience, International Journal of Geographical Information Science, vol 19, edn 5, pp 501-536

Baltensweiler A, Brändli M (2004) Web-based Exploration of Environmental Data and Corresponding Metadata, in Particular Lineage Information. In: Scharl A (ed) Environmental Online Communication. Springer-Verlag, London, pp 127-132

Catarci T, Lenzerini M (1993) Representing and Using Interschema Knowledge in Cooperative Information Systems. Int. J. Cooperative Inf. Syst., vol 2, edn 4, pp 375–398

Fonseca F, Martin J, Rodríguez MA (2002) From Geo- to Eco-ontologies. In: Egenhofer MJ, Mark DM (eds) (2002) Geographic Information Science, Second International Conference, GIscience, USA. Springer LNCS 2478, pp 93-107

Frehner M, Brändli M (2006) Virtual database: Spatial analysis in a Web-based data management system for distributed ecological data. Environmental Modelling & Software. vol 21, pp 1544-1554

Grimm J, Grimm W (1854) Deutsches Wörterbuch von Jacob Grimm und Wilhelm Grimm. 16 Bde. in 32 Teilbänden, Leipzig, Hirzel S (ed) pp 1854-1960

Gruber TR (1993) A translation Approach to Portable Ontology Specifications. Knowledge Acquisition, vol 5, pp 199-220

Holi M, Hyvönen E (2006) Modeling Uncertainty in Semantic Web Taxonomies. In: Soft Computing in Ontologies and Semantic Web, Studies in Fuzzyness and Soft Computing, vol 204, Zongmin Ma (ed), ISSN 1434-992, Springer, pp 31-46

Levi-Strauss C (1962) Das Wilde Denken, Shurkamp Taschenbuch Wissenschaft, ISBN 3-518-27614-X, p 334

Nikravesh M (2006) Beyond the Semantic Web: Fuzzy Logic-Based Web Intelligence. In: Zongmin M (ed) Soft Computing in Ontologies and Semantic Web, Studies in Fuzzyness and Soft Computing. vol 204, ISSN 1434-992, Springer, pp 149-209

Salton G (1970) Automatic Processing of Foreign Language Documents. Journal of the American Society for Information Science, vol 21, edn 3, pp 187-194

Smith B, Varzi A (1999) The Formal Structure of Ecological Contexts. Modeling and Using Context, Second International and Interdisciplinary Conference, 1999

Stoilos G, Stamou G, Tzouvaras V, Pan JZ, Horrocks I (2005) Fuzzy OWL - Uncertainty and the Semantic Web. In: Proceedings of the International workshop on OWL, Experience and Directions (OWL-ED 2005)

Straccia U (1998) A fuzzy description logic. In: Proceedings of AAAI-98, 15th National Conference on Artificial Intelligence, Madison, Wisconsin
Vossen P (2002) EuroWordNet General Document, Version 3, University of Amsterdam, p 108

Data Warehousing with Environmental Data

Ansgar Burmann[1], Jorge Marx Gómez[2]

[1] Otto-von-Guericke-Universität, Magdeburg

[2] Oldenburg University, Department of Informatics,
Ammerländer Heerstr. 114-118, 26129 Oldenburg, Germany,
e-mail: jorge.marx.gomez@uni-oldenburg.de

Abstract: In this paper, the results of the project "Conception of an Environmental Data W" are presented. The actual situation at the Volkswagen AG is shown and the analysis phase is described. After discussing the most important problems that occurred during the project, some of them special to environmental data, it is described what was achieved and what has yet to be done. At the end, the potential of an environmental data warehouse is outlined.

1 Introduction

In an enterprise there are usually several systems that hold data which could be suitable to support management decisions. An often used construct to implement management support is the so called data warehouse. "A Data Warehouse is a subject oriented, integrated, non-volatile and time variant collection of data in support of management decisions" [Inmon 2004]. This definition by Inmon shows the most important aspects of a data warehouse. The data has to be subject oriented, integrated, time variant and non volatile.

In actual literature, one can find slightly differing designs of data warehouses, but the main aspects are mostly the same.

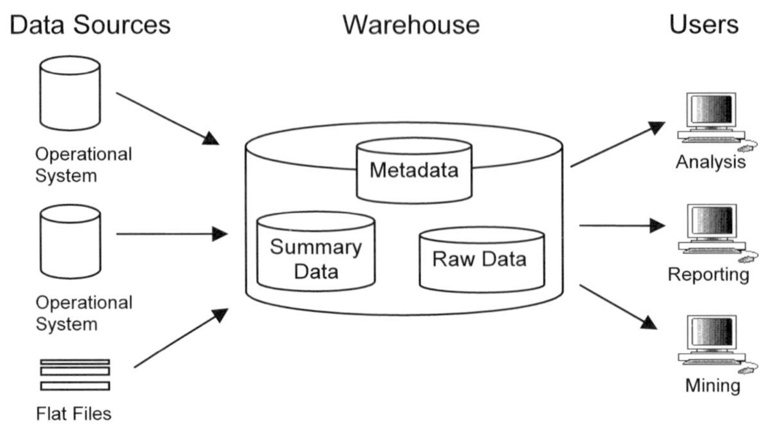

Fig. 1. Architecture of a data warehouse [DWC]

In Figure 1 a simple architecture for a data warehouse is shown, subdivided into the data sources on the left hand side, the warehouse itself in the middle and the users on the right hand side.

Looking at the data sources shows that a data warehouse usually has multiple data sources. That means that there are several operational systems of which data is extracted as well as additional flat files. This is one important point of a data warehouse; it extracts data from several, heterogeneous systems and formats. The warehouse is placed in the middle of Figure 1, containing raw data, summary data and metadata. The raw data which was extracted from the source systems is held within the data warehouse, as well as an additional data type, the summary data. Summaries are very valuable in data warehouses because they precompute long operations in advance. These summaries are also sometimes called materialized views.

One very important aspect of a data warehouse is metadata. Metadata can be used actively, passively and semi-actively. Actively used, metadata like transformation rules can be used to automate the ETL-process. Using metadata passively means using it as a documentation of the data warehouse, which can be accessed by the users for a better understanding of the data. Storing structural information as metadata gives the opportunity to check data for structural correctness, which then is semi-active use of metadata, as it is not actively used for a process.

Metadata is one very important aspect when designing a data warehouse, because the quality and understandability of data is improved.

There are several important aspects that have to be taken into consideration when designing a data warehouse. They can't all be discussed in this paper, so if you have further interest in how to design a data warehouse you can grab one of the many books about data warehouse design.

Dealing with environmental information, the question is if and how a data warehouse can deal with this sort of data. What are the benefits of an environmental data warehouse? Is a data warehouse an option for analyses on environmental data? What are the pros and cons for a data warehouse considering environmental data?

2 Analysis

After taking a look at the theoretical background of the data warehouse and presenting an architecture for a data warehouse, the existing systems where examined to get a complete picture of all the information held within these systems.

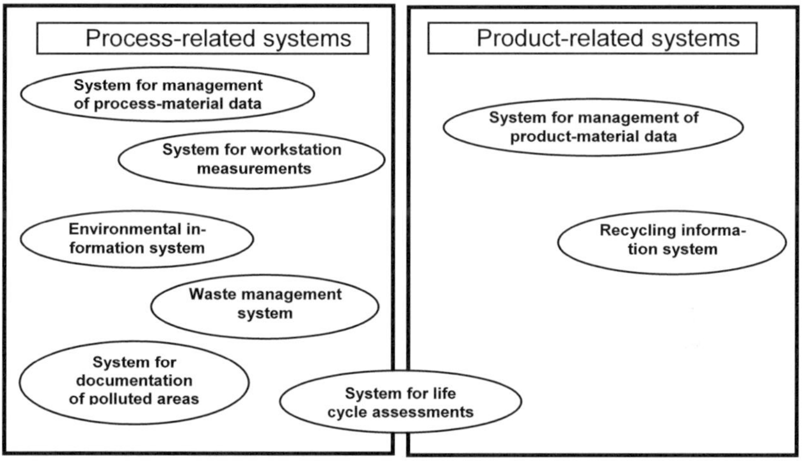

Fig. 2. Given situation of systems

Figure 2 shows the actual situation at the Volkswagen AG concerning the systems holding environmental data. On the left hand side, the process-related systems are presented, on the right hand side the product-related systems. At the bottom in the middle, the system for life cycle assessments is shown. This position between the two categories does not mean that this system actually connects the two types of systems but that it belongs to both sides as it holds both product- and process-related data. Figure 2 actually demonstrates that there is no connection between the two types of systems, which was one cause and motivation for the project.

Having analyzed the different systems and databases, a scheme was derived to represent the possible connections between the different systems. Like that it was possible to discover the missing pieces to connect process- and product-related systems and the data held within these systems.

In addition to that, meetings where held with two different departments of the Volkswagen AG, called "Umweltplanung" (environmental planning) and "Umweltstrategie" (environmental strategy), to find out the needs and requirements of the potential users regarding an environmental data warehouse. As a result of these meetings, some requirements for a data warehouse are collected and potentially implementable reports are presented. The next step was to find out, whether and how a data warehouse could meet these requirements by providing reports on an automated basis.

During the analysis of the different systems, some problems were revealed. First of all, as already described in the introduction, one goal of the project was to connect product- and process-related systems. After looking at the different databases and especially the primary keys used in those databases, it turned out that the systems used different primary keys, making the connection between the systems very difficult. This environmental information system (EIS) holds different information related to manufacturing plants.

In comparison to that, the product related systems use material-numbers as primary keys. So, to connect process- and product-related systems, additional information was needed. So other systems had to be analyzed, searching for a possible connection between the two types of systems. In another department of the Volkswagen AG, the system for working plans (WP) is run. This system contains a key for manufacturing plants as well as a key for material-numbers, both associated to a working plan. A working plan shows which materials are produced on which manufacturing plants in a certain process and creates a connection between process- and product-related systems. Unfortunately, the key for the manufacturing plants in the EIS isn't the same as the one used in the WP system. Nevertheless a preceding data warehouse project already found at least a partial solution to this problem, where 70% of the keys could be mapped automatically.

This is actually a problem that can occur in any data warehouse project and is not specific for environmental data. However during the project, another problem occurred which is more specific.

Analyzing the user demands for an environmental data warehouse revealed different requests for reports to be run on the data warehouse, one of them being the question how many emissions did occur when producing a component. The data needed for those reports wasn't completely available in the analyzed systems, so the question was whether information with an environmental context could be found in other than those systems. After some investigation, a system for process-planning (PP) was found. This system contains all information needed to plan and to run a system, including information about the length of all welded joints made on a component. At first glance, this actually doesn't look like environmental information. It definitely isn't primarily environmental information, but can be used to derive such when looked at it with additional information. During the authoring of the project, emission factors where obtained, giving information about how many emissions occur when doing a 1mm welded joint. Combining this information with those from the PP system gives information about how much emission occur when producing a component, as long as all welded joints are included. It has to be mentioned that of course only the fine particulate emissions produced in the welding processes can be retrieved, as emission factors where only available for this type of emission.

In this context, the data from the PP system actually is of environmental importance. That shows that environmental data or at least data which can be of environmental importance can be found in other than the environmental systems, which is one important conclusion from the analysis phase. As not all companies have a separate department for environmental data, this might even be worse in other companies. The environmental data might be spread over several systems, where there is no intended use for this data in an environmental context. So this is something where a lot of work might be needed when designing a data warehouse using environmental data.

In the case of this project, those two aspects where the main problems that had to be solved – building a connection between product- and process-related systems and gathering all the information needed to be able to do some sample reports on existing data.

Of course those weren't the only difficulties during that project. Retrieving data from other departments is sometimes a quite difficult task, as one has to justify the need for every single data field before being allowed to use the data. This isn't meant as an offence, because those people signing responsible for the systems have the responsibility that their data is used correctly. So complications when being in need of data from other departments are quite common, as the quality of data has to be ensured.

This is where the metadata comes in quite handy. Building up a repository in a data warehouse, which holds all the metadata, improves the quality of the data and makes sure that the users understand the data the right way. Presenting those arguments when in need of data from other departments can help to counter the concerns.

There is one more important thing about the EIS that has to be mentioned. Remember that the focus of the EIS is the manufacturing plants, and all the data found in this system is related to those manufacturing plants. In the EIS, it is possible to create material flows between manufacturing plants. With those material flows, process materials and emissions can be assigned to manufacturing plants. If a possibility existed to connect product- and process-related systems, this would create an opportunity to assign emissions and process-materials to components, which was one of the user demands. The material flows of the EIS have quite a lot potential for a data warehouse, as several connections are possible by those material flows. Not only emissions or process-materials could be assigned to manufacturing plants, this would also be possible for water or energy consumption. The Problem with these material flows is that at present, there is only the option to create them in the EIS but no one is actually using them. This shows that there is still some work to do, but also that there is a lot of potential to be discovered.

3 Results

As described above, the analysis phase revealed some problems and missing data needed to connect process- and product-related systems. Retrieving this missing data would be one of the major tasks when implementing an environmental data warehouse and a way to do so is presented in the project. Nevertheless, the question was whether it is possible to provide reports for the given questions even without this data by using other systems. This goal was achieved by using the data from the PP system so that one practical example could be implemented. Using a practical example developed in cooperation with the Volkswagen department "Umweltplanung", a starflake-scheme is contrived. On the basis of this scheme, a report on component related emissions is run to outline the possible practical benefit of a central environmental data warehouse. To be able to do so, the starflake-scheme was filled with extracts of operational data. Like that, the results of that report are of practical interest as the data is not fictive.

Another area of interest when dealing with environmental data is that of sustainability reports. Among other things, the Volkswagen department environment strategy is responsible for those reports. When creating those reports, a guideline has to be followed. These Sustainability Reporting Guidelines [GRI 2006] contain several measurements for environmental performance, so called performance indicators. Examples for those performance indicators are:

1. Materials used by weight or volume
2. Direct energy consumption by primary energy source
3. Total water withdrawal by source
4. Significant air emissions by type and weight
5. …

Those indicators are just an extract from the guidelines. The reason for presenting those indicators gets clear when remembering the material flows mentioned in the previous chapter. Assuming there would be material flows in the EIS, at least some of the performance indicators could be retrieved from a data warehouse in an automated way, thus making the generation of sustainability reports a lot faster and saving manpower. So this has the potential to reduce costs, too.

The result of the project is that a data warehouse is one possible solution for integrating different systems holding environmental data. Some problems like inconsistent data or different interpretation of data can be solved by a data warehouse, for example with the use of a repository and the metadata held within such a repository. In addition to that, some user demands can already be fulfilled with existing data by using additional systems.

This shows that there might be relevant data in other than the environmental systems as well. Considering the discussed facts a data warehouse seems to be an area of interest concerning environmental data, as for example some aspects of sustainability reports could be processed automatically by such a data warehouse.

4 Outlook

In the project, it was possible to outline the possible benefits from an environmental data warehouse for the Volkswagen AG. The main problems, like the use of different primary keys or the absence of material flows in the EIS are explained. Another problem was that retrieving the data for the practical example sometimes took a lot of time, in some cases about two month. This again isn't a problem specific to the Volkswagen AG but can occur in every company. The larger a company is, the more difficult it might get to obtain the relevant data, as there is a lot of explaining and justifying to do before actually getting access to the data due to security reasons. Despite these complications during the authoring of the project, one practical example was implemented using real data. This was achieved by using other than the environmental systems.

What couldn't be done is the connection between the process- and the product related systems. Nevertheless a solution to this problem is presented as a mapping table between the keys from the EIS and the PP system. This mapping table hasn't actually been created, but this task has been assigned to another project so that within the framework of the project it wasn't sensible to do so.

A second practical example was planned concerning component related use of water and energy, but the necessary data couldn't be provided in time to implement this.

What the project nevertheless shows is that a data warehouse for environmental data has a lot of potential. Not only the demands of some users can be fulfilled, even parts of the sustainability reports can be processed automatically when having access to the right data. Besides the presented examples, there are a lot more that can be processed by such an environmental data warehouse. That is the reason why in the project the development of an environmental data warehouse is recommended.

Taking into consideration the discussed benefits of an environmental data warehouse, this can be a promising project for other companies as well. But the problems discussed above show that several problems might occur, so a lot of work might be needed to accomplish such a data warehouse project. In addition to that, environmental data or data with environmental relevance might be hidden in different systems, so every available system has to be analyzed. Despite these problems, in the opinion of the author, the benefits of an environmental data warehouse outweigh the difficulties.

References

Inmon WH (2004) Building the Data Warehouse, Second Edition, Jon Wiley & Sons, New York

Data Warehousing Concepts (DWC) Oracle9i Data Warehousing Guide

Release 2 (9.2), http://www.lc.leidenuniv.nl/awcourse/oracle/server.920/a96520/concept.htm

Global Reporting Initiative™ (GRI) (2006) Sustainability Reporting Guidelines (Draft), http://www.globalreporting.org, 2007-02-16

Burmann A (2006) Konzept für ein zentrales Umwelt Data Warehouse

New Measures for Evaluating Decision Systems Using Rough Set Theory: The Application in Seasonal Weather Forecasting

Yailé Caballero[1], Leticia Arco[2], Rafael Bello[2], Jorge Marx Gómez[3]

[1] University of Camagüey, Camagüey, Cuba

[2] Central University “Marta Abreu” of Las Villas, Villa Clara, Cuba

[3] Oldenburg University, Department of Informatics,
Ammerländer Heerstr. 114-118, 26129 Oldenburg, Germany,
e-mail: jorge.marx.gomez@uni-oldenburg.de

Abstract: The accelerated growth of the environmental of information volumes on processes, phenomena and reports brings about an increasing interest in the possibility of discovering knowledge from data sets. This is a challenging task because in many cases it deals with extremely large, inherently not structured and fuzzy data, plus the presence of uncertainty. Therefore it is required to know a priori the quality of future procedures without using any additional information. For those reasons, the main goal of this paper is to define and apply new evaluation measures for decision systems by using the rough sets theory in an application of the quality assessment of the decision systems used for learning seasonal weather forecasting. The experimental studies were carried out for demonstrating the feasibility of the proposal.
Keywords: seasonal weather forecasting, rough set theory, evaluation measures

1 Introduction

The likelihood of knowledge discovery from data sets has reached its top interest level nowadays because of the speedy growth of digital information. Automatically processing large amounts of data to find useful knowledge is the primary target of knowledge discovery from databases (KDD) and it can be defined as a non-trivial procedure of identifying valid, novel, potentially useful and eventually comprehensible patterns from data

[Kodratoff 1999]. Machine learning (ML) studies the learning problem in the context of machines, i.e. how machines are able to acquire the knowledge that gets them ready to solve particular problems [Mitchell 1997]. ML is driven to automate the learning process in such a way that knowledge can be attained with a minimum of human dependency. A system is able to learn either by obtaining new information or modifying the knowledge it currently owns so as to make it more effective. The outcome of learning is to outfit the machine (or man) with novel knowledge that allow to embrace (provide solutions to) a wider range of problems as well as to achieve more accurate or cheap solutions or at least, to simplify the knowledge stored.

ML and KDD have a very strong relation. Both of them acknowledge the importance of induction as a way of thinking. These techniques have a broad spectrum of applications. In this work an illustrations of their use in environmental studies is depicted. For so many years, very large amounts of data that comprise the values of indicators used in seasonal weather forecasting have been recorded and gathered. Nevertheless, imprecision in forecasting predictions still arises; therefore a great interest in employing knowledge discovery methods to generate models of forecasting systems exists at the present time. Mining environmental data bases imposes challenges to KDD and ML techniques. This is a defiant task because in many cases inherently non-structured, fuzzy data are handled, there is also some degree of underlying uncertainty, the data sets are extremely large and it is needful to know a priori the quality of a future processing without using any additional information.

Lots of artificial intelligence procedures are applicable to such sort of problems. One of them is the rough set theory (RST) that exhibits great profit in environmental scenarios. Rough sets can be considered sets with fuzzy boundaries – sets that cannot be precisely characterized using the available set of attributes [Pawlak et al. 1995]. The basic concept of the RST is the notion of approximation space. Two advantages of RST can be used for the solution of environmental problems: (i) it does not need any preliminary or additional information about data, and (ii) it is a tool for use in computer applications in circumstances which are characterized by vagueness and uncertainty.

Therefore, the goal of our research is to define and apply new measurements to evaluate the quality of decision systems by using RST in the evaluation of the quality of training sets used by algorithms for learning classification in the seasonal meteorological forecasting problem.

This paper is organized as follows. Section 2 presents the general concepts about RST and some measures of decision systems using RST. In section 3 the description of the seasonal meteorological forecasting problem and the way the RST and the evaluation measures are utilized in the assessment of the quality of decision systems for seasonal weather forecasting are described. The experimental study and results are portrayed in section 4. Finally, section 5 provides some conclusions and future work.

2 Rough Set Theory

Rough set theory, introduced by Pawlak [1982], has often proved to be an excellent mathematical tool for the analysis of a vague description of objects. The adjective vague, referring to the quality of information, means inconsistency or ambiguity which is caused by the granularity of information in a knowledge system. The rough sets philosophy is based on the assumption that with every object of the universe there is associated with certain amount of information (data, knowledge) expressed by means of some attributes used for object description. Objects having the same description are indiscernible with respect to the available information. The indiscernibility relation modeling the indiscernibility of objects thus constitutes a mathematical basis of RST; it induces a partition of the universe into blocks of indiscernible objects, called elementary sets that can be used to build knowledge about a real or abstract world. The use of the indiscernibility relation results in information granulation [Pawlak 1982; Pawlak et al. 1995]. In this section we recall some basic notions related to rough sets and the extension of RST using similarity relations. Also, we mention some measures of closeness of concepts and measures of decision systems. Finally, we propose some new measures of decision systems using RST.

2.1 Indiscernibility, Similarity Relations, Lower and Upper Approximations

An information system is a pair $IS=(U,A)$, where U is a non-empty, finite set called the universe and A is a non-empty, finite set of attributes. Elements of U are called objects. A decision system is a pair $DS=(U, A\cup\{d\})$, where $d\notin A$ is the decision attribute.

The basic concepts of RST are the lower and upper approximations of a subset $X\subseteq U$. These were originally introduced with reference to an indiscernibility relation R. Let R be a binary relation defined on U which represents indiscernibility. By $R(x)$ we mean the set of objects which are indiscernible to x.

In classic RST, R is defined as an equivalence relation (reflexive, symmetric and transitive). R induces a partition of U into equivalence classes corresponding to $R(x)$, $x\in U$. This classic approach to RST is extended by accepting that objects which are not indiscernible but sufficiently close or similar can be grouped in the same class [Slowinski and Vanderpooten 1997]. The aim is to construct a similarity relation *R*' from the indiscernibility relation *R* by relaxing the original conditions for indiscernibility. This relaxation can be performed in many ways, thus giving many possible definitions for similarity. However, this similarity relation *R*' must satisfy some minimal requirements. *R* being an indiscernibility relation (equivalence relation) defined on *U*, *R*' is a similarity relation extending *R* iff $\forall x \in U,\ \ R(x) \subseteq R'(x)$ and $\forall x \in U$, $\forall y \in R'(x)$, $R(y) \subseteq R'(x)$, where $R'(x)$ is the similarity class of *x*, i.e. $R'(x)=\{y\in U\text{: } yR'x\}$. *R*' is reflexive, any similarity class can be seen as a grouping of indiscernibility classes and *R*' induces a covering of *U* [Slowinski and Vanderpooten 1997].

Notice that *R*' is not imposed to be symmetric. Even if most definitions of similarity usually involve symmetry. Notice also that *R*' is not imposed to be transitive. Unlike non-symmetry, non-transitivity has been often assumed for similarity. This clearly shows that an object may belong to different similarity classes simultaneously. It means that the covering induces by *R*' on *U* may not be a partition. The requirement of any similarity relation is reflexivity. *R*' can always be seen as an extension of the trivial indiscernibility relation *R* defined by $R(x)=\{x\}$, $\forall x\in U$.

The rough approximation of a set $X\subseteq U$, using an indiscernibility relation *R*, has been introduced as a pair of sets called *R*-lower and *R*-upper approximations of *X*. We consider here a more general definition of approximations which can handle any reflexive *R*'. The *R*'-lower and *R*'-upper approximations of *X* are defined respectively as:

$$R'_*(X) = \{x \in X : R'(x) \subseteq X\} \tag{2.1}$$

$$R'^*(X) = \bigcup_{x\in X} R'(x) \tag{2.2}$$

When a similarity relation is used instead of the indiscernibility relation, other concepts and properties of RST (approximation measures, reduction and dependency) remain valid.

2.2 New Measures of Closeness of Concepts and Measures of Decision Systems

RST offers measurement techniques for the analysis of decision systems. Accuracy and quality of approximation and quality of classification measures are three representatives of these techniques. Rough membership function is an important function to develop new measures for the analysis of decision systems [Skowron and Stepaniuk 1992].

The accuracy of approximation of a rough set X ($\alpha(X)$) measures how much a set is rough. If $\alpha(X)$=1, X is crisp (exact) with respect to set of attributes, if α(X)<1, X is rough (vague) with respect to set of attributes [Skowron and Stepaniuk 1992]. The quality of approximation coefficient ($\gamma(X)$) expresses the percentage of objects which can be correctly classified into class X. Moreover, $0 \leq \alpha(X) \leq \gamma(X) \leq 1$, and $\gamma(X)=0$ if $\alpha(X)=0$, while $\gamma(X)=1$ if $\alpha(X)=1$ [Skowron and Stepaniuk 1992]. Let $C_1, \ldots, C_m$ the decision classes of the decision system DS. The quality of classification coefficient describes the inexactness of approximation classifications. It expresses the percentage of objects which can be correctly classified in the system. If this coefficient is equal to 1, the decision system is consistent, otherwise is inconsistent [Skowron and Stepaniuk 1992].

The accuracy and quality of approximation are associated to the respective class of a decision system; but in most cases, it is necessary to measure the accuracy and quality of the entire decision system. Thus, two new measures to calculate the accuracy of classification were proposed in [Arco et al. 2006] (see equation 2.3 and 2.4).

Accuracy of classification: It calculates the accuracy average per class [Arco et al. 2006] (see equation 2.3).

$$\mathrm{A}(DS) = \frac{\sum_{i=1}^{k} \alpha(C_i)}{k} \tag{2.3}$$

Because each class has a different influence in the quality of the decision system, the weighted accuracy of classification measure is proposed in [Arco et al. 2006], in the meaning of weighted means of the accuracy per class.

Weighted Accuracy of Classification: This expression calculates the weighted means (considering the cardinality of classes) of the accuracy per class [Arco et al. 2006] (see equation 2.4).

$$A_{weighted}(DS) = \frac{\sum_{i=1}^{k} \left(\alpha(C_i) \cdot |C_i|\right)}{|U|} \tag{2.4}$$

We introduce the generalized variations of the accuracy and quality of classification measures, because the experts can weight classes or use heuristics to define the importance of classes in a lot of applications.

Generalized accuracy of classification: This expression computes the weighted mean of the accuracy per class. The experts can decide the weight of classes or they can use heuristics to define the importance of classes.

$$A_{Generalized}(DS) = \frac{\sum_{i=1}^{k} \left(\alpha(C_i) \cdot w(C_i)\right)}{\sum_{i=1}^{k} w(C_i)} \tag{2.5}$$

Generalized quality of classification: The following expression computes the weighted mean of the quality of approximation per class.

$$\Gamma_{Gemeralized}(DS) = \frac{\sum_{i=1}^{k} \left(\gamma(C_i) \cdot w(C_i)\right)}{\sum_{i=1}^{k} w(C_i)} \tag{2.6}$$

Notice that in both expressions (see equation 2.5 and 2.6), w (C_i) is a value between 0 and 1 representing the weight of class C_i.

There are differences in experimental results among classifiers correlations with measures that distinguish datasets in a general sense, and those describing them by classes; thus, general approximation quotient is proposed in [Caballero et al. 2006] that enables to bound distances among previously mentioned measures.

Generalized approximation ratio: This measure should involve parameters used for general description and by classes in the decision system, without differing for classes [Caballero et al. 2006] (see equation 2.7).

$$\mathrm{T}(DS) = \frac{\sum_{i=1}^{k} \left| R'_{*}(C_i) \right|}{\sum_{i=1}^{k} \left| R'^{*}(C_i) \right|} \tag{2.7}$$

The rough membership function quantifies the overlapping degree between set X and the $R'(x)$ class that x belongs to. It can be interpreted as a frequency-based estimate of $\Pr(x \in X \mid x, R'(x))$, the conditional probability that object x belongs to set X [Grabowski 2003]. The generalized rough membership function is proposed in this paper, considering the combination of the classical rough membership function and the similarity between objects and the classes' centroids.

Generalized rough membership function: Let be object x and the class X. The similarity degree between the object x and the centroid of class X is considered in the generalized function (see equation 2.8).

$$\mu_{Generalized\ X}(x) = \frac{\left| X \cap R'(x) \right|}{\left| R'(x) \right|} \cdot s_X(x) \tag{2.8}$$

where $s_X(x)$ is the similarity of the object x to the centroid of class X. $s_X(x)$ may be computed by means of the expression $s_X(x)$=1-distance(x, center(X)), where center(X) is the mean of all objects belonging to class X.

Taking into consideration the characteristics of the classical and generalized rough membership functions, the rough involvement and generalized rough involvement functions are introduced hereafter (see equation 2.9 and 2.10).

Rough involvement function: The following ratio quantifies the percentage of objects correctly classified into the class X which are related to the object x.

$$v_X(x) = \frac{\left| X \cap R'(x) \right|}{\left| X \right|} \tag{2.9}$$

Generalized rough involvement function: This function is a generalization of the expression above (see equation 2.9). We only have added $s_X(x)$, which allows us to consider the similarity between the object x and the centroid of class X.

$$\nu_{Generalized_X}(x) = \frac{|X \cap R'(x)|}{|X|} \cdot s_X(x) \qquad (2.10)$$

In order to count on a measurement of the membership degree of objects into classes, it is necessary to introduce the "mean of the rough membership per classes" and "mean of the generalized function of rough membership per classes" measures. Thus, to realize on the involvement of objects per classes, the novel measures "mean of the rough involvement per classes" and "mean of the generalized function of rough involvement per classes" have been created. These four new measures consist of calculating the average values of the values reached in rough membership function [Grabowski 2003], equations 2.8, 2.9 and 2.10 per classes.

3 RST and the Assessment of the Quality of Decision Systems for the Seasonal Weather Forecasting

The seasonal weather forecasting demands the application of supervised learning techniques, since the problem is presented as a data set described by features and the classification of those cases by human experts. The aim is to learn from those previously classified cases so as to automatically classify new ones. In this type of learning, there is nearly no direct interaction with the outside word; the universe of learning is the so-called training set. In its most downright formulation learning from examples operates upon a collection of instances, each one being described by a set of features. Such a learning process embodies knowledge generation, and knowledge validation.

The quality of the discovered knowledge mainly depends on two contributing factors: the training set and the learning method being used. Commonly, what is assessed is the quality of knowledge that arises from the application of some learning method, and this evaluation involves the control set, i.e., post-learning assessment. From the data available, different learning methods could be applied to ascertain which one yields better fitting knowledge. Accordingly, being able to estimate data quality before endeavoring in a learning process is a relevant aspect for saving time and computational resources. It is very important when dealing with seasonal weather forecasting problems to perform the edition of the training sets, either for reducing the information volume to be processed or for accomplishing superior outcomes from the classifiers' output.

The proposal is nothing but considering the training set as a decision system, wherein the decision attribute is the classification previously made to the data set and later on applying the introduced measures (see section 2.2 New measures of closeness of concepts and measures of decision systems). Training set quality assessment may serve the way for making decisions as how to develop learning. In other words, given a training set (TS), a function *g* is sought that serves as an indicator of the capability of TS to draw knowledge which is required to build the classifier [Caballero et al. 2006]. Notice in what follows the clear effectiveness of an a priori characterization of samples to forecast the seasonal weather that may be employed as training sets for the classification stage.

3.1 Problem description

There are six meteorological stations in the Province of Camagüey, Cuba. They are located in the following towns: Florida (Station 78350), Santa Cruz del Sur (Station 78351), Esmeralda (Station 78352), Nuevitas (Station 78353), Palo Seco (Station 78354) and Camagüey (Station 78355). The Meteorological Center of Camagüey, in coordination with the Nacional Meteorological Center of Cuba, have provided the data corresponding to the six meteorological stations for the variables Maximum Temperature and Minimum Temperature, as well as other variables that influence temperature variation. Readings for daily values of the data are available for the period from 2000 to 2006, as well as the temperature values forecast by the Weather Forecast Department of Camagüey.

Table 1. Set of classes (ranks in grades) for the temperature features

Ranks	Class	Ranks	Class	Ranks	Class
Temp<12	A	22≤Temp<24	F	30≤Temp<32	J
12≤Temp<16	B	24≤Temp<26	G	32≤Temp<34	K
16≤Temp<18	C	26≤Temp<28	H	34≤Temp<36	L
18≤Temp<20	D	28≤Temp<30	I	Temp≥36	M
20≤Temp<22	E				

After the meteorologists gathered the data, we built 12 case bases for maximum and minimum temperatures at each station; each case base is described by 23 predicting features and one objective feature. The predicting features are: date (year, moth) when the variables were measured; seven maximum and minimum temperatures of the previous week; energetic features like isolation and Wolf number; and weather-conditions related features like wind direction and wind speed (07:00 and 19:00), and relative humidity. Day temperature is the objective feature. It refers either to the minimum or maximum temperature, depending on the problem approached. It may be described either in terms of a range (range-based forecasting) or an integer value (deterministic forecasting); symbolic and integer results, respectively. Thirteen classes emerge from the predicting features for the temperature feature (see Table 1).

4 Experimental study and results

The following steps for the experimental study were applied to each of the 12 case bases used.

1. *Creation of training and control data sets:* For the former, 75% of the cases were chosen, and the remaining 25% were assigned to the latter in a random way. Following the same principle of random choice the process was repeated 10 times so as to obtain 10 training sets and 10 control sets for each base. This was done to implement cross-validation for an improved validation of the results [Demsar 2006].
2. *Classification process:* The classification was performed by applying the algorithms found in the Weka[1] application, we used C-45 (J48 in Weka), MLP and k-NN (IBK). In this classification process we obtain not only the accuracy values for the classification, and the percentage of correct cases found by the classifier but also the Kappa statistics (which is a measure of the quality of the samples); the absolute mean error, the value of TP Rate for each class (which is a behavior conditioned to the class with respect to the expected values) and the values per class of the ROC curve (which consists of a plotting of the false positives and true positives) which allows to make a deeper analysis of the behavior of the classification process. In Table 2 are shown the results of the classification for minimum and maximum temperatures. The only values represented are those obtained for the whole decision system with the classifier k-NN.

[1] Weka - Machine Learning Software in Java http://sourceforge.net/projects/weka/

Table 2. k-NN's classification results for minimum and maximum temperatures

Stations	Minimum temperatures			Maximum temperatures		
	Precision (%)	Kappa Statistic	Error	Precision (%)	Statistic Kappa	Error
78350	63.437	.352	.1468	72.848	.523	.1093
78351	64.559	.378	.1019	71.937	.544	.1129
78352	64.983	.419	.1007	71.937	.544	.1129
78353	74.434	.536	.0738	73.071	.560	.1084
78354	76.031	.498	.0966	72.523	.531	.1105
78355	69.078	.409	.1243	68.554	.479	.3232

3. *Calculation of the inference measures:* For each case base the measures described before were calculated (see section 2.2 New measures of closeness of concepts and measures of decision systems). For this purpose we made use of the MIRST 1.0 system [Caballero et al. 2006]. The weight used to calculate the generalized accuracy of the classification measure (see equation 2.5) is the measure of the mean rough membership per class. The weight considered to calculate the generalized quality of the classification (see equation 2.6) is the mean of the rough involvement per class (see Table 3).

Table 3. RST inference measures' achieved results for the minimum and maximum temperatures

Stations	$\Gamma(DS)$		$A(DS)$		$A_W(DS)$		A_G(DS)		$\Gamma_G(DS)$		$T(DS)$	
	Min	Max	Min	Max	Min	Max	Min	Max	Min	Max	Min	Max
78350	.588	.827	.380	.743	.507	.679	.547	.721	.605	.794	.486	.783
78351	.588	.792	.546	.704	.488	.666	.481	.717	.485	.794	.486	.766
78352	.773	.792	.628	.704	.600	.666	.571	.717	.757	.794	.581	.766
78353	.816	.838	.678	.747	.680	.686	.804	.738	.823	.780	.771	.791
78354	.891	.800	.773	.738	.776	.671	.907	.720	.882	.799	.804	.774
78355	.812	.759	.670	.699	.772	.661	.803	.685	.831	.738	.765	.760

4. *Statistical correlations:* Through the Pearson's correlation method we found that there is a positive correlation coefficient between the measures and the results of the classification process. Table 4 shows only the results of the study of the whole decision system.

Table 4. The Pearson correlation values between the accuracy of the k-NN classifier and the RST measures to the minimum and maximum temperatures

		Γ(*DS*)	A(*DS*)	A_W (*DS*)	A_G (*DS*)	Γ_G(*DS*)	T(*DS*)
Min	Coef.	.838	.824	.809	.928	.785	.916
	Sig.	.019	.022	.026	.024	.032	.005
Max	Coef.	.886	.727	.753	.956	.880	.761
	Sig.	.009	.049	.042	.001	.010	.039

In this comparative study, the performance values with regard to the accuracy achieved by the k-NN classifier (see Table 2) were correlated to the values obtained by the inference measures (see Table 3); likewise, the accuracy per classes and the ROC curve of the classifier were also correlated to those attained by the accuracy of approximation and quality of approximation measures [Skowron and Stepaniuk 1992].

The correlation coefficients computed via the Pearson method are mostly closer to 1 with a significance level less than 0.05, thus claiming that the correlation coefficient is significant and there exists a positive linear relationship between the rough set theory measures and the quality reached by the k-NN classifier. Only the correlation values for the overall decision system are displayed (see Table 4).

5 Conclusions and Future Work

We defined and applied new measurements to evaluate the quality of data sets by using RST to estimate the quality of supervised classifiers' results in the seasonal meteorological forecasting problem. The study showed that the performance values with regard to the accuracy achieved by classifiers were positive correlated to the values obtained by the RST measures; thus measures allow evaluating a training set yielding a quality estimated which approximates the classifier results. It is possible to predict a priori the quality of supervised classification results by means of evaluating the quality of training data sets. It would be nice as a future work to employ statistical methods for outlining the linear models that best fit to these correlations.

References

Arco L, Bello RE, García MM (2006) On clustering validity measures and the Rough Set Theory. In: Proceedings of 5th Mexican International Conference on Artificial Intelligence. IEEE Computer Society Press

Caballero Y, Bello RE, Taboada A, Nowé A, García MM, Casas G (2006) A new measure based in the rough set theory to estimate the training set quality. In: 8th International Symposium on Symbolic and Numeric Algorithms for Scientific Computing. Timisoara, Rumania. IEEE Computer Society

Demsar J (2006) Statistical comparisons of classifiers over multiple data sets. In: Journal of Machine Learning Research, vol 7, pp 1-30

Grabowski A (2003) Basic Properties of Rough Sets and Rough Membership Function. In: Journal of Formalized Mathematics, vol 15

Kodratoff Y, Ras ZW, Skowron A (1999) Knowledge discovery in texts: a definition and applications. In: Proceedings of the 11th International Symposium on Foundations of Intelligent Systems (ISMIS'99). Springer, Berlin, pp 16-29

Mitchell T (1997) Machine Learning. McGraw-Hill Science, Engineering, Math

Pawlak Z (1982) Rough sets. In: International Journal of Information & Computer Sciences, vol 11, pp 341-356

Pawlak Z, Grzymala-Busse JW, Slowinski R, Ziarko W (1995) Rough sets, Comm. ACM vol 38, pp 89-95

Skowron A, Stepaniuk J (1992) Intelligent systems based on rough set approach. In: Proceedings of the International Workshop Rough Sets: State of the Art and Perspectives, pp 62-64

Slowinski R, Vanderpooten D (1997) Similarity relation as a basis for rough approximations. In: Advances in Machine Intelligence and Soft-Computing, vol 4, pp 17-33

A Protocol to Secure Context-aware Service Discovery in Complex Earth Experiments

Ricardo R. M. do Carmo[1], André C. M. Costa[2], José L. Campos dos Santos[3], José G. Pereira Filho[2]

[1] Genius Institute of Technology – Manaus, AM – Brazil,
e-mail: rcarmo@ genius.org.br

[2] Federal University of Espírito Santo – Vitória, ES – Brazil,
e-mail: {andre, zegonc}@inf.ufes.br

[3] National Institute for Amazon Research – Manaus, AM – Brazil,
e-mail: lcampos@inpa.gov.br

Abstract: Ubiquitous Computing allows the use of context-aware applications, a new set of functionalities which adopt contextual information. The structure of such type of information allows the development of more flexible and adaptable services, rich in functions and centered on user demands. The development and management of such services require dynamic mechanisms which consider the nature of contextual information. Service discovery protocols offer such dynamism and ontology is a way to deal with the complexity of the contextual information. Experiments to monitor the environment, like the LBA, due to the diversity of devices, services and contexts, are good examples of ubiquitous environments where service discovery protocols can be a essential tool. SCaSDP is a new service discovery protocol which considers the nature of contextual information and allows a secure management of services and user information. SCaSDP uses a semantic matching algorithm of service descriptions based on ontology's.

1 Introduction

The advantages of mobile computing and the wireless technology lead to the appearance of new paradigms. Ubiquitous Computing [Zhu et al. 2003] is one of the paradigms that allow the use of a new set of applications, known as context-aware, which adopt contextual information (implicit information obtained from software and devices).

The structure of contextual information allows the development of more flexible and adaptable services, rich in functionality and centered on the user demands. This development requires the support of an infrastructure to deal with the characteristics of the contextual information. Some middleware platforms were proposed to achieve this goal. For instance, Dey developed a software architecture to support the building process of context-aware applications and proposes a conceptual supporting framework, named Context Toolkit [Dey 2000; Indulska et al. 2001] proposes an architecture for pervasive enterprise systems which provides an open approach to adaptation; Huang presents a middleware to help developers design applications for *ad hoc* systems, providing a set of toolkits and services for efficient and reliable information discovery and dissemination [Huang 2002]; Costa proposes a generic and configurable service platform architecture, Web Architecture for Service Platforms (WASP), to support context-aware applications [Costa 2003]; and Gu et al. present a middleware, the Service-Oriented Context-aware Middleware (SOCAM), for developing and prototyping context-aware services [Gu et al. 2004]. Those middleware aimed to help the designer to deal with the complexity of the contextual information. The Infraware solution, also based on WASP, uses services and web services to provide users with contextual information [Filho et al. 2006]. To deal with contextual information, Infraware uses ontology's, which give a common understanding within many entities. In Infraware domain, ontology is also used to deal with heterogenic and distributed data acquisition and integration; to coordinate conflicts between applications; to guarantee privacy and data security; and to implement an integrated service management.

The service management in ubiquitous environment requires dynamic discovery mechanisms. There are many approaches of discovering services, although some questions are still opened and requiring more research. The proposed approaches so far, try to cover some important features like user and services privacy and data security, however, all of them fail to consider the nature of the contextual information and do not use semantic to better describe services.

Service discovery has an important role in many scenarios. One of these scenarios is large scale experiments for monitoring the environment, where many procedures need sensors and other devices to collect data and monitor the changes occurred. Usually there is more than one sensor in a huge area and for a researcher it would be desirable to have a tool to automatically discover and collect the data from sensors.

This paper describes a new service discovery protocol that considers the dynamic nature of contextual information. The protocol uses semantic matching algorithm of service descriptions and allows management of information about users, devices and services. The security required by some applications, like medical, financial, military and environment management, is essential to guarantee the data integrity and confidentiality. The protocol's architecture, features and the implementation description are presented and result of tests performed is also presented.

The reminder of the paper is organized as follows. Section 0 discusses some approaches to describe a service. Section 3 classifies some service discovery proposal and identifies some mandatory requirements for a discovery protocol. Section 0 describes the environment in which the protocol can be used. Section 0 presents the service discovery protocol for a context-aware environment. Section 0 presents our concluding remarks.

2 Relevant issues on Service Discovery

Service discovery is a generic term used to describe dynamic mechanisms and protocols which allow devices and services to be aware of the changes in the network. In Ubiquitous Computing, service discovery has an important role, since it allows Context-aware applications to adapt their behavior based on contextual information.

Due to the dynamic characteristic of the environment, automatic adaptation requires applications to find services that can provide the desirable information. Services must fulfill some requirements to guarantee the efficiency during discovering and must be described in a way to allow semantic inferences, which include: selection considering the nature of the contextual information; organization to allow advanced searches; and security and authentication policies to restrict access for certain application classes.

2.1 Service Identification

Services are a piece of software or devices which provide functionalities to users. A software service can be a printer driver and a device service a temperature sensor. In the web world, services are defined as web services and are considered the most important within many other resources in the Internet [Martin et al. 2005].

A service to be discovered has to be identified by some unique way. There are some facets to be considered depending on how the process is done, like duplication and complexity of the identity, limitation imposed by the type of description or even the protocol used to provide the service.

One of the most common ways to identify a service is by name. This allows the service to have more than one identity and does not guarantee the discovery software understands they are similar. The use of numbers is another approach, but besides there are no guaranties they are unique, numbers are not natural to users. The URL description form prevents a service to have more than one identification by the use of "protocol-hostname" pair. The print service *service:printer:lpr://hostname* is unique in the network, but it does not provide information about location, processing capacity or jobs in the queue. Template identification allows the inclusion of attributes, but even been richer than other approaches it is limited by the number of fields supported by the protocol packet, restricting the user to a set of attributes.

Description languages, like XML, allow the use of a variable number of attributes. They, however, do not support semantic which can be achieved by the use of ontology's, giving more precision and dynamicity during the discovery. Ontology's allow the sharing of descriptions between entities and the improving of search mechanisms during the service discovery process. Ontology languages allow services to be semantically described and a domain of services can be formalized through the definition of classes and their properties. Derivation of new information over these classes and over entities to the degree allowed by the language formal semantic is also possible [Buccella et al. 2004]. Automatic location, matching, composition and monitoring are some important functions for the development of context-aware applications and to define them; it can be done by using the advantages of ontology.

2.2 Matching

Service matching can be done by a user when she selects a service from a list or by an application when filters or automatic processes are used in conjunction with implicit and explicit information. Bröens, for instance, proposes a matching method based on similarity degree – a quality measure indicating how close a service description is to a request service [Bröens 2004]. The method uses the set theory to classify the similarity degree, giving applications more precision to discover services.

Service matching is also used by some discovery protocols, which implement mechanisms that do not use contextual information to find services [Zhu et al. 2003]. In ubiquitous computing environment the matching process must use information and mechanisms based on:

- Context Information: machine properties, user profile, temperature, location, activity and time information.

- Service Description: semantic service descriptions based on ontology's and similarity degree.

2.3 Directory Service

Directory service can be defined as software, or a set of integrated applications, which manages information about a network, its resources (printers, switches, etc), users and groups, and allows the sharing of information. Directory services employ classification of information, like yellow, green, or white list. The service descriptions supported by these services are poor in terms of semantic. LDAP, for instance, uses a classification based on attribute-value pair [Johner et al. 1999]. UDDI uses XML to describe objects, but even been a good approach it is not sufficient to guarantee semantic matching. This can be improved by the employing of a semantic description based on service ontology, like Ontology Web Language Service (OWL-S) [Martin et al. 2005].

The need of efficient dynamic discovery mechanisms, required by context-aware systems, justifies the presence of directory services, which have a fundamental role during the discovery process, providing a distributed management model and advanced capabilities for searching and replication of information.

2.4 Security and Authentication

Some types of information are not considered confidential, like temperature and local time. Other types are, like sensitive information. For these types it is very important to have methods to ensure access control. One way to do this is adopting security policies, ciphering data and demanding authentication of users.

In ubiquitous environments some services must have their access allowed to certain groups of applications. For example, medical service, like heartbeat monitoring, must be discovered only by medical applications. This imposes the employing of security policies and authentication methods, which must guarantee data confidentiality, integrity and non-repudiation exchange between context-aware applications and systems.

3 Prior Approaches for Resource Discovery

The first discovery protocol developed and still in use was Service Location Protocol (SLP) describe in RFC 2165 and proposed by the Internet Engineering Task Force (IETF). Following, other protocols were proposed and some of them succeed to be used by the industry like Simple Service Discovery Protocol (SSDP) [Universial Forum UPnP™ 2003], used in Microsoft Universal Plug-and-Play (UPnP); JINI, a java protocol of Sun Microsystems [JINI 2003]; Salutation protocol, proposed by the Salutation Consortium [Pascoe 1998]; and the Secure Service Discovery Service (SSDS), part of the University of California, Berkeley Ninja research project [Czerwinski et al. 1999].

In ubiquitous computing people are surrounded by many devices which provide information and services. Discovery mechanisms play an important role in managing these services when the number grows from a few ones to a huge quantity. To perform this task, however, some requirements must be reached and better discussed. The definition of how services are advertised and can be searched, how the interoperability with other platforms and systems can be achieved, service availability and better use of the network are the main requirements.

3.1 Protocol Requirements

Service discovery has some requirements to perform the management of services in context-aware environments. We identify four requirements, which are discussed bellow.

Advertisement and searching mechanisms are basic requirements in any discovery protocol and they are responsible for making resources available. Advertising and searching can be achieved over unicast, broadcast and multicast. The broadcast and multicast approach are good choices when no computational infrastructure is defined. On the other hand, broadcast usually tend to generate network bottlenecks and income multicast packets, in many networks, are blocked. The unicast is approachable in two forms: requesting many providers; or requesting a directory service. In the two forms only the involved entities are contacted, but the first one imposes a heavy task for poor resource devices like PDAs. The later is the better one because it reduces the network usage.

Depending on how frequent the advertisements are done, the network usage can be very high. Using UDP packets for advertisements is a good approach since it is fast, lightweight and very efficient during message exchange between servers and a huge number of network clients.

The available service discovery protocols do not interoperate with each other. In case of having a common advertising message format, the interoperability could be achieved. Another approach is proposed by Livingstone who implemented a bridge between JINI and Twine protocols to make the interoperability between these protocols to happen [Livingstone 2003].

To guarantee the availability of a service two methods are used: push and pull. In the push method the network usage is not increased because service providers are directly contacted only when a request is made. In the pull method clients wait for the providers' advertisements. With this method the network usage can increase, especially if there are many advertisements of service providers. One good approach to achieve a balance and guarantee the availability of the service is to combine the two methods. This can be done making service providers advertise to a single point, a directory service for instance.

3.2 Service Discovery Protocols and Infrastructures

The available proposals of discovery infrastructures and protocols try to achieve the most essential requirements of a service discovery protocol: automatic discovery and configuration, semantic matching, low human interaction, automatic adaptation and interoperability. But if they support automatic discovery and configuration, they do not have semantic matching; if they have low human interaction, they do not support automatic adaptation; and the interoperability between protocols is not treated. Most discovery protocols do not support data security and user authentication and they just use location as context-aware parameter.

Semantic description, contextual information, security and authentication are other important capabilities not much considered. Within the available proposals only Splendor uses a simple type of contextual information (geographic location) as a parameter to discover services [Zhu et al. 2003]. SLP allows applications to discover the location (not geographic location) and configuration of services; JINI allows richer service description (not semantic) and devices to communicate without human intervention and preliminary planning and setup; UPnP is based on point-to-point network connectivity; and SSDS enforces security. This indicates a lack of service discovery mechanisms for ubiquitous environments.

This work presents a new service discovery protocol which reaches the requirements discussed above. SLP was chosen to serve as a starting point of a new proposed protocol, since it is an IETF standard. Additionally, this work presents some interesting characteristics of service discovery for large scale environment monitoring. Other protocols and infrastructures were also considered focusing some specific purposes.

4 Earth Experiments Context: The LBA Scenario

One of the large scale experiments in progress today is the Large Scale Experiment of the Biosphere-Atmosphere of the Amazon (LBA), a project of the Brazilian govern with the participation of some Latin American countries, the USA and some European countries. LBA is designed to generate new knowledge about the weather, geochemical, hydrological and ecological behavior of the Amazon, the impact of the use of the soil and the interactions between the Amazon and the Earth global biogeophysical system. There are many LBA sites in Brazil and in the countries where the Amazon forest exists. The researchers normally have to drive to these sites and access sensors to collect data and to monitor the experiments.

A service discovery protocol would be an adequate tool to help researchers by the offering of an automatic way to discover the services, in this case sensors. A typical situation is when a sensor fails. The discovery protocol can be used to find a new sensor, according to the capabilities defined by the researcher. The task of finding automatically a sensor within sets of ten types (and this number can increase to hundreds) is not so simple. It is necessary an efficient discovery mechanism which takes into account not only the type of the sensor and the information provided, but also contextual information.

5 SCaSDP: A new protocol for Service Discovery

Secure Context-aware Service Discovery Protocol (SCaSDP), an approach for the Service Manager of Infraware platform, is responsible for registering, advertising and discovering services [Carmo 2006]. It can be used by earth experiments, like LBA, to manage sensors and improve the environment monitoring.

One of the researches of LBA is the investigation of climate change in the Amazon region. A researcher needs to access, for instance, 20 types of sensors in a domain of dozens to collect and to analyze data. Those sensors massively collect data from the environment and store them for post processing. Due to the hostile environment, these sensors normally fail and until the researcher realizes this a huge amount of data can be lost. Finding another onsite sensor with the same characteristics of the failed one can take much time if an automatic mechanism is not used.

The sensors, illustrated in Fig. 1, are considered service providers and to be found through SCaSDP they have to be registered and periodically advertise themselves to a *proxy server*. The protocol *Registration* and *Advertisement* modules are responsible for it. The registered and advertised sensor information (service description) consists of the identification, location, type of service, some attributes and is stored in a directory service, which has classification rules used to improve the service descriptions management.

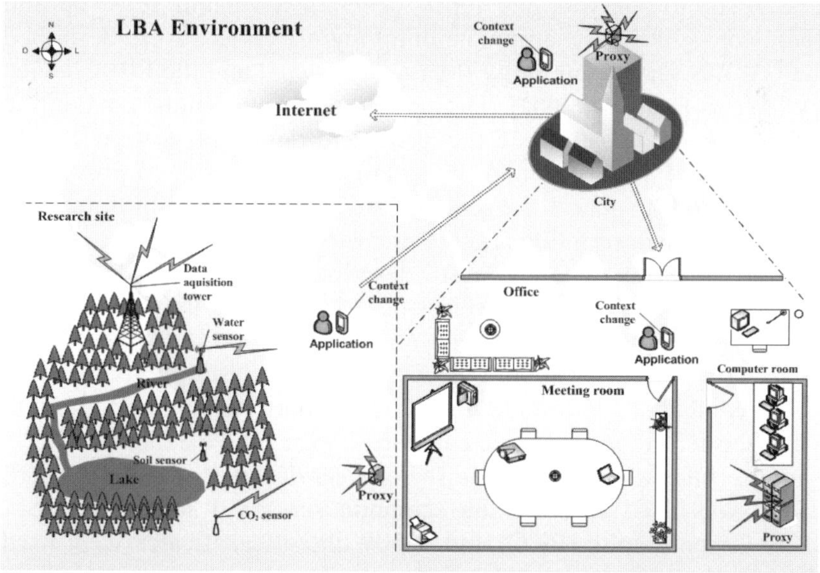

Fig. 1. LBA Scenario

The researcher's client application request service discovery and the protocol use the *Discover* module to find the most relevant services for it. The application provides contextual information, described in an ontology mark-up language. The providers' contextual information includes the type of service, the location and attributes, which are the service's functionalities. The proxy server is a central unit which interconnects service providers and client applications during the discovering phase. Like in SLP, many proxies can communicate with each other to improve the chances to discover a service.

If a researcher context changes to the LBA central office, located in the city, he can also request a discovery because the site's proxy is connected to the office through a communication line. Other services, including the office's services, are registered in office's proxies, which are connected to other research sites of the LBA.

The communication between providers, clients and proxies is authenticated by the *Authentication* module, which accesses the directory service to recover information for the authentication.

5.1 SCaSDP Prototype Implementation

The SCaSDP prototype was developed using Java version 1.5.0_06 and Jena Framework [2005] to manipulate ontology's. Services were described using the OWL-S service ontology [Martin et al. 2005] and the FIPA device ontology [Agents F for I. P. 2002]. The service description must contain the geographic physical location of the provider, the identification of the service; the type of service; which information is necessary to provide to the service; an URL to access the service; and other contextual information to enrich the service description and the precision of the discovery.

To match the services request description provided by client applications with the advertised services descriptions was implemented the *AlComp* algorithm [Costa 2005], based on Bröens proposal [Bröens 2004], which uses service description request, service descriptions, service ontology's, contextual information and a similarity degree method to classify the services found into two categories: exact and inexact match.

Due to the unavailability of a directory service with OWL support, PostgreSQL was used to store the semantic service descriptions. SSL, Public Key Cryptography (PKC) and X.509 digital certificates were used to implement the security and the authentication processes of SCaSDP. Login/password and public keys are used during the authentication phase. Some data, like the providers' MAC address and the mechanisms provided by directory services can be also used to improve the security.

5.2 Experiments

SCaSDP were tested over a wired network using Windows and Linux SuSE 9.1 platform. The machines had at least 512 MB of RAM and 1.4 GHz of CPU. The experiments were based on LBA Earth experiments context described on section 0, with descriptions of services of climatic measurement, simulated by desktop computers. Two factors were considered: precision and recall. Precision were calculated taking the number of relevant items retrieved divided by the total number of selected items. Recalls were calculated taking the number of relevant selected items divided by the total number of relevant items in the set.

The experiments aimed to discover services of speed and direction of the wind. The service description request used was:

- Speed, in km/h, and direction of wind;
- Latitude and longitude and 80 meters far from the user;
- 95% of precision;
- Sensor with memory of at least 64 MB; and

Table 1. Experiments result

SERVICES	QUANTITY	PRECISION	RECAL
Total	23	-	-
Relevant	8	-	-
SCaSDP	8	100%	100%

There were 23 descriptions in the directory service, only 8 of them were relevant and SCaSDP selected the 8 relevant descriptions.

The majority of the providers' advertisements were sent over TCP and not over UDP. The reason for that is justified by the amount of tags needed by OWL to describe a service.

6 Concluding Remarks

SCaSDP is a generic proposal which considers the nature of contextual information. It provides security and authentication through the use of cryptographic keys, digital signatures and certificates and login/password. The protocol uses *AlComp*, a semantic matching algorithm, to give more precision to the discovery phase. OWL mark-up language, OWL-S and FIPA are used to semantically describe services and to also give more precision during the discovery phase.

The LBA environment was an ideal scenario to validate the prototype, due to the diversity of sensors and services and to the different contexts of the research sites. The experiments presented good results in finding specific service description of sensors. However, only a few service descriptions were used and more tests must be performed to verify the protocol performance in a real case with hundreds of service descriptions.

Some improvements were identified to increase the protocol functionalities: a more robust security and authentication method; the advertisement of multiple types of service descriptions by centralizing services in one provider. The *AlComp* matching algorithm needs some improvements to use weighted attributes, services priority and more complete description of the context. A semantic directory service should be developed to better manage semantic service descriptions.

7 Acknowledgements

We would like to thank Amazonas Research Foundation (FAPEAM) for the financial support provided during the development of the research.

References

Agents F for I. P. (FIPA) (2002) Device Ontology Specification. http://www.fipa.org/, access: 2007-02-14

Bröens T (2004). Context-aware, Ontology based, Semantic Service Discovery (Master degree). University of Twente

Buccella A, Penabad MR, Rodriguez FJ (2004) From relational databases to owl ontologies. In: Institute of Mathematical Problems of Biology Ras. Proceedings of Sixth National Russian Research Conference, Russia

Carmo RRM (2006) A Service Discovery Protocol for Context-aware Systems (Master Thesis). Federal University of Espírito Santo, Brazil

Costa ACM (2005) AlComp: A matching algorithm for service Discovery. Federal University of Espírito Santo

Costa PD (2003) Towards a Service Platform for Context-Aware Applications (Master Degree). University of Twente

Czerwinski SE, Zhao BY, Hodes TD, Joseph AD, Katz RH (1999) An architecture for a secure service discovery service. In: Computer Science Division, University Of California, Berkeley. ACM/IEEE Fifth Annual International Conference on Mobile Computing and Networking. ACM Press, pp 24-35

Dey AK (2000) Providing Architectural Support for Building Context-Aware Applications. PhD Thesis Georgia Institute of Technology

Universal Forum UPnP™ (2003) Device Architecture 1.0 Version 1.0.1. http://www.upnp.org/resources/documents/CleanUPnPDA101-20031202s.pdf, access: 2007-02-14

Gu T, Wang XH, Pung HK, Zhang DQ (2004) A middleware for building context-aware mobile services. In: IEEE Vehicular Technology Conference

Huang Q (2002) Supporting context-aware computing in ad hoc mobile environment. In: Saint Louis, MO 63130: Department of Computer Science and Engineering, Washington University

Indulska J, Loke SW, Rakotoniarainy A, Witana V, Zaslavsky, A (2001) An open architecture for pervasive systems. In: CRC for Distributed Systems

Technology/University of Queensland, School Of Computer Science and Electrical Engineering-Monash University. Proceedings of the 3rd. International Working Conference on Distributed Applications an Interoperable Systems (DAIS). Krakow, Poland, pp 175-188

Jena Framework (2005) Jena website. http://jena.sourceforge.net/index.html, access: 2007-02-14

JINI (2003) Jini Introduction. Fakultet for informasjonsteknologi, matematikk og elekroteknikk. TTM47AC Laboratory in construction of self-configuring systems

Johner H, Melot M, Stranden H, Widhiasta P (1999) LDAP Implementation Cookbook. Dept. JN9B, Building 003 Internal Zip 2834, 11400 Burnet Road - Austin, Texas. http://borg.ucsb.edu/aka/Ucdir/sg245110.pdf

Livingstone SR (2003) Service Discovery in Pervasive System (PhD). The School of Information Technology and Electrical Engineering – The University of Queensland

Martin D, Burstein M, Hobbs J, Lassila O, McDermott D, McIlraith S, Narayanan S, Paolucci M, Parsia B, Payne T, Sirin E, Srinivasan N, Sycara K (2005) OWL-S: Semantic Markup for Web Services

Pascoe RA (1998) Salutation Architecture: Enabling Applications and Services. The Salutation Consortium. Senior Technical Staff Consulting

Filho JGP, Pessoa RM., Calvi CZ, Oliveira NQ, Carmo RRM., Barbosa ACP, Farias CRG, Leite MM (2006) Um Middleware de Suporte a Aplicações Móveis Context-Aware. In: SBRC – Brazilian Computer Network Symposium, 2006. Curitiba

Weiser M (1991) The computer for the 21st century. In: Scientific American, edn 265, pp 94–104

Zhu F, Mutka FW, Ni LM (2003) Splendor: A secure, private, and location-aware service discovery protocol supporting mobile services. In: Proceedings of IEEE International Conference on Pervasive Computing and Communications – PerCom, pp 235-242

The Concept of Closed-loop Supply Chain Integration Through Agents-based System

Paulina Golinska[1], Marek Fertsch[1], Jorge Marx Gómez[2], Joanna Oleskow[1]

[1] Poznan University of Technology, Institute of Management Engineering, Strzelecka 11, 60-965 Poznan, Poland, phone: +48 616653401, fax: +48 616653375, e-mail: paulina.golinska@doctorate.put.poznan.pl

[2] Oldenburg University, Department of Informatics, Ammerländer Heerstr. 114-118, 26129 Oldenburg, Germany, e-mail: jorge.marx.gomez@uni-oldenburg.de

Abstract Closed-loop supply chains CLSC) concept is response to a challenge that more and more supply chains meet namely need to include a return flow of materials in they planning and coordination structure. Following paper presents on-going project regarding the application of agent-based systems for supply-chain synchronized production planning including management of raw materials flow as well as flow of returned by customer obsolete products and defected semi-products and products that are refused by quality control within the factory. Authors discuss the main problems that appear by synchronized material flow planning in CLSC. The model for integration of closed-loop supply chain through an agent-based system is proposed.

1 Introduction

The traditional supply chain can be defined as a flexible and cooperative business network of suppliers, manufacturer and distributors through which raw material are acquired, transformed within value adding (VA) processes and delivered as final goods to customers. The material flow in such conditions ends with fulfillment of customer order. The optimization of the traditional supply chain performance in material flow area is regarded as the minimization of product delivery cost to customer, what includes the total cost of manufacturing, distribution and sale. The cost of products repairs and disposal are not usually taken in consideration by business decision making. The new European regulations have imposed on manufactures in number of industry (like for electronics products, automobiles, household electronics equipments) the duty to collect the used products at the end-of-life stage. The legislative regulations for "Extended Product Responsibility" is defined by OECD as "An environmental policy approach in which a manufacturer's responsibility is extended to the post-consumer stage of product life cycle" for recovery and eventually disposal of used goods.

Closed-loop supply chain (CLSC) concept is response for a challenge which more and more supply chains meet namely needs to include a return flow of materials in their planning and coordination structure. In number of industries original equipment manufacturers are nowadays engaged in the remanufacturing, where also production defectives and by-products are reworked. Production planning and material flow management in such a hybrid systems is a real challenge, due the increased uncertainties connected with amount, quality and time of return products flows. Moreover parties involved in the supply chain have their own resources, capabilities, tasks, and objectives. They cooperate with each other autonomously to serve common goals but also have their own interests.

The effectiveness of the supply chain performance might be measured by its ability to provide customer with due-to-date deliveries of final goods and possibility of effortless collection of the used products when they're obsolescent (reverse logistics). In order to meet this goal the effective and integrated management of material flow at a tactical and operational level is of crucial importance. Agent-based system architecture provides an interesting perspective for integration of the supply chain due the fact that they are able to generate, process, store, filter, correlate, broadcast, and route information for real-time coordination and decision making. In following paper an agent-based system is treated as a multi-agent system that's acts as a support tool to company existing business information system (for example ERP) and utilized their databases. Multi-agent system is a collection of heterogeneous, encapsulated applications (agents) that participate in the decision making process [Pechoucek et al. 2003]. In following paper authors apply Jennings and Wooldridge's definition: "an agent is a computer system situated in some environment, and that is capable of autonomous action in this environment in order to meet its design objectives. Agents have the following characteristic [Jennings and Wooldridge 1998]: reactive, pro-activeness, social ability.

2 Closed-loop supply chain

The closed loop supply chain provides the integration of traditional 'forward' supply chain processes and 'reverse' supply chain processes. Guide and Van Wassenhove [2003] identify the common processes required by a closed-loop supply chain as: product acquisition, reverse logistics, inspection, testing and disposition, remanufacturing, and selling and distribution. General overviews of reverse logistics and closed-loop supply chain can be found in Fleischmann [2001], Guide [2003] and Dekker et al. [2004]. The philosophy of closed loop supply chain takes in consideration also the environmental aspects by business decision making.

The ideal closed loop supply chain can be defined as zero-waste supply chain that completely reuses, recycles all materials. However in real life supply chain disposal activities take place the amount of products and components being disposed should be minimized. Figure 1 represents the overall material flow in closed loop supply chain.

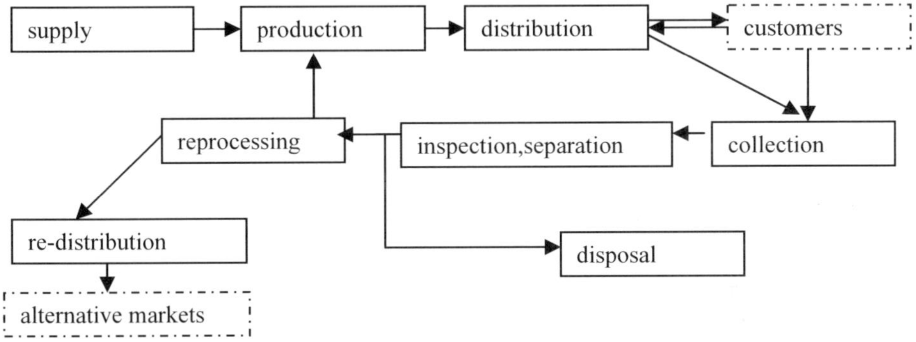

Fig. 1. Material flow in closed loop supply [cf. Bloemhof-Ruwaard et al. 1997]

The forward supply chain processes are as follows:

- Supply includes all activates connected to purchasing, transportation and storing of the raw materials and new components that are input to manufacturing system
- Production/manufacturing enhance all value adding processes that transform the raw materials and components into final products according to actual customers' demand or demand forecasts
- Distribution refers to all activities needed to provide the customers with ordered/demanded products. Distribution usually consists of transportation, storage and sales services. In case of closed loop supply chains the

distribution processes might include the optional collection of end-of-life products from customers.

Reverse processes are [Fleischmann et al.2000] as follows:

- Collection refers to all activities rendering used products available and physically moving them to some point for further treatment
- Inspection/separation denotes all operations determining whether a given product is in fact re-usable and in which way, it results in splitting the flow of used products according to distinct re-use (and disposal) options.
- Reprocessing means the actual transformation of a used product into a usable product /component/ material. This transformation may take different forms including recycling, repair and remanufacturing. In addition, activities such as cleaning, replacement and re-assembly may be involved
- Disposal is required for products that cannot be re-used for technical or cost reasons. Disposal may include transportation, landfilling and incineration steps
- Re-distribution refers to directing reusable products to a potential market and physically to moving them to future users. This may include sales, transportation and storage activities.

In following paper the emphasis will be put on planning and coordination the material flow in CLSC from remanufacturing perspective so the issue related to disposal and re-distribution will be not considered. The characteristics of forward and reverse material flows are presented in Table 1.

Table 1. The characteristics of forward and reverse flows

	Forward flows	Reverse flows
Quality of material /components	Predictable Known in advance and in accordance to given buyers specifications	Unpredictable Known at the moment of disassembly and inspection and depends on stage of products life cycle (degree of wear) and source of returns
Quantity	Constant According to order quantity	Variable Depends on stage of products life cycle and sources returns
Times of deliveries	Predictable According to standard defined delivery cycle for each supplier with some standard deviations	Variable Depends on time of preprocessing processes needed for particular return products like for example dismantling and recycling
Availability	Any time available to order	Variable

3 Related works

The literature research on closed loop supply chain can be grouped in following area:

- the operational aspects of remanufacturing regarding mainly production planning and control and inventory aspects [e.g. Toktay et al. 2000; van der Laan et al. 2003; Gupta 1996; Minner and Linner 2003]
- the strategic issue of CLSC configuration and interaction among the closed–loop supply chain players [Tsay et al. 1998; Fleischmann 2003]
- pricing and cost models for CLSC in order to increase the fraction of remanufacturable products available, or to decrease the costs of remanufacturing [Bloemhof-Ruwaard at el. 1997; Debo et al. 2001] investigate the joint pricing and production technology problem of a manufacturer that offers both new and remanufactured products [Kirkke et al. 2004].

From perspective of synchronized CLSC material flows integration the most relevant are researches regarding the planning, scheduling and control in remanufacturing environment. The following paper authors in elaboration of synchronized production material flow planning and coordination in CLSC refers to previous researches of Guide [1996; 2000] regarding scheduling of remanufacturing activities with Drum-Buffer-Rope concept (for details see section 4.2). There is a number of agent-based systems application for integration of traditional closed-loop supply chain, the overview can be found in Oleskow et al. [2005]. There is a research gap regarding the integration of CLSC regarding both material flows and accompanying them information flows.

4 Synchronized material flow planning and coordination in closed-loop supply chain – problem identification

Parties involved in the supply chain have their own resources, capabilities, tasks, and objectives so there are difficulties in coordination of the constant flows of information, materials, and funds across multiple functional areas both within and between chain members. The better CLSC performance can be achieved by improvement of material flow synchronization and integration of accompanying information flows, where CLSC performance is defined as the ability deliver to due-to-date and cost effectively final goods to customers, by optimization of .forward and reverse material flows.

The benchmark for integration of material flow management in traditional supply chain is the concept of seamless supply chain. The seamless supply chain [Towill 1997] is a state of total integration in which players think and act as one. The following concept is also applicable in conditions of closed loop supply chain. The seamless supply chain is defined by zero defects in material flow and is characterized by Towill [1997] as: no raw material stock, while raw materials arrive at the last moment, no WIP due extreme flexibility and speed and no finish goods because products are in exact order sequence and of good quality. Due the fact that the aim of closed loop supply chain is to reach the goal of zero waste so the optimization of forward and reverse flows is understood as maximization of return and minimization of raw materials and new components flows. Authors by elaborating method for synchronized material flow refers to the concept of synchronous manufacturing which is defined as "an encompassing manufacturing management philosophy that includes a consistent set of principles, procedures and techniques where every action is evaluated in terms of the common goal in organization" [Umble and Sikanth 1990].

It can be assumed that the synchronized material flow planning and coordination can be achieved when 'organization' is perceived as a collaborative network of all supply chain participants and common goal (maximization of CLSC performance) is understood by all CLSC participants, so all action taken benefit to this goal.

4.1 Problem identification and assumptions made

The following problems have been identified by synchronized planning and coordination of material flow in closed-loop supply chain:

- Uncertainty inherent in forecasting of the reverse flow regarding time and amount of returns, due the fact not all products life cycles follow simple stochastic patterns
- Integration of forward logistics flow and reverse logistics flow due the fact that characteristic of raw materials and new components is different then recycled materials and re-used components flows
- Uncertainty inherent in quality of reverse flow depending of product life stage and source of returns [cp. De Brito and Dekker 2003]
- Shorter live cycles of products, what very often means that previous generation products cannot be re-used by manufacturing of next generation of upgraded products.

The mentioned above problems can be transform in following research question:

- How to keep production balanced?

- How to shorten the time between product collection and remanufacturing, so that used-product could be as soon as possible disassembly and preprocessed and include in new products?
- How to be reactive to disturbances in materials flow?
- How provide planners with accurate information about: material resources available which often lead to bloated inventory and inaccurately promised delivery dates to customers?

It is assumed that all final products are manufactured on make-to-order so demand is deterministic. Manufacturer has efficient power over other CLSC suppliers and distributors. Due the fact that the reverse flows of materials require usually special infrastructure investments the participants should cooperate in long term planning horizon. It is assumed that producing a new product from used product is less costly than manufacturing a new one from raw materials, so manufacturer has an interest to collect as much of obsolete products as possible. Production planning is based on Material Requirements Planning (MRP) strategy, so that the production of all parts and raw materials has to match to the Master Production Schedule (MPS) of final goods. The product manufactured in previous period can be reuse in next period.

Effective and synchronized management of materials flow management should be achieved both at a tactical and operational level. Management at tactical level within CLSP refers to planning and scheduling to meet the current demand. Management at operational level refers to plans execution and eventual re-planning when disturbances appear.

4.2 Method for synchronized planning and coordination of material flows in CLSC

The proposed by authors method for synchronized planning and coordination of materials flow management and accompanying information flows in CLSC is based on synchronization at Master Production Schedule level, due the fact that production must be balanced. (MPS) defines the quantity and characteristic of final product, it also points when the manufacturing of particular order should be started and finished. MPS is the main driver and information source for further material flows planning and accompanying calls for supplies and allows making details production schedules for production system resources like machines, internal transportation etc. It helps to preliminary verify whether customers orders can be fulfill due-to-date. It can be assumed that in supply chain where exist a single manufacturer and a single 1-tier supplier/limited number of 1-tier suppliers and the manufacturer has sufficient power over the others parties involved in supply chain a global MPS can be elaborated. Taking in consideration following the synchronized planning can be described as a situation where in centralized process of preparing plans the distributed sub plans for each supply chain's participant are elaborated. The sub-plans for supplier side and distribution side are dependent on manufacturer capacities, so it can be treated as main constrain in planning process. In closed loop supply chain it can be assumed that the main constrain is to balance the production regarding both forward and reverse material flows. However taking in consideration the main goal of closed loop supply chain to be zero waste the used of recycled materials should have the priority over the usage of raw materials.

Authors decided to refer to Drum-Buffer-Rope (DBR) [Shragenheim and Ronan 1990] concept by elaboration distributed planning algorithm due the fact that DBR allows to synchronize the use of resources without having to actively control each resource. The purpose of drum is to exploit the constraint of the closed loop supply chain. Buffers are intended as protection from disturbances and a rope is a mechanism design to force all the parts of the system to work at pace set by drum [Shragenheim and Ronan 1990]. Following concept suits well to conditions of CLSC because entities there are bounded in long term cooperation and manufacturing side can be perceived as a global (in given planning horizon) constrain. Due the fact that all processes are set to the process of most constrained conditions namely bottleneck, it can be assume that the improvement of productivity of all processes at the same speed pace as the improvement of constrained conditions can be expanded to the entire supply chain.

The planning and coordination of material flow management is presented by following algorithm:

1. Define of closed-loop supply chain goal and set of performance indicators
2. Generate a initial MPS for manufacturer taking in consideration customer orders assign to planning horizon plan according to optimization of „potential bottlenecks"
3. Negotiate the initial plan with raw materials supplier and distributors (regarding possibilities of returns flow collection) and find a plan with lower number of constrains among them (so called feasible MPS for supply chain)
4. Decompose the feasible MPS for sub-plans for supplier, manufacturer and distributors
5. Insert synchronization among sub-plans based on concept for time buffers and Drum-Buffer-Line concept where manufacturer sub-MPS is giving pace for supplies and distribution planning activities

6. Allocate sub-plans to agents using task-passing mechanism, if failure come back to previous step or generate new global MPS (step 2)
7. Initiate plan execution and monitoring when plans are executed in DBR green buffer no additional replanning needed when plans are executed in yellow DBR buffer the replanning at local level, if plans are executed in DBR red buffer go to step 2.

The DBR buffer refers to time needed for deliveries of materials and component (raw materials, new components and re-used components and recycled materials). The size of buffer in instable manufacturing system is defined as three times the average lead time from materials release point to the constrained condition [Shragenheim and Ronan 1990]. The size of buffers in CLSC due the uncertainty inherent in amount, time and quality of returns flows should be bigger and can be counted as following [Guide 1996]:

$$BS_i = MULT \sum_{j=1} PT_{ij} \tag{1}$$

Where BS_i is the size of buffer for part i=(1,2,3,..,n), PT_{ij} is the maximum processing time for operation j on part i and MULT is the constant buffer size multiplier. The value of MULT should be asses individually for each CLSC based on historical date regarding the size of disturbances appearing in material flows caused by uncertainty inherent in size, time and quality of return flows. The mentioned above buffer is divided in three area: green (no action needed), yellow (prepare for actions) and red (react). The size of each part of buffer can be equal, but it is recommended to revise the performance of buffer's green, yellow a dread zone on regular basis in order to tune their size to statistical fluctuations appearing in material flows in particular CLSC.

5 The concept of an agent-based system

The CLSC environment is characterized by steady growth in the production and consumption of information, heterogeneous and distributed information sources. Often problem is a lack of coordination and integration between the information systems working within and between chain members to facilitate the flows of materials, information and funds. It can be stated that following characteristics are typical for supply chain also CLSC:

– data, control and resources are inherently distributed,

- the system is regarded as society of autonomous cooperating entities,
- the components must interact with other to reach whole system goals.

Taking in consideration following aspects the agent technology seems to be very suitable to support collaboration in supply chain management. Agent technologies can be used to facilitate and control the scope, the time, and the frequency of information sharing based on specific arrangement and possible impact on each partners. The distributed multi-agent system that performs both inter- and intra-organizational planning and execution tasks has a potential for improvement in the design of CLSC.

Agent technology is the preferable technology for enabling a flexible and dynamic coordination of spatially distributed entities [Villa 2002]. The mentioned above characteristics suits well to requirement for synchronized material flow planning and coordination like common goal orientation, ability to interact with other supply chain participants and reactivity to disturbances appearing in material flows. In an agent-enabled environment, the multiple transactions and exchanges of information that drive material flow can be managed by software agents that maintain visibility across the network [Villa 2002]. Multi-agent systems perfectly suit the demands for global flexibility, cooperation and at the same time, local autonomy. Compared to existing SCM systems, allows the successful integration of both inter- and intra- organizational planning.

The proposed architecture is based of assumption that agent-based system will act as “glue” for existing information systems within each company at global supply chain level and therefore will integrated the whole CLSC. At particular company level system will monitor MPS execution by existing productions management systems and will be responsible for replanning if needed, in order to meet global supply chain goal.

The integration of CLSC is achieved due horizontal synchronized information flow which is achieved by cooperation of following agents:

- Master production planner (MP agent) – is responsible for creation of Master Production Schedules on basis of negotiations with logistics agents, reverse flows agent, if crisis appears on basis of information received from production scheduling agent it changes MPS and inform logistics agent and reverse flow agent about it;
- Logistics agent (M-agent) – is responsible for coordinating plans and suppliers raw materials suppliers, in the enterprise domains to achieve the best possible results in terms of the goals of the closed-loop supply chain, Merges the functions of inventory management and purchasing, and management of inventories of recycled materials and re-used components when the pre-preprocessing activities are finished

- Production scheduling agent (M-agent) – is responsible for scheduling and rescheduling activities in the factory, exploring hypothetical "what-if" scenarios for potential orders, and generating schedules that are sent to the production planning and control system for execution. It dynamically manages the availability of resources so that the schedule can be executed.
- Reverse flow agent (M-agent) – it prepares forecasts of reverse flows regarding time, quantity and quality of materials, the forecast information are received from distribution centers. For forecasting fuzzy logic, and fuzzy-neuro nets will be applied. On the bases of information that are from logistics agent regarding the supply needs in next period for the type of materials and needed amounts prepare re-active simulation (to identify what conditions should be met). It is also responsible for the assignment and scheduling of collection and pre-processing activities. It communicates with production flow control agent in order to update it schedules.

In system AIPPCLUSC (Agent Integrated Production Planning in Closed-Loop Supply Chain) in figure 2 agents are structured in three layers, reflecting supply chain perspective, company coordination level and intra company sub-levels.

The information control agents (Ci Agents) are responsible for monitoring the sub-plan execution within one company on basis of Drum-Buffer-Rope methodology described in section 4.2. On basis of given performance indicators they are reporting to agents in upper layer about buffers status (green, yellow, red). The Ci agents are not allowed to contact Ci agents or M-agents located in another company.

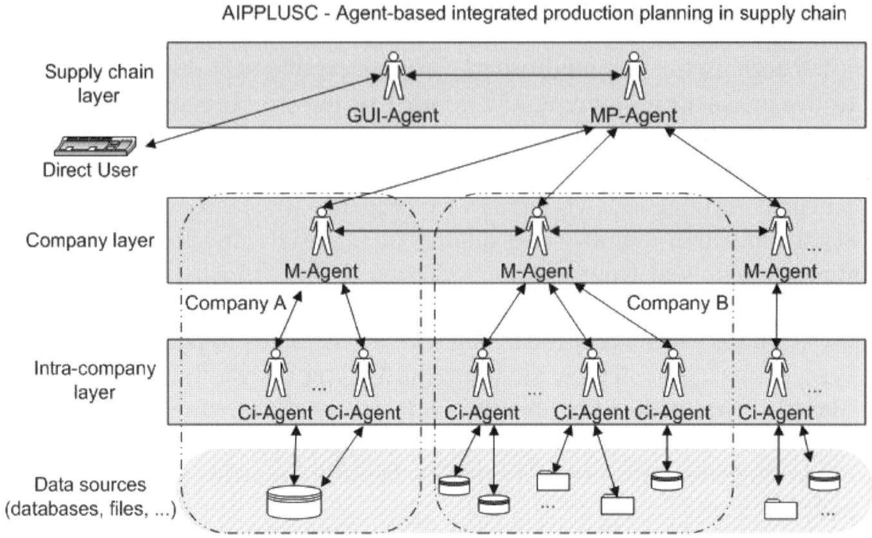

Fig. 2. Architecture design AIPPCLUSC

5 Conclusions and further research steps

The following paper presents the preliminary stage of research regarding the elaboration of agent-based system that might be implemented for improvement of planning and coordination of material flow in closed loop supply chain. The issues related to material flow in closed loop supply chain were discussed and a method for synchronized material flow planning and coordination has been proposed. The next researches steps will enhance the evaluation of proposed synchronized material flow planning method. The further development of the concept agent-based system will include the design of communication and control structure among agents. Perspectives for further research should reflect to elaboration of communication algorithm among agent and structure of standardized input and output data structure for all companies involved in supply chain. The issue related to increasing trust and safety for information exchange between entities involved in supply chain should be examined.

Acknowledgements

The following paper is reporting the preliminary stage of research project financially supported by DAAD in framework of the 5 months research grant for Paulina Golinska at Department of Business Informatics at Carl von Ossietzky University Oldenburg.

References

Bloemhof-Ruwaard JM, Dekker R, van der Laan E, van Nunen JAEE, van Wassenhove LN, Fleischmann M (1997) Quantitative models for reverse logistics. In: Eur. J. Oper. Research, vol 103, pp 1–17

de Brito MP, Dekker R (2003) A Framework for Reverse Logistics. In: Dekker R et al. (eds) Reverse Logistics: Quantitative Models for Closed-Loop Supply Chains. Springer, Berlin, pp 3-28

Debo LG, Toktay LB, van Wassenhove LN (2001) Market segmentation and product technology selection for remanufacturable products, Working Paper, INSEAD. Fontainebleau, France

Dekker R, Fleischmann M, Inderfurth K, van Wassenhove LN (eds) (2003) Reverse Logistics: Quantitative Models for Closed-Loop Supply Chains. Springer, Berlin

Fleischmann M, Dekker R, Flapper SDP (2000) A characterization of logistics networks for product recovery. In: Omega The international Journal of Management Science, vol 28, pp 653–666

Fleischmann M (2001) Quantitative Models for Reverse Logistics. Springer, Berlin

Fleischmann M (2003) Reverse logistics network structures and design. In: Guide Jr. VDR, van Wassenhove LN (eds) Business Aspects of Closed-Loop Supply Chains. Carnegie Mellon University Press, Pittsburgh

Guide Jr. VDR (1996) Scheduling using drum-buffer-rope in a remanufacturing environment for Remanufacturing. In: International Journal of Production Research, vol 34, edn 4, pp 1081-1091

Guide Jr. VDR (2000) Production Planning and Control for Remanufacturing: Industry Practice and Research Needs. In: Journal of Operations Management, vol 18, pp 467–483

Jennings NR, Wooldridge MJ (1998) Applications of Intelligent Agents. In: Agent Technology: Foundations, Applications, and Markets. Springer, Berlin

Krikke H, le Blanc I, van de Velde S (2004) Product modularity and the design of closed-loop supply chains. In: California Management Review, vol 46, edn 2, pp 23-39

Minner S, Lindner G (2003) Lot-Sizing Decisions in Product Recovery Management. In: Dekker R et al. (eds) Reverse Logistics: Quantitative Models for Closed-Loop Supply Chains. Springer, Berlin, pp 157-180

OECD (2001) Extended producer responsibility: A Guidance Manual for Governments. OECD Publications Service, Paris

Schragenheim E, Ronnen B (1990) Drum-Buffer-Rope shop floor control. In: Production and Inventory Journal, vol 32, pp 1867-1877

Olesków J, Fertsch M, Golińska P (2005) The new perspective of supply chain integration through agent-based systems. In: Proceedings of the 2nd International Symposium on Information technologies for Environmental Engineering (ITEE 2005). Magdeburg , Germany, pp 238-252

Pechoucek M, Říha A, Vokrínek J, Marík V, Prazma V (2003) ExPlanTech: Applying Multi-agent Systems in Production Planning. In: International Journal of Production Research, vol 40, edn 15. Taylor and Francis Ltd, pp 3681-3692

Toktay, LB, Wein LM, Zenios SA (2000) Inventory management of remanufacturable products. In: Management Science, vol 46, edn 11, pp 1412-1426

Towill, DR (1997) The Seamless Supply Chain. In: International Journal of Technology Management, vol. 13, edn 1, pp 37-56

Tsay AA, Nahmias S, Agrawal N (1998) Modeling Supply Chain Contracts: A Review. In: Tayur S et al. (eds) Quantitative Models for Supply Chain Management. Kluwer Academic Publishers, Boston

van der Laan EA, Kiesmüller G, Kuik R, Vlachos D, Dekker R (2003) Stochastic Inventory Control for Product Recovery. In: Dekker R et al. (eds) Reverse Logistics: Quantitative Models for Closed-Loop Supply Chains. Springer, Berlin, pp 181-220

Villa A (2002) Emerging trends in large-scale supply chain management. In: International Journal of Production Research, vol 40, edn 15, pp 3487-3498

Integrating Environmental Information with Systems of Factory Planning

Christian Grünwald[1], Jorge Marx Gómez[2]

[1] Volkswagen AG, P. O. Box 1837, 38436 Wolfsburg, Germany,
e-mail: christian.gruenwald1@volkswagen.de

[2] Oldenburg University, Department of Informatics,
Ammerländer Heerstr. 114-118, 26129 Oldenburg, Germany,
e-mail: jorge.marx.gomez@uni-oldenburg.de

Abstract: Before the building of production plants, a detailed planning in view of techniques as well as process is performed in the task of factory and production planning. Production planning under environmental aspects further influences the technical realization of production plants.
An Integration of environmental information into a tool for planning production processes allows the combination of both points of view. In this connection, the generation of substances causing environmental hazards during production can be uncovered even before building an industrial plant.
Actions concerning the protection of employees and environment can early be considered in this way. A later high-cost conversion of a plant is avoidable. The presented concept describes the potential of the integration of both points of view. Moreover, the realization of a prototype evaluation is planned in connection with a case study.

1 Motivation

Increasing competitive pressure, tightened conditions of competition, as well as the target of cost savings in companies often require innovative solutions for products, plants and production methods. The high complexity of new methods and techniques in this connection demands a detailed planning already before realization.

As an example, the term of planning is defined by REFA in form of the systematic search and determination of targets as well as tasks and resources to reach these targets [REFA 1985].

The procedure of planning is supported by specific software solutions [Beck 1990; Fischer 1990]. In this connection the development of the virtual product [Gausemeier et al. 2001] proceeds in Digital Manufacturing [Walter 2002]. The virtual product in this case represents a realistic, integrated computer model, ranging from a singular part to the entire assembly. It further includes all required functions for the support of design, engineering, planning, simulation, production as well as service. Consequently, the term of Digital Manufacturing describes a realistic, integrated computer model of a production site, ranging from singular processes to the entire factory. It contains all necessary functions for the support of planning, simulation, engineering, production and even control and maintenance [Walter 2002].

Task of production planning in this case is to develop not only the product itself, but also the production process [Gausemeier et al. 2001]. A multitude of individual prescriptions, that are valid for running production plants, even have to be considered in the state of planning [Schmigalla 1995]. As an example, plants requiring monitoring are defined in the industrial code. Regulations for the prevention of accidents are specified by the union of industrial trade associations. These terms can even be considered in the state of planning plants and business units. Not to mention last, prescriptions concerning the operation of electrical plants or the use of inflammable liquids represent further factors that influence the planning of plants and production processes [Schmigalla 1995].

Apart from the technical view, production planning proceeds related to environment, regarding existing laws and regulations. A significant part of the german law "Bundes-Immissionsschutzgesetz" contains the aspect of building and running plants that require official authorization [Grünwald 2004; Blumenthal 2002]. Furthermore, the usage of hazardous substances or water-polluting substances is regulated by environmental laws. Examples are the german laws "Gefahrstoffverordnung" (GefStoffV) as well as the regulation for plants concerning water-polluting substances (VAwS). Besides the aspects that are directly related to environment or techniques, further requirements concerning industrial safety and health care have to be considered in the planning process. Industrial safety contains all the actions for protecting life and health of the working employees, to preserve their working capacity as well as to organize the work in a human way. A main target is the prevention of accidents at work as well as occupational illness. This aspect can especially be ensured by a safe design of the technical equipment, for example the machines, and precautions in employing, for example in connection with dangerous plants, material or radiation [Blumenthal 2002].

2 Concept-idea of an environment-related planning system

The described approach is related to practice in connection with a case study in cooperation with the German car manufacturer Volkswagen AG. For the task of process planning, the software Tecnomatix eM-Planner is used in this case. In the planning model of the system production processes are defined by products, resources and operations. Links between these objects are realized via relations. Consequently, product-related data are connected with process-related data in this central planning tool.

As a basis for a possible correlation of environmental information with data of a system for planning, types of environmental information that could influence and optimize the planning of plants and processes have to be determined. A selection of some of that environmental information is shown in Figure 2.1. In the graphic possible central objects of a planning system are colored white: Part, operation and resource. For a correlation to environmental information, an extension of the structure by objects for emissions, process material as well as waste is recommended. In the grey boxes environmental information that can be assigned to the objects in the form of parameters is shown.

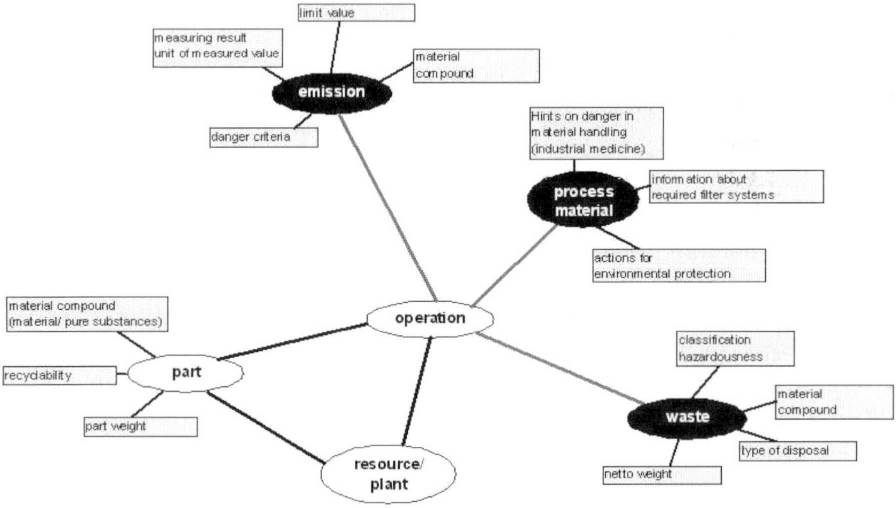

Fig. 2.1. Selection of environmental information being relevant for an integration with a system for production planning

With the help of the described structure several kinds of questions could be answered by a system that combines the technical with the environment related view:

- In which operations do substances appear that are hazardous or dangerous to health? What kind of plants is involved in this case?

A correlation of danger criteria as well as information concerning health care to the specific operation in the planning process allows evaluations of environmental aspects related to production plants.

- At which level has to be the performance of a required filter system in a specific production process to absorb generated substances that are dangerous to health? Which spatial positions of the respective filter system are possible, considering the dimension of this system as well as the respective plant?

The management of quantity data as well as further attributes concerning the used material and the correlation of these data to a specific operation in the planning model makes a calculation of arising emissions in a production process possible. With the help of this information, the planner is able to directly counteract: In this case, for example, a suitable filter system can be planned, and the filter volume suggested by the information of the planning system, will neither be much too high nor too much low. An integrated planning tool connected to a module for planning industrial halls would further consider the spatial position of production plants and filter systems.

- What kind of prescriptions and warnings have to be observed using a certain material in a specific production process? Is the use of an alternative material required at this point, as the original material causes high costs for actions of health care and industrial safety?

On the basis of handling-information for materials that are used in a specific production process, an estimation of costs for industrial safety and health care is possible:

One example is the use of one-way-gloves in a work operation and the related costs.

- Which volume of waste arises in a specific step of processing? Which costs result?

The correlation of waste data with the attributes of an operation makes, among other things, a determination of costs for disposal possible.

- Which amount of emissions arises in a group of plants? Will legal limit values be exceeded? Is it necessary to counteract? Which are the possible effects of these actions?

The calculation of arising emissions allows comparisons with limit values. On this basis, decisions concerning counteractions or alternatives, for example the use of a more environment-friendly plant-type, are possible. Considering environmental aspects, different variants of plants and processes can be modeled in the planning environment.

In a system for production planning, all generated evaluations are even possible in the early phase of planning plants and processes. In this connection a comparison of different types of plants and production processes under aspects of techniques and environment is supported by the system. The consideration of environmental information in this case allows a more holistic point of view in modeling a factory before the phase of realization. Correlating cost factors with the additional environmental data, different manufacturing concepts can be valued under economical aspects. The cost factors are, for example, related to arising waste or additional equipment for industrial safety.

An environment related planning tool in this connection offers a support for decision making in the valuation as well as the selection of alternative production plants and processes.

3 Practice-related conversion

3.1 Initial state of the case study

In connection with the case study, the planning tool eM-Planner is the basis software of this project. This primary standard software is connected with software solutions for planning vehicles, tools and industrial halls in form of an integrated planning system. As an advantage of this kind of data integration, changes of one planning variable, for example the product, are considered in all involved parts of the entire planning process.

Besides the planning systems, several environmental information systems exist, that support tasks of environmental protection. For the integration with a planning system, primary data of emission control and water protection, waste management as well as material data management are relevant.

3.2 Procedure of an evaluation

In the context of the case study an exemplary evaluation based on realistic data is carried out, taking a production process of automobile industry as an example.

As already described before, in this connection an integrated planning system exists. This system, apart from a modification of the process, simultaneously considers the connected effects on the parts of the vehicle, the shape of tools and plants as well as the layout of the working stock in the halls of the factory site. The planning software will be customized to make an integration of environmental information possible. Analogous to the representation of the concept idea, the existing object structure of the system will be extended by classes for emissions, waste, process material as well as environmental factors for parts.

In an extensive and elemental data analysis of the environmental information systems of Volkswagen AG, data fields that are relevant for the integration with the planning system as well as the associated evaluations were identified. These are, for example, danger criteria related to process materials or the substances of parts.

Considering emission factors, additional the amount of arising specific substances can be calculated, if further parameters of the production process are given. With the help of the modified object structure of the eM-Planner, environmental information can be linked to the system-related objects product, resource and operation in the planning model. An example is shown in Figure 3.1.

The example screenshot shows the model of an operating sequence of production. In the customized system an existing operation, displayed in a grey frame in the graphic (operation 4), can be linked to environmental factors, for example emissions and waste. These elements are marked by a black frame in the graphic. The parameters added by the customizing are displayed on the left side of the screenshot in a white frame.

In this case, values for the environmental parameters can either be entered manually or, after extension of the data structure, be read from the database of the planning system.

With the help of the modified tool, the user is able to correlate environmental information with products and operations and to consider this information even in the planning phase.

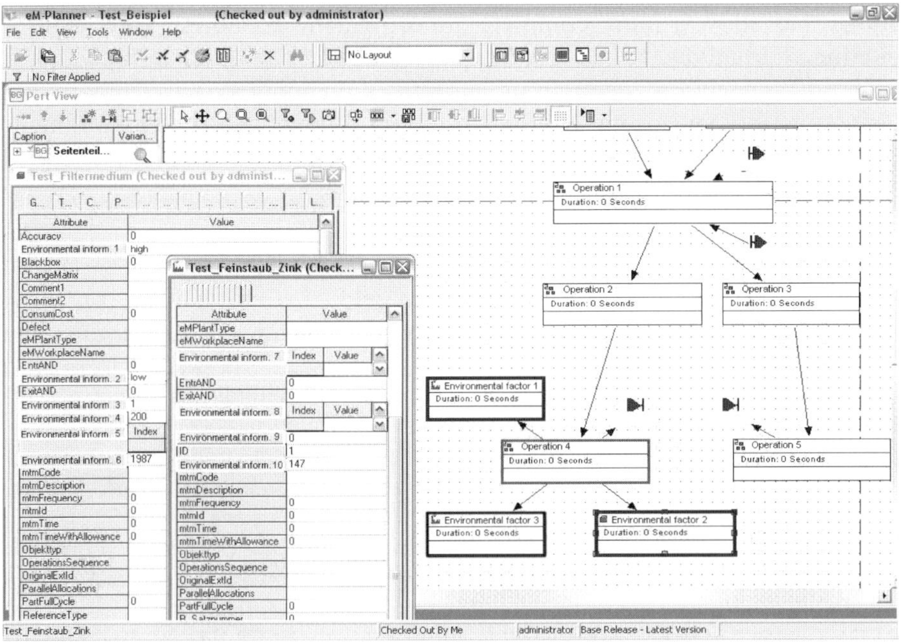

Fig. 3.1 Example of a link between operations and environmental information in the planning tool eM-Planner

4 Conclusions

The starting point of the described project represents the idea of integrating environmental information with systems of factory planning:

The idea is to consider environmental impacts arising in connection with production processes even before the realization of production plants.

At the time of technical planning of plants and production processes already detailed information about types of plants, work operations and products exists. This data is processed in specific planning systems.

Completing the data set of these tools by environmental information that can be correlated to operations, products or resources, hazard potentials can be recognized even in the early phase of planning. Required counteractions like the use of filter systems can be modeled directly in the planning system and can again be viewed under environmental aspects.

The possibilities of correlating costs with the environmental factors or the countermeasures resulting from the environmental impacts over and above that allow a more complete, more realistic economic view of production alternatives.

In this connection, an environment-related planning system supports decision making concerning alternative types of plants and production methods.

References

Beck TM (1990) EDV-gestützte Planungssysteme zur optimalen Belegung von Fertigungslinien. Diss. Karlsruhe

Fischer H (1990) Verteilte Planungssysteme zur Flexibilitätssteigerung der rechnerintegrierten Teilefertigung. Carl Hanser, München

Gausemeier J, Ebbesmeyer P, Kallmeyer F (2001) Produktinnovation - Strategische Planung und Entwicklung der Produkte von morgen. Carl Hanser, München, Wien

Marx Gómez J, Grünwald C, Griese S, Petri T, Rosenau-Tornow D, Eppers R (2004) Konzeption eines integrierten Systems für Umweltinformationen - Fallstudie Volkswagen AG. Papierflieger, Clausthal-Zellerfeld

REFA (1985) Methodenlehre der Planung und Steuerung. 5 Bände. Carl Hanser, München, Wien

Schmigalla H (1995) Fabrikplanung – Begriffe und Zusammenhänge. Carl Hanser, München

Von Blumenthal A (2002) Fabrikplanung Schwerpunkt Recht. Shaker, Aachen

Walter TJ (2002) Einsatz von Methoden der Digitalen Fabrik bei der Planung von Produktionssystemen für die Automobilindustrie. Shaker, Aachen

Designing a Flexible ICT Framework for Processing Remote Gauging-Data

Peter Jakob, Flurin Sutter, Peter Waldner, Gustav Schneiter

Swiss Federal Institute for Forest, Snow and Landscape Research WSL, Zürcherstrasse 111, 8903 Birmensdorf, Switzerland, e-mail: {peter.jakob, flurin.sutter, peter.waldner, gustav.schneiter}@wsl.ch, http://www.wsl.ch

Abstract: In environmental ecosystem research, there is a trend towards sharing research sites infrastructure among project partners when carrying out individual or common projects. As a result a growing number of so-called research platforms are established. In order to take advantage of most of the potential synergies between project partners an adequate ICT infrastructure must be provided. This paper describes an ICT application which fully automates the workflow from measuring parameters in remote gauging stations up to storing validated time-series in a database. The main focus is on designing an appropriate database, but issues of software design and necessary standardization are also included.

1 Introduction

In today's ecosystem research more and more organizations bundle their efforts and use common research platforms for their environmental field research [Thimonier et al. 2001]. Such platforms typically include (i) a measurement network on one or more sites consisting of a conglomerate of basic long-term measurements and different additional temporary campaigns as well as (ii) some agreement among the partners about the division of responsibility for maintenance and for the rights to use the data.

The goal of this paper is to describe an IT solution that enables project partners to share data-collection infrastructure, communication devices and to use a common database. The solution described supports time-series data.

The Swiss Long-Term Forest Ecosystem Research (LWF) is based at the Swiss Federal Institute for Forest, Snow and Landscape Research (WSL) at Birmensdorf, Switzerland [Cherubini and Innes 2000; see http://www.lwf.ch]. The mission of LWF is in agreement with the aims of the "International Co-operative Programme on Assessment and Monitoring of Air Pollution Effects on Forests" (ICP-Forests [see http://www.icp-forests.org]). The aim of LWF is to improve our understanding of how natural and anthropogenic stresses affect forests in the long term, and which risks for humans are involved. The main emphasis is on atmospheric deposition, biogeochemical cycles, climate, soil, ground vegetation and trees [see http://www.icp-forests.org].

LWF in Switzerland consists of a network of 18 research sites located all over the country. Each research site is equipped with two gauging stations for measuring environmental parameters. All data loggers are equipped with a GSM module for digital data transfer. The modular structure of the data logger system makes it easy to add additional sensors and to measure additional environmental parameters.

2 High-level IT requirements

Some of the high-level IT requirements (see Requirement 1) are deduced from the modular, expandable structure of the network:

1. The software must support extensibility: it should be easy to add new sensors in order to carry out additional measurements.
2. Optimized and automated workflows should be established in order to use scarce manpower efficiently.
3. Data should be stored safely and securely in a well-documented structure for a long period of time.
4. Controlled access to data should be possible for different project partners even after decades.

3 Description of the solution

Requirements 3 and 4 can best be fulfilled by using a relational database system (RDBMS) such as, in our case, Oracle. Part of the system can be documented by using the standard features of the RDBMS while the documentation for the rest of the system has to be implemented separately. Data loggers are used to store data in the field (see Fig. 1). Data is transmitted periodically via the GSM network to a PC at the research institute. This step is carried out by a commercial program provided by the logger manufacturer. Another custom-made program validates the data according to specified rules and imports it into the relational database. The workflow is fully automated; manual intervention is only required in cases of faulty devices or a disrupted communication network.

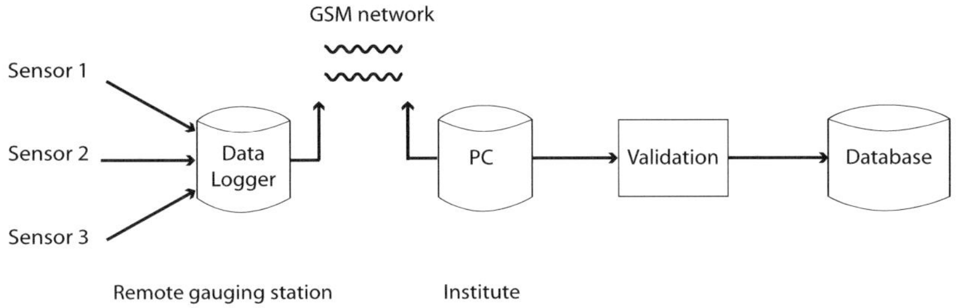

Fig. 1. Data flow from the gauging device to the database

3.1 System architecture of the main data processing module

Requirement 2 must be fulfilled by designing new program modules and using existing ones which offer the necessary workflow functionality. Figure 2 is a more detailed view of the right side of Figure 1 and gives an overview of the structure of the newly designed main data processing module of the whole system. This module is responsible for validating and inserting logger-data into the relational database.

The functions "Format compliance checks (1)", "Data validation checks (3)" and "Process data for database (4)" are part of a custom-made Oracle specific PL/SQL Package residing within the database. The remaining functions of this module are written in "tcsh" script language. Communication between PL/SQL and tcsh scripts is done via "data processing logs" (see Fig. 2). The PL/SQL package is invoked periodically via a tcsh script.

The first task of the PL/SQL package is to carry out the "Format compliance checks (1)". If the format is not correct, the logger file has to be corrected manually (see "Correct data format (2)" on Fig. 2). The next task of this package is to validate data from the files (see "Data validation checks (3)") according to specified rules residing in the database. The invalidation reasons for invalid values are logged in the "Data processing logs" and in the database. Finally, after preparing the data (4) it is loaded into the database.

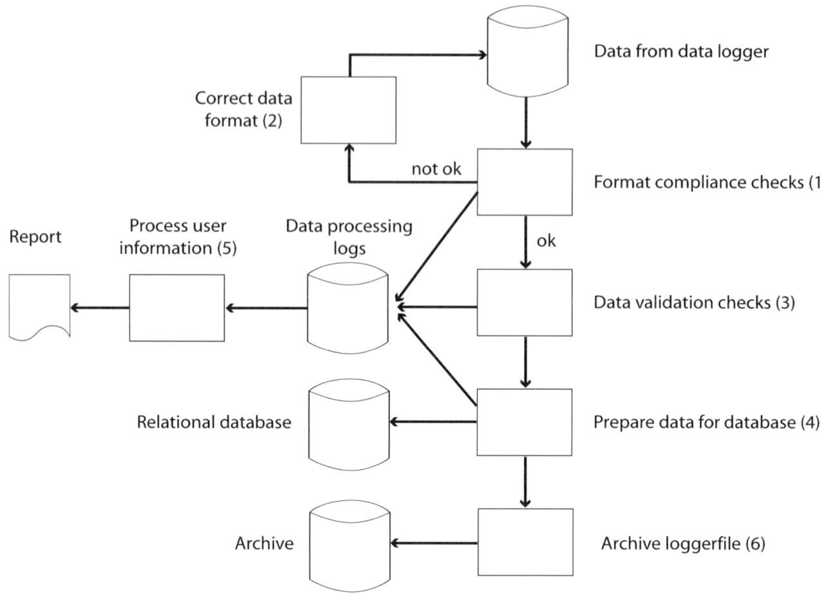

Fig. 2. System architecture of the main data processing module

The "tcsh" script decides according to the information in the "Data processing logs" whether the logger file was processed correctly or not. In the first case (processed correctly) the logger file is archived (6), otherwise the necessary clean-up operations are carried out. The second task of this script is to process the "Data processing logs" and send information from these logs via e-mail to the people responsible for a particular project (see "Process user information (5)").

3.2 Products used

The following products were used for programming the custom-made software:

- Oracle relational database for storing data
- Oracle's PL/SQL programming language for the main data-processing tasks
- Oracle Forms for maintaining the configuration data of the gauging stations
- Oracle Internet Application-Server for enabling Internet access for Oracle Forms
- Tcsh scripts for "gluing" modules together

• different Unix services (mail, file-transfer)

3.3 Required standardization

In order to properly integrate measurement devices and data into the network, certain rules and standards have to be adopted. In the LWF network, for example, Campbell data loggers were used to store the data so all sensors used should be connectable to such a logger. Normally this should not be a problem but sometimes integrated devices which have a data storage module of their own are installed on a research site. Those devices are often difficult to integrate into the existing network and are therefore often used as stand-alone devices.

```
MEASUREMENT PERIOD [MINUTES]
|   STATION_NO
|   | PROJECT_NO
|   | |   DATE (YEAR, DAY OF THE YEAR, TIME)
|   | |   |              VALUE1
|   | |   |              |     VALUE2
|   | |   |              |     |      ...
10,42,2,1998,238,1600,20.35,469.22,...

Values with special meaning:
- Invalid value:         -6999
- Sensor not connected: -999
```

Fig. 3. Data logger file formatting standard

Measurement readings are stored in a file within the data logger. This file has to observe agreed formatting standards. Figure 3 shows the formatting standard of the file within the Swiss LWF network. Each line-header contains the complete information necessary to uniquely identify a line. This redundant information enables the proper processing of incomplete or corrupt files. In addition, it is necessary to agree on a list of values with special meaning (see bottom of Fig. 3).

3.4 Database design

The most obvious way to design the central measurement database is as follows:

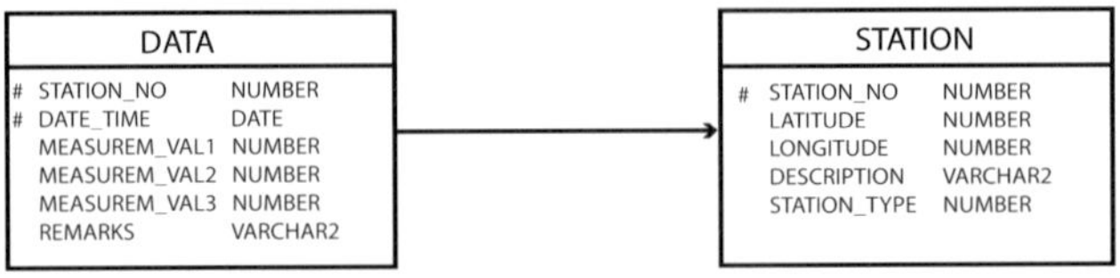

Fig. 4. First, simple ERD of storing measurement data

Unfortunately, this design does not fulfill Requirement 1 (extensibility) very well. Each time an additional sensor has to be added, the structure of the database has to be changed and the associated programs have to be adapted. Different measuring periods would also lead to many NULL values in database records.

Figure 5 shows a more generic or general approach. There is no need to change the database structure when adding a new parameter to be measured; one simply adds an additional record to the "measurement type" table. Requirement 1 is well fulfilled by this design. Simsion and Witt [2005] write about the stability and flexibility of particular data models: a more stable model changes less when changes in the environment of the model occur and a more flexible model is more open to changes in the environment of the model. By generalizing, the second model in Figure 5 becomes more stable and flexible than the first model in Figure 4. On the other hand, the first model is able to support more (specific) "business rules" than the second one. For example, the first model in Figure 4 can support the rule "Values for wind speed cannot be negative" whereas the second model in Figure 5 cannot.

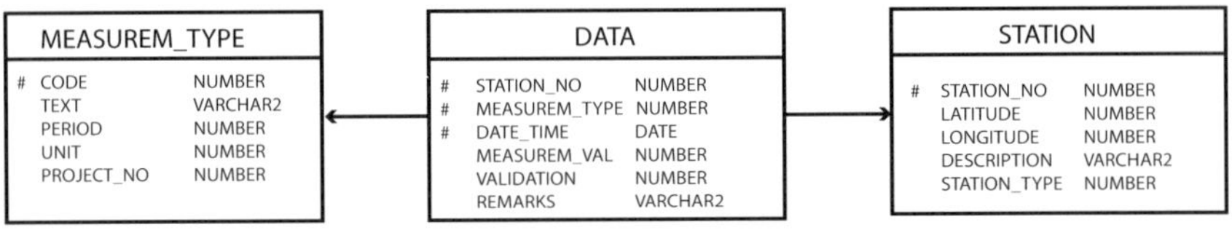

Fig. 5. Second, more generic ERD for storing measurement data

The ERD in Figure 6 shows a model which overcomes this lack of business rule representation in the more generic approach of Figure 5. It enables the representation of more sophisticated validation rules than normal database constraints allow. It is possible, for example, to map time-dependent validation rules like "Minimum value for measurement Type 1 (air temperature) in month 1 (January) for gauging station 3 cannot be below minus 16 degrees Celsius". The following generic rule types are supported: absolute minimum, absolute maximum, changelessness, relative variability and gaps in time-series

When importing the data into the database all values of a time-series are validated according to the specified rules. Values which do not pass this validation test are flagged as "suspicious". The importing process produces a report summarizing all irregularities sent in the form of an e-mail to the people responsible.

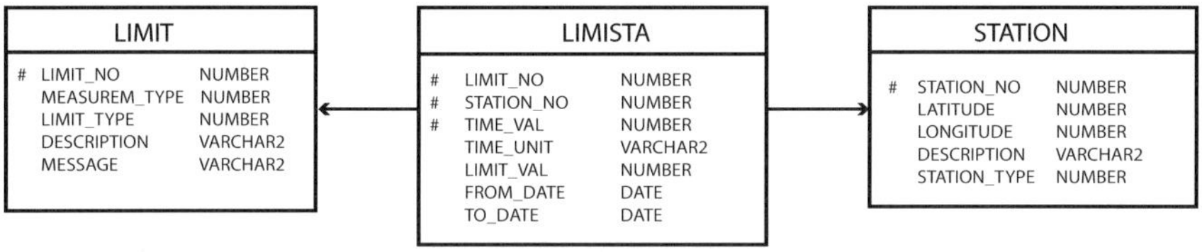

Fig. 6. Generic ERD for storing validation rules

When using the above generic design, the sensor configuration of each gauging station together with the associated validity period has to be stored in the database. The design of this part of the ERD is almost identical to the structure in Fig. 6 for storing validation rules. The sensor configuration data of each gauging station represents important meta-data because it permits the identification of the sensor for every single measurement reading.

3.5 Application for maintaining configuration data

The database design determines some of the requirements for the software design. Figure 7 shows the user interface of the program necessary for maintaining the configuration data of a gauging station. By specifying the sequence number ("Folgenummer") in this program, a unique measurement type can be assigned to each measurement value in the data logger file.

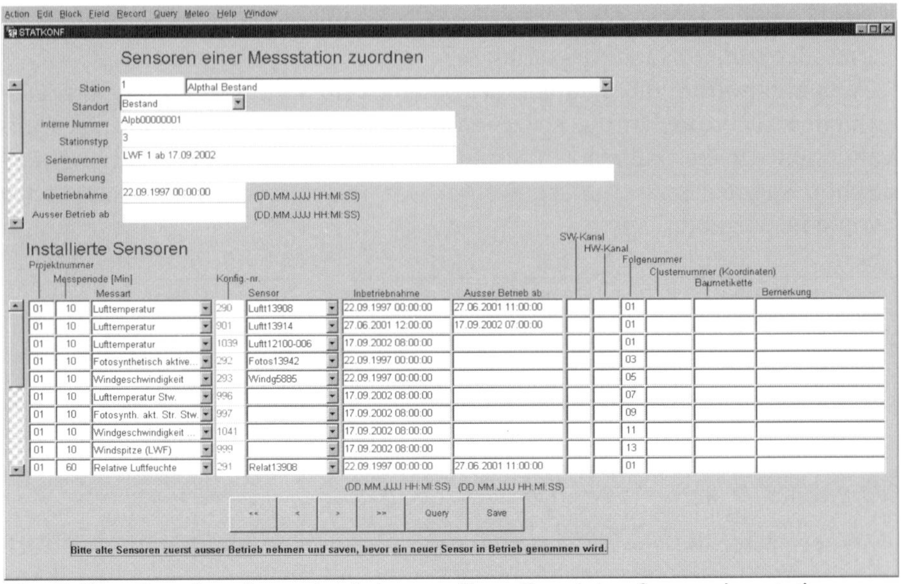

Fig. 7. User Interface for maintaining configuration data of a gauging station

4 Discussion

Since the launch of the generic time-series application in 1998 the number of measurement types that have been added has grown continuously. There are over 50 different types currently in use. In the case of the LWF project, the higher initial effort involved in designing, analyzing, and programming is more than well compensated by the subsequent savings in not having to change the structure of the database and application. In fact, the cost of maintaining a non-generic application would have been prohibitively high [Jakob et al. 2001].

The freely available ICT infrastructure on the LWF research sites and the already existing measurements were strong reasons for project partners with similar research interests to install their own gauging devices on this platform. More then 10 additional projects have integrated their devices into the LWF network and profited from synergies by sharing communication resources, databases and applications [Jakob et al. 2003]. As mentioned earlier, in most cases it was not possible to integrate "turnkey" devices (e.g. devices with data-logging capabilities) into the network. However, such projects also often profited from synergies by using existing datasets at a particular research site.

There are a few examples of environmental databases with generic design. Hoppe and Schulze [1997] describe a comprehensive system which covers deposition fluxes. Because of their broad coverage of topics, their system uses, at least in part, a more generic approach than the LWF system, which has to handle only time-series. Data access is difficult in such highly generic structures and to overcome this problem they introduced another abstraction level into the database (called "virtab"). In the case of the LWF system, access to the data is still easy; therefore it is not necessary to implement an additional abstraction level, which keeps the application relatively simple.

Krause [2005] presents an even more generic and flexible approach. This leads to a very stable design (the structure of the database literally never changes) and it is possible to integrate almost any dataset. One drawback of this approach is the complicated data access. Additionally, part of standard database functionality has to be "reinvented" (e.g. access-rights).

5 Conclusion

The initial effort to realize a generic or general solution for processing time-series data with automated and optimized work-flows for an environmental gauging network is considerably higher than for a specific, ad hoc solution. In addition, finding the optimal level of generalization for the database design is a complex decision-making process. However, there are two main advantages when using a generic approach:

- In many cases, the gauging network needs to be extendable, i.e. new measurement types or sensors will be added frequently. In such cases a specific solution, which would have to be adapted again and again, is much more costly than a generic solution.
- With minimal additional effort a generic solution can be opened to new projects with similar research interests. Such research platforms with good ITC infrastructure represent an attractive asset for new project partners and offer the benefit of potential synergy effects. This advantage is of particular relevance as scientific collaboration becomes an issue of ever-increasing importance.

In many cases the two main advantages outlined above outweigh the initial higher cost and effort involved in realizing a generic solution. For a non-extendable, purely stand-alone device, a specific ICT solution has clear advantages, in particular the smaller overall costs for developing or buying a less complex ICT solution.

Acknowledgement

We thank Norbert Kräuchi, Matthias Dobbertin and his collaborators from LWF for fruitful discussions in an inspiring research environment. We are also grateful to the "Federal Office for the Environment" (FOEN) for financing the LWF and to Ann-Marie Jakob-Murphy for her editorial assistance.

References

Cherubini P, Innes JL (2000) Switzerland: The Swiss Long-term Forest Ecosystem Research Programme. In: Sprott P (ed) The International Long Term Ecological Research Network. Long Term Ecological Research Network Office, Albuquerque, pp 56-59

Hoppe J, Schulze A (1997) ECO Datenbank zur Stoffbilanzierung in Waldökosystemen. Forschungszentrum Waldökosysteme der Universität Göttingen, Göttingen

Jakob P, Bernhard L, Schneiter G, Sutter F (2001) Environmental Data Processing: Science meets Business Engineering. In: Hilty LM, Gilgen PW (eds) Sustainability in the Information Society. Metropolis, Marburg, pp 541-54

Jakob P, Schneiter G, Rösler E, Hug C, Sutter F (2003) IT-replacement investments and sustainability within Swiss forest ecosystem research: the broad and the narrow strategy. In: Gnauck A, Heinrich R (eds) (2003) The Information Society and Enlargement of the European Union. 17th International Conference Informatics for Environmental Protection, Cottbus. Metropolis, Marburg, pp 102-109

Krause R (2005) Development of a Very Flexible Web based Database System for Environmental Research. In: Hrebicek J, Racek J (eds) Networking Environmental Information. 19th International Conference Informatics for Environmental Protection, Brno 2005. Masaryk University, Brno, pp 167-171

Simsion GC, Witt GC (2005) Data Modelling Essentials, edn 3. Morgan Kaufmann Publishers, San Francisco

Thimonier A, Schmitt M, Cherubini P, Kräuchi N (2001) Monitoring the Swiss forest: building a research platform. In: Anfodillo T, Carraro V (eds) (2001) Monitoraggio ambientale: metodologie ed applicazioni. Atti del XXXVIII Corso di Cultura in Ecologia. S. Vito di Cadore, Centro Studi per l'Ambiente Alpino, Università degli Studi di Padova, pp 121-134

Ecological and Economic Effects of Web-based Integrated Management Systems

Horst Junker

FHTW Berlin, Germany, e-mail: horst.junker@fhtw-berlin.de

Abstract: Companies often employ independent management systems for quality, environment as well as work protection and safety. Since these systems are accordingly oriented to the same target group in companies and possess an identical structure for the most part, it is possible to integrate these into a single holistic system. The complexity of those connected via these measures may thereby be reduced, thus such a system may be realized as an intranet solution. Via this web basis, the possibility is thus opened of integrating powerful features such as document management, science management, portal construction and authorization concepts with their capabilities into integrated management systems. Thus users are given the chance of working efficiently and in goal-oriented manner upon applying management systems, thereby in turn leading to an increase in both the economic as well as the ecological effects by realizing these systems.
Keywords: Quality Management, Environmental Management, Safety Management, Intranet

1 Introduction

The term "management systems" was first widely used upon instituting the ISO 900X standard series for Quality Management. Beyond this series of standards, management is of course carried out within companies, in any case a management system requires that the (management) instruments are consciously, systematically and precisely used and especially harmonized among themselves and established. Assuring this is principally the task of a management system.

Besides the ISO 900X Quality Management Standard, two framework rules have existed since passing EU Eco Audit Rule and ISO 14001 for Environmental Management systems. In addition, for example the British

Standard 8800 represents a standard for a management system for work protection and safety.

For both economic and practical reasons, it does not make sense to separately build, implement, maintain, audit and certify specific management systems for all three realms. It is much simpler to elaborate a mutual core of these management systems and only deal with special demands separately.

An integrated procedure differentiates itself principally from a pure extension to an existing management system – e.g. to a Quality Management system – by further aspects. Such an additive extension has the consequence that the systems connected to Quality Management receive less weight, as empirical research proves. Moreover, that the systems connected to Quality Management becomes even more difficult and complicated, as more systems are added. An integrated system thus offers greater flexibility.

An integrated management system achieves the following use:

- Efficiency is greater, since as a whole the quantity of regulations is smaller and therefore its maintenance sinks,
- The risk of contextual contradictions is smaller as for separate systems,
- Both transparency and acceptance are greater.

Management systems are thus justified, if they provide a positive influence to the success of a company. Thus success factors are those, which have a decisive influence upon the sustainability or else the sustainable value in-crease of a company and thereby increase its chances of survival. Implementing certain framework criteria in this regard thus supports successful business dealing. Besides the effective legal requirements, to such frame-work criteria also belong rules, which propose constraints regarding the structure of management systems for the realms of environmental protection, security and quality.

Only well structured management systems are capable of integrating either existing or future demands, which are imposed onto companies by demand groups into a company's functioning and thus precisely eliminate risks. Such a management system is set up in a manner that maintaining both external and internal demands is assured in an economic manner as well as also supporting continual improvement process within companies.

In addition, it is also necessary that employees understand the management system and recognize its use. A requirement in this regard exists that they are empowered with their duties within the system. For this reason and because organizations orient themselves ever increasingly towards customers on processes, these processes should represent the basis of management systems.

In order to provide the required flexibility, durability and acceptance by the employees to a management system, the following procedure is recommended for its construction:

- Identifying the processes occurring within the company,
- Recording the demands to the individual process steps,
- Assigning responsibilities,
- Elaborating understandable instructions from the demands,
- Linking instructions and processes.

Identifying the processes already delivers reference points in this regard; i.e. which regulations are required as the case may be for which processes. In any case, this may only be a rough approach in a process-oriented model. Besides the individual processes it is thus necessary to also record the partial processes and these determining activities. It is therefore necessary to execute a link of regulations and observed unit (process, partial process, activities) at the lowest level. Thereby it will become possible to make relevant regulations, which apply exclusively for this process, partial process or else activities, available to everyone responsible for a process, without his being forced to search for these in non-user-friendly manuals.

By nature, an integrated management system consists of differing contents, which often are processed by various experts. These experts often take over the maintenance of their partial systems. The tendency of isolated processing of partial systems must be counter-acted by organizational measures, e.g. in the form of a management system delegate, in order to thereby assure a continual integration.

Thus the following requirements are to be imposed as a whole onto the construction of integrated management systems:

- The management system is to be adjusted to the individual needs of a company,
- The management system is to be set up in its structure based upon durability; this works best if it is based on the business processes,
- The management system is to be conceived in such a manner that new requirements, which are posed by demand groups onto a company, may be integrated with minimal complexity.
- The management system is flexible regarding consideration of process changes and thereby supports a continual improvement of the processes.

The execution of tasks influenced by the management systems is determined by the regulations. Principally, such regulations serve to implement external demands and internal goals. In addition, regulations are made to provide a general basis for managing a company.

Regulations allow themselves to be classified according to two criteria, first according to the level of concretion, second according to the object of regulation. Regarding the level of concretion, procedural directions, work and company instructions are differentiated. They have the character of requirement regulations, since they present requirements to the workers for executing duties. Under the aspect of the object of regulation, it is subsequently differentiated, if process flows are either adjusted or set down firmly.

Regulations such as e.g. process flows contain the following elements:

- Description of the regulation,
- Purpose and application area,
- Terms,
- Responsibilities,
- Flows,
- Further applicable documents,
- Documentation,
- Attachments.

A guideline for documenting regulations should be that they not only clarify and specify tasks to those executed, i.e. the employees, rather also clarify their causes – e.g. a legal requirement – and give advice regarding processing methods and tools.

The goal of a process-oriented management system consists among other things of assigning the existing regulations—procedure, work and company directions – to the processes, partial processes and activities, in order to achieve goal-oriented informing of employees.

However, regulations are not only based directly on a process, rather also on equipment and material (e.g. maintenance, air purification and handling toxic compounds). Since however equipment and materials are in turn linked with processes, these regulations at least directly also affect processes.

In the case of high regulation density, an (direct) assignment to processes is only economical and practically possible by implementing a suitable database. In this manner, a detailed job description of the relevant functional executor as well as proof of control responsibilities of management is obtained; consequently, a meaningful clarity is thereby created about responsibilities.

2 Integrated Management System

In the past individual management systems for quality, environmental, work and security were separately implemented within companies; which were applied more or less process-oriented with complex and non-harmonized documentation "side by side". The experience was therefore acquired that for practical reasons only management and therefore a management system may result.

Integrated management systems are illustrated in the form of a management manual, which is applied for the following purposes:

- In order to make policies, procedures and requirements known to a company,
- In order to achieve better controlling flows and facilitate tasks,
- In order to protect the continuity of the management system under dynamic circumstances,
- In order to educate personnel regarding the management systems and
- In order to externally expose the management system.

The layout of such a manual is principally set up according to the individual needs of a company. It should however be set up in the following manner:

- Title, purpose and application area,
- Index,
- Policies and goals of the organization,
- Description of the organization, responsibilities and competencies,
- Description of the elements of the management system, if necessary with reference to procedural instructions,
- Instruction about handling the management manual as well as
- Attachments with complementary documents.

3 Web-based Integrated Management System

For The goal of realizing an integrated management system on a web-basis consists of:

- Facilitating handling the management system to user,
- Being able to import modification of elements of the management system in a timely manner,
- Only making those portions of a management system available to users that they require for performing their duties.

Thus the central system task consists of optically presenting the contents (of integrated management systems) to users on an information carrier.

Web-based integrated management systems are challenged to treat contents as a sum of important individual information; these are especially structure, display form and text or graphic contents. Furthermore, the requirement of reusing and recycling leads to analyzing individual information and separately storing them.

Typically, the required functionality of "classical" document management systems is realized here.

Moreover, integrated management systems demonstrate themselves since their software solution must be capable of structuring and managing an extremely extensive and complex database. Therefore, attention must be paid to principles and procedures of modeling data in a special manner.

In order to open preferably both powerful and elegant access possibilities to the contents of integrated management systems to users, an intranet solution is offered, into which a portal construction is integrated.

Usually, companies choose the intermediate step of an information portal upon developing an employee portal. The latter show the traditional form of the intranet, sometimes extended to the function of knowledge management. Normally, aggregating contents is dealt with for these portals. Employee portals are connected with significant process improvements and additionally offer functionalities in the area of Content Management, thus managing contents.

In order to realize an increase in employee productivity and an improvement of the efficiency of business processes for the purpose of goals of integrated management systems, employees must have timely access to necessary information. In any case, the availability of a current and complete overview is required, in order to harmonize the respectively executed activities with the goals of the integrated management system. The up-to-dateness of the contents required here can only be granted via a holistic portal concept.

Since the information is of differing meaning for each employee, the main duty of the portal consists of processing the information specifically for each employee. This personalizing is the key, in order to fully exploit possible productivity via portals.

On a whole, a portal solution means not only technical modification, rather modifications in the organization and process adjustments are necessary, in order to achieve desired success. The company culture also plays an important roll in establishing a portal solution. As previously stated, the contents are a further critical success factor. It must be presented in the context and be current. The presentation structures may be standardized and the design must display a clear recognition effect and clear navigation structure.

Personalized employee portals demonstrate that a database adjusted to the task realm and information needs is made available to each employee. For this purpose, for each user or else for each user group, the respectively specific and necessary data volume must be set and assured, so that only these defined data are accessed. Conversely, it must be avoided that users access database contents, which are neither necessary nor suitable for their tasks. In order to pay special attention to these requirements, the development of an authorization concept is urgently needed.

4 Ecological and Economic Effects

In By means of the outlined solutions of a web-based integrated management system, all data and information are stored on an internet platform and precisely assigned to individual business realms – if necessary subdivided onto the workplace levels. Hence they may be retrieved at any time. Making the information available involves both those applying to the entire company, as well as those, which are tailored to groups of workplaces or else to an individual workplace.

The first group of information deals primarily with structural organization, the second addresses itself primarily to the organizational workflow question.

In the realm of processes of workflow organization, above all information are processed or else made available that are related to material and information flows of a company. Hence various advantages are realized, for example:

- Increased quality assurance via steady process monitoring and assurance in all company departments.
- Saving resources by optimizing all business processes e.g. increased equipment efficiency or optimizing the transport division.
- Minimizing risks by constantly providing information, e.g. regarding the technical inspection of equipment. This thereby ensures improved worker safety for employees.
- Via consequent and integrated application of standards:
- ISO 9001:2000 (Quality Management)
- ISO 14001:1996 or else EMAS II (Environmental Management)
- British Standard 8800 (Work Protection and Safety Management)

A technologically holistic new development of integrated management systems as real-time multi-user system is realized, which permits all users to simultaneously work on either one or numerous tasks. The results of each individual are immediately visible and available to all.

Finally, the web-based integrated management system consists of a type of archiving procedure, enabling the duties and tasks of a company's employees to adjust to the management system's goals. The system supports the users upon coordinating their work and upon following rules and regulations of management standards.

The web-based display enables preparing all company criteria such as, e.g.:

- Assuring quality,
- Saving resources,
- Avoiding risks or
- Preventing work accidents

for an intranet solution, in order to process running procedures more simply, quickly and safely.

5 Conclusions

Empirical studies prove that integrated management systems will establish themselves in an increasing manner in the future. The connected ecological implications will also be used as a possibility for increasing economic efficiency. In any case, besides quality and environmental aspects, social and ethical viewpoints will be increasingly addressed in the future.

Competitive advantages exist there, where companies do not bet on island solutions, rather recognize the use of synergies of holistic and web-based integrated systems.

Finally, processes need to be planned, controlled and improved in order to efficiently create quality-justified production not just environmentally, but also safely and responsibly conscious as well.

The second step of the strategic action planning aims at the generation of project ideas for the elimination of the strategic gaps. This includes a check, which information systems are available in the enterprise and what kind of data these information systems include. For this purpose, an overview of the different architectures of an enterprise is necessary in order to avoid redundancies or, if redundancies can not be avoided for technical reasons, to minimize these. During this step, the existing information systems were analyzed. The focus was set on the following data which is relevant for EIS [Lang and Jürgens 2003]:

- Material master data
- Structure oriented data
- Process data
- Energetical and material flow data
- Organizational data

The analysis made clear, that numerous environmentally relevant data is available throughout the enterprise. But that data was distributed across six different information systems.

The third step results in a selection of information system categories, which can be used to close the strategic gaps. The results of the environmental analysis and the therein identified information systems serve as a starting point. The following information system categories were identified:

- Material and energy controlling systems
- Audit support systems
- Document Management Systems
- Environmental Controlling Systems
- Reporting systems
- Material Flow Analysis Systems

On the basis of the identified categories, commercial information systems can be identified or the necessity for individual development can be uncovered, in order to close the gaps in the environmental management.

6 Conclusion

Environmental management is an accepted object of business organization. Nevertheless, environmental management is highly operational aligned. This might even result from the heavy operational alignment of the used environmental management systems. The operational alignment leads, among other things, to the missing consideration during the performed IT planning. So far there are only few large enterprises in Germany, which recognize the necessity for a holistic approach considering information systems in environmental management. Thereby it becomes obvious, that the complexity in the area of environmental management is widely underestimated. In a rather not irrelevant number of individual development projects that underestimation leads to serious problems. This is the reason why custom software was excluded during the process of project idea generation for the closing of the strategic gaps. This approach offers different advantages. On the one hand, the available information systems are already proven. The vendors, on the other hand, already showed some experience in the industry of the viewed enterprise.

The concrete arrangement of the information systems portfolio led to a system portfolio, which can close the identified strategic gaps and restore scope for design for future strategic decisions, as the identified information systems will lead to an enhanced informational quality for decision makers in the enterprise. Above all the securing of legislational compliance and with the help of controlling instruments identified cost optimization potentials will guarantee the basis for economic operation of the enterprise.

References

ASU/UNI Unternehmerinstitut e.V. (1997) Öko-Audit in der mittelständischen Praxis

Beucker S, Jürgens G, Lang C, Rey U (2003) Betriebliche Umweltinformationssysteme (BUIS) im Umweltcontrolling: http://www.bum.iao.fraunhofer.de/intus/ergebnisse/BUIS-Umfrage.pdf, access: 2007-05-01

Computerwoche (2006) Palms Smartphone "Treo 650" ist für Europa zu giftig http://www.computerwoche.de/knowledge_center/wireless/578331/index.html access: 2007-01-07

Dyllick T, Hamschmidt J (2000) Wirksamkeit und Leistung von Umweltmanagementsystemen, vdf Hochschulverlag AG

Heinrich L, Lehner F (2005) Informationsmanagement. edn 8. Oldenbourg

Pietsch T, Martiny L, Klotz M (2004) Strategisches Informationsmanagement. vol 4. Erich Schmidt Verlag, Berlin

Krcmar H (2005) Informationsmanagement. edn 4. Springer, Berlin

Lang C, Jürgens G (2003) Aufgaben Betrieblicher Umweltinformationssysteme In: Kramer M, Brauweiler J, Helling K (eds) Internationales Umweltmanagement Band II: Umweltmanagementinstrumente und –systeme. Gabler, Wiesbaden

Niemann KD (2005) Von der Unternehmensarchitektur zur IT-Governance. vieweg, Wiesbaden

Porter M (1998) Competetive Strategy, Free Press

Rautenstrauch C (1999) Betriebliche Umweltinformationssysteme. Springer, Berlin

Schwegler R, Schmidt M (2003) Lücken im Umweltmanagement: Forschungsansatz für ein rationales Umweltmanagement auf der Basis der St. Galler Management-Lehre In: Schmidt M (ed) Umweltschutz und strategisches Handeln. Schweg

Modeling of a Corporate Environmental Management Information System (CEMIS) for Production Processes

Corinna V. Lang

FHTW Berlin, Germany, e-mail: c.lang@fhtw-berlin.de

Abstract: Corporate environmental protection has developed into an elementary building block of corporate policy in recent years and has now become necessity and opportunity for companies. It is not longer a question of simply protecting the environment. It is increasingly becoming a matter of social and economic aspects as well, of companies' global responsibility for future generations; to put it briefly, a matter of sustainable development.

Against this background, companies have to meet two sets of demands. They have to satisfy the ecological and economic efficiency of the products and production processes. In addition to this, they contribute towards conserving resources.

In order for an environmentally oriented management to be able to gain influence in time, problems and mistakes must be recognized quickly. However, it is problematic that the variety and complexity of the existing environmental data and information are relatively high. To the recording, analysis and treatment of these data and information, efficient and extensive information systems have an essential importance.

Corporate environmental management information system (CEMIS) as an organizationally technical system offers the possibility of systematically covering, analyzing, processing and appraising as well as to archiving environmentally relevant information [cp. Page and Hilty 1995]. These information systems support strategic and operative management with planning, control and transaction of environmental-issue measures [cp. Rautenstrauch 1999].

1 Impact of environmental management systems for production processes

Through intensive social discussions and statutory provisions about the protection of the environment, corporate environmental management has become increasingly important. Environmentally oriented ordinances and laws, like the DIN EN ISO 14001, EMAS or EMAS II, are designed to motivate companies to show an environment-conscious behavior. They are about strategic decisions and instructions and increasingly about targeted control of relevant material and energy flows.

An environmental management system is an important component of an extensive management system in ecologically oriented businesses. It includes the organizational structure, planning tasks and responsibilities concerning the development, implementation, fulfillment, evaluation and maintenance of environmental politics. This requires an ecological analysis and evaluation of the applied methods, procedures, processes and resources.

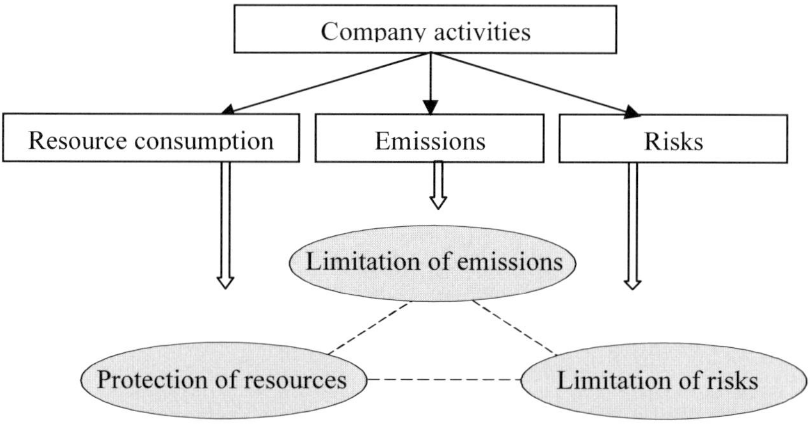

Fig. 1. Goals of corporate environmental management [cp. Dyllick 1990]

Corporate environmental management helps to make the environmental activities transparent and to document them. The goal is to reduce risks and emissions as well as the resource consumption in companies (Fig. 1). The main focus is on the continuous reduction or avoidance of environmental burdens for all processes in connection with product manufacture [cp. Rautenstrauch 1999].

A long-range integration and extensive variation of the procurement, the production, the distribution and the disposal can only be realized by means of integrated and complex corporate environmental management information systems. Each department must be supported with process information. By providing the necessary information at the right place at the correct time, it becomes possible to reduce the complexity, variety and low transferability of the tasks involved.

2 Approaches of CEMIS

Information systems that thus have a direct reference to corporate environmental protection measures are also called CEMIS. Such a system is defined as an organizational-technical system for the systematic recording, processing and preparation of environmentally relevant information in companies. It should be designed to record industrial environmental burdens and to support environmental measures [cp. Hilty and Rautenstrauch 1995]. CEMIS support the management and the operative divisions with the planning, control and realization of environmental protection measures [cp. Rautenstrauch 1999].

The tasks and objectives of CEMIS result from the intersection of the realms of business information systems (BIS) and environmental management information systems (EMIS). The complexity of such systems is extremely large so that modeling is very extensive. Analyzing individual business processes serves the purpose of reducing the complexity and thereby simplifies the consequential creation of software.

It is necessary to either link the isolated systems by modifying the system structure by interfaces or else to replace those isolated with integrated systems [cp. Hassis et al. 1995]. Both solutions require appropriate construction models for applying CEMIS.

Consequently the possibility is created by reducing, optimizing and restructuring company internal material flows of identifying existing cost-saving potentials in connection with environmental relief, thereby achieving targeted influence.

The Production department connects the departments of procurement, distribution and disposal (Fig. 1). An integrated company environmental protection unites the individual integration arrangement of the departments into a common concept for a holistic CEMIS. In an ideal scenario, procurement, production, distribution and waste disposal form a closed material cycle.

The concept of this process model for CEMIS is based of production processes of in-house logistics. The process model contains also parts of procurement, production, distribution and in-house disposal.

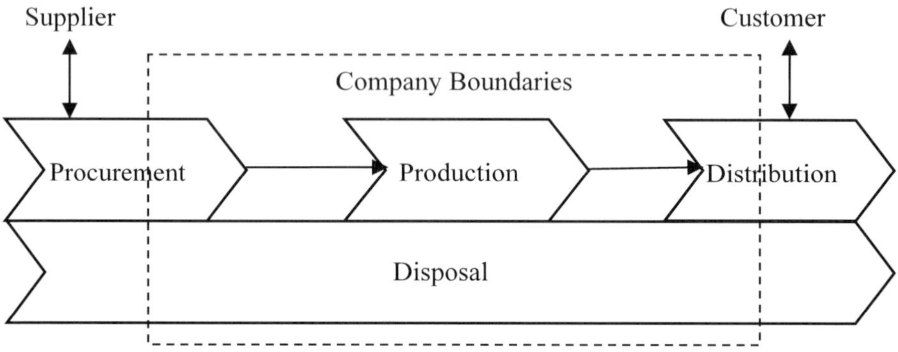

Fig. 2. Main departments of production companies

The processing of environmentally relevant data to environmental data is still a problem due to the particular requirements, functions and application areas of business information systems. Therefore the preparation of information for environmental management in the company is difficult.

CEMIS offer a possible solution. In contrast to business management information systems, CEMIS pursue economic and ecological objectives. On the one hand this increases the functional scope; on the other hand very heterogeneous data need to be processed. These social-technical systems are therefore very complex. There are still no universal principles for the practical procedure of modeling such an extensive information system.

In order to better integrate the user, the modeling is closely related to business managements processes.

The main objectives of this integrated approach to environmental protection in industrial processes are a reduction of environmental impact and cost reduction. Environmental protection measures should be integrated into the production processes since it is during the production planning because there are included great potentials for reduction of costs. The substitution of traditional production procedures by technical and organization methods appears particularly promising. Estimation of the benefits of such procedures critically depends on information instruments such as eco-balances in combination with analysis of economic efficiency.

Different strategies exist for the organizational structure of CEMIS. With additive environmental protection efforts, particular environment-oriented authorities are merely added to the already existing organizational units. The organizational restructuring involved is relatively modest with this integration, since the existing structures and tasks remain largely untouched. The function and process-oriented strategies require the integration and penetration of the entire production processes with environmental activities.

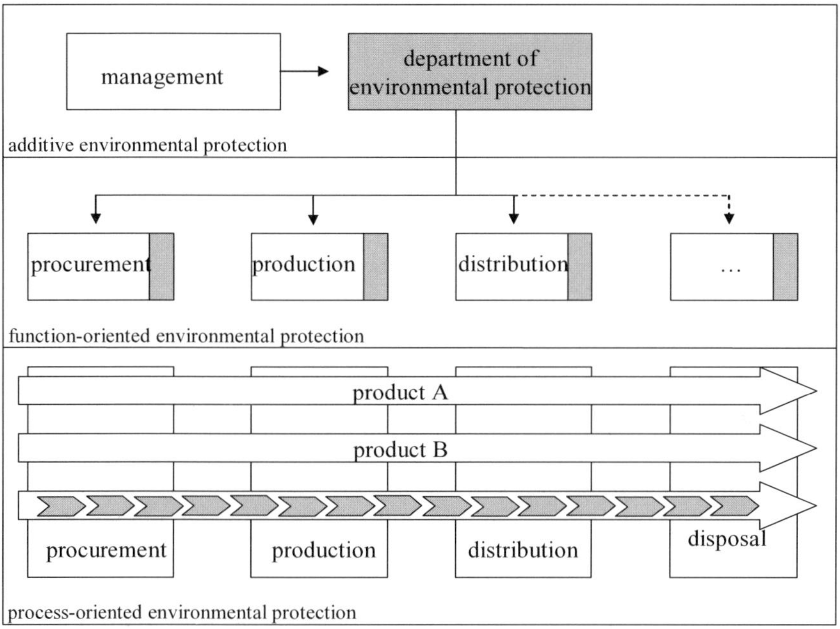

Fig. 3. Strategies for environmental protection in companies [cp. Kraus 1997]

It makes sense to integrate the environmental management activities into the individual department areas. The ecological requirements can lead to selective or even extensive changes of the tasks of the respective functions.

3 Model of CEMIS

Even at an ecologically-oriented production processes do not function automatically. The environmental management, which takes over the planning, leadership, organization and supervision, is responsible for this.

It is possible to integrate environmental aspects into the production processes in companies starting with material provision, product manufacture and customer supply up to material and product recycling.

3.1 Procurement

The individual elements and the relationships within ecologically-oriented procurement can be shown with the help of modeling. Knowledge of the functional processes makes it possible to improve the communication. A business management process of procurement will be described and modeled, which is directly related to material and remains as well as information flow in the business and in cooperation with the suppliers. It is about activities, which change the condition and the situation or generate new objects.

Procurement influence the decision about a supplier, the time of order, the quantity ordered and the delivery date. The management of warehouses also is a task of procurement. Its responsibility finishes when the materials leave the warehouse.

For the environment-oriented procurement in companies, the following requirements exist:

- Selection of materials according to recyclability and environmental compatibility,
- Cooperation with suppliers, for example concerning the taking back of package material and remains,
- Cooperation with the product development on the development of easy-to-recycle product compositions,
- Presentation and analysis of all the material and energy flows within the company by means of material and energy balances,
- Cooperation with production in selecting procedures targeted at enforcing low waste and economical coupled production,
- Avoidance of waste covering everything from material procurement and manufacture to application.

Materials are planned and acquired upon considering economical and ecological objectives. Ecological objectives are protecting resources, relieving the environment and avoiding waste. Consequently modifications and extensions of tasks in this department and during the resulting business processes are realized.

3.2 Production

By integrating environmental protection measures into production processes, companies may also realize an offensive environmental management. Negative environmentally impacts resulting during production must be associated to the individual processing procedures.

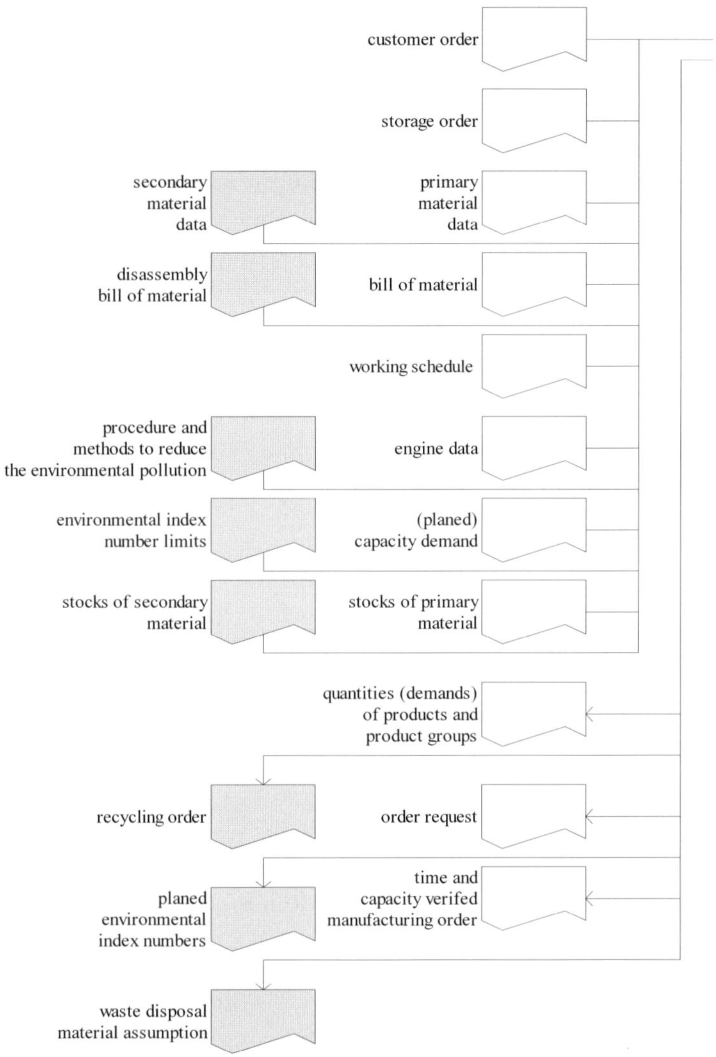

Fig. 4. In- and output objects of ecologically-oriented production planning [Lang 2007]

Integrating environmental objectives into production requires modified, environmentally oriented management and new methods and modulations within the functions of planning (Fig. 4) and control (Fig. 5). Environmentally oriented production have tasks such as material-need planning upon considering recycled products and pre-products as well as materials, reducing material and energy consumption by selective lot-size determination, assuring continuous production processes, storing data and analyzing material and energy flows, avoiding or else reducing undesirable output, moreover registering emission, finally remains as well as waste quantities within production.

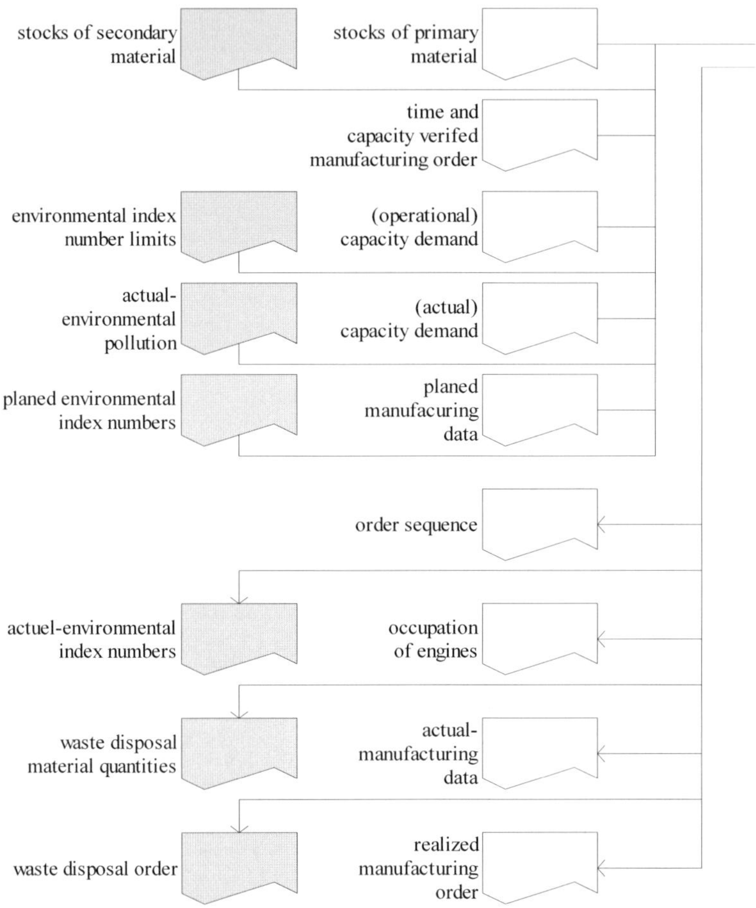

Fig. 5. In- and output objects of ecologically-oriented production control [Lang 2007]

3.3 Distribution

In order to provide customers with offers of environmentally friendly products and indications regarding their components, data regarding ecological product attributes, product life expectancy, recycle-ability, environmental tolerability criteria must be additionally stored.

To the extended tasks of distribution count henceforth also monitoring usage of secondary materials, substituting or else minimizing environmentally detrimental components, products and packaging for environmental tolerability and recycling-ability, the return of used products and packaging material, checking disposal possibilities of remains and wastes as well as recording material and energy flows.

The requirement for supporting an environmentally oriented distribution is a holistic information system with which both customer and company-internal data and information may be analyzed in an objective-oriented manner, evaluated and processed. Nonetheless, it is problematic that despite environmental tolerability claims, environmental effects from products and production procedures often lie in the distant future and therefore are not completely assessable.

3.4 Disposal

By legally induced and proactive environmentally oriented responsibility consciousness of companies, disposal has developed as an additional department. Processes for getting rid of disposable goods (waste, remains, emissions and used products) are thereby integrated into company processes. If by structural measures no avoidance may be achieved, disposable goods are created during procurement, production and distribution processes and must therefore either be dealt with disposal or else substantially reused.

Disposal must entirely structure their tasks so that wastes are reduced and so that secondary materials are as much as possible added to the production process.

The emphasis of disposal is that of reducing remains and recycling. Dealing with remains upon considering these emphases poses both broad economic as well as ecologic demands upon in-house disposal processes. The in-house recycling processes can be strengthened if this data is supplemented by information concerning the disassembly and salvaging of products.

Important tasks of disposal are controlling disposal orders regarding recycling possibilities within the company, substituting or else minimizing environmentally detrimental materials, controlling materials regarding environmental tolerability and recycle-ability, investigating materials regarding ecologically optimal order quantities, checking return possibilities for packaging as well as disposal possibilities of remains and waste.

4 Conclusion

CEMIS are necessary for companies for legal compliance, continual improvement and interested parties as well as Corporate Sustainability Reporting. There are very complex, caused of integration of economical and ecological objectives.

This concept presented a model of CEMIS for production processes limited to analyzing business processes of production companies. It describes the value-added chain of procurement, production, distribution and inhouse disposal. This model offer a way for the production integrated environmental protection.

Already upon planning and deciding procurement, production and distribution, it is necessary to consider the consequences of possible disposal activities. In addition, besides both technical and economic, also disposal-oriented data are to be provided [cp. Kreis 1997].

It is primarily important that models and available data correspond with the user's understanding of the processes. Only then, the user can understand and correctly interpret the available data. For this purpose the need for information in the respective areas and departments of the companies needs to be limited and possibly suitable codes are to be defined. In the development of a CEMIS the focus must be on the intended data utilization of the users.

References

Hassis HD, Hilty LM, Hundscheid J, Kürzl H, Rautenstrauch C (1995) Umweltinformatik in der Produktion. Marburg

Hilty LM, Rautenstrauch C (1995) Betriebliche Umweltinformatik. In: Page B, Hilty LM (eds) Umweltinformatik: Informatikmethoden für Umweltschutz und Umweltforschung. edn 2. München

Kraus M (1997) Informationsmanagement im betrieblichen Umweltschutz – Strategien und Architekturen betrieblicher Umweltinformationssysteme. Universität des Saarlandes, Saarbrücken

Kreis H (1997) Umweltgerechte Gestaltung von Produkten als ganzheitliche Aufgabe. UmweltWirtschaftsforum, vol 5, edn 4. Berlin, pp 42-45

Lang CV (2007) Konzeption eines Referenzmodells für betriebliche Umweltinformationssysteme im Bereich der innerbetrieblichen Logistik. Aachen

Rautenstrauch C (1999) Betriebliche Umweltinformationssysteme – Grundlagen, Konzepte und Systeme. Berlin

Petri Net Based EMIS-Mappers for Flexible Manufacturing Systems

Andreas Moeller

University of Lueneburg, Germany

Abstract: The key obstacle to environmental management information systems (EMIS) is how to efficiently integrate them into enterprise application environments. While there is considerable work on technical aspects of the problem, there are few suggestions of how to solve the problem on the conceptual level, in particular with dynamic business environments, mass customization and flexible manufacturing systems in mind. In this contribution, the Petri net theory is applied as an innovative conceptualization for providing transformation capabilities between widely-used enterprise resource planning (ERP) systems and material flow oriented EMIS.

1 Introduction

From the earliest days of environmental management information system research, the integration into existing corporate IT infrastructures has driven research and development. Today's technical support ranges from java and .NET connectors to extensible markup languages (XML) and service oriented architectures (SOA). However, integration is still one of the most challenging R&D tasks. The standards say little about the conceptual integration of EMIS into enterprise application environments. To address this problem, this paper proposes conceptual mappers as one of the major design patterns for EMIS – mappers not in terms of transformation of documents, rather in terms of a simulation model that helps to join different perspectives. The mapper takes available data in corporate IT infrastructures as input parameters and generates data in a form that fits to the conceptual demands of EMIS. Particularly with regard to flexible manufacturing systems, the communication between widely-used ERP systems and material flow oriented EMIS serves as a concrete example. It reveals that the communication itself is a modeling and simulation challenge; coherent and appropriate modeling concepts are required.

2 Conceptual Mappers

In EMIS, three phases of data processing can be distinguished: data collection, data storage – implementing a consistent accounting system – and data evaluation. After that, integration has in principle the objective to avoid manual data collection effort: data collection should be replaced by an access to available data in corporate IT infrastructures. However, as existing enterprise information systems become the data source of first choice for EMIS, requirements of EMIS are poorly addressed in these systems. In order to overcome this shortcoming, adaptation of data is required. In principle, three approaches are possible:

- All relevant business applications provide appropriate export components;
- The EMIS comprises a set of import components for all relevant enterprise applications;
- Integration servers facilitate independent mappers between the applications [cf. Lee et al. 2003].

Following Fowler, a mapper can be defined as "an object that sets a communication between two independent objects" [Fowler 2003, p 473]. Mappers have some advantages compared to the other alternatives. Fowler mentions two of them: It can be undesirable to modify the two systems. And often dependencies should not be created between the two systems [Fowler 2003, p 473]. Another point is that the demands on the component can result in a quite complex software solution. In such a case, the mapper should be based on a coherent methodology that interconnects the different perspectives and resulting requirements. Customizing such a mapper becomes more and more a modeling activity. In the following, such a mapper is presented relating to the integration of ERP systems and material and energy flow oriented EMIS.

3 EMIS-Mappers for flexible manufacturing Systems

Common basic concepts for EMIS are materials flow analysis (MFA) and life cycle assessment (LCA). These approaches tend to focus on fairly simple manufacturing systems: One or a few products, single-purpose processors, simple production lines and so on. They are in line with strategies for industrial-age make-sell businesses that aim at eliminating variation [cf. Ramanathan 2005, p 51]. But in today's dynamic business environments "selling a single product is not the goal for most manufacturing organizations. Rather, most enterprises are selling a complete portfolio of products and providing related services and operations over the entire life cycle of those products" [Jones and Deshmukh 2005, p 48]. Recent enterprise applications support these organizations.

Furthermore, the approaches like LCA assume that it is not relevant in product-oriented assessments to distinguish between processes (actions, operations, phases, activities, transformations, transports, etc.) and systems (locations, plants, manufacturing systems, etc.).

As a result, underlying concepts of EMIS like LCA or MFA make it difficult to represent and to analyze the material and energy flows of today's flexible manufacturing systems. In typical cases of manufacturing organizations, the databases of ERP systems include orders of customers, materials master data, production jobs, bill of materials (BOM), process plans, production batches and so on. Obviously, all these data categories are relevant determinants of material and energy flows and stocks. But it is also obvious that direct data mappings are not possible. It is not surprising then, that more ambitious, model-based mappers are required. In this contribution, the Petri net theory and in particular object systems are applied as an innovative conceptualization for mapping ERP data to EMIS data objects.

But what makes Petri nets so attractive in this context? – The concept of Petri nets has played a fundamental role in analyzing material and energy flows in organizations, value chains and product life cycles by providing a consistent accounting system on the level of material and energy flows as well as stocks [cf. Möller et al. 2001]. Hence, with Petri nets, there is a close relationship between the environmental accounting system and data collection. In this setting, Petri nets serve as an overall modeling language in EMIS.

4 Petri Nets as a Conceptualization for EMIS-Mappers

The origins of Petri net theory can be traced to Carl Adam Petris "Kommunikation mit Automaten" in 1962 [Petri 1962]. In distinguishing two types of nodes in a graph – transitions and places – it was possible to represent and analyze complex concurrent systems. In the first phase of Petri net development mathematical foundations and logical analysis of concurrent computer systems were in the main focus: liveliness, deadlocks, synchronization, etc. However, the analysis of dynamics of complex systems came from a different context – simulation of flexible manufacturing systems: throughput, bottle necks, conflicting requirements, etc. A new category of Petri nets, in development since the early 1980s, combine the different approaches. "Time can be incorporated into Petri nets by introducing a delay after a transition is enabled. This results in a timed transition that will have the ability to model tasks or activities. Such a Petri net is known as a Timed Petri net (TPN)" [Sawhney 1997; cf. Gile and DiCesare 1997] or, more precisely, a timed-transition Petri net. In real systems not all events require delays. To model these events in Petri nets, one would use classical, so-called immediate transitions. Petri net models which consist of exponentially timed transitions and immediate transitions are called generalized stochastic Petri nets [cf. Marsan et al. 1995, p 131]. Main attention in the analyses is directed on the dynamics of complex concurrent systems and quantitative performance measures, without disregarding previously researched mathematical foundations.

Focusing on the most basic aspect of MFA-oriented EMIS – material and energy flows and stocks – timed Petri nets cannot serve directly as an underlying concept for the mapper. How can the conceptual ideas behind workshop models, process plans and BOMs be reformulated in Petri net languages? Valk presented one possible way by introducing object systems. He defines object systems as a pair of two types of Petri nets: the system net and the object net. The object nets replace anonymous tokens in common Petri nets [cf. Valk 1996].

This new kind of Petri nets allows an explicit disjunction of processes and systems. This, in turn, facilitates the application of Petri nets in today's organizational structures. With respect to flexible manufacturing systems, the system nets can serve as a conceptual foundation of workshop models, and the object nets can be applied to represent process plans. In such a way, it is possible to bring together two different directions of abstraction: process orientation and systems orientation.

One remaining question to ask when considering object systems is simply "What's with material and energy flows?" – There is a quite simple answer. BOMs are assigned to processes in the process plan or object net respectively [cf. Busse 2006]. In this setting, the process plan is a dynamic object within a system net. It is possible in such a model to separate material and energy flow specifications from the location, where they possibly occur. Transition can be specified not only in the system net but also in the object nets, primarily on the basis of BOMs. Firing of transitions in the Petri net simulation is associated with an application of BOMs.

Data input of the object system based simulations include:

- A system net – this Petri net represents the workshop model, value chain or similar. The system net consists of several functional units (e.g. processors), paths and stocks. To facilitate future-oriented analyses, in particular timed Petri net approaches come into consideration.
- At least one object net – the object nets represent the corresponding process plans in the ERP system.

- Loads – the loads stand for production job, the parts to be produced (products or subassemblies), an arrival object, that specifies the release time of the load, the entry points etc. Each load is associated with an object net as its process plan.
- BOMs – BOMs are assigned to single transitions (processes) in the object net (process plan).

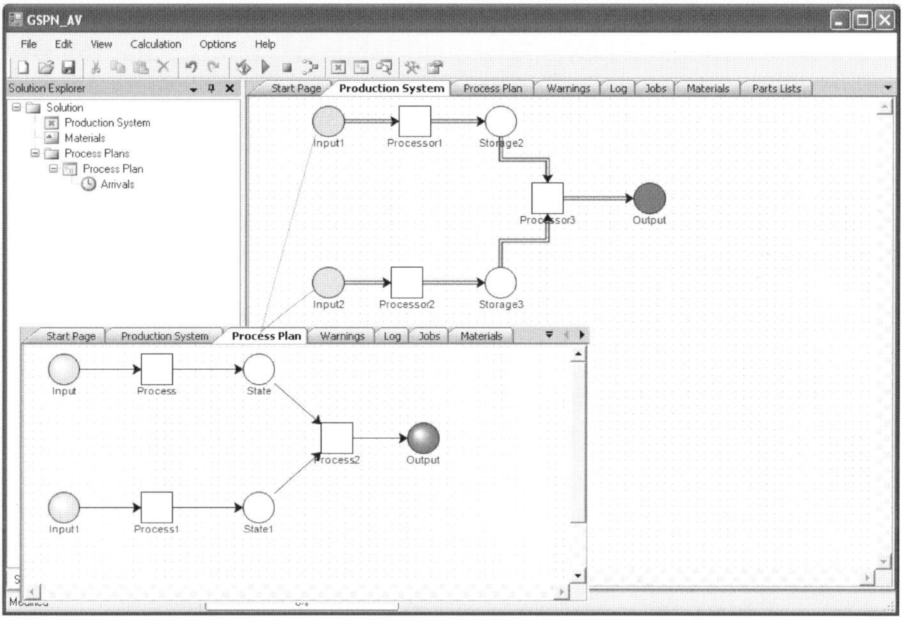

Fig. 1. Object system of a simple manufacturing system

Petri net simulation is applied to trace the paths of process plans in the workshop model. Firings of conjoint transitions within the simulation runs facilitate the application of BOMs and therefore the calculation of material and energy flows. The resulting aggregated data sets serve as a direct data input of MFA-based EMIS. After the simulation run, still missing data in the material flow network can be determined by using common calculation procedures in the EMIS, e.g. in case of pre chains and post chains [cf. Möller 2000, p 105].

5 Examples

Figure 1 depicts a simplified implementation of a Petri net-based mapper. The objective of the prototype is to demonstrate that it is possible to apply object systems to transform data between the two information systems. The first example (Fig. 1) provides a simple model of a production system in the for instance chemical industry. The system net and the object net have the same structure.

The figure shows that instances of the object net "Process Plan" are tokens of the places "Input1" and "Input2" in the system net. If the pair (Processor1, Process1) fires, the object net will change from "Input1" to "Storage2". At the same time the token is taken away from place "Input1" of the object net and a new token occurs on place "State". After firing of the transitions (Processor2, Process2) the pair (Processor3, Process3) is activated. And after firing of these transitions, the process plan reaches the output place in the system net. The production job is finished.

The processes in the process plan are associated with BOMs; the conjoint firing of transitions makes it possible to apply the BOMs. It is possible to calculate the job-related material and energy flows.

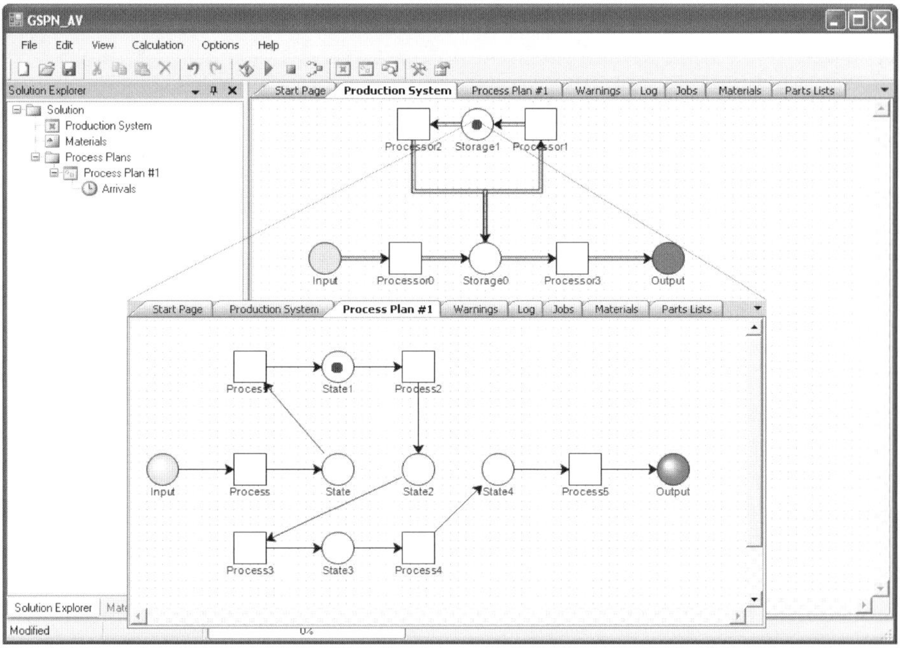

Fig. 2. An object net ("process plan #1") as a token in a system net ("production system")

Note that in constellation like in example 1 EMIS-mappers based on object systems are not required. Normally, it is possible to apply build-in calculation routines in EMIS.

The second example (Fig. 2) shows a more complex situation. The model consists of one system net ("Production System") and a different production job – an instance of the "Process Plan #1" – as a token within the system net. Enabled are the conjoint transitions "Processor2" in the system net and "Process2" in the object net. The conjoint firing of these two transitions results in a new state ("State2") in the object net, and the job becomes a token in place "Storage0". In common Petri nets this causes a so-called conflict in the system net (Processor 1 and Processor 3 are enabled), but in object systems there is a little bit different situation. The object net specifies that the job has to pass through processor 1 and processor 2 again. Further process plans can provide other ways through the system net. This makes the mapper highly flexible: the differentiation of systems and processes.

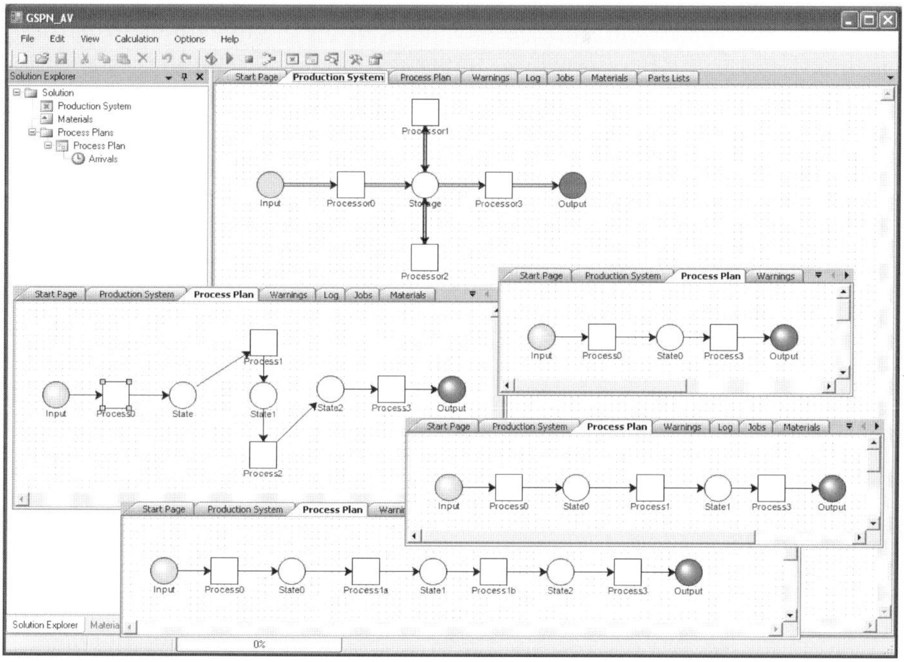

Fig. 3. One system net (manufacturing system) and four different object nets (process plans)

The third example demonstrates the potentials of the differentiation of systems and processes: Figure 3 shows one manufacturing system and four different process plans. The systems and the processes can result from different data sources. Graphical editors can help to model the manufacturing system (to some extent, the editor helps to customize the mapper), and ERP systems can serve as a data provider of process plans and the associated BOMs. In other words, the process plans and all related data objects are imported from the ERP system. It is not required to use a software tool to model the process plans.

Think about furniture industry. Such an EMIS mapper makes is possible to calculate the material and energy flows of chairs, tables and whatever the factory produces, provided that the company collects all relevant data.

6 Conclusion and Outlook

The question of how to conduct efficient material flow analyses in organizations is in essence dependent on the question of how to collect the required data efficiently. From there, it is obvious to think about interfaces to other computer-based information systems in an organization. In particular, interfaces to ERP systems are under development. But the issue of interfacing ERP systems and EMIS is often cut down to technical aspects.

Employing simulation techniques like timed Petri nets and object systems to deal with conceptual gaps between the different perspectives will enable MFA-oriented EMIS to address basic requirements to IT systems in dynamic business environments. The components apply Petri net theory and as well discrete event simulation. This makes it possible to utilize the underlying approaches as well in future-oriented EMIS and particularly in Product Lifecycle Management frameworks. Input parameters of the components come from product design and process design. In this arrangement, the components can be used to interface eco-design and life-cycle assessment.

References

Busse T (2006) Adaption des Erzeugnisstrukturkonzepts aus ERP und PPS-Systemen auf Stoffstromnetze. Diploma thesis, Department of Informatics, University of Hamburg (in German)

Fowler M (2003) Patterns of Enterprise Application Architecture. Addison-Wesley, Boston, San Francisco, New York

Jones A, Deshmukh, A (2005) Test Beds for Complex Systems. Communications of the ACM. vol 48, edn 5, pp 45 – 50

Gile MR, DiCesare F (1997) Towards Distributed Simulation of Complex Discrete Event Systems Represented by Colored Petri Nets: A Review. In: Manufacturing and Petri Nets – A workshop within the XVIII International Conference on Applications and Theory of Petri Nets. Toulouse

Lee J, Siau K, Hong S (2003) Enterprise Integration with ERP and EAI. Communications of the ACM. vol 46, edn 2, pp 54 – 60

Marsan MA et al. (1995) Modelling with Generalized Stochastic Petri Nets. John Wiley and Sons, Chichester, New York

Möller A (2000) Grundlagen stoffstrombasierter Betrieblicher Umweltinformationssysteme. Projekt-Verlag, Bochum (in German)

Möller A, Pager B, Rolf A, Wohlgemuth V (2001) Foundations and Applications of Computer Based Material Flow Networks for Environmental Management. In: Rautenstrauch C, Patig S (eds) Environmental Information Systems in Industry and Public Administration. Idea Group Publishing, Hershey London

Petri CA (1962) Kommunikation mit Automaten. Schriftenreihe des Instituts für Instrumentelle Mathematik, Bonn (in German)

Ramanathan J (2005) Fractal Architecture for the Adaptive Complex Enterprise. Communications of the ACM. vol 48, edn 5, pp 51 – 57

Sawhney A (1997) Petri Net based Simulation of Construction Schedules. In: Andratottir S et al. (eds) Proceedings of the 1997 Winter Simulation Conference. Atlanta, pp 1111 – 1118

Valk R (1996) On Processes of Object Petri Nets. Report No. 185/96 of the Department of Informatics. University of Hamburg

Environmental Management Systems and Information Management – Strategic-Systematical Integration of Green Value Added

Ian Molloy

IMBC GmbH, Berlin, e-mail: ian.molloy@imbc.de

Abstract: Environmental management is becoming a commodity. Nonetheless it is characterized by stringent operative processes and a missing strategic integration. The strategic integration is only achievable if necessary information is quickly available and is of high quality. Hence a paradigm shift away from operative processes pursued by employees to automation is necessary. Therefore it is necessary to include environmental management in the IT planning processes to foster an integrated information systems architecture. An integrated approach supports an enhanced legal compliance and the identification of cost optimization potentials.

1 Introduction

The finiteness of the natural environment is known at the latest since the reports to the club of Rome[1] .The knowledge resulting from it has led to a considerable support and spreading of the environmental management thought in politics and economy. In the meantime, a considerable range of theoretical and empiric knowledge is available concerning environmental management, which proves not only its relevance for eco-logical aspects, but also concretizes the exceptionally positive consequences on the economic parameters of an enterprise [Dyllick and Hamschmidt 2000]. Regardless of the economic relevance, enterprises have adopted environmental guidelines in their corporate visions, since the protection and sustainment of the environment is an important request for customers in industrial nations [Dyllick and Hamschmidt 2000]. The integration of environment-oriented guidelines in the corporate vision clarifies that enterprises view environmental management or the protection of the environment as an equivalent formal goal besides profit achievement (Fig. 1).

---

[1] http://www.clubofrome.org/

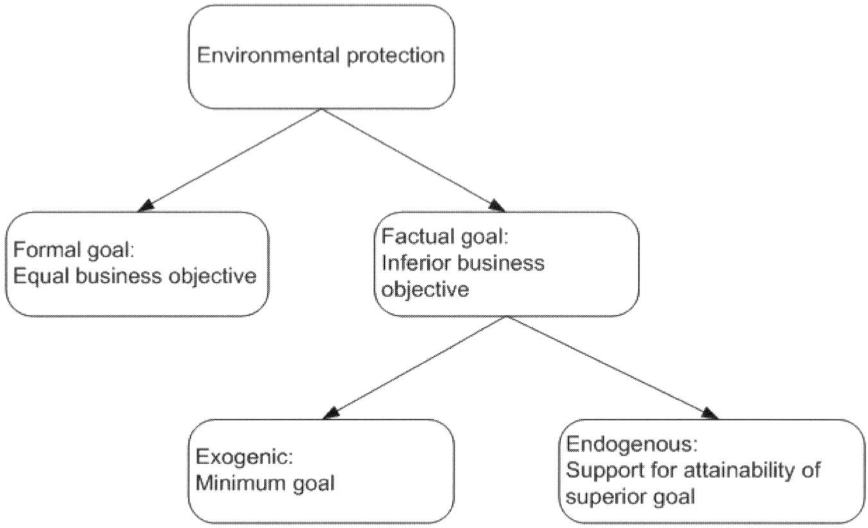

Fig. 1. Type of goals

This leads to the assumption, that enterprises stopped viewing environmental management as a burden and begin to realize the benefit of environmental management. Nevertheless, most of the enterprises realize environmental management primarily using operative processes with the help of very simple software products like MS Office. This approach leads to increasing difficulties for enterprises, because time and effort for the handling of environmental management will not decrease but increase. The acceptance of the status quo leads to the fact that the IT planning entirely or partially ignores environmental management, in spite of the demand for a holistic approach and the avoidance of isolated solutions [Heinrich and Lehner 2005; Rautenstrauch 1999; Beucker et al. 2003]. This is explained by the fact that the integration of environmental information systems (EIS) is of considerable complexity [Beucker et al. 2003]. Therefore there is a significant difference between the governmental demands to environmental management and its IT support and the actual organizational and informational state of the art [Rautenstrauch 1999]. The lacking integration of EIS in the enterprise IT-architecture leads to the point that the environmental management can not fulfill his cross section task.

To solve these problems different actions are necessary. On the one hand, suitable technical conditions must be implemented by the manufacturers of EIS to allow the integration in already available heterogeneous IT architectures. On the other hand it is suitable if the strategic IT planning accepts environmental management as a strategic designer and creates guidelines and a framework to accelerate the process of integration.

2 Strategical Information Management

Strategical information management (SIM) is a cross sectional task [Heinrich and Lehner 2005; Krcmar 2005]. Krcmar defines SIM as the management of informational economy, information systems, information and communication technology as well as spanning leadership tasks [Krcmar 2005]. According to Heinrich and Lehner SIM is called the management action in an enterprise concerning information and communication, consequently all leadership tasks which deal with information and communication in the enterprise [Heinrich and Lehner 2005; Pietsch et al. 2004].

Therefore, SIM is a management-centered approach, which plans and coordinates the allocation of the resource "information" according to the logistical principle [Heinrich and Lehner 2005; Rautenstrauch 1999].

Goal of the logistical principle is the allocation of the right information, in the right time, in the right amount, in the right place and in necessary quality [Krcmar 2005].

Certain conditions must exist or have to be created in an enterprise to guarantee an adequate supply of the resource information.

The different in enterprises mostly organically grown architectures have to be structured and methodically analyzed. Because without a consistent overview about available

- data,
- information systems,
- infrastructure components and
- organizational structures

no goal-oriented actions can be planned for the improvement of informational supply.

However practical experience shows, that architectural transparency is rather the exception. Thus 50% of the accomplished as-is analyses' of IT structures uncover redundancy, which leads to considerable waste of financial resources [Niemann 2005]. The removal of these redundancies leads to relevant savings considering personal (10-20%) and license (5-15%) expenditure [Niemann 2005]. This consideration however ignores the costs, accumulating along the life-cycle of information systems, software products and infrastructure components. Hence it is to be assumed, that the actual saving of financial resources excesses the actual numbers.

However, the downstream identification of redundancies or saving potential is not to be preferred, because relevant costs have already incurred for the implementation, use and maintenance of the suitable components.

Instead, a planning instrument should be used to realize value generating IT projects on the basis of transparent enterprise architectures. This planning instrument should be installed on a strategic level to en-sure a holistic approach of IT development in an enterprise. It is necessary to use a methodical approach to control the IT of an enterprise objectively. It is an inevitable necessity for useful and transparent results. Therefore a strategic IT planning process, outlined by Heinrich und Lehner, can be used. This process consists of the following sub processes [Heinrich and Lehner 2005]:

- Strategic Situation Analysis
- Strategic Target Planning
- Strategy Development
- Strategic Action Planning

3 Environmental Management Systems

For the implementation of environmental management as well as analysis and promotion of resource efficiency in enterprises, there are various instruments and methods. Environmental management systems like EMAS or ISO 14001 have reached an extensive degree of popularity in the meantime. These systems are integrated into the organizational and operational structure. A goal of these standardized management systems is to achieve an improvement of the environmental impact of an enterprise by substantially improving the administrational processes regarding environmental aspects. The improvement of the environmental impacts as a continuing process is optional in the ISO 14001 and a must-have in the EMAS to ensure (re-)certification.

The motivation for using an environmental management system varies widely [Dyllick and Hamschmidt 2005]. Figure 2 provides a selected overview of the different motivations for the use of the ISO 14001.

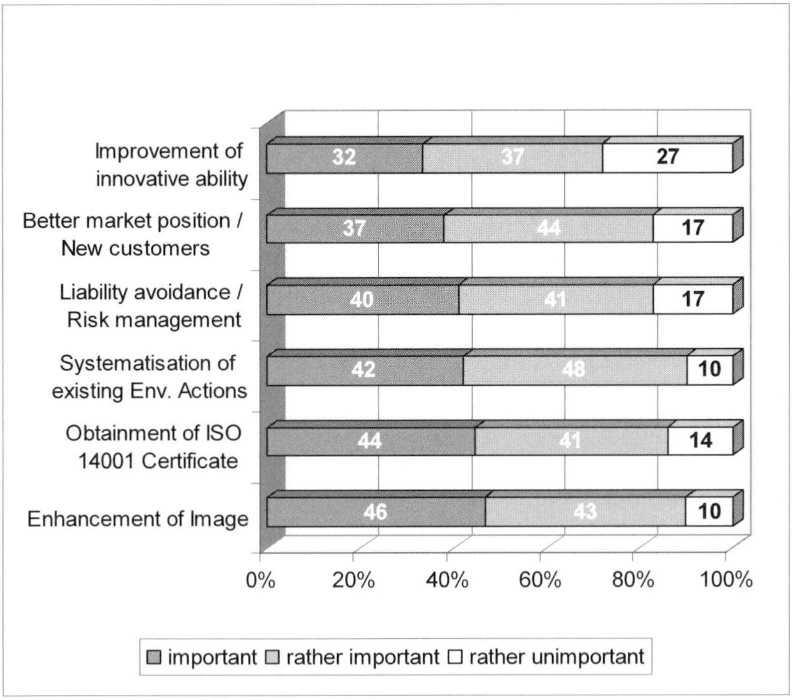

Fig. 2. Motivation for the implementation of an ISO 14001 environmental management system [Dyllick and Hamschmidt 2000]

Figure 2 illustrates that the motives for the implementation of an environ-mental management system are manifold and not always cost-oriented. But according to an ASU/UNI study, environmental management systems can be viewed as very interesting investment, as these systems tend to amortize within 1.6 years [ASU/UNI 1997]. Regardless of these realizable cost effects, an integration of environmental management on the strategic level is missing [Schwegler and Schmidt 2003]. It can be assumed that an integration of environmental management systems in the strategic processes of enterprises would lead to substantial and sustainable improvements of their economical and ecological progresses. However, until today there is no relevant empirical data available.

4 Strategic IT planning for the Environmental Management of a selected Enterprise

In order to guarantee the information supply within an enterprise, a holistic approach is necessary for IT planning. If environmentally relevant information is to be available during important decision processes, an integration of environmental management should be conducted with the operational, administrative and strategic information systems. If an enterprise refrains from such integration, a substantial fixation of employees and uncertain decisions has to be accepted. The character of environmental management as a cross sectional function leads to the necessity to regard environmental management as integral aspect within an enterprise and to use systems for redundancy-free processing of environmental relevant information. This is relevant insofar, as global changes have to be anticipated and integrated in an increasing pace.

Therefore it is necessary to expand the systematical approach of strategic IT planning on environmental management and to make integrated information systems available. This ensures that an enterprise can act proactively and leads to an improvement of legal compliance and security and the identification of cost reduction potentials. In the following, a strategic IT planning process is shortly described and discussed.

5 Strategic Situation Analysis

The situations analysis in the enterprise was limited to the range of the environmental management, in order to limit the expenditure for the collection and analysis of data. The first step of the situation analysis is an analysis of the IT in the environmental management. The result of the analysis was that only MS Office products are used. Despite the absence of information systems in the environmental management it could be determined, that the used ERP system stores substantial quantities of environmentally relevant data. The data however is in no consistent form available, which leads to a substantial binding of employees.

The next step of the situations analysis is the determination of critical success factors for the environmental management. On the basis of interviews and discussions, different critical success factors have to be determined. Two of these are exemplary represented here. The first critical success factor concerns Compliance Management. The enterprise had substantial problems to follow legislation requirements. It is to be expected that future developments within the legal range will lead to a further degradation of the competitive ability. Data quality can be viewed as a second critical success factor, since enterprises miss significant information regarding environmentally relevant processes. That led to the problem, that concrete statements towards the current condition of different media could not be made.

The environmental analysis, as last step of the situation analysis, is focused on the determination of technologies and information systems, which should be used in the problem afflicted areas. The analysis revealed that relevant and usable information systems are already available on the market. A successful implementation would not only reduce the operational load of the employees, but would also improve strategic scope for design by providing data of higher quality.

6 Strategic Target Planning

The phase of strategic goal planning has to be used to transfer the results of the strategic situation analysis into goals for the administrative and operational level. As starting point for the development of the strategic goals the critical success factors are selected. An example of a strategic goal is the collection and generation of environmental key performance indicators (see Table 1).

Table 1. Exemplary strategic goal

Target content	Generation of key performance indicators
Target measure	Aggregation and structuring of necessary data
Degree of target attainment	Supply of key performance indicators with an actuality of at least five working days
Time-frame of target attainment	Number of months for target attainment after availability of information systems

Apart from the strategic goal outlined here, another seven goals for the environmental IT were generated. Thus a rough target structure for the integration of information systems in the environmental management is available.

7 Strategy Development

The analysis of the competitive position of the examined enterprise showed, that the enterprise followed the strategy of cost leadership [Porter 1998]. Outgoing from this strategy, a strategic direction for the entire IT can be formulated, which can be partitioned into sub-strategies for single highly relevant objects of the information infrastructure. For the area of environmental management this partitioning was abdicated, as the information systems implemented for environmental management should completely support the strategic direction of the enterprise. Thus the following factors must be supported by information systems in the environmental management:

- Enhancement of transparency in the enterprise
- Identification of cost advantages and disadvantages

Beyond that, a framework for the realization of the different IT development projects was made, which serves as guideline for the implementation of the different information systems. In the course of this elaboration the following aspects were concretized among other aspects:

- Acquisition and usage of standard software
- Development and usage of custom software
- Integration of information systems
- Criteria for sequencing of projects in the project portfolio

This manual has to guarantee that a consistent validation basis is available for the evaluation of current and future projects.

8 Strategic Action Planning

Strategic action planning must lead to the generation of a concrete project portfolio, which must be implemented within a certain period of time. The first step for the generation of the portfolio is the identification of strategic gaps. Therefore an analysis of the as-is and the to-be situation is necessary, which should be advantageously supported by key performance indicators. As there was no relevant data available in the area of the environmental management, the goals, identified during the target planning were used as fallback solution. These goals were identified as strategic gaps, which are to be closed by information systems.

The second step of the strategic action planning aims at the generation of project ideas for the elimination of the strategic gaps. This includes a check, which information systems are available in the enterprise and what kinds of data include these information systems. For this purpose, an overview of the different architectures of an enterprise is necessary in order to avoid redundancies or, if redundancies can not be avoided for technical reasons, to minimize these. During this step, the existing information systems were analyzed. The focus was set on the following data which is relevant for EIS [Lang and Jürgens 2003]:

- Material master data
- Structure oriented data
- Process data
- Energetic and material flow data
- Organizational data

The analysis found out, that numerous environmentally relevant data is available throughout the enterprise. But that data was distributed across six different information systems.

The third step results in a selection of information system categories, which can be used to close the strategic gaps. The results of the environmental analysis and the therein identified information systems serve as a starting point. The following information system categories were identified:

- Material and energy controlling systems
- Audit support systems
- Document Management Systems
- Environmental Controlling Systems
- Reporting systems
- Material Flow Analysis Systems

On the basis of the identified categories, commercial information systems can be identified or the necessity for individual development can be uncovered, in order to close the gaps in the environmental management.

9 Conclusion

Environmental management is an accepted object of business organization. Nevertheless, environmental management is highly operational aligned. This might even result from the heavy operational alignment of the used environmental management systems. The operational alignment leads, among other things, to the missing consideration during the performed IT planning. Up to now there are only few large enterprises in Germany, which recognize the necessity for a holistic approach considering information systems in environmental management. Thereby it becomes obvious, that the complexity in the area of environmental management is widely underestimated. In a rather not irrelevant number of individual development projects such an underestimation leads to serious problems. For this reason custom software was excluded during the process of project idea generation for the closing of the strategic gaps. This approach offers different advantages. On the one hand, the available information systems are already proven. The vendors, on the other hand, already showed some experience in the industry of the viewed enterprise.

The concrete arrangement of the information systems portfolio led to a system portfolio, which can close the identified strategic gaps and re-store scope for design for future strategic decisions, as the identified information systems will lead to an enhanced informational quality for decision makers in the enterprise. Above all the securing of legislation compliance and cost optimization potentials identified with the help of controlling instruments will guarantee the basis for economic operation of the enterprise.

References

ASU/UNI Unternehmerinstitut e.V. (1997) Öko-Audit in der mittelständischen Praxis

Beucker S, Jürgens G, Lang C, Rey U (2003) Betriebliche Umweltinformationssysteme (BUIS) im Umweltcontrolling: http://www.bum.iao.fraunhofer.de/intus/ergebnisse/BUIS-Umfrage.pdf, access: 2007-01-05

Computerwoche (2006) Palms Smartphone "Treo 650" ist für Europa zu giftig http://www.computerwoche.de/knowledge_center/wireless/578331/index.html, access: 2007-01-07

Dyllick T, Hamschmidt J (2000) Wirksamkeit und Leistung von Umweltmanagementsystemen, vdf Hochschulverlag AG

Heinrich L, Lehner F (2005) Informationsmanagement. edn 8. Oldenbourg

Pietsch T, Martiny L, Klotz M (2004) Strategisches Informationsmanagement, vol 4. Erich Schmidt Verlag

Krcmar H (2005): Informationsmanagement, edn 4. Springer, Berlin

Lang C, Jürgens G (2003) Aufgaben Betrieblicher Umweltinformationssysteme In: Kramer M, Brauweiler J, Helling K Internationales Umweltmanagement Band II: Umweltmanagementinstrumente und –systeme. Gabler, Wiesbaden

Niemann KD (2005) Von der Unternehmensarchitektur zur IT-Governance. Vieweg Verlag, Wiesbaden

Porter M (1998) Competetive Strategy, Free Press

Rautenstrauch C (1999) Betriebliche Umweltinformationssysteme. Springer

Schwegler R, Schmidt M (2003) Lücken im Umweltmanagement: Forschungsansatz für ein rationales Umweltmanagement auf der Basis der St. Galler Management-Lehre. In: Schmidt M, Schwegler R (eds) Umweltschutz und strategisches Handeln. Gabler, Wiesbaden

Integration of MRP II and Material Flow Management Systems

Claus Rautenstrauch

Otto-von-Guericke-Universität Magdeburg, Faculty of Computer Science, Department of Business Informatics, PO Box 4120, 39016 Magdeburg, Germany

Abstract: Nowadays, Environmental Management Information Systems are not integrated into business information systems, in particular ERP-Systems. The transition to a completed integration is neither done nor on the horizon. On the first view, this is somehow astonishing. A closer look at the basic data of Manufacturing Resource Planning (MRP II) and material flow management, for example, shows that there are quite clear similarities in data structures. In principle, bills of materials, operation plans and production orders contain exactly the data, which are useful for the design of material flow networks. Material transformations are represented through operation plans and input/output equations can be derived from bills of materials. And indeed it is possible in practice to generate material flow networks from MRP II data. Therefore, in this paper current approaches for the integration of material flow management systems and manufacturing resource planning (MRP II) and control will be presented.

1 Current State of Integration

During outgoing 20th century the environmental informatics community was rather enthusiastic. The idea of sustainable development appeared as an upcoming hype and was announced to be the driving paradigm of future business behaviour [Forum info 2000, 1998]. Consequently it was expected that the integration of functionality of Environmental Management Information Systems (EMIS) into Business Information Systems becomes somehow obvious. For example, Bruckmann and Weinert suggested in [1998] to add material flows as a fifth view to the ARIS architecture (ARIS = Architecture of integrated InformationSystems – [Scheer 1998]) and Rautenstrauch predicted that EMIS are only a transition phenomenon on the way to a sustainable society [Rautenstrauch 1999, p 162].

The new millennium "surprised" the computer science community first with the Y2K problem and then with introduction of the Euro. Thereafter, the E-Business hype moved EMIS out of the view. In the field of Business Information Systems applications from different business units were integrated to complex ERP systems (ERP = Enterprise Resource Planning). Since mid of the nineties ERP systems became a global de facto standard in large enterprises [Rautenstrauch and Schulze 2003, p 313ff]. Although it should be a "natural" development that such systems include EMIS as integrated component, this is only partially the case. For example, one can find a component called Environment, Health and Safety (EH&S) in my SAP® PLM (Product Lifecycle Management), but this offers mainly functions for document management in the field of waste management following national laws and regulations and for the handling of hazardous goods [SAP 2005]. Especially material flow management systems are still systems of their own which are loosely coupled with ERP systems though more or less standardized interfaces.

This cannot be accepted as a satisfying solution, because material flow models need input from accounting, manufacturing resource planning (MRP II), which today is a standard component of ERP systems, and EH&S [Schmidt 1995]. Furthermore, measures for the improvement of material flows cause changes to production processes. Therefore also a feedback towards MRP II and EH&S appears.

2 Integration of MRP II and Material Flow Management

2.1 Data Structures

Data integration is determined to be the precondition for the integration of business functions [Becker 1990, p 9], because functions can only interact properly if they access a joint data base. Therefore, as a first step data structures of MRP II and material flow management have to be analyzed.

Basic data structures of MRP II systems are bills of materials (BoMs), process plans and production orders [Kurbel 2005]. BoMs can be depicted as so called gozinto graphs, where nodes represent parts and edges represent "goes into"-relationships. Edges are marked quantity coefficients, which show how many lower parts are demanded to produce one unit of the upper part. Process plans describe for each assembly or product, which operations in which sequence demanding which resources have to be executed. Alternatives regarding different operations or sequences are taken into account. For manufacturing execution production orders have to be generated from process plans, which can be understood as templates to be enhanced with concrete quantities and scheduling data (cf. Fig. 1). Since the availability of all assemblies and materials is the precondition for the production of a part, the production orders are embedded into a production order network (cf. Fig. 2). The dependencies between productions orders are derived from the dependencies of upper and lower parts in the BoM.

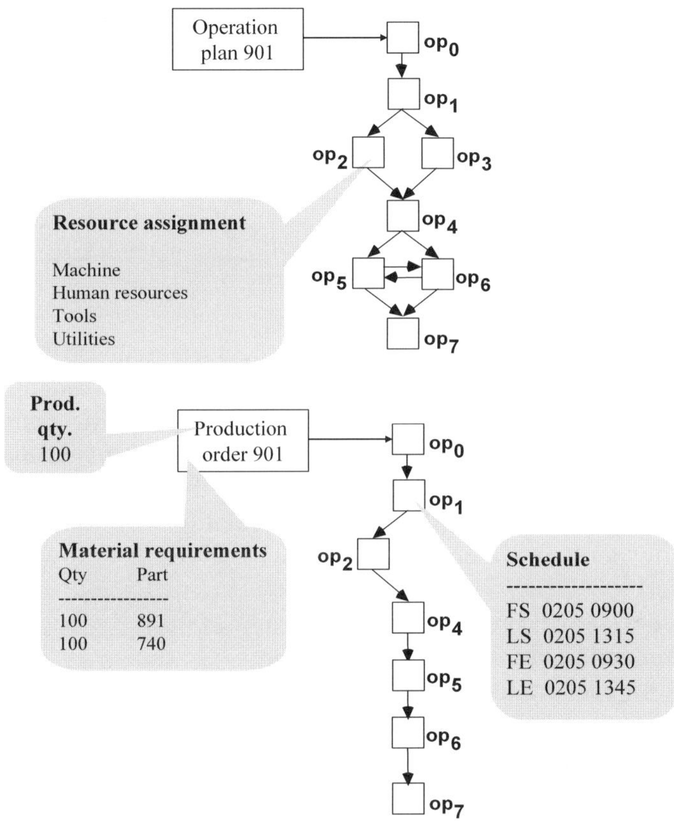

Fig. 1. Process plan and production order

A common data structure for the modeling of material flows are material flow networks [Möller 1994]. Material flow networks are higher Petri nets, where places represent buffers for material and energy and transitions represent material or energy transformations. A transition is specified through an input/output equation, summarizing the inputs (of material and energy) on the left side and the outputs on the right (cf. Fig. 3).

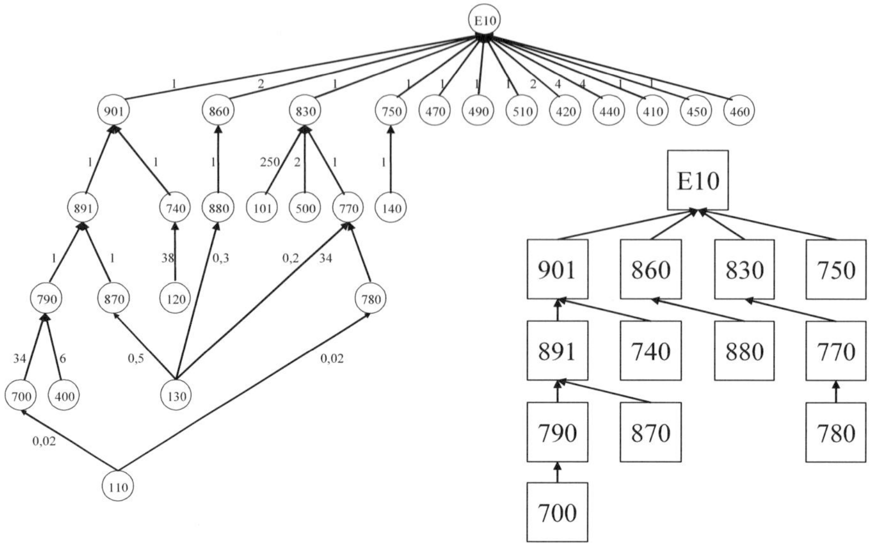

Fig. 2. Gozinto graph and production order network

Now, production orders can be understood as specifications of input/output equations. Material requirements determine the input and required quantity the output. The algorithm for transformation of production order to material flow networks in bottom-up manner is given as follows:

1. level = max (level)
2. Repeat until level = 1 {final product}
   a. Choose one production order from lowest level.
   b. Create a place for each input material.
   c. Create a transition for production order.
   d. For each input material i
      i. Input = quantity(i)
   e. Output = Output (quantity(production order))
   f. Create a place for output material
   g. Connect inputs and outputs as given in production order structure
3. level = level – 1

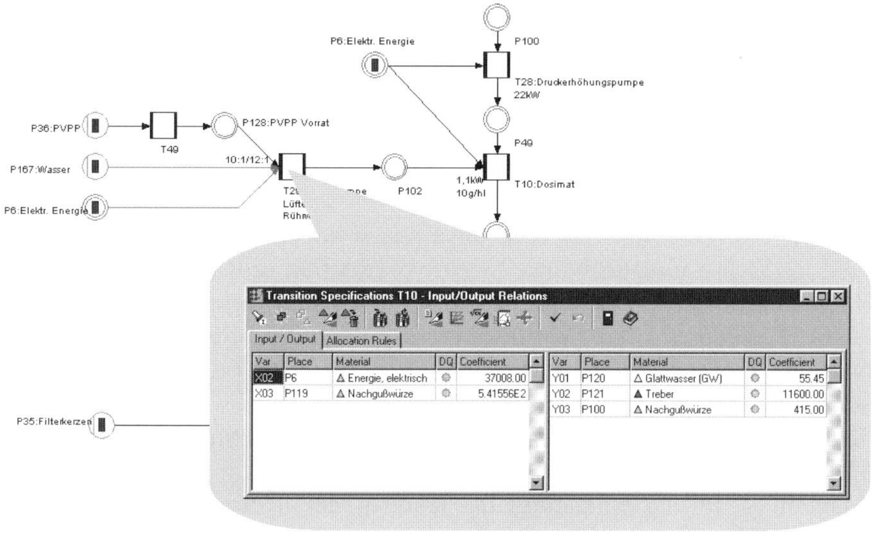

Fig. 3. Material flow network with input/output equation

Figure 4 shows an example for a production order network and the generated material flow network. Place P120 represents the material buffer for part 120, which is a purchased material. T2 represents a production order for manufacturing of part 740, which thereafter, together with part 891 goes into part 901. This becomes an assembly for the final product E10. The method is reversible that means production orders can also be reconstructed from material flow networks.

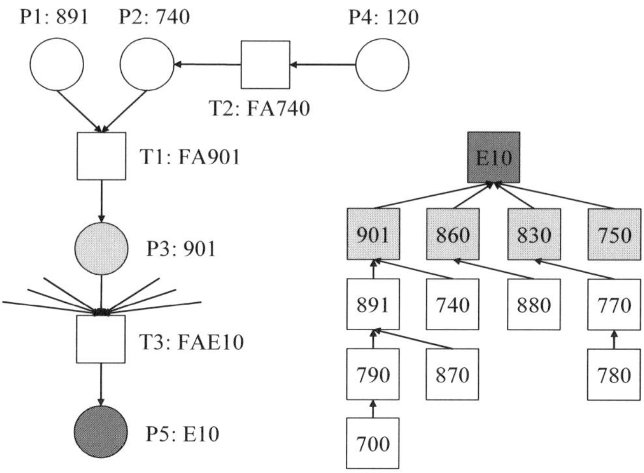

Fig. 4. Generation of material flow network from production order network

With this it looks as if the problem of integration of data structures from MRP II and material flow management is solved, but a closer look at the problem shows that some detail are not taken into account.

2.2 Problem Fields and Approaches

Eco balances are related to a balance space and period. The method depicted in the last chapter works for one production order only. Since it has to be taken into consideration that during the balancing period several lots of a final product will be produced, all production orders for such parts have to be aggregated. Therefore, for material inputs and outputs, all quantities have to be summarized over the period.

The second problem is the level of detail of information. In MRP II only such parts are taken into account which are relevant for material and time management. In eco balancing, the level of detail depends from kind of balance. While enterprise eco balances are on the level of material classes, product eco balances go down to the level of materials and substances. This level is out of the view in MRP II. Furthermore, in MRP II normally energy consumption, auxiliary materials, operating supply items and transportations and unwanted outputs are not observed. Consumption of energy, auxiliary materials and operating supply items can be derived from master data of resources assigned to production orders, if such data are maintained in MRP II data base. For a precise allocation of quantities of energy, auxiliary materials and operating supply items, variants of process executions and resource assignments have to be considered. If for the eco balance data of higher granularity are required then offered by MRP II systems, the information on substance level have to be inquired from the vendors and maintained by hand.

Since data in MRP II systems are not maintained in a way that they provide all information regarding material and energy transformation without gap, one can act on the assumption that the derived material flow networks are defective. Defects can be classified as missing data, wrong or improper data or missing elements or subnets [Marx Gómez et al. 2004]. If the whole network contains sufficient information, such defects can be repaired through data balancing computation methods based on neural nets and case based reasoning [Arco et al. 2005]. The basic idea is to derive the structure of the multi-layer structure of the neural net from a Petri net applying the method of [Yu et al. 1995]. Then the learning results of the neural nets will be adjusted by assignment of nearby cases.

The last problem to be discussed here is data redundancy in MRP II and material flow management systems. Data redundancies are problematic, because thy cause inconsistencies when same data in different systems have a different grade of up-to-dateness. This problem can be solved with a joint data model for MRP II and material flow management. Since MRP II systems today are embedded into ERP systems, their data structures are closely interconnected with data structures of other business areas. Therefore it is useless to isolate them from ERP systems and to move them to data structures of material flow management. Instead of that data structures and perhaps also data base from MRP II have to be enhanced in a way that they can cover data for material flow management, too:

- *Inclusion of substance level in BoMs*: Since materials purchased from the market form the lowest level of BoMs in MRP II, a level of substances has to be added on bottom of the data structure. This also means that information about substances has to be included in the master parts file and marked as substances. This necessary, because they are not to be taken into account in methods of BoM explosion for material requirements planning. Furthermore it might be necessary to insert a reference to hazardous material management in the area of EH&S.
- *Introduction of unwanted outputs to BoMs*: Co-production and recycling planning provide an idea how to deal with unwanted outputs though deployment of negative quantity coefficients [Rautenstrauch 1997, pp 162ff]. An unwanted output goes into an upper part with a negative quantity. This method allows that all conventional algorithms for BoM explosion can be applied without changes and unwanted output can be identified immediately by its negative quantity. Unwanted outputs have to be added to the master parts file and have to be linked to waste management data in EH&S. Figure 5 depicts the enhanced BoM for assembly 740. It shows that if one unit of part 740 is produced, also 20 units of unwanted material (waste) 089 appear.

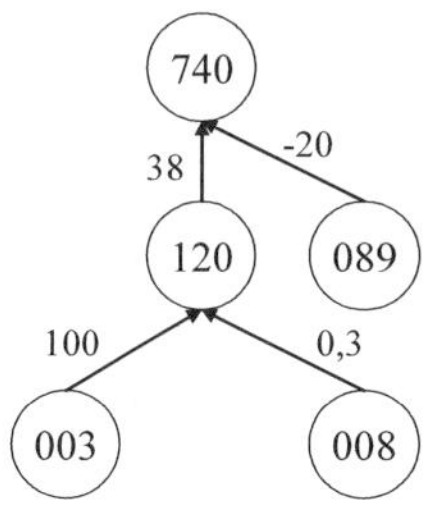

Fig. 5. Enhanced BoM

- *Transportation processes* have to be taken into consideration in operation plans.

- *Consumptions of energy, auxiliaries and operating supply items* per period have to be added for each resource.

All this items can be implemented without changing the data model of ERP systems. The special requirements can be met through customizing measures. Therefore, realistic and suitable preconditions for a smooth integration of material flow management are created.

3 Summary and Outlook

This paper describes the theoretical foundations and solutions for the integration of ERP and material flow management systems as an example for EMIS. It shows that such integration possible without substantial changes of ERP data structures and that material flow networks can be derived from MRP II data (as part of ERP data), if the data base is extended to fulfill some special demands.

However, material flow management is not the one and only area of EMIS to be integrated. Other examples can also be found on higher management level like environmental data warehouses or sustainability reporting. Therefore, two prospects on further research can be identified: First, the theoretical models have to be approved through their implementation and application in software prototypes. Second, other areas of EMIS have to be identified to follow up integration with ERP environments.

References

Arco L, Bonet I, Garcia L, Marx Gómez J, Rautenstrauch C (2005) A New Approach to Solve Data Defects in Material Flow Networks. In: Leal W, Marx Gómez J, Rautenstrauch C (eds) ITEE 2005: Information Technologies in Environmental Engineering (2nd International ICSC symposium Magdeburg, Germany, September 25-27, 2005). Shaker, Aachen, pp 515-524

Arndt H-K (1997) Betriebliche Umweltinformationssysteme. Wiesbaden

Becker J (1990) Das CIM-Integrationsmodell. Berlin

Bruckmann C, Weinert M (1998) Die fünfte Sicht – ARIS, EPK und Stoffstromnetze. In: Rolf A Einführung in die Organisations- und Wirtschaftsinformatik. Berlin, pp 288-289

Clausen J, Loew T, Klaffke K (2002) Nachhaltigkeitsberichterstattung. Berlin

Forum info 2000 (ed) (1998) Herausforderungen 2025 – Auf dem Weg in eine weltweite nachhaltige Informationsgesellschaft. Ulm

Kurbel K (2005) Produktionsplanung und –steuerung im Enterprise Resource Planning und Supply Chain Management, vol 6. München, Wien

Marx Gómez J, Prötzsch S, Rautenstrauch C (2004) Data Defects in Material Flow Networks – Classification and Approaches. In: Cybernetics and Systems: An International Journal, vol 35, edn 5-6, pp 549-558

Marx Gómez J, Rautenstrauch C (eds) (2001) Von der Ökobilanzierung bis zur automatisierten Umweltberichterstattung mit Stoffstrommanagementsystemen – eine Fallstudie. Aachen

Möller A (1994) Stoffstromnetze. In: Hilty LM, Jaeschke A, Page B, Schwabl A (eds) Informatik für den Umweltschutz, 8. Symposium, Hamburg. Band II: Anwendungen für Unternehmen und Ausbildung. Marburg, pp 223-230

Mertens P (2005) Integrierte Informationsverarbeitung 1. vol 15. Wiesbaden

Rautenstrauch C (1999) Betriebliche Umweltinformationssysteme. Berlin

Rautenstrauch C (1997) Fachkonzept für ein integriertes Produktions-, Recyclingplanungs- und -steuerungssystem (PRPS-System). Berlin, New York

Rautenstrauch C, Schulze T (2003) Informatik für Wirtschaftswissenschaftler und Wirtschaftsinformatiker. Berlin

SAP AG (2005) Integrated Industrial Hygiene and Safety with SAP® Environment, Health & Safety. Walldorf

Scheer A-W (1998) ARIS – Vom Geschäftsprozeß zum Anwendungssystem. vol 3, Berlin

Schmidt M (1995) Stoffstromanalysen und Ökobilanzen im Dienste des Umweltschutzes. In: Schmidt M, Schorb A (eds) Stoffstromanalysen in Ökobilanzen und Öko-Audits. Berlin, pp 3-13

Spatial Inventory of Greenhouse Gases on Regional Level

Rostylav Bun, Khrystyna Hamal, Mykola Gusti, Andriy Bun, Olga Savchyn

Lviv Polytechnic National University, Lviv

Abstract: The principals of spatial inventory of greenhouse gases on regional level which are based on IPCC methodologies and digital maps are considered. Mathematical models for spatial inventory using elementary plots are proposed. A geoinformation approach for creation of thematic maps of greenhouse gas emissions in MapInfo format is developed. The mechanism is illustrated on example of Energy sector of Lviv Region of Ukraine.
Keywords: greenhouse gas, spatial inventory, digital map, geoinformation technology, energy sector, mathematical model

1 Introduction

Spatial inventory of greenhouse gases (GHGs) on regional level is important for a number of reasons, particularly, it gives local government valuable information on location and magnitude of GHG sources helps to identify cost effective ways of GHG emission reduction and involve local people into GHG reduction measures. An interest in regional GHG inventories exists especially in large countries [see RCCA 2004; LCS 2002; AQPMDPPD 2006]. Besides, regional detailed GHG inventories allow improving national inventory if used as a part of multilevel model [Bun et al. 2005].

Development of the spatial regional inventory requires solving a few specific problems: lack of information on spatial distribution of GHG sources and absence of data necessary for direct estimation of GHG emissions in every point, in general.

In the paper we further develop the geoinformation technology for GHG inventory, which is described in [Bun et al. 2005], and apply it for spatial GHG inventory in the Lviv Region of Ukraine.

2 Geoinformation technology for GHG inventory on regional level

Developed geoinformation technology for GHG inventory on regional level links digital maps, geographical information system and IPCC methods [IPCC 1997a] programmed in Microsoft Excel [IPCC 1997b]. Realization of the technology for spatial inventory consists in carrying out the inventory for each elementary plot (region territory is divided into elementary plots) one-by-one following the IPCC methods. The digital maps are used for forming input database and visualization and interpreting the inventory results. Summing up the results of the elementary plot inventories yields regional inventory.

Structurally the geoinformation technology is composed of two main software modules: Mod1Inventaryz and Mod2Maps.

Functions of *Mod1Inventaryz* module are recording input data into IPCC Excel tables and running the inventory for selected economy sector. Information from respective layers of digital map, statistic data and results of researches are used for forming georeferenced database corresponding to sections of the IPCC methods (energy industries, manufacturing industries and construction etc.). Data for each elementary plot from the georeferenced database is recorded into respective Excel software [IPCC 1997b] cells using OLE technology and MapBasic queries for running the inventory. Also the module picks up the inventory results from the Excel soft-ware and records them into output georeferenced database corresponding to emissions of GHGs (carbon dioxide, methane, etc.) in selected economy sector.

Functions of *Mod2Maps* are forming queries to tables with the inventory results for the elementary plots, which are created by *Mod1Inventaryz*, forming new layers and reflecting them in the digital map of the region. The module also uses topographic information from the digital map of the region as input data. For visualization an IDW-interpolation (averages with weights, which are inverse proportional to distance from the object) can be used.

A digital map layer corresponding to total GHG emissions in CO_2-equivqlent is formed. Also a layer for each economic sector used in the IPCC methods is created. All layers contain data on specific GHG emissions (relating to an area unit). Such an approach provides good transparency of the inventory process and possibility of easy visualization and analysis of the inventory results.

3 Models of spatial inventory and results for the energy sector of the Lviv Region

3.1 A model and GHG inventory schemes for the energy sector

We describe development and application of the geoinformation technology on the example of the Lviv Region in Ukraine. The Lviv Region is located in the west of Ukraine. The region has the largest population and the most developed industry in western part of the country. The population and the industry are distributed non-uniformly in the region. The region consists of 20 administrative districts and 7 cities of direct subordination to the region, for which separate records in the statistical yearbooks exist.

A general model for estimation of GHG emissions in the energy sector can be presented in the following form

$$\Delta Y_n = \sum_{m=1}^{M} a_{nm} \Delta x_{nm} , \quad n = 1,\ldots,N , \tag{1}$$

where a_{nm} s the coefficient for calculation of emissions for the m-th economic activity of the energy sector in the n-th elementary plot, N is the total number of the elementary plots, Δx_{nm} are the data on economic activity for the n -th elementary plot and the m-th economic activity, ΔY_n is the emissions from the n-th elementary plot.

According to the IPCC methods [IPCC 1997a] the energy sector is divided in five subsectors ($M = 5$): energy industries; manufacturing industries and construction; transport; other sectors (including commercial/institutional, residential and agriculture/forestry/fishing). Georeferenced data on each type of the economic activity in each elementary plot are input data for the model. We consider only the most significant fuel types (according to results of economic activity of the region in 2004): oil and oil products (48.0% of standard fuel), natural gas (40.2%) and coal (8.5%). Other fuel types (e.g. fuelwood) comprise only 3.3% of the total consumed fuel and are not accounted in the model. Also a number of coefficients (emission factors, calorific values for the fuels, fraction of non-oxidized carbon etc.), which are taken from special reference books or IPCC Workbook [IPCC 1997a], are used in the model.

Inventory schemes, which are used for numerical experiments in the subsectors energy industries, manufacturing industries and construction, transport, other sectors are described in detail in [Bun et al. 2004]. Inventory of CO_2 emissions and emissions of other GHGs (CH_4, N_2O, NO_x, CO and NMVOC) were done separately.

3.2 Creation of elementary plots for spatial analysis

In order to perform spatial GHG inventory first of all one must create the elementary plots, in which the GHG emissions are estimated. The elementary plots are objects of size lxl km, which are bordered with the administrative district boundaries. That is each administrative district is divided in a number of elementary plots of size lxl km. Some of the elementary plots are complete squares of area l^2 km^2, but some of the elementary plots are cut by the administrative boundaries of the districts. Besides, elementary plots for the towns and cities of direct administrative subordination to the region should be formed. Total number of the elementary plots (N) is defined using the formula

$$N = \sum_{r=1}^{R} N_r + N_M, \tag{2}$$

where r is the number of administrative district, $r = 1,..,R$; R is total amount of the districts; N_r is the amount of elementary plots in the r-th district; N_M is the amount of elementary plots corresponding to the towns and cities of direct administrative subordination to the region.

Let v_{ri} is the i-th elementary plot (an elementary object of the digital map) of the r-th district and d_{ri} is its area. The area of the r-th administrative district can be defined with the formula

$$D_r = \sum_{i=1}^{N_r} d_{ri}, \tag{3}$$

and the set of all elementary plots of the Lviv Region V can be presented as a union of sets of the elementary plots of the districts and the elementary plots of the towns and cities of direct administrative subordination to the region v_m

$$V = \left[\bigcup_{r=1}^{R} \left\{ v_{ri}, i = \overline{1, N_r} \right\} \right] \bigcup \left\{ v_m, m = \overline{1, N_M} \right\}. \tag{4}$$

Taking into account written above we divide the territory of the Lviv Region into elementary plots of size 10x10km using MapInfo Professional 6.5.

3.3 Formation of secondary map with spatial distribution of population in the region

When doing modeling and spatial analysis of a number of processes of green-house gas emissions it was assumed that the intensity of anthropogenic activity is proportional to the population of a certain territory. Thus, a parameter 'population' is calculated for each member of the set V (a set of elementary objects on a digital map of the region). This is done using data of the last population census (2001) and the following algorithm.

1. For elementary plots v_m, $m = \overline{1, N_M}$, representing towns and cities of direct subordination to the region, population z_m is taken from the results of the population census.
2. For each administrative district $r = \overline{1, R}$ for elementary plots $k = \overline{1, N_{Mr}}$, representing urban areas (N_{Mr} is the number of such inhabited localities in r-th region), the population $z_{rk}^{(m)}$ is taken from the results of the population census.
3. The population z_{ri} of rural areas, i.e. the rest of the elementary plots of each region v_{ri}, $i = \overline{1, N_r}$, is calculated as follows

$$z_{ri} = \frac{d_{ri}\left(Z_r - \sum_{k=1}^{N_{Mr}} z_{rk}^{(m)} \right)}{\left(D_r - \sum_{k=1}^{N_{Mr}} d_{rk} \right)}, \tag{5}$$

where Z_r is the population of r-th region according to the population census. In other words, in accordance with proposed algorithm, rural population is 'distributed' uniformly in the territory of an administrative district. At the same time total population of the region Z can be expressed as

$$Z = \sum_{k=1}^{N_M} z_M + \sum_{r=1}^{R} \left(\sum_{k=1}^{N_{Mr}} z_{rk}^{(m)} + \sum_{k=1}^{N_r} z_{rk} \right), \qquad (6)$$

and it must correspond to official results of the population census. The algorithms mentioned above are realized using MapBasic queries for each elementary plot.

3.4 Spatial analysis of emissions resulting from fuel processing and electricity generation

Use of fossil fuels in the energy sector reflects expenditure of fuel resources for converting fuels into other kinds of energy or other types of fuels, i.e. fuel expenditure for production of coal and peat briquettes, different types of gas, oil expenditure and expenditure of other components for production of oil products, fuel expenditure for electricity generation in power plants, use of fuels in boiler-houses [IPCC 1997a].

1. Formation of geographically distributed input data for analysis of greenhouse gas emissions due to fuel processing and electricity generation. Inventory results. According to the proposed approach, for each fuel type $j=\overline{1,K_p}$ (where K_p is the number of considered fuel types) number ($N_{kp,j}$) of the most influential enterprises in the region for this activity type, which consume $p_{kp,jk}$ ($k=\overline{1,N_{kp,j}}$) tone of j-th fuel type, is found. Values of consumed fuel in these enterprises are assigned to the elementary plots in the digital map in which these enterprises are located.

The rest of consumed fuel (in practice these amounts are not considerable) is distributed among the elementary plots proportionally to the population. The following dependence is used

$$p_{jri} = \frac{z_{ri}}{Z} \left(P_j - \sum_{k=1}^{N_{kp,j}} p_{kp,jk} \right), \quad j=\overline{1,K_p}, \; r=\overline{1,R}, \; i=\overline{1,N_R}, \qquad (7)$$

where p_{jri} are the amounts of fuel of j-th type consumed in i-th plot of r-th administrative district, P_j are the regional totals for consumed amounts of fuel of j-th type.

Activity data for fuel processing in 2004 (the latest available statistical data) in administrative districts are extracted from statistical yearbooks.

Moreover, activity data from the most influential enterprises, considering greenhouse gas emissions due to fuel processing and electricity generation, are used.

According to the algorithm mentioned above, some data are distributed in the elementary plots containing refineries, public power plants, power plants of enterprises, heat and power plants and other enterprises which use fuel for converting it into other types of energy and fuel products. The rest of fuel (it is fuel used in boiler-houses, power plants of enterprises and heat and power plants) is distributed in the elementary plots, which correspond to urban areas, proportionally to the population. Respective queries are formed using MapBasic.

Then greenhouse gas inventory according to the IPCC methods [IPCC 1997a] is carried out for each elementary plot in the set V using the activity data derived on the previous step.

2. *Formation of geographically distributed input data for analysis of methane emissions due to coal mining and activities related to oil and gas.* Data on coal mined in the region are extracted from statistical yearbooks. Also data on total amount of crude oil processed in refineries are used. These data are distributed in the elementary plots containing the refineries (Drohobych, Lviv); plots in which coal (Chervonohrad, Hirnyk, Sosnivka), natural gas (Krakovets, Rudky, Dashava and others), oil (Drohobych District) are produced. Then methane emissions are estimated according to the IPCC methods [IPCC 1997a] using the derived activity data and a digital map is formed.

3.5 Total emissions from the energy sector in the Lviv Region

Approaches similar to the described above were used for estimation of emissions resulting from fuel combustion in the residential and transport subsectors. The activity data are distributed in the region territory proportionally to the population.

For the subsector 'manufacturing industries and construction' the data on fuel consumption are distributed among the elementary plots proportionally to gross product of the administrative districts in a certain year. Gross products of towns and cities of direct subordination to the region are taken from statistical yearbooks. Such parameter for other areas is assumed to be proportional to the population.

Figure 1 shows final map of total greenhouse gas emissions in CO_2 equivalent for the Lviv Region according to activity data in 2004. The results of spatial inventory are summed up for administrative districts for a better clarity.

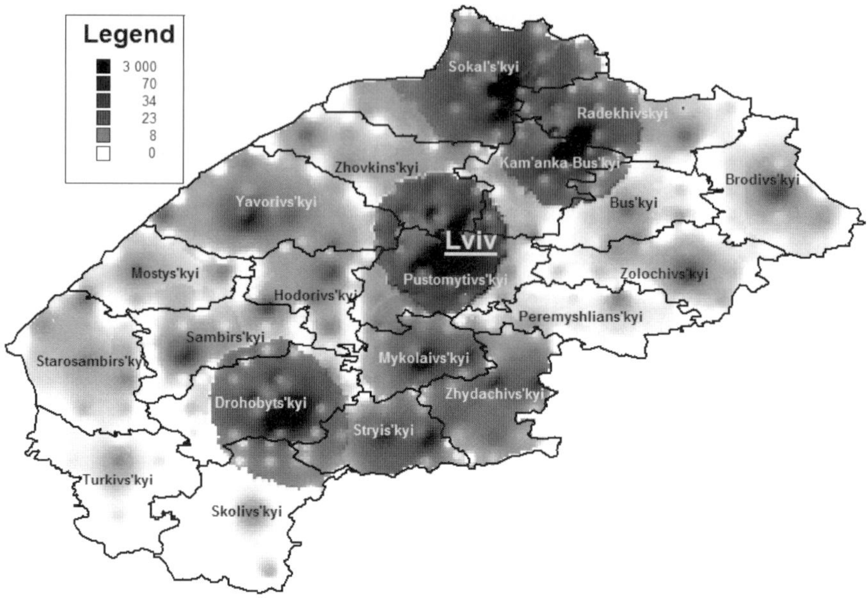

Fig. 1. Total greenhouse gas emissions in CO_2 equivalent, Gg/100km^2

4 Concluding remarks

A geoinformation technology for spatial GHG inventory on regional level is developed and applied in Lviv Region in Ukraine. The geoinformation technology uses IPCC methods and software for GHG estimation in every elementary plot that makes the inventory compatible with national GHG inventory and UNFCC requirements. Utilization of the developed geoinformation technology improves quality of GHG inventory, particularly, because of using location-specific emission factors and additional direct and indirect information on GHG source locations.

Numerical experiments on spatial GHG inventory show considerable variability in GHG emissions in Lviv Region. According to the results of activity in 2004, Lviv gives the largest amount of emissions – 3,770 Gg of GHG in CO_2 equivalent, including 3,208 Gg of CO_2 itself. Kamyanka-Buzka District which is not very well developed industrially is the second largest GHG source. It emitted 1,966 Gg of GHGs (basically it is because of location of a big power plant in this region). Agglomeration of towns Drohobych-Boryslav-Truskavets, which emitted 1,427 Gg of GHG in CO_2 equivalent, is the third.

References

Air Quality Policy and Management Division Policy and Planning Department (AQPMDPPD) (2006) Developing a Greenhouse Gas Reduction Strategy for Greater Vancouver

Bun R, Gusti M, Dachuk V (eds) (2004) Information Technologies for Greenhouse Gas Inventories and Prognosis of the Carbon Budget of Ukraine. Lviv, pp 376

Bun R, Gusti M, Oleksiv B (2005) Greenhouse gas inventory on the base of multilevel model. In: Second International ICSC Symposium on Information Technologies in Environmental Engineering (ITEE 2005), Magdeburg. Shaker, Aachen, pp 294-308

Low Carbon Spaces (LCS) (2002) Area-Based Carbon Emission Reduction: A Scoping Study. In: A report to the Sustainable Development Commission prepared by the Tyndall Centre for Climate Change Research. Tyndall Centre for Climate Change Research, UK

Report Card on Climate Change Action (RCCCA) (2004) First Assessment of the Region's Progress Towards Meeting the Goals of the New England Governors/ Eastern Canadian Premiers Climate Change Action Plan of 2001. http://www.masspirg.org/reports/2004GlobalWarmingReportCard.pdf, access: 2007-02-15

Intergovernmental Panel on Climate Change (IPPC) (1997a) Revised 1996 IPCC Guidelines for National Greenhouse Gas Inventories (1997a) Reporting Instructions. The Workbook, Reference Manual, vol 1-3, IPCC/OECD/IEA, Working Group I (WG I)

Intergovernmental Panel on Climate Change (IPPC) (1997b) IPCC Greenhouse Gas Inventory Software for the Workbook

Weblogs, Legitimacy and Organizations

Markus Glötzel, Martin Müller

Oldenburg University, Chair for Production and the Environment, Germany, e-mail: {markus.gloetzel, martin.mueller}@uni-oldenburg.de

1 Introduction

The last decades have seen a movement of production to less economically developed countries. These countries often don't meet Western social and environmental standards. Non-governmental organizations (NGOs) have challenged poor conditions such as child labor, discrimination or environmental pollution caused by suppliers. Purchasers are criticized in the public realm, which leads to damage to their reputation [Hansen and Schrader 2005; Gilbert 2003]. Nike, Dole Foods and GM have experienced such campaigns [Lawrence 2002]. Focal companies with direct contact to the end customer and big labels are most vulnerable to this strategy. A loss of reputation may weaken their ability to compete and result in decreasing sales volume [Hansen and Schrader 2005].

In the past, coverage in the mass media such as newspapers and television was the main vehicle for stakeholder-driven campaigns. Today, however, several companies are confronted with a relatively new phenomenon: weblogs or personal online journals. Bloggers (blog authors) can document nuisances, spread news through their own network, set agendas, and in general be a problem to companies or other organizations within the public perception. On the one hand, the inclusion of the blogosphere can amplify the effect of NGO campaigns. On the other hand, as will be shown, companies and organizations in general haven't yet developed corresponding behavior patterns for an adequate reaction to bloggers.

An attack on the reputation of a name-brand company means questioning its legitimacy. Nevertheless, the term is a complex and controversial one [Dingwerth 2004]. An agreement exists in the way that we differentiate between a normative and an empirical based understanding of legitimacy. Empirical legitimacy relies on stakeholder acceptance or aims at the effective societal acceptance of rules and structures, while normative legitimacy is based on the compliance of procedures to normative concepts such as formal and transparent ways of decision making or rule setting [Dingwerth 2004]. This context distinguishes between input, throughput and output legitimacy [Scharpf 1999; Zürn 1998]. Input legitimacy relies on the acceptance by all parties involved. Throughput legitimacy means fair procedures of decision making, while output legitimacy is maintained through decisions that meet the respective demands, e.g. they contribute to the common welfare.

With this contribution, we will build on both of the concepts. Two selective case studies will show how fast the legitimacy of an organization can be questioned. As a result, maintaining legitimacy gets harder and harder. The inclusion of parts of the normative concept of legitimacy can help organizations to develop adjusted strategies.

2 Weblogs

A weblog is a website organized like a personal journal by time and date. New entries show up on top of the start page. The start page usually shows a certain amount of entries (usually between 10 and 15), or it shows all entries that have been written within the last few weeks. In the first case, the oldest entry on the start page disappears when a new one is published.

Typically, older entries aren't erased. They are always available in the archives. In most cases, weblogs are driven by special web content management systems (WCMS).

An entry may consist of a single link and an annotation, or can just be a chunk of plain text. Any other web compatible content can be embedded along with it, such as a movie clip shown through a flash player.

2.1 History

Berners-Lee, creator of many important standards for the web, maintained a list of links to new websites and servers in 1992. Andreesen, developer of the first web browser mosaic, had a similar list. John Barger probably coined the term 'weblog' in 1997 [Blood 2000].[1]

Until 1999, bloggers formed a very small community. In July 1999 the first commercial blog hosting service became available (http://pitas.com),

---

[1] http://web.archive.org/web/19991011003943/
http://www.robotwisdom.com/ shows the blog robotwisdom as archived on October 11th, 1999

followed by blogger.com, which today is a big player in the market, and was acquired by Google Inc. in 2003. Since then, basic blogging functionality has been available free of charge, and the number of blogs has been increasing steadily.

The growth of the blogosphere was accompanied by the development of an appropriate format. While commented link lists are still popular, some blogs *"eschewed this focus on the web-at-large in favor of a sort of short-form journal. These blogs, often updated several times a day, were instead a record of the blogger's thoughts: something noticed on the way to work, notes about the weekend, a quick reflection on some subject or another"* [Blood 2000].

After the September 11, 2001 attacks on the World Trade Center, blogs began to be used by an increasingly broader public. In 2003, 4% of the Internet users in the USA went to blogs for information and opinions [Rainie et al. 2003].

2.2 Anatomy

Basic features

Weblog software enables users to publish new entries through a simple web interface (backend). There may or may not be other ways to put new content on the blog, e.g. through a desktop client which interacts through XML-RPC[2] with the server.

New entries get a unique and stable URI called *permalink* (short for "permanent link"). This implies the generation of a new entry not only on the start page of a blog but also in the archives. Archives are often organized by months (one page per month). Most weblog CMS even generate an archive page for each entry. Stable URIs are very important for cross linking between entries in several discrete weblogs.

More features

Blog visitors may want to leave comments on certain entries, and blog software usually supports this. With the rise of commercial comment spamming, comments are often moderated by the owner of the blog, or the blog software engages automatic or semi-automatic counter measures.[3]

---

[2] The XML-RPC specification allows remote procedure calls via http; see http://www.xmlrpc.com/

[3] For example, word filters or Turing tests where commentators are asked to type a number displayed in an image.

Another common feature is the *trackback* functionality. If person A writes an entry on her blog about another entry on person B's blog and links it to B's entry, she can telecommand B's blog to link back to her posting. This backlink appears within or near the comments on B's entry. Trackbacks enable distributed discussions where one can follow the discussion by following the trackback links on the initial entry. Automatic backlinking is, of course, another invitation to commercial spammers.

Weblogs often provide a *newsfeed*, an XML version of the blog with the latest entries lacking any presentational information. While there are different competing standards in use, the data format is always quite strict and allows easy content syndication in web and desktop applications.

The ability to '*ping*' other services is perhaps the most important advanced feature of weblog software. The default behavior of common blog software is to send a small data package (XML-RPC request) to some central services like technorati.com or weblogs.com whenever a new entry is published. This request contains the name and URI of the requesting site (or its newsfeed). The remote services then check the calling weblog for new entries and index them. Those services can be used like search engines or directories but are much more up-to-date because they don't have to crawl billions of pages, as would be the case with e.g. Google. As we will discuss later, these services play an important role for the diffusion of topics through the blogosphere.

2.3 The Blogosphere today

Metrics

In 2006 8% (12 Million) of online Americans maintained a blog, while 39% read weblogs [Lenhart and Fox 2006]. These figures are based on a telephone survey. Technorati.com tracked 35.3 million weblogs worldwide in April 2006 based on the ping-technique described above [Sifry 2006]. However, technorati.com has a vested interest in keeping such figures high – the growth rate of the blogosphere has a direct impact on their company value.

The German speaking part of the blogosphere is quite small. According to an estimate by [Schmidt 2006], there were 40,000 blogs in Germany at the beginning of 2006. Schmidt also discusses the problems of different estimation methods quite thoroughly.

The quantitative aspects of the distribution of attention among the bloggers are much more important than the exact number of active blogs. If one measures attention by the number of incoming links, it seems to be distributed following the power law where the poor stay poor and the rich always get richer.[4] In November 2006 the German blog "Industrial Technology and Witchcraft" had 3,100 incoming links while the vast majority of weblogs (the 'long tail') only have a few or even no incoming links. Highly networked blogs are important hubs for the propagation of themes among the blogosphere – nevertheless new themes can be invented in the poorer long tail and then shoot to fame, as we will show later.

Development of a format

Aside from the continuous growth of the blogosphere, the use of blogs diverges steadily. While many blogs are read by only a few readers who know the author personally, and a few famous, often tech-centric blogs cannot be ignored by any serious blogger (like the German "Schockwellenreiter"[5]), other and perhaps more sophisticated forms emerge. For example, scientists use blogs as a kind of slip box and also as a way to make contacts with other scientists in the same field. Companies experiment with blogs as an instrument for PR or simply to get feedback from their customers. There are also organizations that try to leverage blogs as an internal knowledge management and networking tool. Importantly, blogs focused on certain areas with niche content cultivate niche markets, which are not profitable for conventional media.

In the last two years, the inclusion of multimedia content in blogs has become very popular. The fact that most mobile phones are also digital cameras with Internet access enables people to blog pictures when they are on the go.

Some bloggers try to make money with their blogs, typically by paid advertisement, although only a few can make a living with their blogs. These bloggers are either so-called "A-list" bloggers with high traffic, or they belong to the aforementioned niche markets.

3 Weblogs and the legitimacy of organizations

3.1 Coining search terms

Until the end of 2006, the Google search for the phrase "miserable failure" returned the biography of George W. Bush as its first result. This is caused by a collective linking strategy of hundreds of bloggers who linked the

[4] See [Barabasi 2002] for more information on power law distributions

[5] http://www.schockwellenreiter.de, access: 2007-02-13

page with the words "miserable failure". According to [Tatum 2005] this practice ('Google Bombing') can be seen as a case of online collective action, and under certain circumstances the participating people are even considered to form a social movement.

The practice of Google Bombing clearly indicates a shift in the distribution of media power. However, such coordinated action among a larger group of bloggers is quite unlikely and does not happen very frequently.

What *is* more frequently happening is the uncoordinated evolvement of certain themes within the blogosphere. A famous example is the Kryptonite Lock controversy. In 2004, someone found out how to open a Kryptonite bicycle lock with a ballpoint pen. Kryptonite ignored the problem for a time, but then started a product exchange program. If you search for Kryptonite today, Google delivers a pointer to the weblog www.engadget.com[6] as the fifth result. The third and fourth result point to wikipedia.org, where the ball point pen incident is discussed as well. The brand label is now damaged, and even today, more than two years later, the incident has probably caused fewer sales because many consumers tend to search the Internet before they decide to buy something expensive. This incident shows the development of what we call a *blog swarm*. We will now analyze two similar episodes that happened in the German part of the blogosphere.

3.2 Blog Swarms

Jamba

Jamba is a company that sells ring tones and games for mobile phones, had a sales volume of nearly € 75 million in 2004, mostly on the German domestic market, and currently belongs to Rupert Murdoch's 'News Corp' and Verisign.

On December 12th, 2004, the German blogger and journalist Johnny Häusler published an article about Jamba's business practices on his weblog 'Spreeblick'. Häusler imitated the phrasing of a well-known German infotainment telecast for children, 'Sendung mit der Maus'. Here, he stated that Jamba sold to children who just wanted a single ring tone, but were then charged for an entire subscription to the service. The story about the rip-off with the small print began to diffuse through the German blogosphere. Meanwhile, employees of Jamba attempted to help their company by commenting on the issue on Häusler's blog – without revealing

[6] http://www.engadget.com/2004/09/14/kryptonite-evolution-2000-u-lock-hacked-by-a-bic-pen/, access: 2007-02-13

themselves. The swindle was detected soon after, since most of the comments originated from an IP address belonging to Jamba. This behavior annoyed several bloggers so much that the issue was pushed even more – several A-list bloggers commented and linked to Häusler's entry.

After the holidays, Spiegel Online, a news portal run by the well-known German magazine *der Spiegel* picked up the story. A link to the original article induced more traffic than Häusler's web server could process. On January 10th, the story made it to the printed *Spiegel* magazine. This was followed by features on TV (SAT.1, one of Germany's most important non-public TV programs) and on several radio stations [Fischer 2006].

The coverage in the conventional media again fuels the debate in the blogosphere, according to Fischer. The distributed debate leads to excessive linking and back linking between several blogs. As an outcome, a Google search for 'Jamba' lists Häusler's article on Jamba at the 5th position even today (January 2007), while on the 3rd position there is a Wikipedia article covering the issue of ring tone subscriptions. And, the issue wasn't stopped by the language barrier: In April 2005 *The Mail on Sunday* printed an article titled 'Save our children from this ringtone rip-off'.[7]

Transparency International

Transparency International (TI) is an UN-accredited NGO whose goal is to fight corruption. TI is organized into local chapters which work very independently of one another. The following case also occurred in Germany and has nothing or little to do with other chapters of the NGO.

On January, 13th Moni blogged about a friend of hers who used to work for TI and was fired within her trial period after failed salary negotiations.[8] More than two months later, on March 25th Transparency International's legal adviser asked her to delete the entry within two days. In the legal advisor's opinion, Moni's article was biased and violated the personal rights of TI. He threatened to take legal action. Moni put the email on her blog.[9]

Until then, Mona's blog had had about 100-150 page impressions per day. But now, other bloggers with more readers picked up the story. One day later, 'Transparency International' became the top search phrase on Technorati.com. So the directory itself became an important hub, perpetu-

7 http://www.mailonsunday.co.uk/pages/live/articles/news/news.html?in_article_id=345213&in_page_id=1770, access: 2007-02-13

8 According to Moni's entry her friend would have agreed to stay with the old arrangement.

9 http://wasweissich.twoday.net/20060324/, access: 2007-02-13

ating the diffusion of the issue. Moni took the relevant article down on the due date. Some other bloggers reacted immediately by retrieving copies of the article from Google's cache and documenting it on their own blogs.

On March 27th, Transparency's lawyer asked Moni to delete the blog entry where she documented his email, arguing that she would otherwise infringe his copyright. One hour later a blogging lawyer picked up the issue,[10] finding Transparency's reasoning to be quite obscure. A few hours later, he plead Moni's case. His first letter to Transparency International was published on the 'lawblog'[11].

One day later, the story diffused to the mainstream media. Several newspapers reported it in their online editions. Even tagesschau.de, the website of the newscast 'Tagesschau' picked it up. However, the issue was not covered in printed media.

On March 29th, according to Moni's lawyer, Transparency International made a 'satisfactory statement', giving Moni legal security not to be sued. Meanwhile, bloggers x-rayed the financial situation of the NGO and the past of TI's legal adviser (who was a show host in the former East Germany).

Today the search phrase 'Transparency International' on google.de still shows the impact of this incident. The first five hits are related to official TI websites and to wikipedia.org. Result #6 is a blog, followed by the aforementioned article at tagesschau.de. #8 points to a search at technorati.com for the same phrase. #9 links to an article at focus.de, the online edition of another important German magazine. The article is titled 'Der selbstgemachte PR-GAU' (The Self-Made Public Relations Disaster).[12]

3.3 Analysis of different strategies

From a marketing point of view, the questioned organization has four strategic options when an issue arises. *First*, one could ignore the affair completely. *Second*, one could confront the group which broaches the issue. This approach involves actively stating one's own view. The *third* option is to adapt, either in a symbolic way or by eliminating the cause of the criticism. *Finally*, there is a strategy of innovation. Innovation means to be one step ahead, with an open-minded attitude towards already-identified

---

[10] http://www.lawblog.de/index.php/archives/2006/03/27/die-gutmenschen/, access: 2007-02-13

[11] http://www.lawblog.de/index.php/archives/2006/03/27/gestortes-verhaltnis/, access: 2007-02-13

[12] http://www.focus.de/digital/netguide/transparency-international_nid_26808.html, access: 2007-02-13

stakeholders. This option also requires an active assessment of processes respective to social or environmental aspects [Meffert and Kirchgeorg 1998].

In both cases the attacked organizations chose to confront the bloggers. In the first case it is not clear if the commentators on Häusler's weblog were instructed to do something or if they acted on their own account. In any case, their behavior seriously annoyed many bloggers. In the second account, the confrontation was the driving force behind the following events. Moni's blog was quite poor (as indicated by the lack of incoming links), and her entry about her friend who was fired legally but not treated fairly had already been posted for months. Almost nobody would have noticed if Transparency International had opted for the "ignore" strategy, but the threat to sue her made the story interesting to some A-list bloggers. In the first case study, Jamba later switched to the "ignore" strategy. Transparency International at least decided not to pursue Moni with legal action, which allowed the situation in the blogosphere to cool off. Yet Moni's friend wasn't hired again. So it can be stated that TI partially chose the strategy of adaptation.

Both cases somehow recount the narrative of David and Goliath. With this in mind, the first lesson for organizations dealing with bloggers could be 'Don't let biblical stories be revisited where the little guy gets all the sympathy, and don't sue them!' In fact, there are a few cases in the German blogosphere where companies sued bloggers and then were surprised by the rising countervailing power. However, reliable figures do not exist about the cases where bloggers were under legal pressure and nothing happened.

While ignoring an issue sometimes might be the best option, it also means taking the risk to be left behind with a loss of reputation and therefore a loss of legitimacy in the view of important stakeholders.

At first sight, the strategy of innovation seems to be the most promising approach for organizations that don't want to be a target of coordinated campaigns or rather uncoordinated blog swarms. However, following this strategy would require an active monitoring of the blogosphere as well as other places on the Internet and maintenance of an ongoing dialogue with identified stakeholders. Along with this, the challenge probably is to configure an organization in such a manner that enables it to deal actively with criticism and therefore sustain its ability to compete in the future.

3.4 Deductions from the normative concept of legitimacy

Expression of opinion in blogs or even coordinated campaigns by NGOs concerning unethical practices of a company, be it non-sustainable exploitation of the environment or bad conditions of employment, forces a company to legitimize itself. According to [Dingwerth 2004], normative concepts of legitimacy have to

- be inclusive by integrating all involved parties and stakeholders
- ensure the quality of discourse
- enable democratic control through a high level of transparency and responsibility.

Campaigns or uncoordinated blog swarms target the empirical legitimacy of organizations (especially in the case of name-brand companies). Independent of the chosen strategy, the standards of normative concepts of legitimacy can be helpful here. First, the dialogue should include anyone who demands something or who is affected by the expressions in the weblogs in question. Then, the quality of the discourse can be enhanced if no party tries to run down other people involved. Besides that, adhering to legal demands is sometimes not a sufficient argument because even legal practices can seem unfair or unreasonable when issues are discussed in the context of sustainable development. Lastly, the company in question can try to maintain transparency by explaining their approach in the past and how they are going to deal with the respective issue.

4 Conclusions

4.1 Research questions

To begin with, it can be valuable to better understand how issues diffuse from weblogs to the conventional media. Also, one could analyze the development and also the downturn of blog swarms – both in a quantitative approach (modeling the diffusion of themes) and in a qualitative approach (based on content analysis: Why are some stories more likely to emerge than others?).

The concept of empirical and normative legitimacy and their role in an increasingly connected society should be refined and developed. The partial shift of power away from the conventional mass media has to be considered here.

4.2 Practical implications

Organizations in general and especially name-brand companies have to live with the fact that there is now a disruption of media power through weblogs. Moreover, the probability grows of aroused public interest if there are unsolved sustainability issues, be they of an environmental or social nature.

The increasing availability of Internet access in economically less-developed countries can put these companies at additional risk. Pictures taken in a sweatshop and immediately published in a blog could trigger a landslide in the blogosphere. Companies are therefore required to have an idea about how to act when an issue arises. Besides this, a proactive innovative approach should be seriously considered.

NGOs can try to leverage the power of the blogosphere to set an agenda. The redistribution of media power leaves them less dependent on the conventional mass media. However, they should not try to exploit bloggers, and should act carefully and fair-mindedly. A first step could be to join the conversation in the blogosphere. Some NGOs already try to do this[13], but we have yet to see cross-media campaigns that include weblogs.

Scientists who research in the context of a sustainable development could start blogging as well. In their blogs they could explain actual findings in an easy-to-understand language. NGOs could use these articles as a resource to develop and fuel campaigns. Journalists who pick up a theme would than have accessible and independent sources for their research and inquiries.

References

Barabasi A-L (2002) Linked: The new Science of Networks, Cambridge

Blood R (2000) Weblogs: a History and Perspective. http://www.rebeccablood.net/essays/weblog_history.html, access: 2007-02-13

Dingwerth K (2004) Effektivität und Legitimität globaler Politiknetzwerke. In: Brühl T, Feldt H, Hamm B, Hummel H, Martens J (eds) Unternehmen in der Weltpolitik: Politiknetzwerke, Unternehmensregeln und die Zukunft des Multilateralismus. Bonn

Fischer T (2006) Die Entwicklung von Weblog Issues am Beispiel des Klingeltonanbieters Jamba. In: Picot A, Fischer T (eds) Weblogs professionell. Grundlagen, Konzepte und Praxis im unternehmerischen Umfeld. Heidelberg, pp 247-252

Gilbert DU (2003) Instutionalisierung von Unternehmensethik in internationalen Unternehmen. Ein Ansatz zur Erweiterung der Zertifizierungsinitiative Social Accountability 800. In: Zeitschrift für Betriebswirtschaft, vol 73, edn 1, pp 25-48

[13] e.g. http://weblog.greenpeace.org/, access: 2007-02-13

Hansen U, Schrader U (2005) Corporate Social Responsibility als aktuelles Thema der Betriebswirtschaftslehre. In: Die Betriebswirtschaft (DBW), vol 65, edn 4, pp 373-395

Lawrence A-T (2002) The Drivers of Stakeholder Engagement. Reflections on the case of Royal Dutch/Shell. In: Andriof J, Waddock S, Husted B, Sutherland Rahman S (eds) Unfolding Stakeholder Thinking. Theory, Responsibility and Engagement. Sheffield

Lenhart A, Fox S(2006) Bloggers. A Portrait of the internet's new storytellers. http://www.pewinternet.org/pdfs/PIP%20Bloggers%20Report%20July%2019%202006.pdf, access: 2007-02-13

Meffert H, Kirchgeorg M (1998) Marktorientiertes Umweltmanagement: Konzeption – Strategie – Implementierung mit Praxisfällen. Stuttgart

Rainie L, Fox S, Fallows D (2003) The Internet and the Iraq War - how online americans have used the internet to learn war news, understand events, and promote their views. http://www.pewinternet.org/pdfs/PIP_Iraq_War_Report.pdf, access: 2007-02-13

Scharpf FW (1999) Regieren in Europa: Effektiv und Demokratisch? Frankfurt a. M.

Schmidt J (2006) Weblogs. Eine kommunikationssoziologische Studie. Konstanz

Zürn M (1998) Regieren jenseits des Nationalstaats: Globalisierung und Denationalisierung als Chance. Frankfurt a. M.

Unifying Architecture for Industrial Ecology Communication and Education

Ralf Isenmann

University of Bremen, Institute for Project Management and Innovation (IPMI), and Research Center for Sustainability Studies (artec), Wilhelm-Herbst-Strasse 12, 28359 Bremen, Germany, phone: +49 421 218 8229, fax: +49 421 218 8222, e-mail: ralf.isenmann@innovation.uni-bremen.de, http://www.innovation.uni-bremen.de

Abstract: In this contribution the emerging body of Industrial Ecology theory is organized in a unifying architecture. This basic architecture is illustrated in the form of a pyramid and built upon four basic layers (I-IV) or contexts respectively: (I) context of statements, (II) context of phenomena, (III) context of instruments, and (IV) context of basics. The architecture has its methodological basis in tools of philosophy. It is particularly based on a framework used in epistemology. This generic framework is further conceptualized through a document analysis identifying certain issues that are prototypical for the Industrial Ecology community and literature. Due to its solid structure, the basic architecture is proposed for Industrial Ecology education and communication what is distinctive for the field.

1 Introduction

Since its launch nearly two decades ago, Industrial Ecology has started from a smart "idea", becoming a "somewhat fuzzy concept", to give rise to a professional international society. Now it constitutes a "powerful prism" with numerous tools, studies, publications, resources of research, and other characteristics that make it a discipline.

The overall object of Industrial Ecology is to study industrial systems and their fundamental linkage with natural ecosystems, with the aim to contribute to a more sustainable future [e.g. Isenmann and von Hauff 2007]. As Industrial Ecology is developing such a degree of institutionalization, hence, it may be the right time to describe Industrial Ecology's disciplinary contours and constitutive characteristics in a unifying architecture. Such an effort is particular essential for the field as issues of theory building are becoming a topic of increasing importance, not just inside the community, but also from an outside perspective, especially for education like university curricula, course syllabi, training tools, and development of instructional materials.

2 Discourse on what makes Industrial Ecology unique

The discourse on what makes the field distinctive usually requires considering a number of crucial topics, e.g. philosophical basis, common language, definitions and goals, concepts, tools for applications, and no less important, aspects of institutionalization and historical development [e.g. Weston and Ruth 1997; Lifset and Graedel 2002].

Setting the boundaries is essential to any field of research. This is true for well-established entities like physics, ecology, and economics, but in particular for emerging entities evolving from diverse intellectual roots. Hence, in the Industrial Ecology community there is no clear picture how it is primarily described, yet: Some see Industrial Ecology as a field of research, some term it academic branch, others prefer school of thought. Some others again announce it as scientific community, and some call Industrial Ecology an "umbrella approach" merely covering and/or providing a helpful toolset but without features of uniqueness. The different perspectives to the field are usually accompanied by attributes like "multidisciplinary", "interdisciplinary", "transdisciplinary", or even "metadisciplinary". All these different terms and various labels may be a sign that Industrial Ecology has several scientific roots and thus makes use of various intellectual resources [Fischer-Kowalski 2003].

As a field "with a name as provocative and oxymoronic as industrial ecology" [Lifset and Greadel 2002, p 3], the discourse on a distinctive "self-image" in Industrial Ecology is certainly an ongoing process. Lively dialogues cover various perceptions from different backgrounds [e.g. Lifset 1998; Allenby 1998, 1999, 2006; Boons and Roome 2001; Allen 2001; Bey 2002; Bourg 2003; Ehrenfeld 2001, 2004, 2006; Keitsch 2006]. For example, Lifset [1999, p 1] is aware that "(q)uestions of definition and identity will not completely take care of themselves". As an early perception of how Industrial Ecology would evolve, he maintains, its scientific profile could develop like a monolithic body of theory in the sense of a discipline or a modular structure seen as a complement to other fields [Lifset 1998, p 1].

Research papers, journal articles, and textbooks are the main part of the literature in the field, so far. Another part deals with education issues, including the works of Ehrenfeld [1994], Lynch and Hutchinson [1992], Starr

[1992], Troxell [1992], and more recently Staniskis and Arbaciauskas [2003] and Cockerill [2004]. Despite these useful proposals, Industrial Ecology is not clearly defined. Hence, a unifying architecture illustrating what is distinctive for the field is still needed: Clarifying its identity and uniqueness will not just support Industrial Ecology's intellectual and institutional development. Further, it will improve communication, particularly since the field has become under fire, be it through sceptics inside the community or critics outside the community [e.g. O'Rourke et al. 1996; den Hond 2000; Roundtable discussion on 'industrial ecology' 2001]. Moreover, a unifying architecture will put forward Industrial Ecology education [e.g. Leal 2002]. Lastly, such an effort helps finding a place within the scientific communities, in particular: to define its basics, to highlight distinctive assumptions, and to communicate its unique features.

The early stages in Industrial Ecology have served their purpose well in ensuing years because they helped to facilitate community engagement and stimulated a wide range of projects. Too rigid or prescriptive a definition could have stifled interested members in science, government, business, and education from this development path. As moved forward, however, establishing Industrial Ecology will require greater clarity in communication if it is to maintain any substantial meaning in contrast to other fields.

Clearer definition what Industrial Ecology should be and what not becomes crucial as more communities position themselves as disciplines focused on sustainability and claim to provide relevance for approaching this ambitious goal. As Industrial Ecology matures, features of identity, differentiation, and uniqueness will become more important. For a comparison with Industrial Ecology, Ecological Economics, Environmental Engineering, Cleaner Production, and similar approaches to Environmental Management see e.g. Ehrenfeld [1997], Jackson [2002], Tilley [2003], Seuring [2004], and Kronenberg [2006].

3 Unifying architecture for Industrial Ecology

An analysis of Industrial Ecology from an epistemological perspective is considered a proper method for organizing its emerging body of theory and describing the characteristics of the field. Such an effort seems particular useful to provide a unifying and methodologically sound architecture for Industrial Ecology. Obtaining order in the obvious disorder of assumptions, principles, concepts, tools, and research objects is not as simple a process as it may appear at first glance. As a young and ambitious field, this is particular true for Industrial Ecology that makes use of various intellectual roots originating from different theoretical backgrounds.

Usually all research fields share a common "perspective", no matter of certain roots or specific origins. Put differently, any form of intellectual arrangement could be characterized through a specific "view". In terms of epistemology, such a "perspective/view" is understood as a specific way to treat and deal with "problems". These problems are phenomena that are regarded as being of relevance.

For the clarification of Industrial Ecology's theoretical foundation with the help of tools of philosophy, here a basic architecture is used, represented in the form of a pyramid. This architecture serves as a rough framework for examining epistemological issues when portraying the contours of a scientific profile along four basic layers (I-IV) or contexts respectively (Fig. 1): (I) context of statements, (II) context of phenomena, (III) context of instruments, and (IV) context of basics:

- Layer I: A statement or a system of statements, that covers insights, information, expertise and knowledge, documented in articles, textbooks, journal publications etc. (context of statements)
- Layer II: Different phenomena that could be considered or observed, an object or a class of objects, e.g. design of products, processes, wastes and services as well as certain fields of study (context of phenomena).
- Layer III: Methods, instruments, calculations etc., e.g. material flow analysis (MFA), life cycle analysis (LCA) etc. that are part of the industrial ecology toolbox (context of instruments).
- Layer IV: Basic ideas, theories, principles, and axioms, particularly the underlying assumptions of understanding: human, nature, relationship human-nature, and the role economy and technology should take. These assumptions represent the philosophical fundamentals that are underlying any conceptualization in a scientific sense.

The epistemological framework described here has its methodological basis in the philosophy of science [Zwierlein 1994] and sociology of science [Krüger 1987]. It is especially useful to illustrate the "inter-disciplinary" common ground of various disciplines, analyze different fields of research, and compare diverse schools of thought.

The distinction between the layers/contexts (in the framework above) could be traced back even to earlier works in philosophy [e.g. Rickert 1926/1986], and in economics [e.g. Amonn 1927].

Amonn [1927, pp 21-22] introduced the terms "object of experience" (German "Erfahrungsobjekt") and "object of cognition" (German "Erkenntnisobjekt") to distinguish between

- on the one hand a phenomenon that is considered, observed, or treated as an object of research, and
- on the other, the distinct method of consideration, observation, or treatment as the manner of providing insights.

According to Amonn, a field is defined through a combination of certain objects of research and specific objects of cognition for which a set of methods, instruments, and tools could be employed. It is this approach proposed by Amonn that serves as the basic method primary used in modern economics and management sciences in order to define what to research and thus to determine how to prepare or manipulate certain research objects.

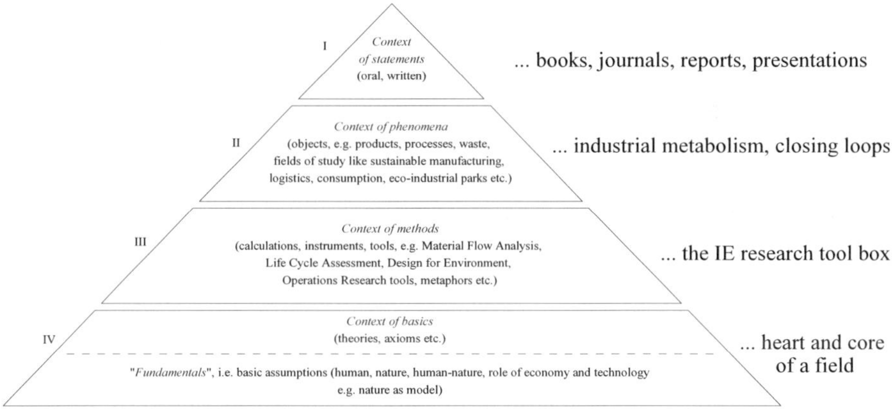

Fig. 1. Architecture of Industrial Ecology

The architecture above (Fig. 1) indicates a schematic model only. For example, layer III includes several contexts such as the context of discovery, description, explanation, justification and application [Isenmann 2003; Bey and Isenmann 2005]. Further, a set of basic assumptions is situated typically at the bottom of layer IV, carrying an understanding on humanity, nature, and the relationship human-nature. The viewpoint from which humans interpret nature notably depends on underlying presuppositions which indicate the human observer's subjective perception employed for explaining and making sense of all that is happening on the Earth.

As the assumptions necessarily contain experiences, final interests, and usually (implicit) valuations and value judgments, they constitute a normative compass. Consequently, the anthropological basis of understanding the human (what or who is man?) is the ultimate key defining the importance of nature and influencing both the purpose and the value that nature should have. In the sense of Kant [1787/2000, edn 3, 443 B 699], these assumptions serve as "regulative ideas". They play a dominant role for research, arranging the way of thinking, organizing imagination of phenomena, and governing decision-making.

Kant made it clear that we are not able merely to extract knowledge from nature as passive observers through our senses, but always do so in a more active manner by constructing knowledge out of human experience. The intellect, as Kant said, does not draw its laws from nature, but imposes its laws upon nature. Comparable with Kantian interpretation, the significant influence of such underlying meta-ideas to modern economic theory and different schools of thought was also described by a string of famous economists. For example, Schumpeter [1954] and recently Daly [1991] termed Kant's regulative idea a "preanalytic vision", whereas Boulding [1971] called it an "image". Despite conceptual differences and though the terms used are different, the similarity is obvious.

In the sense of the example above the epistemological framework comprises (I) a statement about (II) something that is (III) represented, interpreted or manipulated by something (IV) in the light of something. This generic framework has been conceptualized by certain issues that are widely discussed in the Industrial Ecology community.

These issues can be taken as prototypical for the current state of the field. For example, material flow analysis (MFA), life cycle analysis (LCA), and design for environment are some of the methods that comprise the Industrial Ecology toolbox. The issues listed here are resulting from a document analysis, covering all oral and poster presentations that were held during the international Industrial Ecology conferences, including

- the inaugural conference "The Science and Culture of Industrial Ecology" in Leiden 2001,
- the second conference "Industrial Ecology for a Sustainable Future" in Ann Arbour 2003, and
- the recent conference "Industrial Ecology for a Sustainable Future" in Stockholm 2005.

The issues again have been grouped in terms of the four levels above (i.e. statements, phenomena, methods, and basics).

4 Conclusions

The unifying architecture for Industrial Ecolgy presented here has some links to previous works:

- For example, it looks similar to a concept proposed by Vellinga et al. [1998] who made a distinction between: "intuitions and beliefs", "operational principles and concepts", and "practical applications".
- Further, the architecture corresponds to a proposal provided by Weston and Ruth [1997]. They argue that a new science usually requires: a "consistent philosophical base", a "common language", a "set of concepts", and "methods for empirical application".
- Moreover, den Hond [2000] arranged his review to Industrial Ecology along the key terms: "vision", "research field", and "tools". These key terms help to structure the way of thinking about industry-ecology relations and describe the profile of Industrial Ecology as a whole.

Despite the usefulness of all proposals above, however, what the unifying architecture makes essential and unique is – among others – its sound methodology, clear structure, and ease of visualization. The architecture surely supports further conceptual developments and promotes theory building. Based on such architecture, discussions on theoretical dimensions of Industrial Ecology, inside and outside the community, could be made clearer. Further, discussions on emerging issues like objectivity, normativity, epistemological standards, values, and value judgments could be brought to the surface and explained in a transparent manner. No less important, the architecture helps to find other fields perhaps for institutional co-operation, e.g. to the LCA community (SETAC), the International Society for Ecological Economics (ISEE), and the Environmental Informatics community as some shared features could be identified.

The aim of this contribution is to stimulate the ongoing discussion on Industrial Ecology education and the dialogue what is key for the field: What are the features of its identity, and what makes this field of research unique compared to other scientific communities striving for sustainability? When considering a unifying "self-image", scholars in the field will necessarily come across such issues, which are of relevance for theory building but in particular for education and communication, too.

It is argued here that a proper method would be to analyze Industrial Ecology from an epistemological perspective, exposing the emerging body of theory and describing the contours of its scientific profile in a basic unifying architecture: While some aspects of Industrial Ecology may even have attained a status of well-defined relationships, other issues still seem to be in a status nascendi. This interim status of in-between is not seen as a weakness in principle but rather as an opportunity for fruitful development. For that purpose, an epistemological architecture is proposed that highlights the features of Industrial Ecology's identity.

Matters of clear argument, proper vocabulary, and fair communication are considered as of increasing importance for Industrial Ecology both, in terms of the intellectual heritage of its proponents, the development of the community on the whole, and in terms of its reputation as the "science of sustainability" [Ehrenfeld 2004].

Even if some may find these issues a rather useless task compared to "real research" like empirical studies, such discussions are crucial to recognize that these issues have consequences: They shape concepts, instruments, and vocabularies that are used to approach research problems. For this reason, if no other, Industrial Ecology's intellectual heritage needs to be examined.

References

Allen D (2001) A set of core principles and tools? In: Journal of Industrial Ecology, vol 4, edn 4. MIT Press Journals, Cambridge, pp 1-2

Allenby BR (1998) Context is everything. In: Journal of Industrial Ecology, vol 2, edn 2. MIT Press Journals, Cambridge, pp 6-8

Allenby BR (1999) Culture and industrial ecology. In: Journal of Industrial Ecology, vol 3, edn 1. MIT Press Journals, Cambridge, pp 2-4

Amonn A (1927) Objekt und Grundbegriffe der theoretischen Nationalökonomie, edn 2. Franz Deuticke, Leipzig, Wien

Bey C (2002) Quo vadis industrial ecology? Realigning the discipline with its roots. In: Greener Management International, vol 34 (summer). Greenleaf, Sheffield, pp 35-42

Bey C, Isenmann R (2005) Human systems in terms of natural systems? In: International Journal of Sustainable Development, vol 8, edn 3. Inderscience, pp 189-206

Boons F, Roome N (2001) Industrial ecology as a cultural phenomenon. In: Journal of Industrial Ecology, vol 4, edn 2. MIT Press Journals, Cambridge, pp 49-54

Boulding KE (1971) The image. University Michigan Press, Ann Arbor

Bourg D (2003) Introduction. In: Bourg D, Erkman S (eds) Perspectives on industrial ecology. Greenleaf, Sheffield, pp 13-18

Cockerill K (2004) Innovative industrial ecology education can guide us to sustainable paths. In: Sikdar SK et al. (eds) Technological choices for sustainability. Springer, Berlin, pp 77-90

Daly HE (1991) Steady-state economics, edn 2. Island Press, Washington

Ehrenfeld J (1994) Industrial ecology and design for environment: The role for universities. In: Allenby B, Richards D (eds) The greening of industrial ecosystems. National Academy Press, Washington, pp 228-240

Ehrenfeld J (1997) Industrial ecology: A framework for product and process design. In: Journal for Cleaner Production, vol 5, edn 1/2. Elsevier, Heidelberg, pp 87-95

Ehrenfeld JR (2001) Industrial ecology begets a society. In: Journal of Industrial Ecology, vol 4, edn 3. MIT Press Journals, Cambridge, pp 1-2

Ehrenfeld JR (2004) Can industrial ecology be the "science of sustainability? In: Journal of Industrial Ecology, vol 8, edn 1/2. MIT Press Journals, Cambridge, pp 1-3

Ehrenfeld JR (2006) Advocacy and objectivity in industrial ecology. In: Journal of Industrial Ecology, vol 10, edn 4. MIT Press Journals, Cambridge, pp 1-4

Fischer-Kowalski M (2003) On the history of industrial metabolism. In: Bourg D, Erkman S (eds) Perspectives on industrial ecology. Greenleaf, Sheffield, pp 35-45

Hond F (2000) Industrial ecology: A review. In: Regional Environmental Change, vol 1, edn 2. Springer, Berlin, pp 60-69

Isenmann R (2003) Further efforts to clarify industrial ecology' hidden philosophy of nature. In: Journal of Industrial Ecology, vol 6, edn 3/4. MIT Press Journals, Cambridge, pp 27-48

Isenmann R, Hauff von M (eds) (2007) Industrial Ecology: Mit Ökologie zukunftsorientiert wirtschaften. Elsevier, Heidelberg

Jackson T (2002) Industrial ecology and cleaner production. In: Ayres RU, Ayres LW (eds) A handbook of industrial ecology. Edward Elgar, Cheltenham, pp 36-43

Kant I (1787/2000) Kritik der reinen Vernunft (1 und 2).In: Weischedel W (ed) Kritik der reinen Vernunft, vol 3 and 4, edn 14. Suhrkamp, Frankfurt am Main

Keitsch M (ed) (2006) Special issue "theoretical dimensions of industrial ecology". In: Progress in Industrial Ecology, vol 3, edn 1/2. Inderscience

Kronenberg J (2006) Industrial ecology and ecological economics. In: Progress in Industrial Ecology, vol 3, edn 1/2. Inderscience, pp 95-113

Krüger L (1987) Einheit der Welt – Vielheit der Wissenschaften. In: Kocka J (ed) Interdisziplinarität: Praxis – Herausforderung – Ideologie. Suhrkamp, Frankfurt am Main, pp 106-125

Leal Filho W (2002): Towards a Closer Integration of Environmental Education and Industrial Ecology. In: International Journal of Environment and Sustainable Development, vol 1, edn 1. Inderscience, pp 20-31

Lifset R (1999) On becoming an industrial ecologist. In: Journal of Industrial Ecology, vol 2, edn 3, pp 1-3. MIT Press Journals, Cambridge

Lifset R (1998) Setting the boundaries? In: Journal of Industrial Ecology, vol 2, edn 2. MIT Press Journals, Cambridge, pp 1-2

Lifset R, Graedel TE (2002) Industrial ecology: Goals and definitions. In: Ayres RU, Ayres LW (eds) A handbook of industrial ecology. Edward Elgar, Cheltenham, pp 3-15

Lynch DR, Hutchinson C (1992) Environmental education. In: Proceedings of the National Academy of Sciences, vol 89, edn 3, pp 864-867

O'Rourke D et al. (1996) Industrial ecology: A critical review. In: International Journal of Environment and Pollution, vol 6, edn 2/3. Inderscience, pp 89-112

Rickert H (1986) Kulturwissenschaft und Naturwissenschaft. Reclam, Stuttgart

Roundtable discussion on "industrial ecology 2000: Maximizing shareholder value, lessons from the natural world". 2001. Business as a living system: The value of industrial ecology. In: Transcript of the roundtable discussion. Conference held at the Haas School of Business, 5-8 October 2000. California Management Review, vol 43, edn 3, pp 16-25

Schumpeter JA (1954) History of economic analysis. Oxford University Press New York

Seuring S (2004) Industrial ecology, life cycles, supply chains: Differences and interrelations. In: Business Strategy and the Environment, vol 13. Wiley, pp 306-319

Staniskis J, Arbaciauskas V (2003) Industrial ecology in university curriculum. In: Clean Technology and Environmental Policy, vol 5. Springer, Berlin, pp 92–94

Starr C (1992) Education for industrial ecology. In: Proceedings of the National Academy of Sciences, vol 89, edn 3, pp 868-869

Tilley DR (2003) Industrial Ecology and Ecological Engineering. In: Journal of Industrial Ecology, vol 7, edn 2. MIT Press Journals, Cambridge, pp 13-32

Troxell DE (1992) Educating and training industrial ecologists for the 1990s. In: Proceedings of the National Academy of Sciences, vol 89, edn 3, pp 862-863

Vellinga P et al. (eds) (1998) Managing a material world: Perspectives in industrial ecology, vol 1. Springer, Berlin

Weston RF, Ruth M (1997) A dynamic, hierarchical approach to understanding and managing natural economc systems. In: Ecological Economics, vol 21, pp 1-17

Zwierlein E (1994) The paradigm of self-organization and its philosophical foundations. In: Mishra RK. et al. (eds) On Self-Organization. An interdisciplinary search for a unifying principle. Springer, Berlin, pp 288-298

SmartLIFE: Toward European Approaches to Sustainable Building

Walter Leal Filho

TuTech Innovation GmbH, Harburger Schlossstrasse 6-10,
21079 Hamburg, Germany, e-mail: leal@tutech.de

Abstract: The experience gathered from both the academic world and the industry sector is that there is a strong business argument for more sustainable construction. Being sustainable is as much about efficient profit-orientated practice and value for money as it is about helping the environment. Yet, despite the great potentials and advantages seen in the development of approaches towards sustainable building, progress in this field has been rather slow.

In order to address the need for catalyzing initiatives in sustainable building, the project SmartLIFE, an European initiative aimed at surveying provisions and needs as well as fostering awareness on sustainable construction among students across Europe, has been created.

This paper presents the experiences from SmartLIFE and the rationale behind the project, which is the need to develop approaches and methods which may further the implementation of sustainable construction into practice. In addition, the paper describes the project structure and methods to be developed as part of the project, which caters for the introduction of a sustainability dimension as part of construction projects. Finally, the paper considers the extent to which sustainable construction may play its part in delivering sustainable communities.

1 Introduction

Much has been said and written about sustainability and how to apply it to different areas and sectors [e.g. Leal Filho 1999; Leal Filho 2000; Leal Filho 2002]. Yet, the debate has largely been theoretical, with very few examples of concrete action taken in order to integrate sustainability into mainstream areas such as the construction sector. Indeed, although there has been on the one hand a growing awareness about the need for a "restructuring" of sustainable development [Ayres and Simonis 1994] and of the importance of sustainable construction [e.g. Laing 2002; DTI 2002; DTI 2004] the incorporation of concepts of sustainable construction around the globe for the last few decades, on the other hand, has not been as widely spread as it should be and there is much room for improvements.

As outlined by Young and Sachs [1994] the creation of a sustainable materials economy requires not only good will, but also a market demand which the construction sector is uniquely well placed to catalyze.

This paper, which has a very practical nature, will present a case study from a project where the links between sustainability, innovation and construction ate explored.

At the outset it should be pointed out that the Project SmartLIFE is a partnership between three EU regions - UK, Germany and Sweden, which aims at delivering solutions to some of the common sustainability problems faced in European growth areas. It is funded by the Interreg IIIB Programme (North Sea), over a period of 42 months (Jan 2004 – June 2007) and with a budget of approximately Euro 9 million, being the largest pilot project of this scale in Europe today.

Bodies working together in SmartLIFE include central government agencies, local authorities charged with spatial planning in the partner countries, local training providers and private bodies such as developers and building systems suppliers. It is an innovative pilot project led by Cambridgeshire County Council, with partners in the cities of Hamburg (Germany) and Malmö (Sweden). Each city faces very similar growth related challenges. The project's main aim is address three challenges of housing delivery in growth areas:

- affordability
- sustainability/energy efficiency
- skills/capacity shortages in the construction industry

To this purpose, partners work closely, sharing expertise and overcoming obstacles to innovative sustainable construction.

The project's vision is to unlock strategic solutions to sustainable growth area management and sustainable communities, combining the set up of transnational infrastructures with a cross-sectoral approach linking public awareness, education and training in respect of modern methods of construction and environmental protection.

In addition, the strategic aim of SmartLIFE will be to develop and implement an innovative approach to facilitating sustainable communities in the North Sea Region. Transnational polycentric development and spatial planning thinking will be advanced at all levels of governance by encouraging a grassroots interaction of various stakeholders to influence business practices and policy makers. SmartLIFE will recognize and incorporate various planning policy and practices within the European partnership and identify a route through planning regimes. It will also bring together educators and planners to challenge traditional (national) methods of delivery by demonstrating SmartLIFE transnational approaches.

By promoting social inclusion and community development, along with the other aims, we believe that SmartLIFE can be elevated to a transnational recognized position to promote sustainable spatial planning and transfer knowledge, foster economic and socially cohesive development in the North Sea region. Most importantly, the project aims to deliver effective, practical and concrete outputs from raising the skills level of the targeted population, increasing the acceptability of modern construction methods, promoting sustainability in growth areas, and providing increasing capacity to deliver EU growth area development targets.

In concrete terms, the project aims to:

1. Develop a coherent and coordinated approach to sustainable growth area management.
2. Establish a transnational network of SmartLIFE business centers that will act as European delivery and dissemination hubs for the project.
3. Provide environmental, technical support and financial advice to local people who are engaged in community self-build and in sustainable approaches to construction.
4. Increase skills and capacity in the local construction workforce by developing and implementing transnational programs that provide training in modern methods of construction, thus contributing to the reduction of unemployment in the partner regions.
5. Utilize transnational experience in various aspects of training, construction and development to promote the adoption of best practice among the partners.
6. Design an effective communications campaign to promote and disseminate the SmartLIFE approach, the cross-sectoral partnership and the Interreg programme.
7. Develop a comprehensive strategy for fortifying and extending Smart LIFE beyond its Interreg lifespan.

The subsequent parts of this paper will provide further details on the scheme.

2 Project Activities and the Partnership

It is often emphasized that sustainability is best delivered by involving the various stakeholders who have an interest in a particular issue or who may be motivated to work towards solving a given problem. This premise has been taken seriously in the context of SMARTLIFE. The partnership is composed by the following agencies:

UK Partnership

- Cambridgeshire County Council
- Fenland District Council
- Learning & Skills Council
- Cambridgeshire Jobcentre
- CITB-Construction
- SkillsFoundation for the Built Environment (FBE)
- Self-Build Agency
- English Partnerships
- The Housing Corporation
- Cambridgeshire Centre of Vocational Excellence for Construction (CoVE)
- Warden Housing Association
- Office of the Deputy Prime Minister (ODPM)

German Partnership

- TuTech
- G19 (Gewerbeschule 19):
- Institute for Construction Training
- Hamburg Technical University
- Hamburg City Council
- ZEB (Zentrum fur Exzellentes Bauen)
- Centre of Construction Excellence
- AZB (Ausbildungszentrum Bauen)

Swedish Partnership

- Malmö Stad LIP – kansli m.m. (City of Malmö – Environmental Department)
- Sustainable Business Hub
- Ekologiskt marknadscentrum
- Agenda 21 forum
- Skane Milore Centre
- Green Roof
- Institute Stadsekologisktkunskaps-ochutvecklingscentrum

The elements of the project are delivered by means of work packages, which are led by a given partner. This approach allows a more elaborated handling of the various project matters, bearing in mind that each work package is processed by an international team. The work packages are as follows:

Work Package One: SmartLIFE European Business Centres

The building or fortification of business centers in each of the partner countries will establish a transnational network of sustainable development hubs. These EU facilities will give support to the SmartLIFE projects and associated initiatives within each country. SmartLIFE business centers will be multi-functional and built using sustainable construction methods. They will serve as Community icons for sustainability and social inclusion whilst demonstrating the creative application of European funding. All of the training programs marketing and communications, media events, partnership interaction, conferences and self-build support will be run from these comprehensive centers.

Additionally, it is intended that in the longer term the expertise and know-how gained will be easily transferred to other interested regions in the North Sea Region.

Work Package Two: Community Training and Education

By determining the strengths and weaknesses of each partner country in respect of their local labor supply this work package encompasses the establishment and delivery of transnational training programs that offer additionality to programs already being offered within the partner countries.

It is envisaged that training programs, for both modern methods of construction and self-builders, will be run nationally on a day-to-day basis with frequent international meetings and seminars spread throughout the duration of the project. These meetings and workshops will represent an exciting opportunity for local people (both teachers and students) to exchange their views on various approaches to training, experience different training cultures and as a result adopt the best training practice. This is connected to the wider agenda of community outreach education programs promoting sustainable living and careers based on sustainable development. This will give the wider local community access to this European facility. Furthermore, this work package will support a scheme offering easy access for local people to these training programs and development sites in particular from economically deprived and rural areas.

Work Package Three: Planning Policy and Practice

By recognizing and understanding respective spatial planning and sustainable development policies within the European partnership, a clearer picture will be presented in terms of goals and ambitions of current leaders which the SmartLIFE projects seeks to apply and implement. This work package involves:

1. Planning and Investigation. Research will determine the similarities and differences between the various relevant planning policies and practices within the SmartLIFE partnership. This will aid the partnership in defining the most appropriate solutions and applying continuity to SmartLIFE principles and approaches to foster sustainable development within the partner countries.
2. Local Stakeholder Conferences. Each partner country will bring together and exchange views from those parties with a vested interest in growth area demands and outputs.
3. Expert Panel. Consultation Expert panels will be brought together and asked a similar set of questions to those of the local opinion polls, comparing at the end of the session the standpoint of local population and the experts. This will set the stage for the international seminar.
4. International Seminars. Following on from the results of the expert panels and telephone consultations, representatives from each partner country will take it in turns to facilitate seminars, which will address growth area management concerns and bring forward possible solutions in a structured approach.

Work Package Four: Communications and Marketing

The idea is to deliver a robust and transnational communications campaign that will not only highlight the project itself, but the dynamic partnership and the Interreg programme as well. This work package entails a comprehensive communications and marketing strategy for the business centers, for self-build and for modern methods of construction. In doing so, the project will determine and utilize all available IT solutions to foster social inclusion by realizing and communicating sustainable building and living to both private and public actors alike. A web site is already available (Fig. 1).

Other elements are also included in the work packages but the ones here listed provide a rough overview of what the project intends to achieve.

The approach used in the project shows that there are clear benefits for transnational cooperation on spatial development issues in the growth areas. All the partner regions involved in SmartLIFE encounter similar problems associated with growth such as population density, urban pressures, lack of affordable housing and inadequate house-building rates due to a skills and capacity shortages in the construction industry.

SmartLIFE represents a much-needed cross-sectoral and transnational partnership as a possible solution to some of these common problems. The work package elements and organization of SmartLIFE have been designed to enable expertise to be shared and transnational work to take place on common issues, as all partners are participating in all work packages. The nature of the project is also such that cross-sectoral approach leads to greater co-operation and communication between the various partners and sub-partners and more specifically with the targeted population, resulting in the most effective realization of the aims and objectives of the project.

Over the period of the project it is expected that other regions in Europe will encounter similar issues associated with growth in the relatively near future and it is hoped that the SmartLIFE project or specific elements of it will be applicable to other EU regions and beyond the present life span of the project. The project will deliver internationally applicable solutions to a currently crucial problem, namely the lack of means to foster sustainable spatial planning via environmentally sound and socially acceptable housing, which is of national significance in all the participating partner countries.

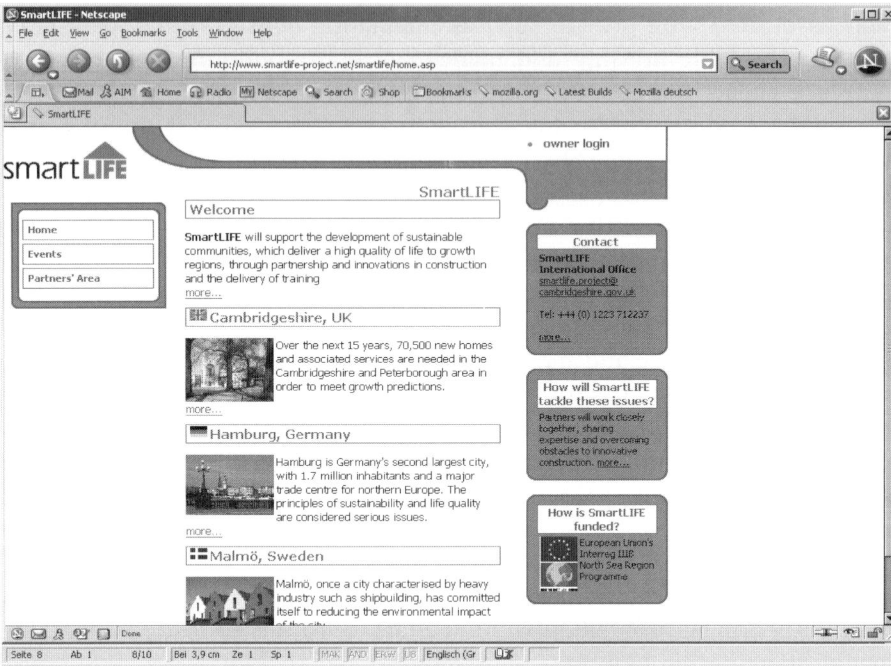

Fig. 1. Overview of the project home page

3 Conclusions

Within the European partnership the SmartLIFE project runs in parallel with many domestic/ national initiatives and partnerships. These national partnerships are largely involved in the planning and delivery of affordable housing. The project will deliver international solutions to difficult domestic/national problems.

SmartLIFE also seeks to go beyond inter-regional cooperation between the main partners and sub-partners and in addition aims to establish a strong and on-going interaction at the grass-root level. In practice, this will include various international workshops and seminars for developers, off-site manufacturers, self-builders, construction trainees and other relevant experts where the participants will be able to view and discuss sustainable construction alternatives and learn from each other's expertise.

The chief aim of this interaction is to encourage intra-sectoral partnership between German, British and Swedish interested parties and to promote the transnational adoption of the best practice in sustainable development and environmentally friendly construction in the whole of the North Sea region. Overall, the international partners and their sub-partners believe that this unique project combining development and furthering of local construction skills with the delivery of affordable and sustainable housing through an innovative transnational partnership will offer a significant contribution to spatial development of the North Sea region.

References

Ayres RU, Simonis UE (1994) Industrial Metabolism – Restructuring for Sustainable Development. UN University Press, Tokyo

Department of Trade and Industry (DTI) (2002) Sustainable Construction Report. Department of Trade and Industry, London

Department of Trade and Industry (DTI) (2004) Sustainable Construction Brief 2 (April 2004). DTI, London

Laing M (2002) Reputation, Risk and Reward: the business case for sustainability in the UK property sector (http://projects.bre.co.uk/rrr/)

Leal Filho W (1999) Sustainability and University Life. Peter Lang Verlag, Frankfurt

Leal Filho W (2000) Communicating Sustainability. Peter Lang Verlag, Frankfurt

Leal Filho W (2002) Teaching Sustainability – towards curriculum greening. Peter Lang Verlag, Frankfurt

Young J, Sachs A (1994) The Next Efficiency Revolution: Creating a Sustainable Materials Economy. World Watch Paper 121, p 41

The Effect of ICTs on Mobility and Environment in Urban Areas

Ji Young Song[1], Bong Gyou Lee[2], Koen Steemers[1], Nick Baker[1]

[1] The Martin Centre for Architectural and Urban Studies, Department of Architecture, University of Cambridge, 6 Chaucer Road, Cambridge CB2, e-mail: {jys22, kas11}@cam.ac.uk, nickvbaker@aol.com

[2] The Graduate School of Information, Yonsei University, 134 Shinchon-dong, Seodamun-gu, Seoul 120-749, Korea, e-mail: bglee@yonsei.ac.kr

Abstract. With the IT (Information Technology) Revolution in the 21th century, many cities have developed strong infrastructural communication networks. It is expected that well-built communication networks will diminish people's need for traveling. Subsequently, the decreased traffic would lead to reduced energy consumption in transport. The purpose of this study is to investigate the environmental impact resulting from changes in traveling activities as a result of an enhanced communication infrastructure supported by ICTs. We have made a preliminary study to examine the relations between the use of ICTs and people's mobility. The impact on the environment was further analyzed using regression analysis. This study aims to provide some information about the potential of ICTs on the environment.

Keywords: Information and communication technologies (ICTs), Mobility, Environmental impact, Energy consumption

1 Introduction

According to the advances on the developments of ICTs (Information and Communication Technologies), people's lifestyle and urban structure have been changed. People in their home, office buildings and many other places have been assisted unawares by ICTs. Mitchell [1996] named this highly-connected society by ICTs in near future 'networked city'[1]. He ar-

[1] Mitchell referred the networked cities to advanced cities using ICTs actively in the near future in his book series "City of bits (1996)", "e-topia (1999)" and "Me++ (2003)".

gues that physical distance and location will be weakened in networked cities and thereby cities could provide us with green environment. In other words, people have alternative way to replace physical movements, e.g., using conventional vehicles in networked cities where were equipped with strong communication infrastructure such as telephony and internet.[2]

In networked cities, can actually ICTs contribute to solve environmental problems as well as to decrease mobility of people? How closely do existing methods such as statistical analyses investigate these kinds of questions? Unfortunately, there are very few empirical studies regarding these issues with emerging ICTs until recently [Ferrell 2004; Kakihara and Sorensen 2002].

To examine how the enhanced communication infrastructures in urban areas have an effect on people's mobility and also on the environment, we have made a preliminary study using regression analyses. This study provides some information about the potential of ICTs on the environment as the first step for further research.

2 Study design

We assume that the advanced communication infrastructure supported by ICTs might decrease people's mobility by substituting digital communication for physical travel needs especially for business purpose. Thereafter, energy consumption (CO_2 emission) for traveling would also be reduced according to the change in travel demand.

Multiple regression models in this study focuses on the relation between the communication networks and mobility and further environmental impacts. Internet connectivity[3], the number of households with access to the internet and the number of teleworkers[4] have been chosen as variables to explain the influence of ICTs. The mobility of people has been estimated by the road traffic and the trip length. In addition, to investigate environmental impacts, we have considered the overall energy consumption in road traveling.

---

[2] "Travel is no longer the only way to go [Mitchell 1999, p147]."

[3] Office for National Statistics developed the Index of Internet Connectivity as part of a package of measures to help monitor the UK's use of the Internet and the growth of e-commerce. Publication of the index commenced in 2001. The publication shows the change in the overall level of connections to the Internet.

[4] According to the report of National Statistics, teleworkers are defined as people who work in their own home and who use both a computer and a telephone for their work. This includes people who:

- mainly work from home in their main job – "teleworker homeworkers"
- work in various locations but use their home as a base – "home-based teleworkers"
- don't usually work at home but do so for at least one day per week – "occasional teleworkers".

To examine the correlation between communication networks and mobility, road traffic of all vehicles in the UK (1998-2004) has been used as a dependent variable in the estimated multiple linear equation. Internet connectivity index (1998-2004), the number of households with access to the internet (1998-2004) and the number of teleworkers (1998-2004) in the UK have been used as independent variables. The trip length for business (1998-2003) has been used as a dependent variable to observe relation between the communication networks and the mobility in business sector using the same independent variables. Furthermore, to investigate the relation between mobility and its environmental impacts, the annual overall energy consumption by transport (1998-2003) has been used as a dependent variable and Internet connectivity, the number of households with access to the internet and the number of teleworkers in the UK have been used as independent variables. Data have been collected from reports of National Statistics [2003, 2005], Department for Transport [2003, 2004, 2005], Department of Trade and Industry [2004], and National Environmental Technology Centre in the UK (2004).

Table 1. Model design for analyses

		Model 1
Dependent variable	y	The road traffic by all vehicles
Independent variables	x_1	Internet connectivity
	x_2	The number of households with access to the internet
	x_3	The number of teleworkers
		Model 2
Dependent variable	y	The trip length for business
Independent variables	x_1	Internet connectivity
	x_2	The number of households with access to the internet
	x_3	The number of teleworkers
		Model 3
Dependent variable	y	The overall energy consumption by transport
Independent variables	x_1	Internet connectivity
	x_2	The number of households with access to the internet
	x_3	The number of teleworkers

3 Statistical analysis

For multiple regression analyses, Microsoft Office Excel (2003 Edition) has been used following linear equation. Table 2 summarizes the results of regression analyses of four models.

$$y = \alpha + \beta_1 x_1 + \beta_2 x_2 + \beta_3 x_3 + \varepsilon \quad (3.1)$$

Table 2. Summary of results

Independent variable	Road traffic by all vehicles	Trip length for business	Overall energy consumption by transport
	OLS	OLS	OLS
Constant	431.317527 (8.648841)	708.511370 (6.723102)	50020.058187 (1497.792342)
Internet connectivity	0.029808 (0.047001)	-0.240282[a.] (0.033051)	-10.287869 (8.488610)
Number of households with internet access	-0.000000 (0.000001)	0.000008[a.] (0.000000)	
Number of teleworkers	0.000032[b.] (0.000009)	-0.000042[b.] (0.000007)	0.003777[b.] (0.001244)
Adjusted R^2	0.960523	0.967111	0.783469
Observations	7	6	7

[a] Significant at 1 percent
[b] Significant at 5 percent
[c] Significant at 10 percent.
Standard errors are in parentheses

The result of model 1 illustrates that the number of teleworkers has the positive relations with road traffic by all vehicles with the significance level 5% (P-value of the variable x_3: 0.04) i.e., road traffic increases as the number of teleworkers grows against the assumption that enhanced communication networks would decrease the mobility of people by substituting road traffic.

For the model 2, with the high reliability (the adjusted R-squared value of 0.96) and the high significance level (significance F: 0.01 < 0.05), all three variables are statistically significant at the 5 % level (P-values of x_1, x_2 and x_3 are 0.01, 0.00 and 0.02). The result shows that the number of households with access to the internet has an effect on the business trip length positively however, internet connectivity and the number of teleworkers affects the business travel negatively i.e, the business trip length decreases according to the increase of internet connectivity and teleworkers while it increases slightly as the number of households which can access to the internet grows.

In the model 3, the number of teleworkers has positive relations with the overall energy consumption by transport with the significance level 5% (P-value of the variable x_3: 0.03). This means energy consumption for travelling goes up as the number of teleworkers increases.

On the contrary our early assumption, the results show that the overall road traffic and thus energy consumption for traveling increase as teleworkers grow. This could result from the overall road traffic includes all traffic as well as traffic for commuting purpose thus, traffic for other purposes e.g., home-delivery might grow even though traffic decreases due to teleworking. This argument is supported by the result of model 2. For the business trip, the travel length is lessened as teleworkers grow and internet connectivity rises consistent with the assumption. Especially, between internet connectivity and business trip length, we can see stronger correlation (coefficient: -0.24) compared with other two variables.

4 Conclusion

In order to explore environmental impacts from changing pattern in traveling resulting from well-equipped information and communication network, we made a preliminary study using existing data and method.

The results show that road traffic increases as teleworkers grows and thus energy use for traveling also increases against the assumption that advanced communication networks would decrease the mobility of people by substituting digital communication for physical traveling needs. However, the trip length for business decreases according to the increase of internet connectivity and the number of teleworkers while it increases slightly as the number of households which access to the internet grows.

The study has found out several interesting results as the first step to examine how ICTs affect traveling behavior and consequently environment, even though there are limitations e.g., small sized-sample data and missing dependent variables. However, this study provides some information about the potential of ICTs on the environment as a pilot study.

References

Baxter A (2001) Infrastructure and cities. In: Echenique M, Saint A (ed) Cities for the new millennium. Spon press, London & New York, pp 105-114

Briggs DJ, de Hoogh C, Gulliver J, Wills J, Elliott P, Kingham S, Smallbone K (2000) A regression-based method for mapping traffic-related air pollution: application and testing in four contrasting urban environments. Sci Total Environ. Department of Epidemiology and Public Health, Imperial College of Science, Technology and Medicine, London

Cervero R (2003) The Built Environment and Travel: Evidence from the United States. In: EJTIR, vol 3, edn 2, pp 119-137

Corporation of London (2001) ICT Infrastructure Review

Department for Environment, Food and Rural Affairs (2005) National Statistics 2005: e-Digest of Environmental Statistics. Defra Publications

Department for Transport (2003) Focus on freight. The Stationery Office Books

Department for Transport (2004) Transport Statistics Great Britain 2004. The Stationery Office Books

Department for Transport (2005) Transport Statistics Great Britain 2005. The Stationery Office Books

Department of Trade and Industry (2004) Energy consumption in the United Kingdom. The Stationery Office Books

Douglas AA (1973) Home-based trip end models: A comparison between category analysis and regression analysis procedures. In: Transportation, vol 2, edn 1, pp 53-70

Ferrell CE (2004) Home-based teleshoppers and shopping travel: Do teleshoppers travel less? In: Transportation research record, edn 1894, pp 241-248

Frank LD, Stone B, Bachman W (2000) Linking land use with household vehicle emissions in the central puget sound: Methodological framework and findings. In: Transportation Research Part D, vol 5, pp 173-196

Kakihara M, Sorensen C (2002) Mobility: An extended perspective. The Hawaii' international conference on system sciences, January 7-10, 2002, Big Island, Hawaii

Lass H, Gottlieb P (1971) Probability and statistics. Addison-Wesley publishing company

Mitchell WJ (1996) City of bits: space, place, and the Infobahn. The MIT Press

Mitchell WJ (1999) e-topia: urban life, Jim-but not as we know it. The MIT Press

National Statistics (2003) First release: Internet Access, Households and Individuals. National Statistics

National Statistics (2005) The Internet Connectivity First Release.National Statistics

Owens S (1986) Energy, planning and urban form. Pion Limited

Ruiz Y, Walling A (2005) Home-based working using communication technologies. In: National Statistics feature

Signor R, Westphal FS, Lamberts R (2001) Regression analysis of electric energy consumption and architectural variables of conditioned commercial buildings in 14 Brazilian cities. In: 7th International IBPSA Conference, Rio de Janeiro, Brazil, 13-15 August, 2001

Wu C-H, Ho J-M, Lee DT (2004) Travel-time prediction with support vector regression. In: IEEE Transactions on Intelligent Transportation Systems, vol 5, edn 4, pp 276-281

Data Mining as a Suitable Tool for Efficient Supply Chain Integration - Extended Abstract

Joanna Oleskow[1], Marek Fertsch[1], Paulina Golinska[1], Katarzyna Maruszewska[2]

[1] Poznan University of Technology, Poland

[2] Poznan School of Banking, Poland

1 Introduction

Emerging technologies have enabled many companies to collect a tremendous amount of data on their customers, suppliers and partners. How to deal with the very large database in supply chain management is a very important problem. A large amount of data is likely to be gathered from the many activities along supply chains. This data must be warehoused and mined to identify patterns that can lead to better management and control of supply chain. One of the solution is using of data mining. Data mining is a process that uses a variety of data analysis tools to discover relationships in data that may be used to make valid predictions.

A supply chain can be defined as a series of activities that are involved in the transformation of raw materials into the final product. A supply chain consists of all stages involved, directly or indirectly, in fulfilling a customer request. The supply chain not only includes the manufacturer and suppliers, but also transporters, warehouses, retailers, and customers themselves. If the supply chain process is looked at from a cycle view, all supply chain processes can be broken down into the following four process cycles: customer order cycle, replenishment cycle, manufacturing cycle and procurement cycle [Raisinghani et al. 2005].

The keys to success in the supply chain management should be focused on integration of activities, cooperation, coordination, and information sharing throughout the entire supply chain. With information sharing, demand information at a downstream site is transmitted upstream in a timely fashion and thus, facilitate decision making. Within the context of supply chain logistics, integration can occur at different levels: physical integration, information integration, control integration, structure integration [van Gor 2001].

2 Supply chain management and integration framework based on data mining

Enterprises forming supply chains rely on vast amounts of data and information that is located in large databases. This information is a valuable resource, but its value can be increased if additional knowledge can be gained from it. The exploration of database information, to identify and extract deep and hidden knowledge, is made possible by data mining (DM) techniques.

The existing databases of manufacturing enterprises, or indeed of most large organizations, are huge, but largely untapped sources of information, since they contain valuable records of operational and market history. DM techniques can be used to improve strategic and operational planning activities, as databases can be explored to gain feedback on the past performance and business behavior of the enterprise [Neaga and Harding 2005].

Data mining is a form of information extraction activity whose goal is to discover hidden facts contained in databases; the process of using various techniques (i.e. combination of machine learning, statistical analysis, modeling techniques, and database technology) to discover implicit relationships between data items and the construction of predictive models based on them [Wang 2006].

Data mining process is the core of the knowledge discovery process and its relationship to data cleaning and integration of databases into data warehouses, and the subsequent selection of appropriate task-relevant data and appropriate data mining techniques to yield pattern evaluations (see Fig. 1).

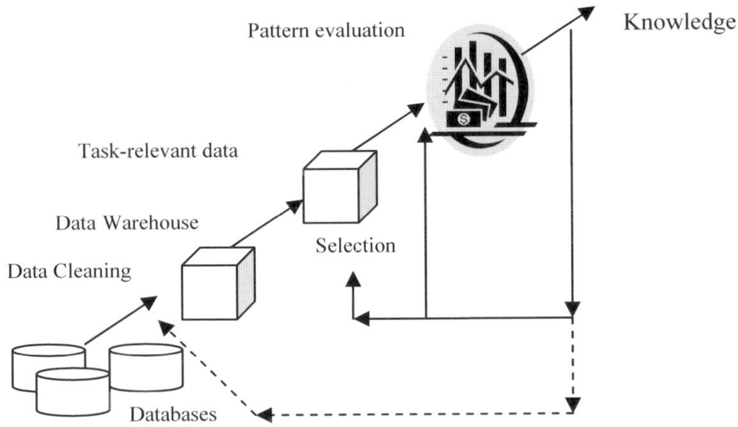

Fig. 1. Data mining: a knowledge discovery process [Han and Kamber 2001]

To data mine effectively, data need to be preprocessed in a variety of steps that include removing duplicates, performing statistical data editing and imputation, and doing other cleanup and regularization of the data.

The fundamental steps of mining process shall be as follows:

- *Establish mining goals* – consider the cost-benefit tradeoff and the expected accuracy and usefulness of the results.
- *Select data* – decide which data is useful, which attributes are worth considering, and how big the sample size should be.
- *Preprocess data* – filter out noisy, erroneous, or irrelevant data, and handle missing data.
- *Transform data* – where possible, reduce the number of data attributes or extract new ones from existing data attributes.
- *Store data* – integrate and store data at a single site under a unified scheme.
- *Mine data* – perform appropriate data mining functions and algorithms according to mining goals.
- *Evaluate mining results* – perform various operations such as knowledge filtering from the output, analyzing the usefulness of extracted knowledge, and presenting the results to the user for feedback. The feedback from this step can prompt changes to earlier steps.

Depending on specific mining goal it is necessary to determine one general orientation of our research and furthermore select a proper approach and methods of analysis (see Table 1).

Table 1. Data mining approaches and methods [Samli et al. 2002]

General orientation	Specific approach	Method of analysis
Exploration	Classification Clustering Association Sequential analysis	Decision trees Genetic algorithms Neural networks Fuzzy logic K-nearest neighbor
Prediction	Regression Time series Association Sequential analysis	Decision trees Genetic algorithms Neural networks Fuzzy logic K-nearest neighbor Linear discrimant

In the paper we want to point out the key application areas of data mining in supply chain integration domain. There are different sources to collect data i.e. consumers, manufacturers, suppliers, retailers, wholesalers. Data from consumers are the core data that affect all other data items along supply chains. The types of items that are appealing to consumers can be determined by various data mining techniques. Information collected from the consumers at the point of sale include data items regarding the sold product (e.g., type, quantity, sale price, and time of sale) as well as information about the consumer (e.g., consumer address and method of payment). These data items are analyzed often.

Data-mining techniques can be also used to: (1) Identify patterns in returns so that retailers can better determine which type of product to order in the future and from which supplier it should be purchased, (2) Better promote their own product and establish a competitive advantage, (3) Identify political and economic conditions in supplier countries, (4) Identify patterns in the defects of products, their components, or the manufacturing process.

Manufacturers collect data regarding a) particular products and their manufacturing process, b) suppliers, and c) the business environment. Data regarding the product and the manufacturing process include the characteristic of products and their component parts, the quality of products and their components, and trends in the research and development of relevant technologies. Data mining techniques can be applied to identify patterns in the defects of products, their components, or the manufacturing process. Data regarding suppliers include availability of raw materials, labor costs, labor skills, technological capability, manufacturing capacity, and lead time of suppliers.

Retailers, manufacturers, and suppliers are all interested in data regarding transportation companies. These data include transportation capacity, prices, lead time, and reliability for each mode of transportation.

3 Conclusions

Global competition and distributed resources make it necessary for the extended enterprise to create and use a framework that enables the association of product development, supply chain management activities and manufacturing strategies. The application of the common knowledge model based on DM methods/approaches to the extended enterprise is directed at solving the above issue.

Future work should be focused on development mining tools and its application for particular supply chain issues such as wholesalers and retailers selection or customers behavior pattern identification.

References

Han J, Kamber M (2001) Data Mining: Concepts and Techniques. Morgan Kaufmann, San Francisco, CA

Mathieu R, Lavary RL (2006) Data warehousing and mining in supply chains. In: Wang J (ed) Encyclopedia of Data Warehousing and Mining. Idea Group Reference, Hershey, PA, p 323

Neaga EI, Harding JA (2005) An enterprise modeling and integration framework based on knowledge discovery and data mining. International Journal of Production Research, vol 43, edn 6, 15th March 2005, pp 1089–1108

Raisinghani MS, Medea LL (2005) Strategic decisions in supply-chain intelligence using knowledge management: an analytic-network-process framework. Supply Chain Management: An International Journal 10/2, pp 114–121

Samli AC, Pohlen TL, Bozovic N (2002) A Review of Data Mining Techniques as they Apply to Marketing: Generating Strategic Information to Develop Market Segments. The Marketing Review, 2002, edn 3, pp 211-227

Van Gor AR (2001) Demand & Supply chain management: A logistical challenge. 17th International Logistics Congress, Thessaloniki

Wang J (2006) Encyclopedia of Data Warehousing and Mining, Idea Group Reference, Hershey, PA, p 1096

A Pilot Project for Eco Industrial Parks in North Rhine-Westphalia - Extended Abstract

Veronika Wolf

Ministerium für Umwelt, Naturschutz, Landwirtschaft und Verbraucherschutz des Landes Nordrhein-Westfalen, Ref. VII-2,
e-mail: veronika.wolf@munlv.nrw.de, http://www.munlv.nrw.de/umwelt/wirtschaft_umwelt/gewerbegebiete/

Abstract: The state of North Rhine-Westphalia's pilot project for Eco Industrial Parks aims to integrate ecological, economic, and social considerations into the planning, development and operation of new and existing industrial and commercial parks in the state. Toward this end, the Ministry is helping eight of the state's communities to realize sustainable planning for pilot projects in various stages of development involving an extremely broad range of commercial and industrial park typologies and developmental phases. The project stakeholders are municipalities, economic development agencies, enterprises, and private investors. As part of a project-related kickoff event on April 11, 2006, the state of North Rhine-Westphalia granted participating municipalities the status of Eco Industrial Park partners for the project.

The structural change we have seen in North Rhine-Westphalia over the past decades has brought small businesses from the chemical, machine manufacturing, food processing, electronics, media and IT sectors – a total of 600,000 enterprises, all of which have their head offices in North Rhine-Westphalia. And they are joined by 50 of Germany's 100 largest corporations. North Rhine-Westphalia currently has approximately 1,300 commercial and industrial parks with a surface area of 10 or more hectares (24.71 acres) each. The state of North Rhine-Westphalia itself has a disproportionately large amount of residential and commercial areas: 22 percent versus 12 percent for Germany as a whole. Moreover, the state has a population density of 523 persons per square kilometer, which far exceeds the nationwide average of 228. And the march of development continues to encroach upon open space, with more than 100 hectares (247 acres)—and approx. 15 hectares (41 acres) in North Rhine-Westphalia—being lost to residential and road transport development every day in Germany.

In light of shrinking world resources and the considerable body of knowledge and practical experience that are available today with respect to economic development and its impact on human health and the environment, the question arose as to whether there might be potential for sustainable development of North Rhine-Westphalia's present and future industrial/commercial parks. Sustainable development of these locations from an ecological, economic and social standpoint could render them more efficient, appealing and ecologically compatible, while at the same time avoiding the ecological risk entailed by the potential development of new commercial areas, thus protecting the environment, the landscape, recreational areas and human beings.

As part of a pilot project that was initiated earlier this year, the North Rhine-Westphalia Ministry for Nature Protection, Agriculture, and Consumer Protection is currently helping eight communities to achieve sustainable development for their respective commercial/industrial parks. The facilities in these municipalities reflect the range of commercial/industrial park developments that are currently operating in North Rhine-Westphalia. I will now briefly describe these eight facilities:

- A new 12 hectare (30 acre) *ecological commercial park in the city of Kürten* in the Bergisches Land region.
  This facility features optimized development practices; compact buildings; structures realized along slope profiles; buildings oriented to the direction of the sun and wind; energy supplied by a neighboring lumbermill; innovative and sustainable water and rainwater management; involvement of the tenants in facility planning processes.
- Refurbishment and consolidation of an existing 15 hectare (37 acre) commercial park in the *city of Petershagen*, where the following activities are being realized:
  elimination of residual pollution, refurbishment of the existing buildings, structural consolidation, energy use optimization, development of sustainable water and wastewater management systems.
- Restructuring of an existing 20 hectare (49 acre) industrial district in the *city of Remscheid* involving the following activities:
  reuse of old factory buildings (some of which protected landmarks); consolidation and reuse of the district as a whole; realization of sustainable water and wastewater management systems; construction of new thoroughfares; elaboration of an energy plan based on the use of renewable resources; optimization of traffic access to the downtown area; bio-

tope protection and networking; integration of residential use; enhancing the overall appeal of the area and its buildings.

- Establishment of a 10 hectare (24.7 acre) "residential and occupational village" (*LuADo*) in the town of Oppenwehe featuring intelligent ecological planning and management of a village district that will be used for businesses, housing, and leisure time activities. Local businesses will also be involved in project planning and management.
- Sustainable development of two adjoining commercial and industrial areas (125 hectares; 309 acres) in the city of Bottrop (*Am Kruppwald* and *An der Knippenburg*) that were built in the 1960s.
  In this project, sustainable site development, traffic planning, water supply and wastewater management, construction, landscaping and energy supply activities will be realized in collaboration with the facility's envisaged 250 businesses.
- The *Campus Fichtenhain* building complex in Krefeld is a landmark site in a protected park adjacent to Europa Park Fichtenhain. The project involves building refurbishment, development and implementation of water and wastewater systems, waste management, and occupancy of the resulting commercial space by companies from the IT industry, as well as media and other design oriented sectors. One of the most exciting features of the project is the daycare center (located in Europa Park) that will provide up to 12 hours of daycare at a stretch for the children of individuals that are employed at the facility. In addition, Fichtenhain will become the first commercial park to provide year-round daycare for the children (including infants) of individuals that work nights or on weekends.
- Ecological construction and space use planning for *Alfred Judokus Kwak Haus* in Goch.
  In this project, the Hermann van Veen foundation will construct and operate a recreational facility for extremely needy families or families where one or more members have a chronic or acute medical disorder. The facility will meet state of the art standards in terms of its architecture, ecological sustainability, and clinical facilities.
- Development of a *virtual commercial space pooling resource* for the Kleve region.
  This project will enable the communities in the Kleve region to jointly manage commercial and industrial space use through the creation of a virtual space "pool." The project will elaborate the legal, economic and organizational framework for the realization of the space "pool" in a sustainable manner. Businesses that rent pool space will be legally required to meet North Rhine-Westphalia sustainability requirements.

The following *supplemental* and highly innovative projects are also being realized as an "extended wing" of the pilot project:

1. A project known as*: metabolon* involves the development of a sustainable 45 hectare (111 acre) commercial area on the site of a former refuse dump in Engelskirchen, with a view to converting the site into a competency and innovation center, as well as an educational facility, for substance conversion and environmental technology. The new facility will mainly focus on the processing of renewable raw materials using available landfill gas and biogas, as well as innovative photovolatic technologies.
2. In the *neue bahn stadt:opladen* project, the city of Leverkusen is reactivating a centrally located former rail depot with a view to sustainable realization of one of North Rhine-Westphalia's largest urban development projects, which will be used for commercial, industrial, residential and recreational purposes. A center for ecological construction has already been established on the site as an initial catalyst for sustainable development.

Scenario specific solutions and objectives will be elaborated by the participating municipalities and investors with the assistance of state/regional agencies and the North Rhine-Westphalia Ministry of the Environment, Nature Protection, Agriculture and Consumer Protection and state agencies.

Envisaged project milestones:

- Recruiting and involving project actors and stakeholders
- Realization of location and project analyses
- Obtaining funding by documenting the need for the project and realizing feasibility studies for it
- Determination of development objectives
- Determination of support requirements (troubleshooting potential weak points)
- Researching and employing specialized knowledge related to the various development objectives
- Establishment of a core group of expert advisors
- Realization of investigations and experts reports to support project implementation
- Development of scenario specific ecological, economic and social solutions
- Elaboration of scenario specific criteria and indicators
- Development of operational solutions

- Project implementation
- Project evaluation and documentation
- Implementation of sustainable management for project facilities

This will lead to the development of sustainable concepts in the following arenas:

- Recreational area management
- Land use management
- Building construction management
- Energy resource management
- Traffic management resource management
- Waste management
- Water
- Business models
- Social models

One focus of these efforts is the establishment of a network that will link the various pilot projects via innovative solutions.

The pilot projects will be supported by the establishment of the requisite communication platform; the realization of feasibility studies; analyses of potential problems; site analyses; the integration of lifecycle factors into planning processes; and the elaboration of sustainable facility operation and management models. The overarching goal is to elaborate proposals for the development of sustainable commercial facilities in North Rhine-Westphalia. Doing so will improve the employment situation in the state and enhance its citizens' quality of life. At the same time, it will help to achieve viable environmental stewardship.

Knowledge Acquisition and Risk Analysis in Material Recovery Facilities by a Virtual Team

Ioannis Dokas[1], Konstantinos Kitis[2], Christos Dodis[2], Demetrios Panagiotakopoulos[2], Demetrios Karras[3]

[1] University of Paderborn Institute of Informatics, Fürstenallee 11, 33102 Paderborn, Germany and Cork Constraint Computation Centre, University College Cork, Cork, Ireland.

[2] Democritus University of Thrace, Civil Engineering Department, 67100 Xanthi, Hellas.

[3] Chalkis Institute of Technology, Department of Automation, Hellas

Abstract[1]: A knowledge acquisition process that has been implemented by a virtual team for the development of a knowledge based early warning system in Material Recovery Facilities (MRFs) operations will be presented. During the process, hazard and risk analysis techniques have been used extensively. In this paper emphasis will be given on the: a) activities and workflow, b) communication methods and tools used, c) the model of collaboration The result was to acquire and codify the complicated picture of the events, which are acting as triggering mechanisms to a number of operational problems, incidents and accidents during MRFs operations.

1 Introduction

The recycling of waste became a major strategic goal in many countries, resulting to the design, construction, and operation of numerous facilities that sort and process household and commercial waste, as well as construction and demolition waste. This paper is dealing with the operational problems of the facilities that sort and process household and commercial waste

---

[1] This work has been supported mainly by the EU 5th FP project ADVISES RTN (Contract No HPRN –CT –2002-00288) and by the Hellenic Operational Programme for Education and Initial Vocational Training PYTHAGORAS II. During the last stages this work has been supported by the Marie Curie Fellowships for the Transfer of Knowledge (ToK) Grand No MTKD-CT-2006-04563.

which are commonly known as MRFs and are defined as [Gladding 2002]: A central operation where source segregated, dry, recyclable materials are sorted, mechanically or manually to market specifications for processing into secondary materials.

During the operations of solid waste treatment facilities, a considerable number of problems, faults, incidents and accidents can occur leading to direct and indirect consequences ranging from citizen complaints and increased operational cost to human lives losses. Statistics have shown that the overall accident rate for the waste industry in the UK during 2001/02 was estimated to be around 2,500 per 100,000 workers [HSE 2004]. This rate was about four times that year's national average. In particular, for scrap and MRFs the rates of incidents and accidents are not encouraging. In the 2004/05 UK statistics of fatal injuries [Hertel et al. 2005], the industry with the highest rate of fatal injury to employees was the recycling of waste and scrap, where the rate was approximately 27 times the national average. These statistics are revealing the size of occupational health and safety problem in the recycling industry, and point out the need for better and safer practices during the operational phase.

Public and occupational health and safety are topics of great importance in the field of solid waste management. Nevertheless, and in contrary to the large amount of research on the design and construction of safer solid waste treatment facilities, the research on their safe and invulnerable operation is very low, scattered, and mainly is focused in landfill operations. These attempts have been made by a small number of waste management associations like ISWA [1999] and SWANA [1998] by practitioners [Bolton 1995], and by researchers [Dokas and Panagiotakopoulos 2006, p 11].

The amount of analysis and research on MRFs operational problems is very limited as well. The UK DEFRA, after reviewing the relevant literature, concluded that a national estimate of emissions couldn't be made because of the absence of quantifiable environmental emissions [DEFRA 2004]. Gladding has conducted studies on the risks and hazards of MRFs operations [Gladding 2002, p 5]. She grouped the hazards in physical, chemical, and biological types, and focused mainly on the impact of aerosols, in the workers health and on the determination of occupational exposure limits for microorganisms, and biaerosols. These studies are highlighting a number of hazards on MRFs operations, as well as some of their adverse effects. Nevertheless, these sources of information do not show the entire picture of the coincidences that can trigger operational problems and thus the mechanisms that can trigger operational problems in MRFs are still not known.

The research goal is to develop a web - knowledge based - early warning system in MRFs operations which it will be able to estimate the occurrence possibility of operational problems during MRFs operations and to provide advice on how to prevent them. A high priority objective towards achieving this goal is to acquire and codify the expert knowledge about safe and invulnerable MRFs operations. Therefore, a team of researchers implemented hazard identification and risk analysis techniques within a knowledge acquisition process in two MRFs in Greece, under the guidance of their managers.

A characteristic that made this process challenging was that the members of the knowledge acquisition team were physically separated and had to interact, and coordinate their work electronically using media such as e-mail, telephone, internet telephony, video conference etc. as well as digital photographs and videos. In essence, the knowledge acquisition team was a virtual team. The result was to acquire and codify the complicated picture of the events, which are acting as triggering mechanisms to a number of operational problems, incidents and accidents during MRFs operations.

2 Virtual Team Description

As minimum consensus virtual teams consist of two or more persons who collaborate interactively to achieve common goals, while at least one of the team members works at a different location, organization, or at a different time so that communication and coordination is predominantly based on electronic communication media [Hertel et al. 2005]. Considering the work of Powell et al. [2004], there are several key issues that affect the virtual team performance. Some that are relevant to the virtual team of this study are listed in brief below. Some others are described in detail in the next paragraphs.

- *Project Duration*: 6 months from June 2005 to December 2005.
- *Members, Roles and Locations*: The virtual team had been composed of three members. One had been in Germany and had the role of the manager. The other two had been in Greece and had the role of the knowledge engineer / risk analyst (KE/RA) on the MRFs sites.
- *Skills*: The virtual team manager has had experience from real cases on knowledge acquisition and risk analysis by previous projects and background knowledge in solid waste management. The other two members have had background knowledge relevant to solid waste management but not experience from real cases on performing knowledge acquisition and risk analysis techniques. All members had been using electronic communications techniques before, each one for different purposes.

None was a member of a virtual team before, and none had previous real experience on MRFs operations.

- *Relationship, Cohesion and Trust*: The members had been working together as a traditional team in other projects before the formation of the virtual team. They already knew each other, had mutual respect and high level of trustworthiness.
- *Cultural Differences*: There was no cultural difference between the team members.
- *Face-to-Face Meetings*: One at the beginning of the project.

3 Knowledge Acquisition Process: Tasks of Team Members, Interaction and Communication

The knowledge acquisition process had three main activities that had to be performed. These were the design and preparation, the knowledge acquisition on site, and the evaluation and validation. The process is briefly described in Figure 1. Detailed description of each activity is given bellow.

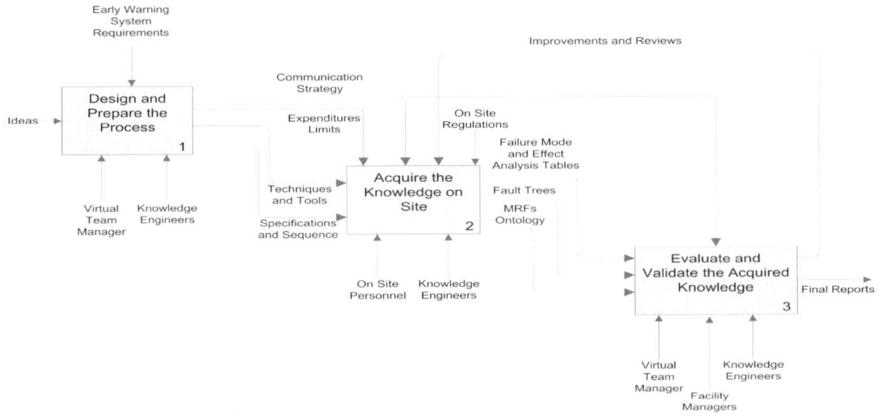

Fig. 1. The knowledge acquisition activities described with IDEF0 diagram

Design and Preparation:

During this activity the virtual team manager had defined the:

1. Analysis techniques that would be used later by the KE/RA on site, after he first considered the objectives and the requirements of the early warning system and the potential difficulties in communication between the virtual team members.
2. Acceptable level of expenditures.

3. Specifications and the sequence of applying the analysis techniques during the next phase of the knowledge acquisition (i.e. which will be described in detail in the next paragraph).

He also provided the KE/RA with necessary documents and tools to start familiarizing themselves with the analysis and communications techniques.

The selected techniques were the Failure Mode and Effect Analysis (FMEA), and the Fault Tree Analysis (FTA). Both are widely known risk and reliability analysis techniques and many published works describes them. These characteristics were important because the KE/RA did not have previous experience with risk analysis techniques. In addition, both techniques were considered very effective from a communication point of view because the acquired knowledge and information can be displayed through tables and logic diagrams (i.e. fault tree diagrams) respectively. Moreover, the virtual team manager identified the need of a common vocabulary between the team members. Therefore, the development of an ontology that describes MRFs was considered very important for the success of the entire knowledge acquisition process.

The KE/RA familiarized themselves with the analysis techniques and the corresponding tools. They also made binding agreements with two MRFs managers for help and guidance during the on site knowledge acquisition phase. Both KE/RA and the virtual team manager reviewed the literature for documents that describe MRFs operations and operational problems and made some initial virtual communications. Through this a general communication strategy was established which was based on different communication scenarios. Based on these scenarios the virtual team manager:

1. Requests status report from the KE/RA.
2. Reports any identified error and provide advice on how to be corrected.

The KE/RA:

1. Request advice on how to perform a task or they request additional information from the manager.
2. Submit the acquired knowledge to the manager.
3. Communicate between them for experience sharing and for comparing the operational procedures in different MRFs.

By performing these scenarios the communication media that were preferable in each occasion and for each scenario were identified. For example, the videoconference was identified as the preferable media for the case of the 1st scenario, while the combination of e-mail, telephone and internet telephony, for the rest cases, depending on the emerging need for communication and on the available resources.

Knowledge Acquisition on Site and Validation of Knowledge:

During the on site knowledge acquisition phase the KE/RA worked separately in the Athens and Patras MRFs in Greece. At first, a KE/RA has been occupied in the MRF of Athens for a period of two months and right after the second KE/RA was occupied in the MRF of Patras for another two months. That strategy was chosen in order to have the second KE/RA learned from the experience of the first. The KE/RA interacted with the facility personnel, following the on site regulations, in order to develop and deliver the MRFs ontology, the FMEA tables and the fault trees diagrams. The KE/RA were collaborating with the virtual team manager and with the corresponding MRFs managers. The collaboration model is illustrated in Figure 2.

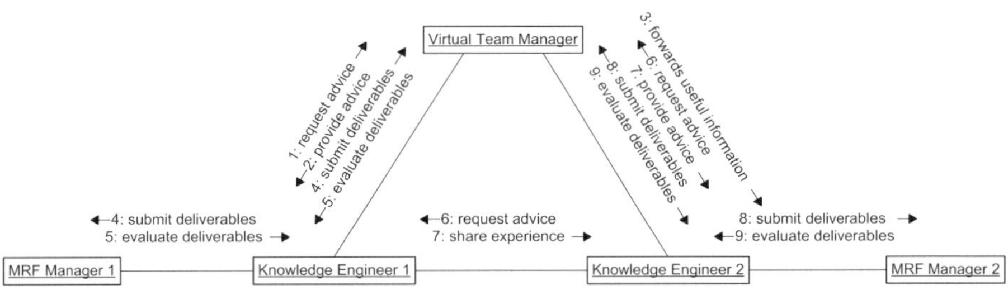

Fig. 2. The collaboration model of the virtual team including the MRFs managers

By developing the ontology, the KE/RA managed to record and hierarchically describes the physical items of the MRFs like the personnel, the infrastructure, the equipment etc. as well as the workflow, providing simultaneously useful definitions. Afterwards, the KE/RA applied the FMEA analysis for each and every physical object within the ontology. During this stage the following steps were followed:

1. The possible modes of failure of each object within the MRFs were identified, along with the possible causes.
2. The likely effect of each failure was determined along with the frequency of each failure mode, the level of its severity, and the degree to which the effect of each failure is detectable.

Finally, the KE/RA performed the FTA following the process that has been described in [Dokas and Panagiotakopoulos 2006]. The FTA is a graphical representation that relates, via logic gates, the failure modes -or the top events- of a system with its causes. The aim was to connect all the previously identified causes via AND-OR logic gates with the top event (i.e. the failure mode) forming a fault tree and simultaneously to enrich, and validate, the results of the FMEA.

Periodically, the KE/RA were submitting the deliverables to the MRFs managers and to the virtual team manager for evaluation. The MRFs managers had to validate the acquired knowledge. That was not an easy task at first because MRFs managers did not have the appropriate background to evaluate the ontology, the FMEA and the FTs. In addition, the virtual team manager had to evaluate the correctness of the FMEA tables and FT diagrams. The deliverables they were only considered acceptable when the managers approved their validity and correctness. In order to reach that point several iterations had to be made. At the end, each KE/RA made a final report illustrating the acquired knowledge about MRFs operational problems.

4 Results

During the preparation stage and based on the literature review, eight operational problems have been recognized. During the risk analysis phase in the facility, the operational problems that have been analyzed were 34. These problems can be triggered by around 250 events/causes. Ontology development helped the analysts to fully comprehend the facility, understand the workflow, and to record the majority of physical objects that can contribute to failures. FMEA tables have been developed for the majority of the physical objects identified and recorded in the ontology. In addition 34 fault trees have been developed in which all the identified causes are related with the top events.

The 250 MRF operational problems causes that have been detected can be categorized as follows:

- 9% are events that cannot be controlled by the personnel like extreme weather conditions, strikes of municipal workers, improper recycling habits by organizations and citizens etc;
- 9% of the failure modes causes had to do with human errors and human shaping factors like low performance due to lack of light, bad calculations and judgments etc;
- the failure modes causes that were related to bad facility design were 3.5%;
- 3.5% was also the numbers of failure causes that have organizational-managerial nature like bad estimation of workers that must work in a particular shift and
- the rest 75% is related with equipment malfunctions and maintenance and with the nature of the incoming recyclable refuse.

Next goal is to use this knowledge to develop the knowledge base of the early warning system.

5 Conclusions

This paper described a knowledge acquisition process applied by a virtual team in two MRFs in Greece. The main goal was to identify and analyze MRFs operational problems. This work has proofed that virtual teams can be used to perform the task of knowledge acquisition. However, in order to work effectively the virtual team must have a manager/coordinator with previous experience in knowledge acquisition. In addition, the development of ontology and the use of diagrams and tables helped significantly the communication of the members of the virtual team.

References

Bolton N (1995) The handbook of landfill operations: A practical guide for landfill engineers, owners and operators. Blue Ridge Solid Waste Consulting, Atascadero

Department for Environment, Food and Rural Affairs (DEFRA) (2004) Review of environmental and health effects of waste management: Municipal solid waste and similar wastes. The University of Birmingham. http://www.defra.gov.uk/environment/waste, access: 2006-02-07

Dokas I, Panagiotakopoulos D (2006) A knowledge acquisition process to analyse operational problems in solid waste management facilities. In: Waste Management Research, vol 24, pp 332 - 344

Gladding TL (2002) Health risks of materials recycling facilities. In: Hester RE and Harrison RM (eds) Environmental and health impact of solid waste management activities. Issues in environmental science and technology, vol 18. Royal Society of Chemistry, Cambridge, pp 53-72.

Gladding T, Thorn J, Stott D (2003) Organic dust exposure and work related effects among recycling workers. In: American Journal of Industrial Medicine, vol 43. Wiley, pp 584-591

HSE (2004) Mapping health and safety standards in the UK waste industry RR240. http://www.hse.gov.uk/research/rrpdf/rr240.pdf, access: 2006-02-07

HSC, Statistics of fatal injuries (2004/2005). http://www.hse.gov.uk/statistics/overall/fatl0405.pdf, access: 2006-002-07

Hertel G, Geister S, Konradt U (2005) Managing virtual teams: A review of current empirical research. In: Human Resource Management Review, vol 15. Elsevier, pp 69-95

ISWA Working Group on Sanitary Landfill (1999) Operations guidelines. ISWA

Powell A, Piccoli G, Ives B (2004) Virtual teams: A review of current literature and directions for future research. In: The DATA BASE for Advances in Information Systems, vol 35, pp 6-36

Rodić L (2000) Reliability of landfill technology. TU Delft, Holland

SWANA (1998) Training sanitary landfill operating personnel. SWANA

Early Warning and Risk Management – An Interdisciplinary Approach

Walter Hürster[1], Thomas Wilbois[2]

[1] T-Systems GEI GmbH, Ulm, Germany

[2] T-Systems, Ulm, Germany

Abstract: An interdisciplinary approach to Early Warning and Risk Management is described in this paper as well as the general technical implementation of Early Warning and Risk Management Systems. Based on this approach, a concept has been developed for the design of an Integrated System for Coastal Protection. In addition to this, as a prototype implementation of a modern environmental monitoring and surveillance system, a system for the remote monitoring of nuclear power plants is presented here in more detail, including a Web Portal to allow for public access. The concept, the architectural design and the user interface of Early Warning and Risk Management Systems have to meet high demands. It will be shown that only a close cooperation of all related disciplines is able to fulfill the catalogue of requirements and to provide a solution which is suitable for environmental monitoring and surveillance, for early warning and for emergency management.

1 Introduction

The frequency and the severity of environmental disasters seem to increase continuously: almost every day TV news reports on disastrous events from somewhere in the world. Earthquakes, tsunamis, volcanic eruptions, hurricanes, heavy thunderstorms, floods and avalanches alternate with tanker collisions, oil spills, coastal pollution, wildfires, accidents in chemical or nuclear plants (related to the emission of toxic gases or radioactive nuclides) and terrorist attacks, affecting human and animal populations, surface infrastructures, atmosphere and oceans.

The loss of human life and the tremendous damages caused by those catastrophes as well as the increasing sensitivity of the general public make it reasonable to protect the population and the environment by means of a new generation of intelligent surveillance, information, early warning and emergency management systems. This includes a highly sensitive monitoring, fast and reliable prognostic calculations, but also a timely dissemination of the relevant information to the general public within the endangered area and in adjacent regions. This challenge calls for a cooperation of academia, industry and public administration, for interdisciplinary approaches involving physicists, chemists, biologists, computer scientists and application engineers.

2 Interdisciplinary Approach and Integrated Concept

In a first step, the relevant stakeholders and experts from all related disciplines will have to meet and to produce the catalogue of requirements, as a prerequisite for a more detailed functional specification and for the technical design of the solution. Facing the complexity of the challenge, it is obvious quite from the beginning that only a network of computers with dedicated individual tasks and appropriate communication structures will provide a promising approach to solve the problem of monitoring, surveillance, threat prediction, decision support, early warning and emergency management. However, defining an adequate logical structure for those networks, a variety of subtasks and prerequisites have to be fulfilled in each case in order to include intelligence in various ways and to reach the goal of an integrated operational system:

- Development/integration of adequate sensor systems and sensor networks (autonomous or remotely controlled) providing data and background information. This may include mobile sensory platforms and remote sensing systems (air space surveillance and satellite systems). The variety of sensor types to be used clearly depends on the disaster type and the involved scientific disciplines.
- Advanced modeling, i.e. development/improvement of scientific prediction models for prognostic calculations of each disaster type supporting interpretation and extrapolation of data, e.g. calculation of the atmospheric dispersion in case of NBC releases, taking into account the current meteorological situation and the expected forecast values
- Integration of scientific prediction modules into operational systems and definition of adequate interfaces to allow for fast response actions even under critical conditions
- Integration of GIS functionality for an online situation display, using various types of geographical maps to facilitate the recognition of affected areas

- Development and implementation of risk analysis and decision support systems based on sensor data, prognostic calculations and disaster specific experience
- Development/improvement and inclusion of knowledge management components (in general and disaster specific) to integrate the long-term experience of experts and observations made in the past – a challenging task for computer scientists and artificial intelligence.
- Definition of adequate information, alarm and warning strategies, using the full range of state-of-the-art technology, such as satellite communications, web technologies, radio broadcast, TV, telephone (fixed network and mobile systems), siren systems and loudspeakers
- Definition of appropriate interfaces to catastrophe handling and resource management subsystems (transport facilities, shelters, medical care, etc.)

Figure 1 shows an example for the integration of remote sensing systems (radiological surveillance, helicopter flight path).

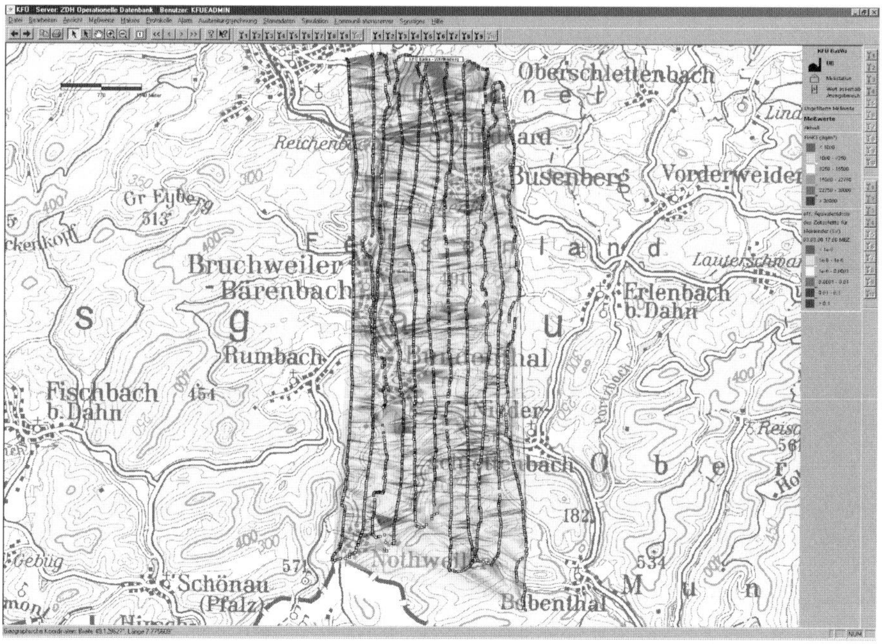

Fig. 1. Radiological Surveillance, Helicopter Flight Path

3 Application of the Conceptual Approach: An Integrated System for Coastal Protection

The present situation with respect to coastal protection is characterized by a lack of superimposed intelligence, recognition of the situation, co-ordination, communication and integration of the existing subsystems and sources of information. Therefore, it is indicated to choose an integrated, essentially centralized approach for a possible solution.

One essential component of such an integrated system for coastal protection will be a central data and communication platform where all relevant data streams converge and which can act as an information turntable and distributor for the connected subsystems. It can also serve as a knowledge base for spread information and for the sources that can be accessed, if required.

Then, there are the main functional components or subsystems (see Fig. 2):

- Surveillance
- Threat analysis and situation display
- Early warning and alarm raising
- Emergency management

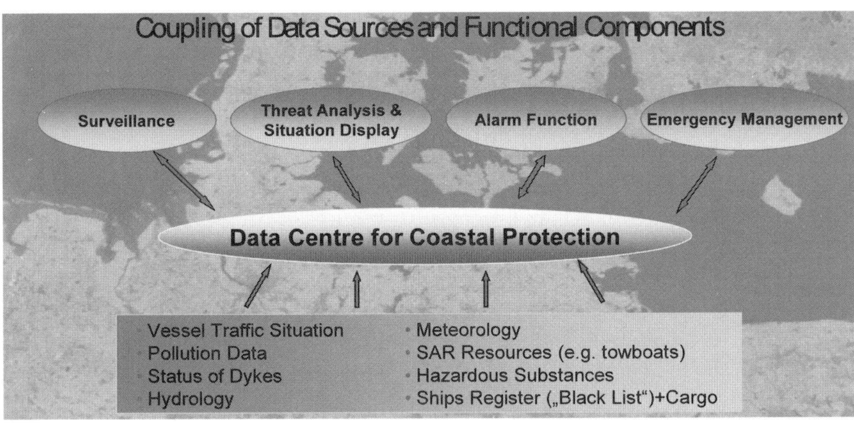

Fig. 2. Integrated System for Coastal Protection

Surveillance here primarily means surveillance of vessel traffic, but also surveillance of the air space and of maritime pollution (e.g. oil slicks). Various types of sensor systems, including mobile sensor platforms, are to be used for this purpose (helicopters, marine patrol aircraft, and satellites).

In addition to online representation of the current situation, a threat analysis has to be accomplished if indicated by the danger situation at hand. This could, e.g. require collision calculations and collision warnings for vessel traffic, as well as calculations for the dispersion of harmful substances in the water, on the surface and in the atmosphere. In doing so, the current meteorological and hydrographical data will have to be taken into account as well as their forecast values. Contributions from biology, chemistry and medicine are very essential at this step.

With respect to early warning, alarm raising and information of the public, all usable and widespread systems of telecommunication and media technology have to be applied.

Computer supported emergency management means in particular digitized emergency protocols with a suitable representation on geographical maps as well as numerous status overviews (dykes, embankments, evacuation areas, emergency sheltering capabilities, traffic routes, means of transportation, hospitals, ambulances, mission and SAR vehicles, etc.)

An additional and essential feature of computer support is simulation, i.e. the possibility to prepare mission scenarios, to modify them and to run them as a training exercise for emergency management.

Finally, there remains to be mentioned the fact that effective coastal protection is only possible if international data exchange takes place among the different systems of neighboring coastal countries, fulfilling the requirements of early recognition of danger situations and mission coordination across national boundaries.

4 The Elements of the System Structure

It is strongly recommended to use a communication server to handle the input data streams from various sensor systems and sensor networks and to convert different data formats, if necessary. This server may also handle the connection to external information networks. In case of higher system loads or larger systems, a specific server computer may be used for this purpose, e.g. a dedicated web server.

Since most of the data are usually needed for central access, the concept of a central database has been confirmed by practical experience. The distribution of specific subsets of the data for decentralized use (e.g. remote application servers) can be achieved by means of dedicated replication functions.

Prediction models and prognostic calculations, such as those for the atmospheric dispersion of toxic gases or radionuclides, tend to need large sets of data and therefore are also best served by a central access. Agent technologies and service access layer ontology's have proven to be most suitable for the integration of the various scientific components into an operational system.

Decision support systems, based on data describing the actual situation as well as the prognostic development, will reflect the implications for the population in the affected areas and thus help the crisis squad to make their decisions. These systems will have to include advanced knowledge management components and sets of metadata providing access to the long-term disaster specific experience of experts and to observations made in the past.

The increasing sensitivity of the general public and the enhanced public awareness of environmental threats call for Web-based information systems. Therefore, and in order to allow for public access to the system, it is recommended to provide a Web Portal.

5 The Web Portal

It is obvious that various user groups and stakeholders have their specific needs and therefore emphasize different aspects of the system. The following user groups can be identified and categorized (see Fig. 3):

- Administrative Sector
- Operational Sector
- Restricted Public Sector
- Public Sector

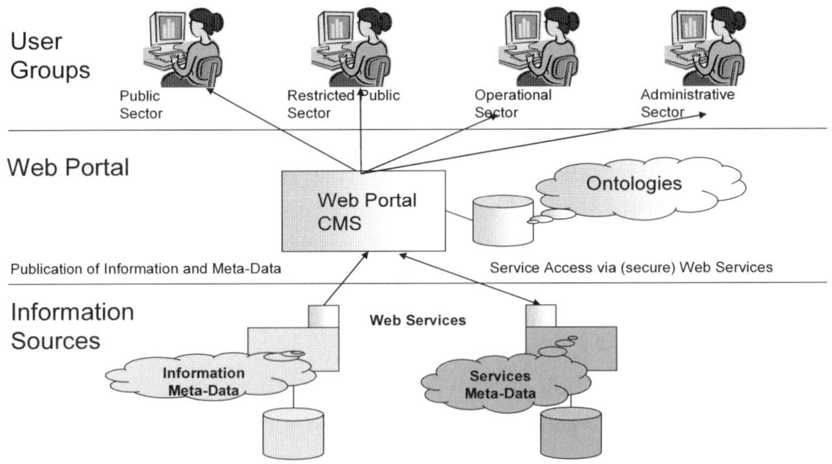

Fig. 3. Basic concept, overview and structure

The administrative sector covers the system administration, maintenance of configuration lists, adaptation and optimization of the system itself and of the related workflows.

The operational sector deals with the main task of the system, i.e. surveillance and monitoring functions, display of the current and prognostic situation, risk assessment and decision support.

The restricted public sector will provide the necessary information for the crisis squad, for public services (the staff of rescue forces, fire brigades etc.) and all other authorities responsible for civil protection. This may contain confidential information or security related orders which are not foreseen for public disclosure, e.g. in order to avoid panic reactions and pillage.

Finally, the public sector will serve as an information platform for the general public, giving an overview about the current threat situation, exposure risks and the development of these risks. The public sector will also provide general and specific recommendations in case of an imminent dangerous situation.

The large extent and the complexity of the available information combined with various views of diverse user groups call for specific selection and preparation of the data for display in graphical and/or tabular form (depending on the user group). This is the core point for the design and implementation of the Web Portal: to provide for each user group a specific set of web pages which contain all information that is needed to achieve the assigned tasks in the best possible way.

By analyzing the required functionalities and the customer needs, a set of requirements can be derived. The main aspects are:

- Harmonization and matching of the heterogeneous sets of information
- Electronic situation display, incl. animation features
- Simplicity of the user interface
- Modern display capabilities, especially for graphical representations
- Possibility to combine various representations
- Easy-to-use approach by offering well structured information
- Definition of user groups by means of hierarchical access privileges
- Well targeted preparation of the presentations ("generated by experts, to be viewed by anybody")
- Automated, timely publication of (selected) information and metadata
- Publication of reports via secure web services (e.g. in alarm situations)
- Implementation of an "intelligent" public warning system
- Improvement of the emergency management capabilities by introducing workflow tools and corresponding templates
- High quality standards with respect to safety, security and system availability – even under emergency conditions

Further requirements can be deduced from the demand for reliability and high performance of the portal. In any case, a highly reliable network infrastructure with good performance will have to be provided.

A typical publication of web reports may a.o. include the following information:

- Protective measures, including their representation on geographical maps
- Evacuation areas/safe areas and their current availability status
- Overview of arterial roads, traffic flow and means of transportation (incl. status information)
- Layout plans for assembly and collecting points in case of collective transports
- Emergency stations for medical care, including hospital trains and hospital ships
- Overview of regular hospitals and medical centers (incl. status and capacity information)
- Layout plans of schools, kindergartens and retirement homes
- Layout plans of emergency sheltering capabilities, such as gymnasiums, roof covered stadiums

6 The Reference Implementation: A System for Radiation Protection

As a prototype example for a modern environmental monitoring and surveillance system and for public access to it, a system for the remote monitoring of nuclear power plants (RM/NPP) will be presented here in more detail.

This system includes the collection of radiological and meteorological variables that have an influence on the diffusion and deposition of radioactive nuclides. A central role of the monitoring system is the use of these variables in the calculation of radiation exposure values and areas. These results are used for decision support, dissemination of information and the issuing of public warnings.

In the event of an imminent, occurring or already terminated release of radioactive nuclides, safety measures pertaining to disaster control and the provision of radiation protection could be required. For instance, the distribution of iodine tablets or precautionary evacuations is included among these measures.

In its role as a supervisory authority for the nuclear facilities (Obrigheim, Philippsburg and Neckarwestheim) in the Federal State of Baden-Württemberg, Germany, and for foreign facilities close to the German border (Fessenheim/France and Leibstadt/ Switzerland), the Ministry of Environment in Baden-Württemberg has been operating such a remote monitoring system for nuclear power plants for almost 20 years. Recently, the system has been completely renewed using modern hardware platforms and software technologies [Obrecht et al. 2002; Hürster et al. 2005].

As described by Hürster et al. [2005], the RM/NPP is a complex measuring and information system which records and monitors approximately 20 million data sets per day. The actual operational state of the nuclear facilities including their radioactive and non-radioactive emissions is automatically recorded around the clock, independently of the operator of the nuclear power plant. In addition, the RM/NPP system continuously collects meteorological data at the sites and also receives data from external measuring networks (national and international). It provides numerous possibilities to visualize the data and to check them against threshold values and protection objectives. In the case of a radioactive leak, potentially affected areas can be determined at an early stage by a transport calculation [Schmidt et al. 2002] and protective measures can be adopted by the Ministry in cooperation with the authorities responsible for civil protection.

In order to allow for a broader but selective access to the information kept within the Operational System, the decision was taken by the Ministry to establish a web access function by means of a dedicated Web Portal [Hürster et al. 2006]. Similar applications are envisaged by the Federal States Baden-Württemberg and Saxony-Anhalt in order to open the access to general environmental information, as imposed by legislation [Schlachter et al. 2006].

Both, the operational system and the web portal have been designed and developed in accordance with the logical structures described above [see Land Baden-Württemberg 2004; Wilbois and Chaves 2005]. The client software offers numerous possibilities to visualize the data by means of a modern graphical user interface with GIS functions. Also, it provides standardized export interfaces to office and graphical applications.

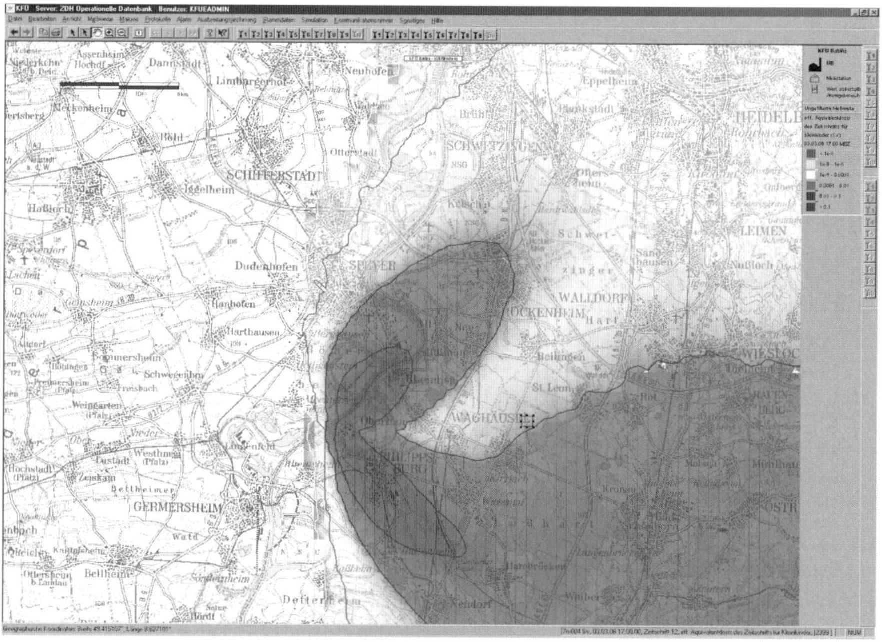

Fig. 4. Propagation cloud on the background of a topographical map

For demonstration purposes, a first prototype version of the Web Portal has been implemented which is being intensively used and is therefore considered to be highly accepted by the user groups. Actually, an animated presentation of a propagation cloud has been selected thus illustrating the results of a Dispersion Modelled Transport Calculation (DMTC) for radio nuclides (Fig. 4). This type of calculation has to be carried out in case of a radioactive incident or accident and the result is of greatest importance for radiological protection and emergency management.

6.1 The Importance of Meteorological Data in the Remote Monitoring of Nuclear Power Plants

The collection of the significant meteorological variables (with influence on diffusion and deposition of radioactive nuclides) and the calculation of radiation exposure in the vicinity of nuclear power plants constitute a central responsibility of a remote monitoring system for nuclear power plants, especially with regard to the determination of the affected area and for decision support.

For instance, the as-precise-as possible determination of precipitation pattern is an essential factor in the wet deposition of radioactive nuclides, which (as a logical consequence) has a considerable influence on radiation exposure ("hot spots"/ see Fig. 5). The precipitation events in the southern part of Germany during the Tchernobyl nuclear accident are a prominent example of this. By precipitant washout, these rainfalls contributed to a considerably stronger deposition of radioactive nuclides south of the Danube River in Bavaria and Upper Swabia than in the rest of Germany. Therefore taking into account an inhomogeneous distribution of precipitation considerably increases the quality of dispersal calculation results.

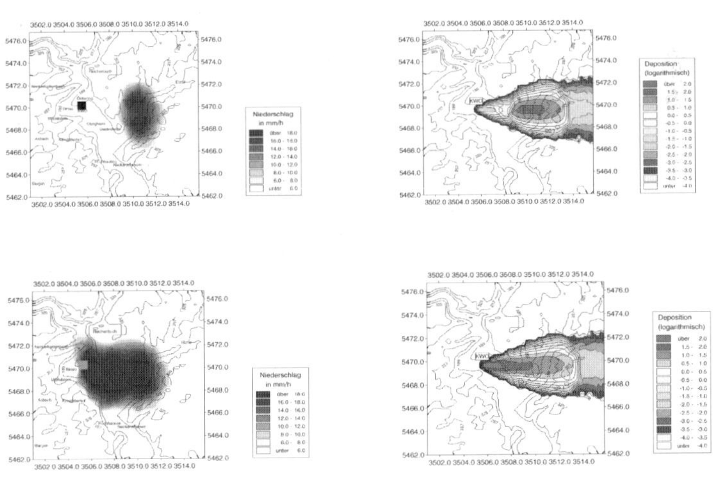

Fig. 5. Precipitation and radioactive deposition ("hot spots")

In order to be able to carry out a reliable risk assessment it is even necessary to expand the meteorological basis — in terms of space (supraregional data) and time (prognostic data). These data are made available in Germany by the German Meteorological Service (Deutscher Wetterdienst, DWD). Here, the "DWD Local Model" (LM) is applied for the utilization in the prognostic dispersal calculation. The LM is the most precise and most crucial DWD weather forecast model and it encompasses a time domain of up to 72 hours ahead. Radar precipitation data and forecast data as provided by the Local Model allow for a more realistic and accurate calculation of radiation exposure.

7 Conclusion

Due to the increasing importance of early warning and emergency management systems and recognizing the great attention paid to the subject by a sensitive general public, a large number of initiatives and projects on national, international and even global scale are searching for adequate solutions. It has been shown that, based on an interdisciplinary approach and on a detailed requirements analysis, an integrated concept for Early Warning and Risk Management Systems can been derived. In a logical sequence, an IT concept has been produced in accordance with the integrated concept and with the aim to fulfil the identified requirements to the largest possible extent. The feasibility of the concepts has been proven by the application to an Integrated System for Coastal Protection and by the prototype implementation for the Remote Monitoring of Nuclear Power Plants.

The reference implementation of the operational system and the pilot installation of the Web Portal received a great deal of interest from the user groups. The good cooperation with all of them produced an optimistic view for further developments and implementations.

Having started with coastal protection and with an improvement of radiation protection (including the related emergency management), we are confident that the system presented here can significantly contribute to finding a general solution to the indicated problems. The proof will be left to international multi risk scenarios and corresponding across border exercises, supported by the system capabilities described above.

Acknowledgments

The system "Remote Monitoring of Nuclear Power Plants" was contracted to T-Systems by the Federal State of Baden-Württemberg, Ministry of Environment, as a turnkey system, with the integrated service "DMTC" provided by the Institut für Kernergetik und Energiesysteme (Institute for Nuclear Energetics and Energy Systems) of the University Stuttgart (IKE). The research work related to the development and integration of the DMTC was supported by the Ministry of Environment within the framework "Environmental Information System" Baden-Württemberg. The underlying program modules of the DMTC were taken from the library of the OECD Nuclear Energy Agency.

References

Chaves F, Wilbois T, Grinberg E (2005) IT-Konzept für die Erstellung eines KFÜ-Portals. Fraunhofer IITB, Karlsruhe (Germany)/ T-Systems GEI GmbH, Ulm (Germany)

Hürster W et al. (2005) Remote Monitoring of Nuclear Power Plants in Baden-Württemberg, Germany, In Hilty L, Seifert E, Treibert R (eds) Information Systems for Sustainable Development. Idea Group Publishing, Hershey

Hürster W et al. (2006) A Web Portal for the Remote Monitoring of Nuclear Power Plants. In: Tatnall A (ed) Encyclopaedia of Portal Technology and Applications. Idea Group Publishing, Hershey

Land Baden-Württemberg (2004) Das Corporate Design im Internet, Styleguide für das Landesportal und die Ministerien-Websites, mit allgemeinen Regelungen zur Gestaltung weiterer Auftritte. Land Baden-Württemberg, 07/2004, Stuttgart (Germany). http://www.baden-wuerttemberg.de/internetstyleguide

Moßgraber J et al. (2005) Entwicklungshandbuch für WebGenesis Version 7.10. Fraunhofer IITB, Karlsruhe (Germany)

Obrecht R et al. (2002) KFÜ BW - Erneuerte Kernreaktorfernüberwachung in Baden-Württemberg. In: Mayer-Föll R, Keitel A, Geiger W (eds) (2002) Projekt AJA, Anwendung JAVA-basierter Lösungen und anderer leistungsfähiger Lösungen in den Bereichen Umwelt, Verkehr und Verwaltung – Phase III 2002. Wissenschaftliche Berichte FZKA-6777, Karlsruhe (Germany), Forschungszentrum Karlsruhe, 2002. http://www2.lfu.baden-wuerttemberg.de/lfu/uis /aja3/index1.html, access: 2007-02-13

Schlachter T et al. (2006) Environmental Portals of the federal states of Baden-Wuerttemberg and Saxony-Anhalt with access to the administrative environmental information. In: Tatnall A (ed) Encyclopaedia of Portal Technology and Applications. Idea Group Publishing, Hershey

Schmidt F et al. (2002) KFÜ-ABR – Weiterentwicklung des Dienstes Ausbreitungsrechnung in der Kernreaktor-Fernüberwachung Baden-Württemberg.

In: Mayer-Föll R, Keitel A, Geiger W (eds) Projekt AJA, Anwendung JAVA-basierter Lösungen und anderer leistungsfähiger Lösungen in den Bereichen Umwelt, Verkehr und Verwaltung – Phase III 2002. Wissenschaftliche Berichte FZKA-6777, Karlsruhe (Germany), Forschungszentrum Karlsruhe. http://www2.lfu.baden-wuerttemberg.de/lfu/uis/aja3/index1.html, access: 2007-02-13

Wilbois T, Chaves F (2005) Fachkonzept für die Erstellung eines KFÜ-Portals. T-Systems GEI GmbH, Ulm (Germany)/ Fraunhofer IITB, Karlsruhe (Germany)

W3C (2002) Web Services Description Language (WSDL). World Wide Web Consortium (W3C)/ Web Services Description Working Group. http://www.w3.org/2002/ws/desc/, access: 2007-02-13

W3C (2005) The Synchronized Multimedia Integration Language (SMIL 2.0) – Second Edition. World Wide Web Consortium (W3C), Recommendation 07 January 2005. http://www.w3.org/AudioVideo/, access: 2007-02-13

Development of a GIS-based Risk Assessment Methodology for Flood Pollutants

Axel Sauer, Jochen Schanze, Ulrich Walz

Leibniz Institute of Ecological and Regional Development (IOER), Weberplatz 1, 01217 Dresden, Germany

Abstract: The article presents the current development of a methodology to assess flood risks due to contaminants with a high spatial resolution. The integrated approach combines methods from flood risk analysis with those from contaminant risk assessment. It focuses on toxicological receptors of urban areas represented by land-use data. For each receptor risks are analyzed considering the hydrodynamic contaminant distribution and the exposure of specific receptors. These risks cover both the recurrence probability of certain flood events with their matter fluxes and the exposure of receptors with dose-response relationships. GIS is used for data management, calculation of algorithms and display of the results in a DSS.

1 Introduction

Research on the consequences of extreme flood events has usually focused on damages to buildings and infrastructure caused by water, whereas only little interest has been put on pollutant impacts. Especially following the 2002 flood in the Elbe River basin, negative effects due to contaminants transported by flood waves could be observed and became of increasing meaning [cp. Geller et al. 2004]. Most of the substances were deposited bound to sediments, thus polluting children's playgrounds, allotments, fields, and meadows. People, animals, and plants as well as soils were affected by the contaminated sediments. Management of flood risk should therefore integrate procedures to assess the multiple risks caused by different substances which influence different receptors.

At present, methods can be found for the analysis and assessment either of flood risks caused by surface and groundwater [e.g. Merz 2006; Schanze 2006] or risks to humans stemming from contaminants [Swartjes 2002; Ferguson et al. 1998; Mathews et al. 1998; Mathews et al. 1999]. Most studies dealing with both contaminants and flooding focus on the physical and chemical processes and their simulation [e.g. Stuyt 2003]. Integrated approaches for the assessment and management of the multiple contaminant risks due to extreme floods – notably taking the spatial dimension into consideration – are still missing.

This gap should be filled by the modeling and assessment framework of the SARISK research project: "Development of a pollutant fate and transport model for the risk analysis and assessment of contaminants due to extreme flood events taking the city and the district of Bitterfeld as an example" [von Tümpling et al. 2006]. The project is being funded under the German RIMAX research programme (Risk Management of Extreme Flood Events) of the BMBF (German Federal Ministry of Education and Research). In the following, the state of development of the integrated assessment methodology contributing to the above mentioned project is presented.

2 Risk Assessment Methodology

The development of the integrated risk assessment methodology consists on the combination of methods from flood risk analysis and contaminant risk assessment. To facilitate the understanding of the new approach, relevant aspects of the underlying methods are briefly described.

2.1 Methods for Flood Risk Analysis

Flood risk in general can be defined as the probability of negative consequences due to flooding. The occurrence probability of flood events with certain magnitudes represents the hazard. Negative consequences of such hazard depend on the vulnerability of specific receptors, e.g. damage potentials of buildings [cp. Schanze 2006].

Hydrological and hydraulic models are used to quantify the discharge, water level, flow velocity etc. of a flood hazard. Their results are overlayed with investigations on flood vulnerability like economic classifications of the damage functions of building types. Risks are finally calculated as probable damages for a certain time span, often displayed in risk maps.

The approach of SARISK consists on hydraulic modeling of flood and matter-flux propagation based on pre-calculated flood waves and flood frequencies. The models provide a 2d-inundation simulation and a fate, transport and sedimentation simulation of selected substances. They lead to the probability of the spatial deposition of certain substances [von Tümpling 2006].

2.2 Methods for Contaminant Risk Assessment

A common approach for the analysis of risks from contaminants defined as negative consequences for man and the environment is the quantitative Human and Environmental Risk Assessment [e.g. Paustenbach 1989; Paustenbach 2002], here in particular the Quantitative Risk Assessment (QRA) for human health. This method predicts the exposure of a person to a contaminant by modeling the Daily Intake Rates (DIR) via the ingestion, inhalation and dermal adsorption route, taking multiple pathways and different contact media into account. Examples are the direct ingestion of polluted soil by children through hand-to-mouth contacts [Stanek 1995] or the inhalation of contaminated house dust.

This step usually follows a risk assessment which is carried out by comparison of the DIR with the tolerable daily intake rate (TDIR), i.e. the daily dose over lifetime without observable adverse health effects. These substance-specific dose-response relationships are determined by toxicological studies and can be accessed through public databases (e.g. US Environmental Protection Agency's ECOTOX, World Health Organisation's EHCs) [Hassauer et al. 1993].

At present, several software tools for exposure analysis as part of QRA exist, amongst others CLEA, UMS, CSOIL, CalTOX, and SADA [cp. Swartjes 2002]. Neither those systems are specified for contaminant impacts due to flooding nor do they include the spatial dimension, except SADA.

2.3 Integrated Risk Assessment Methodology for Flood Pollutants

The methods from the above mentioned flood risk analysis and contaminant risk assessment are combined in an integrated approach. To explain the way of linkage, the Source-Pathway-Receptor-Consequence (SPRC) model is used [ICE 2001]. Figure 1 gives an overview of the whole approach, indicating a two-tired SPRC chain. Sources and pathways are mentioned twice, firstly representing the hazard determination as part of the flood risk analysis, secondly displaying the initial steps of the contaminant risk analysis. All steps designate principle features and do not specify the scale of time, space and process. A further description of the methodological elements can be found in Table 1.

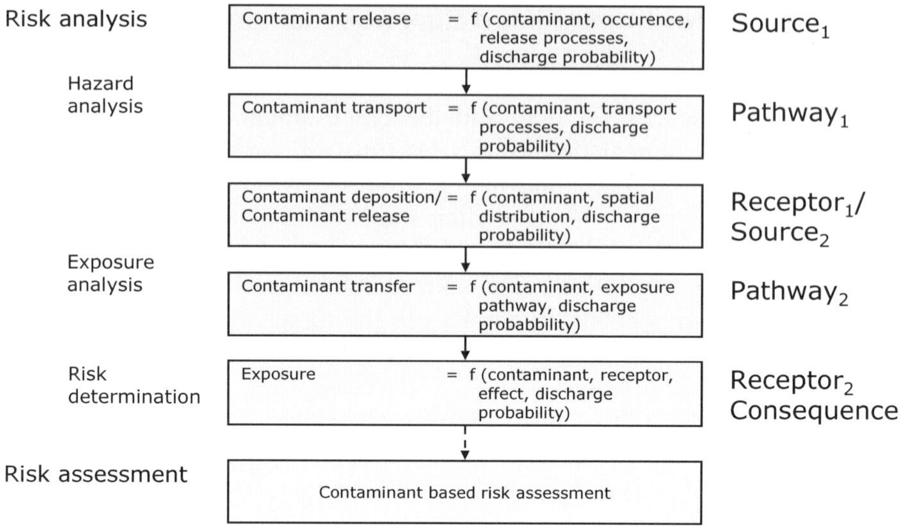

Fig. 1. Overview of the general methodology

At first, the hazard analysis (i) describes the release, transport and accumulation of substances based on flood characteristics and substance properties. It covers the identification of pathways from the sources to the interim deposition. Results are spatially distributed substance concentrations in environmental media, e.g. sediments.

The following exposure analysis (ii) is based on this spatially distributed substance depositions. It consists of a receptor analysis which indicates and characterizes potential receptors taking their localization and properties into account. In essence, it performs the linkages between the contaminant sources to the receptors via the pathways.

In a next step, the exposure analysis than quantifies the receptors' exposition to a certain substance by calculating the transfer, deposition and intake rates. Results are amounts of absorbed contaminants, e.g. for the receptor human Daily Intake Rates (DIR).

The contaminant risk determination (iii) describes the effects (consequences) on the receptors by dose-response relationships, forming the basis for the final substance-based risk assessment. Simpler approaches are used in cases where no dose-response relationships for certain receptors exist [e.g. Blume and Brümmer 1991].

Table 1. Two-tiered SPRC chain for analysing and assessing flood-pollutant risks

Method	SPRC-Element	Description
Flood hazard analysis	S_1	Contaminated land and areas with high soil background levels. Probability of contaminant release during flood events.
	P_1	Distribution of contaminated sediments in flooded areas with respect to the frequency and magnitude of a certain flood event.
Contaminant risk analysis and assessment	S_2	Contaminated sediments deposited during and after a flood event.
	R_1/S_2	Deposition of contaminated sediments (depending on P_1).
	P_2	Contaminant intake for exposed receptors from different contact media via direct and indirect pathways.
	R_2	Receptors, like humans, animals, plants, soils or surface water.
	C	Adverse (health) effects for receptors leading to disturbances of their functions; assessment of the tolerability of disturbed functions based on dose-response-relationships.

3 Operationalization for the Study Area

The methodology is being developed and tested in the catchment of the Mulde River, taking the city and the district of Bitterfeld (Germany) as an example. Here, we focus on the four heavy metals Arsenic, Lead, Cadmium and Mercury as relevant inorganic contaminants. These substances are primarily originating from abandoned mining dump sites in the upper reaches of the Mulde (Saxonian Ore Mountains). During flood events, deposits in the lower reaches are eroded, transported and sedimented and form secondary sources of contamination (S_2). Risks from these flood pollutants have to be expected for the following receptors of the study area: Humans, animals, plants and soil.

3.1 Methodological Implementation

The flood hazard analysis as the first step consists of hydraulic modeling and coupled fate and transport simulations which are carried out by the UFZ Centre for Environmental Research Leipzig-Halle and the USF-

Institute of Environmental Systems Research Osnabrück [cp. Tümpling et al. 2006]. Results are dynamic simulations and maps of heavy metal distributions for certain flood scenarios (e.g. magnitude for return periods, dike breaches), constituting the contaminant flood hazard (S_1, P_1). In the following contaminant risk analysis a GIS-based receptor analysis is carried out. It identifies and characterizes receptors and exposure pathways for different land uses. The latter are derived from ATKIS data (see below).

Table 2. Selected pathways [and processes] for the receptor human

Media	No	Pathways
Sediment	1	Sediment–[Ingestion]–Human
Plant	2	Sediment–[Adhesion]–Plant–[Treatment]–Plant Products–[Ingestion]–Human
	3	Sediment–[Dissolved Transport]–Soil (Water)–[Absorption]–Plant–[Treatment]–Plant Products–[Ingestion]–Human
	4	Sediment–[Adhesion]–Plant–[Treatment]–Plant Products–[Ingestion]–Animal–[Treatment]–Animal Products–[Ingestion]–Human
Air	5	Sediment–[Particulate Transport]–Outdoor Air–[Inhalation]–Human
	6	Sediment–[Particulate Transport]–Outdoor Air–[Particulate Transport]–Indoor Air–[Inhalation]–Human
	7	Sediment–[Particulate Transport]–Outdoor Air–[Particulate Transport]–Indoor Air–[Deposition]–House Dust–[Ingestion]–Human

Due to the particular relevance of the receptor human, its risk analysis and assessment methodology is presented in more details. Land-use types where the occurrence of humans can be assumed are linked to tables containing the relevant pathways and parameters necessary for the calculation of human exposure. The basic algorithms are derived from Ihme [1994] who developed a comprehensive and general exposure model, comparing numerous calculation procedures.

Table 2 lists the selected pathways including contact and transfer media and processes under consideration. For these elements the relevant parameters and exposition factors are assigned. They can be distinguished into receptor-specific factors such as inhalation rate or body weight as well as pathway-specific factors like e.g. concentration in the contact media, transfer rates, and enrichment factors. These factors are derived from exposure factor standards such as AGLMB [1995], ExpoFacts or the soon released RefXP database [cp. Okken et al. 2004]. The latter is based on fate and transport simulations as well as on receptor specific databases, e.g. the TRANSFER database of the UBA (German Federal Environmental Agency) containing soil-to-plant transfer factors [Knoche et al. 1999].

As a result of the calculation, human exposure is spatially explicitly presented for each model patch. Exposure maps combined with TDIRs lead to maps displaying the risk index (DIR/TDIR). Values greater than 1 indicate a risk for adverse health effects, whereas values less than 1 show the opposite.

3.2 Spatial Data and Technical Implementation

To develop the methodology in a way which is applicable throughout Germany, generally available data sets are used. The basic geometry stems from the ATKIS Basis-DLM (Digital Landscape Model). Additional data are biotope and land-use types (Biotop- und Nutzungstypen BNT), statutory land-use plans (Flächennutzungspläne FNP) for land-use functions and soil maps provided by the soil surveys of the federal states (Bodenkarte 1:50 000 BK50).

The ATKIS objects and their attributes are described in a Microsoft-Access database. Relational tables link the objects to receptors, pathways and the corresponding parameters. Similar constructs are used for the FNP and BNT data, as well as for the BK50. Calculations of exposure are implemented within ESRI's Arc Macro Language (AML) of ArcGIS and within the MS-Access database using SQL and Visual Basic.

4 Conclusions and Outlook

The methodology presented is dedicated to an integrated assessment of flood pollutant risks caused by different substances. Due to the fact that multiple receptors and different flooding scenarios are under consideration, highly integrated results covering a broad range of possible situations will be achieved. The spatially high resolved approach will allow an identification of concrete contaminant flood risks with a practical relevance for flood risk management. It may contribute to an improved land-use planning as part of long-term flood risk management and to risk mitigation during running events as part of the operational flood risk management.

The generic approach and its foundation on commonly available data sets such as ATKIS will allow the transfer of the methodology to other flood-prone areas in Germany. The regular maintenance of this data set provides results that are spatially up-to-date.

Further developments of the methodology need to adapt and broaden the approach in order to cover the specific properties of other groups of contaminants, e.g. organic substances such as Hexachlorcyclohexanes (HCH) which also play an important role as flood pollutants.

Another important task of the future work of SARISK will be the accessibility and visualization of the results for decision makers and the general public. Therefore, a web-based decision support system (DSS) will be built. The tool will be tailored according to the needs of local decision makers and the citizens being directly affected. Meeting these requirements will be realized by the use of a participatory approach including actor analysis and expert interviews as well as (actor) workshops.

Acknowledgement

The project is being funded by the German Federal Ministry of Education and Research (BMBF) under the RIMAX research programme.

References

AGLMB Arbeitsgemeinschaft leitender Medizinalbeamtinnen und -beamter (1995) Standards zur Expositionsabschätzung (Standards for Exposure Assessment). Hamburg

Ferguson C, Darmendrail D, Freier K, Jensen BK, Jensen J, Kasamas H, Urzelai A, Vegter J (1998) Risk Assessment for Contaminated Sites in Europe. vol 1. Scientific Basis, Nottingham

Geller W, Ockenfeld K, Böhme M, Knöchel A (eds) (2004) Schadstoffbelastung nach dem Elbe-Hochwasser 2002 (Contaminant Pollution following the Elbe River Flood 2002). Umweltforschungszentrum Halle-Leipzig GmbH, Magdeburg

Hassauer M, Kalberlah F, Oltmanns J, Schneider K (1993) Basisdaten Toxikologie für umweltrelevante Stoffe zur Gefahrenbeurteilung bei Altlasten (Basic Toxicological Data for Substances Relevant to the Environment in the field of Hazard Assessment for Contaminated Land). Umweltbundesamt, Berlin

ICE Institution of Civil Engineers (2001) Learning to live with rivers. Final report of the Institution of Civil Engineers' Presidential Commission. Institution of Civil Engineers, London

Ihme W (1994) Allgemeines Modell zur Quantifizierung der Exposition des Menschen durch kontaminierte Böden (General Model for Quantifying the Human Exposure to Contaminated Soil). Shaker, Aachen

Knoche H, Brand P, Viereck-Göttle L, Böcken H (1999) Schwermetalltransfer Boden-Pflanze: Ergebnisse der Auswertungen hinsichtlich der Königswasser- und Ammoniumnitrat-Extraktion anhand der Datenbank TRANSFER (Heavy-metal transfer soil-plant: Results from analysis of aqua regia and ammonium nitrate extraction using the TRANSFER database). Umweltbundesamt, Berlin

Mathews T, Fischer AB, Exner S, Eikmann T (1999) ECORISK-Verfahren zur Expositions- und Risikoabschätzung in der Gefährdungsabschätzung und Sanierungsplanung von Altlasten. Teil II: Ableitung von Sanierungszielwerten mittels probabilistischer Methoden (ECORISK-procedure for Exposure and Risk Assessment in Hazard Assessment and Remediation Planning of Contaminated Land. Part I: Derivation of Remediation Target Values with Probabilistic Methods). Umweltmedizinische Forschung & Praxis. vol 4, edn 5, pp 289–300

Mathews T, Fischer AB, Exner S, Eikmann T (1998) ECORISK-Verfahren zur Expositions- und Risikoabschätzung in der Gefährdungsabschätzung und Sanierungsplanung von Altlasten. Teil I: Allgemeiner Überblick über die Systematik des Verfahrens (ECORISK-Procedure for Exposure and Risk Assessment in Hazard Assessment and Remediation Planning of Contaminated Land. Part I: General Overview of the Systematic). Umweltmedizinische Forschung & Praxis. vol 3, edn 3, pp 156–167

Merz B (2006) Hochwasserrisiken: Grenzen und Möglichkeiten der Risikoabschätzung (Flood Risks: Limitations and Potentials of Risk Assessment). Schweizerbart'sche Verlagsbuchhandlung, Stuttgart

Okken P-K, Mekel O, Mosbach-Schulz O, Wintermeyer D, Schümann M, Herrmann J, Fehr R, Hehl O, Peters C, Bubenheim M, Timm J (2004) Xprob – Forschungsprojekt zur bevölkerungsbezogenen Expositionsabschätzung (Xprob – Research Project for Population-based Exposure Assessment). Umweltmedizinischer Informationsdienst. vol 3, pp 41–44

Blume H-P, Brümmer G (1991) Prediction of Heavy Metal Behavior in Soil by Means of Simple Field Tests. Ecotoxicology and Environmental Safety, vol 22, pp 164–174

Paustenbach DJ (ed) (1989) The Risk Assessment of Environmental and Human Health Hazards: A Textbook of Case Studies. Wiley, New York, Chichester, Brisbane

Paustenbach DJ (ed) (2002) Human and Ecological Risk Assessment: Theory and Practice. Wiley, New York, Chichester, Brisbane

Schanze J (2006) Flood Risk Management – A Basic Framework. In: Schanze J, Zeman E, Marsalek J (eds) Flood Risk Management: Hazards, Vulnerability and Mitigation Measures. Springer, Dordrecht (NATO Science Series IV, Earth and Environmental Sciences, vol 67, pp 1–20

Stanek EJ, Calabrese EJ (1995) Daily estimates of soil ingestion in children. Environmental Health Perspectives. vol 107, edn 3, pp 276–285

Stuyt LCPM, Reinders JEA, van der Hoek EE, Hermans AGM, de Munick Keizer M, Kamperveen, JP, Hoogvliet MC, Gerritsen AL, Hounjet MWA, Bontenbal MP, Hoefsloot FVJ, Icke I (2003) Environmental impact of flooding. Delft Cluster, Delft (Delft Cluster Publications DC1-233-8)

Swartjes FA (2002) Variation in calculated human exposure: Comparison of calculations with seven European human exposure models. RIVM, Bilthoven

von Tümpling W, Rode M, Büttner O, Baborowski M, Böhme M, Gläßer C, Sonnabend B, Matthies M, Schulz M, Schanze J, Walz U, Sauer A (2006) Entwicklung eines Schadstoffausbreitungsmodells zur stoffbezogenen Risikoanalyse und -bewertung extremer Hochwasserereignisse am Beispiel des Landkreises und der Stadt Bitterfeld (Development of a pollutant fate and transport model for the risk analysis and assessment of contaminants due to extreme flood events taking the city and the district of Bitterfeld as an example). Umweltwissenschaften und Schadstoff-Forschung, vol 18, pp 129–131

Modeling Arsenic and Oil Contamination After High Water Events in the Town and Floodplain of Bitterfeld (Germany)

Marcus Schulz[1], Olaf Büttner[2], Michael Matthies[1], Michael Böhme[2], Wolf von Tümpling[2]

[1] Institute of Environmental Systems Research (USF), University of Osnabrück, Barbarastrasse 12, 49076 Osnabrück, Germany

[2] Helmholtz Centre for Environmental Research – UFZ, Brückstraße 3a, 39114 Magdeburg, Germany

Abstract: Applying the two-dimensional model system Telemac2D, the transport of arsenic and oil during extreme flood events was modeled for the region Bitterfeld (Germany). Telemac2D includes sub-routines that describe hydrodynamics, transport and first-order partitioning of soluble and particle-bound pollutants. Oil spills were numerically treated as particulate pollutants with zero settling velocity. Simulations of the inundation of a small region after a dam-break revealed distinct small-scale contamination patterns and partly exhibited severe contamination. Comparison with measurements from the Mulde flood in August 2002 gives reasonable agreement. The model will be calibrated and validated by several measurements with sediment traps during flood events.
Key words: Pollutant contamination, two-dimensional model, arsenic, uncertainty analysis, oil spills

1 Introduction

In industrial regions, extreme high waters may cause severe contamination of inundated areas with organic and inorganic pollutants, as it was observed during the flood event of the Rivers Elbe and Mulde (Germany) in August 2002 [Krüger et al. 2005]. Potential risks for downstream areas arise from the mobilization of highly contaminated soils, but also from point sources, such as manure and oil tanks and pipes, deposits of pharmaceuticals, and chemical industries. In addition to the mechanical damages of private and public households caused by the flood wave, severe health problems of citizens in flooded areas may result from contact with contaminated floodwater and sedimentary deposits [Mielke et al. 2006].

Therefore, there is a need for preventive measures against the flooding of settled areas, which requires the knowledge of geo-referenced potential risks by pollutant exposure to humans and ecosystems. In turn, this knowledge can be derived from empirical observations. However until present, physical, chemical, and biological measurements during extreme flood events have seldom been performed, but numerical models represent an efficient alternative to predict hydrodynamics during and contamination by flood events [Ashauer et al. 2006].

For these purposes, we simulated a dam-break scenario in the region of Bitterfeld (Germany) and predicted the contamination with arsenic and from oil spills by using the numerical two-dimensional model system Telemac2D [Hervouet and van Haren 1996]. We intended not only to calculate pollutant concentrations in floodwater at the downstream boundary of the flooded area, but also to compute geo-referenced deposition fluxes of sediments and pollutants.

Telemac2D seemed to be a suitable tool for our intentions, as it was previously confirmed for flooding of terrestrial areas and wetlands [Hervouet 2000]. Various scenarios assuming different dam-breaks and water routes of the flood wave could be calculated to identify areas of high contamination in the region of Bitterfeld, feeding into a risk analysis and a decision support system (DSS) applicable to Bitterfeld and other industrialized floodplains.

2 Methods

2.1 Study site

The region of Bitterfeld and Wolfen is situated in the federal state Saxony-Anhalt in the central eastern part of Germany. Since the beginning of industrialization in the 19th century, Bitterfeld experienced severe pollution by local chemical industries, lignite open pits in the lower and ore mining in the upper catchments of the River Mulde. Highly concentrated residues, deriving from periodic arsenic and heavy metal exports during high waters of River Mulde, are evenly distributed within the floodplain, whereas other pollutants, such as HCH, dioxins, and mercury, are locally enriched in the vicinity of former and present chemical industries [Franke et al. 2005].

Closing the lignite open pit east of Bitterfeld and after the Mulde Dam had been built in 1975, the River Mulde bed moved through the mining hole. From that time on, the dam has acted as a sink of particulate pollutants transported by the River Mulde retaining up to 90% of seston load [Junge & Jendryschik 2003]. South east of Bitterfeld after ending the lignite mining activities in the so-called Goitzsche, the flooding with water from the River Mulde downstream the dam started in May 1999 and brought to a preliminary halt early 2002, at a water level of 71.5 m a.s.l.. In August 2002, a flood disaster raised the water level by 7 m within 2 days. (>90 Mio m^3 water). Afterwards, the water level lowered to the envisaged final level of 75 m. Partly downstream areas could be prevented from inundation [Wycisk et al. 2005]. However during the extreme high water in 2002, Bitterfeld was flooded to a great extent and experienced high contamination with chemicals and oil from fuel tanks.

2.2 Model system Telemac2D

The two-dimensional model system Telemac2D consists of a hydrodynamic (Telemac-2D), a transport (Subief-2D), and a water quality module (wq2subief). A detailed description of Telemac-2D is given elsewhere [Hervouet and van Haren 1996].

The hydrodynamic module enables to calculate flow velocity components and water depth with high temporal resolution ($\Delta t = 10$ s) and is based on a spatial mesh grid. At each node, the numerical code solves depth-averaged free surface flow equations as derived by Barré de Saint-Venant, and considers friction on the bed, turbulence, dry areas such as floodplains, entrainment and diffusion of a tracer by currents, particle tracking, and drag forces by vertical structures.

Subief-2D relies on the results of Telemac-2D and solves equations for transport, deposition, and erosion of several soluble or particulate tracers. Deposition and erosion fluxes are calculated by setting modeled flow velocities into relation to critical flow velocities for erosion and deposition.

The water quality module can be generated by the user and may describe the transformation of tracers, such as partitioning between the soluble and particulate phase. In our study, we used the following first-order equation to describe the partitioning via sorption and desorption between soluble and particle-bound arsenic:

$$\frac{d[As_{aq}]}{dt} = k_1 \cdot [As_s] - K_D \cdot [SM] \cdot [As_{aq}] \tag{1}$$

where $[As_{aq}]$ is the soluble arsenic concentration [g l^{-1}], $[As_s]$ is the particle-bound arsenic concentration in the free water [g l^{-1}], [SM] is the concentration of suspended matter [g l^{-1}], K_D is a coefficient of partitioning between the soluble and particle phase [l g^{-1}], and k_l is a kinetic coefficient [s^{-1}]. Changes in [As_s] due to partitioning were modeled to the negative rate of temporal changes in [As_{aq}].

Oil [kg m^{-2}] was treated as a particulate tracer with zero settling velocity, to account for the fact that oil from fuel tanks and upstream waters becomes distributed as a thin film on the water surface and settles, when the water level in the flooded area is decreasing to zero again.

2.3 Simulations

We re-calculated a hydrodynamic [Büttner et al. 2006] and two transport scenarios (arsenic and oil) of the flood event in August 2002 for a mesocale area situated between the Mulde Dam and the town Bitterfeld (Fig. 1). A regular grid of this area comprised 4,384 elements, each of approximately 1,250 m^2 size. Surface roughness was derived from a biotope survey. We simulated a dam-break at the western margin of the lake Goitzsche with subsequent flooding of the simulation area. Prior to flooding, the simulation area was completely dry. Water level at the dam-break was kept constant over the entire simulation period of 12 hours. Table 1 provides water discharge and loads of arsenic and oil at the dam-break. For the parameters K_D and k_l implemented in the transport and water quality model, we used generic values. Boundary loads and concentrations of soluble arsenic were taken from measurements in the flood wave in August 2002 [Junge et al. 2004; Tables 1 and 2]. Boundary particle-bound arsenic concentrations were calculated from seston concentrations measured in the flood wave and particulate arsenic concentrations in the high water layer of 2002 in the Mulde dam. For boundary oil concentrations and loads we assumed generic values.

Table 1. Water discharge, and loads of arsenic and oil to the simulation area at the dam-break

t [h]	Discharge [m^3 s^{-1}]	As_s [g As s^{-1}]	As_{aq} [g As s^{-1}]	Oil [kg s^{-1}]
1	226	28.3	4.5	753
2	226	28.3	4.5	753
4	226	28.3	4.5	753
6	195	24.4	3.9	650
8	6	0.75	0.12	20
12	0	0	0	0

Table 2. Input data for the transport scenarios of arsenic and oil.

Parameter	Value	
	Arsenic scenario	
K_D	31.0 l g^{-1}	
k_1	0.0001 s^{-1}	
Settling velocity of [As_s]	0.1 cm s^{-1}	
Settling velocity of [SM]	0.1 cm s^{-1}	
	Initial value	Boundary value
[As_{aq}]	0.0 mg l^{-1}	0.02 mg l^{-1}
[As_s]	0.0 mg l^{-1}	0.125 mg l^{-1}
[SM]	0.0 mg l^{-1}	0.5 g l^{-1}
	Oil spill scenario	
Settling velocity of oil	0.0 cm s^{-1}	
	Initial value	Boundary value
Oil concentration	0.0 kg m^{-2}	1.0 kg m^{-2}

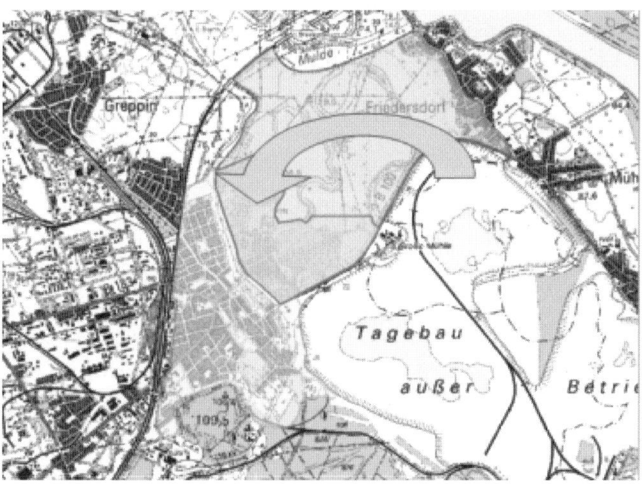

Fig. 1. Topographical map and dam-break scenario of the simulation area north-east of Bitterfeld.

2.4 Uncertainty analysis

An uncertainty analysis was conducted, calculating 1,000 Monte-Carlo simulations of the partitioning equation implemented in wq2subief. An uncertainty analysis was necessary, because we used a generic value of K_D, which under natural soil conditions can vary over several orders of magnitude for the same substance. Values of the input parameters were equal to the boundary concentrations given in Table 1, k_l was set to 0.02 s^{-1}, and K_D was normally distributed between 0.1 and 100 l g^{-1}. We computed 50 time steps of 1 s in order to simulate a complete reaction.

3 Results and Discussion

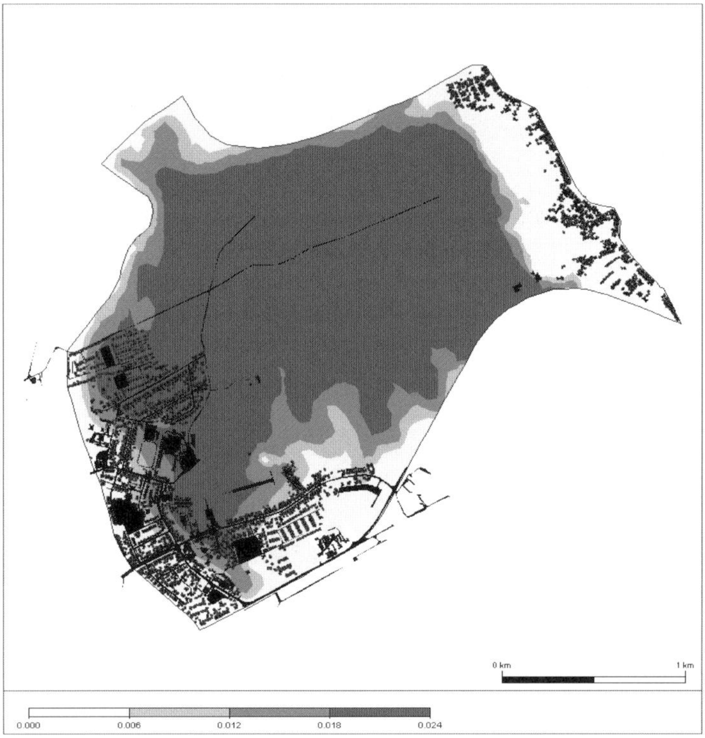

Fig. 2. Soluble arsenic concentration [mg As l^{-1}] in the simulation area at hydrodynamic steady state, 10 hours after start of flooding.

The simulated course of the flood event was in good agreement with observations of the high water in August 2002. Seven hours after dam break, the simulation area was nearly filled with flood water, and a hydrodynamic steady state was reached. Therefore, flow velocities amounted to 0.0 m s^{-1} in the entire basin 10 hours after dam-break. The fate of soluble arsenic (As_{aq}) was mainly influenced by advection and dispersion, rather than by partitioning with As_s (Fig. 2). Sedimentation of particle-bound arsenic (As_s) and suspended matter (SM) increased because of the quiescent conditions. Accordingly after 10 hours, deposition fluxes of As_s approximated to zero, because in the free water, As_s was scarcely available for settling.

However, higher deposition fluxes of As_s were calculated for the first six hours of simulation, especially at the margin of the flood wave, where the flow regime was supposed to be very turbulent and heterogeneous. Accumulated sediments revealed high spatial patchiness after 10 hours (Fig. 3) and were in good agreement with observed sedimentation patterns after the high water in August 2002.

In accordance with the two-dimensional simulation of As_{aq}, the uncertainty analysis exhibited that partitioning exerted little influence on As_{aq}. After complete reaction, As_{aq} showed a total range from 15.9 $\mu g\ l^{-1}$ to 34.6 $\mu g\ l^{-1}$. Mean As_{aq} and mean percentage error of As_{aq} amounted to 24.1 $\mu g\ l^{-1}$ and 17.8%, respectively. Probably, sorption and desorption had minor impact on As_{aq}, because the flood water already had traveled for several hours and therefore was in phase equilibrium.

Fig. 3. Sediment accumulation [m] in the simulation area at hydrodynamic steady state, 10 hours after start of flooding.

A simple equation was implemented to account for partitioning. Soil science has produced far more elaborate sorption models, which account for substrate heterogeneity in soils and suspended matter [Dijkstra et al. 2004]. However, our simple model, based on a K_D value, filled the need to describe sorption and desorption in a highly turbulent system, where reaction velocity is high, because of high activation energy and maximum particle surface as potential sorption sites.

The transport of oil was solely affected by advection and dispersion (Fig. 4), because the implementation of partitioning would not have been reasonable and any settling of oil did not occur. Until present, we cannot state if the oil spill scenario is plausible, and calibration and validation of the model for the region Bitterfeld is difficult without high-resolution aerial photographs of the high water 2002 and direct measurements of oil concentrations. Other oil spill models revealed a good performance, but needed far more computational effort [Al-Rabeh et al. 1992; Venkatesh and Murty 1994]. However, our model will be improved by considering cohesion and adhesion of oil films, which could improve the performance of the settling of oil during flooding.

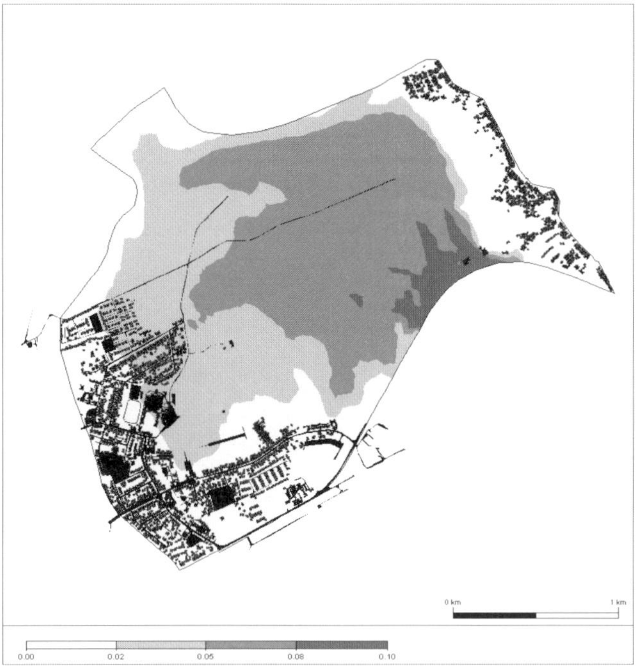

Fig. 4. Oil concentration [kg m^{-2}] in the simulation area at hydrodynamic steady state, 10 hours after start of flooding.

The simulations of arsenic and oil contamination in a small area showed plausible results with reasonable computational efforts. The model still has to be calibrated and validated, which will be done by using sediment traps during high waters. Overall, we succeeded in detecting areas of high deposition as potential hot spots of particulate pollutants, and realistic arsenic concentrations could be calculated, which again might display an increased risk for downstream areas.

Future simulations will be calculated for large areas in the region of Bitterfeld and will include point sources, such as fuel and manure tanks. The oil spill model will be extended allowing for simulating adhesion of supernatant oil films to vertical surfaces. The latter could be performed by setting the critical deposition velocity of oil close to zero. Thereby, we would suggest that adhesion occurs when flowing oil becomes decelerated to zero flow velocity by vertical structures, such as buildings. Furthermore, the processes of emulsification and evaporation will be considered within Telemac2D, partly accounting for losses of volatile oil components to the atmosphere.

Acknowledgements

We kindly acknowledge F. Walkow (Environmental agency of the district Bitterfeld) for the delivery of chemical data. M. Schulz and M. Böhme were funded by the German BMBF project SARISK.

References

Al-Rabeh AH, Cekirge HM, Gunay N (1992) Modeling the fate and transport of Al-Ahmadi oil spill. In: Water Air Soil Pollution, vol 65, edn 2. Springer, Berlin, pp 257-279

Ashauer R, Boxall A, Brown C (2006) Predicting effects on aquatic organisms from fluctuating pulsed exposure to pesticides. In: Environ Toxicol Chem, vol 25, pp 1899-1912

Büttner O, Böhme M, Sonnabend B, Gläßer C, Rode M (2006): Hydrodynamic flood simulation in urban areas in the region of Bitterfeld, Germany. In: Contribution to the conference " Strategies and Instruments for Improving Flood prevention". 23.-25.11.2006, Tangermünde, pp 213-216

Dijkstra JJ, Meeussen JCL, Comans RNJ (2004) Leaching of heavy metals from contaminated soils: An experimental and modeling study. In: Environ Sci Technol, vol 38, pp 4390-4395

Franke S, Heinzel N, Specht M, Francke W (2005) Identification of organic pollutants in waters and sediments from the Lower Mulde River area. In: Acta hydrochim hydrobiol, vol 33. Wiley InterScience, pp 519-542

Hervouet JM (2000) A high resolution 2-D dam-break model using parallelization. In: Hydrological Processes, vol 14, edn 13, pp 2211-2230

Hervouet JM, van Haren L (1996) Recent advances in numerical methods for fluid flows. In: Floodplain Processes, Anderson MG, Walling DE, Bates PD (eds). Wiley InterScience, Chichester, pp 183-214

Junge FW, Hanisch C, Zerling L, Czegka W, Arnold A, Jendryschik K, Morgenstern P, Treutler HC, Scharf B (2004) Investigations on the consequences of the high water in 2002 for heavy metal contents in seston and sediments within the Mulde Dam. In: Geller W, Ockenfeld K, Böhme M, Knöchel A (eds) Pollutant contamination after the Elbe flood, pp 136-158

Junge FW, Jendryschik K (2003) Investigations into the distribution of element concentrations in a recent dam and their seasonal and hydrographical correlation (Bitterfelder Muldestausee, Saxony-Anhalt). In: Acta hydrochim hydrobiol, vol 31. Wiley InterScience, pp 378-390

Krüger F, Meissner R, Gröngröft A, Grunewald K (2005) Flood induced heavy metal and arsenic contamination of Elbe River floodplain soils. In: Acta hydrochim hydrobiol, vol 33. Wiley InterScience, pp 455-465

Mielke HW, Powell ET, Gonzales CR, Mielke Jr. PW, Ottesen RT, Langedal M (2006) New Orleans soil lead (Pb) cleanup using Mississippi River alluvium: Need, feasibility, and cost. In: Environ Sci Technol, vol 40, pp 2784-2789

Venkatesh S, Murty TS (1994) Numerical simulation of the movement of the 1991 oil spills in the Arabian Gulf. In: Water Air Soil Pollution, vol 74. Springer, Berlin, pp 211-234

Wycisk P, Neumann C, Gossel W (2005) Flooding induced effects from the mining lake Goitzsche on groundwater and land-use in the Bitterfeld area. In: Acta hydrochim hydrobiol, vol 33. Wiley InterScience, pp 507-518

EcoClass – An Easy-to-access GUI for LCA Data to Assess Product Utilization Schemes

Volker Barth[1], Jens Krefeldt[2], Jan-Christian Marinesse[3]

[1] GELENA, Oldenburg University, Faculty 2, 26111 Oldenburg, Germany, e-mail: volker.barth@uni-oldenburg.de

[2] TOJAQ GmbH, Breite Str. 39, 38159 Vechelde, Germany

[3] OpenKnowledge GmbH, Bismarckstr. 13, 26122 Oldenburg, Germany

Abstract: We present experiences from using the EcoClass life cycle assessment tool in participatory product development. EcoClass is designed as an easily accessible, user-friendly tool providing data on three emissions categories during production and utilization of urban transport vehicles. In contrast to most other LCA models, EcoClass allows online modifications of vehicle utilization parameters to reflect changes in behavior. Its use with lay people revealed that the model in its current version is well suited to raise awareness and provide information, whereas it still needs improvement before it can be used for detailed production decisions.

1 Introduction

In recent decades, life-cycle assessment (LCA) has become a frequently used method in product development to assess ecological impacts of products or production methods. Addressing the ecological impacts of all processing steps from resource extraction to disposal/recycling requires large databases of technologies, inputs and by-products, and their associated emissions. Currently available software products to work on these databases are highly specialized tools that require expert knowledge and training in order to produce meaningful results.

Insights from innovation research indicate, however, that inclusion of non-experts like consumers during product development may be advantageous: companies take customers' needs and environmental concerns more seriously [van den Ven 1986], and combining the technical knowledge of companies with the contextual every-day knowledge of customers may lead to products that fit better to customers' needs [Lester et al. 1998]. However, consumer participation is less ambitious than political participation and aims mainly at increasing consumer satisfaction, commitment to the company, and market success [Hansen and Raabe 1991].

With rising environmental concerns among average consumers, the necessity to address environmental impacts already during the design phase becomes an important issue. LCA models offer the data and the methodology to assess these impacts in a comprehensive manner. However, the integration of LCA models in participatory product development processes is still challenging for mainly two reasons: consumers usually do neither have the expert knowledge to run usual LCA models nor are they used to use computer models at all, while existing models are neither designed nor being flexible enough to be used by non-experts in participatory processes. (This of course applies not only to LCA models; see e.g. Kasemir et al. [2003] for experiences with climate change models.)

We present the EcoClass model that has been used by lay users for participatory development of climate-friendly products according to the INNOCOPE guidelines in a pilot study in 2005 [Hoffmann 2006]. These guidelines feature model support to address and assess the ecological impacts of the product to be developed. The paper describes the main model features and reflects on its use during the INNOCOPE process.

2 The INNOCOPE product development process

The INNOCOPE process (INNOvation through COnsumer-integrated Product dEvelopment [Hoffmann 2006]) consists of a series of three workshops within approximately six months. Participants are consumers as well as company representatives from different divisions. When recruiting consumers care is taken to ensure diversity of social and demographic background and of affinity towards the product to develop, which was determined by the company in advance. During the first workshop the participants generate product ideas, which are evaluated, substantiated and refined in two subsequent meetings by using professional creativity and moderation techniques. The company is expected to take up ideas from the participants and work on their realization in between the workshops to provide feedback on workshop results from the production side, and to allow for reflection and discussion among the participating consumers.

The INNOCOPE method was first employed in a pilot study in 2005, where some 15 consumers and 4 company representatives developed a pedelec ('pedal electric cycle', a bicycle where pedaling is supported by an electric motor). In the first workshop, awareness for environmental issues was raised by an expert talk focusing on climate change and the environmental impacts of traffic. In the second workshop, participants used the EcoClass model (see section 3) to evaluate the environmental impacts of their current traffic behavior, and to compare the impacts of different means of transportations in more detail. It was also possible to evaluate the effects of substituting one or more vehicles in a mobility chain.

3 The EcoClass model

The EcoClass model (ECOlogical and CLimatic ASsessment Software, available at http://www.gelena.net) in its current version 0.2 is a tool to assess the environmental impact of vehicles used in urban personnel transport. Its purpose to visualize alternative ways to build and operate the product developed in the INNOCOPE process.

3.1 Database

To allow for greater flexibility to apply the final software we use a broad database covering lots of products and processing steps, which can easily be extended with few additions once a particular product is selected. We use the GEMIS database included in the LCA model GEMIS [Fritsche et al. 1989, 1994] in version 4.2 (http://www.oeko.de/service/gemis/).

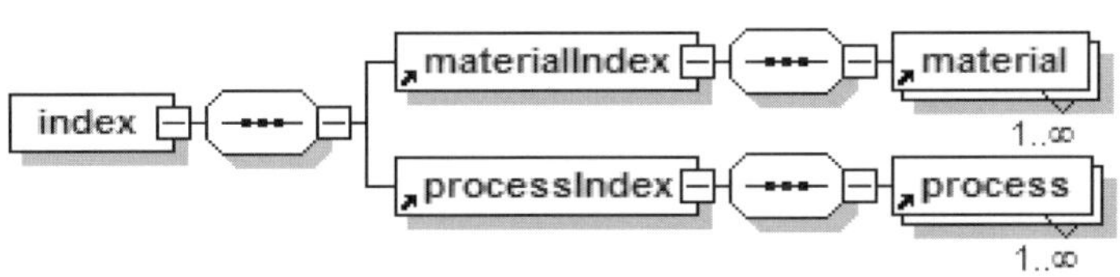

Fig. 1. An index XML file is loaded at application startup and acts as a memory-saving proxy to materials and processes. The material and process proxy elements have XML attributes (not shown here for complexity reasons) with values "pointing" to XML files containing the actual data. Their structure is shown in Figure 2.

EcoClass is written in Java to ensure availability on most major computer platforms. The proprietary database format of the Windows-based GEMIS model was converted to XML by exporting the database to the Microsoft Access format and extracting this using Java's ODBC interface. During this process care was taken to preserve the structure of the GEMIS database, although few ambiguities had to be resolved manually, and extraction of cost information has been skipped. Figures 1 and 2 show the structure of the XML database elements forming a meta model.

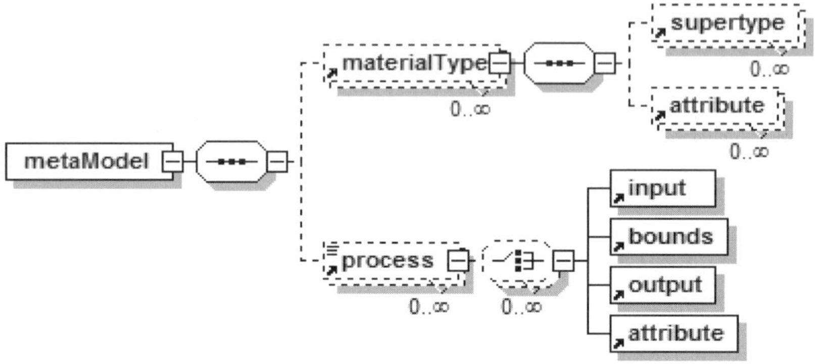

Fig. 2. The materialType and process elements including their subelements represent the meta model for EcoClass' product system (XML attributes not shown here for complexity reasons).

3.2 User Interface

The graphical user interface (GUI, Fig. 3) of EcoClass is a plugin for the Eclipse framework[1]. Main area of interaction is a graphical editor that employs the Graphical Editing Framework of Eclipse. Within this editor, users construct so-called product systems by selecting predefined production or utilization processes from the database and connecting the outputs of one process with the inputs of other processes in order to build a production chain. Inputs and outputs are materials or assembled products. The central process is a utilization process combining an input (e.g. car) with the energy input to operate it (e.g. gasoline) to produce one service unit as output, e.g. one person-kilometer (pkm). The amount of service units to be delivered in the given setup is defined separately.

Utilization processes extend standard production processes (from the GEMIS database) to provide a user interface to specify the operation mode of the product in a parameterized manner. In EcoClass 0.2, vehicles may differ in the passenger count, the road mix, or the driver's style of driving. Manipulating these specifications affects the energy use (and thus emissions) per service unit, the actual relation being either derived exactly (passenger count), or based on simple extrapolations (driving style, road mix). This feature allows the user to directly monitor different usage con-

[1] see http://www.eclipse.org

ditions (although at reduced accuracy), whereas in most other LCA models this can only be achieved by setting up a completely new process offline. Of course, this latter method can also still be employed in EcoClass.

Utilization processes in EcoClass have pre-selected input processes, so that it is usually only necessary to select a utilization process and specify the amount of service units that shall be delivered. It is, however, possible to manually exchange inputs of electric energy, e.g. to compare emissions of a pedelec whose batteries are charged with average household electricity with one that is running on green electricity. It is also possible to display all input processes of a given process and the input processes of the input processes and so on, so as to retrace the whole production chain as defined in the GEMIS database.

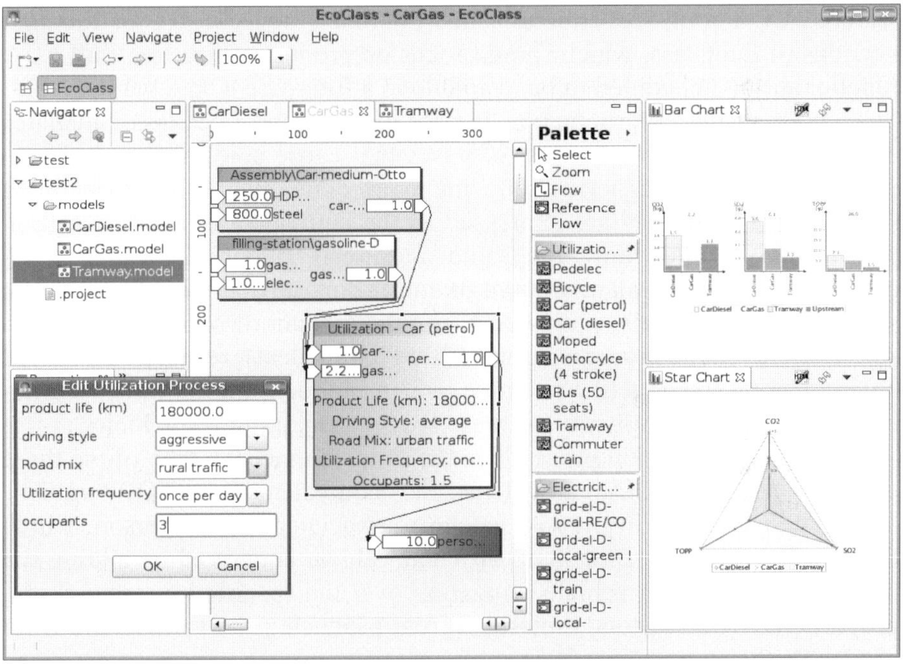

Fig. 3. Screen shot of the EcoClass GUI. The central part shows the graphical editor containing a product chain with one utilization process. The two smaller boxes above represent the processes that provide the inputs of the utilization process. The little box right below the utilization process is used to define the amount of service units to be used. Some of the utilization parameters have just been changed in the dialog box in the lower left of the screen shot and will be transferred to the utilization process after clicking "Ok". Graphical visualization of results is shown on the right. The user may rearrange all parts of the GUI.

3.3 Calculation of Results

The GEMIS database provides two sets of emissions associated to a given process: those actually emitted during execution of that process (direct emissions), and the total emissions that have been set free over the whole process chain. Using this second set allows to calculate the emissions "from the end" of a process chain and to skip the tedious explicit setting up of all steps that are required to build the final product. This is helpful when only the utilization phase shall be examined since the user can concentrate on product operation. On the other hand, this restricts the model to the process chains (and the emissions), that are predefined in the database; it is not possible to assess production alternatives during the construction of the product. Therefore, the model loses much of the characteristics of a fully featured LCA tool. Nevertheless, this approach was chosen for the first versions of EcoClass, which should focus on product utilization. Full LCA functionality is scheduled to be available in a future EcoClass version 1.0.

To address the environmental impact of products, EcoClass calculates emissions of greenhouse gases, of gases that cause acid rain, and of precursor gases of tropospheric ozone, the major component of photochemical smog. These quantities are expressed as the equivalent amount of carbon dioxide that would cause the same additional greenhouse effect (CO_2 equivalent), the equivalent amount of sulfur dioxide that would lead to the same acidification (SO_2 equivalent), and the amount of ozone that will result from the ozone precursor gases (TOPP equivalent), respectively. Other environmental impacts like resource use, waste, or water pollution are neglected to reduce complexity, as these are less important for vehicles.

The actual calculation proceeds in the same manner for each of the three emission categories CO_2 equivalents, SO_2 equivalents, and TOPP equivalents. The emissions in a given category associated to one person, which drives a specified distance in a particular vehicle is derived by adding the direct emissions from vehicle utilization over the specified distance to the proportional share of total emissions from producing the vehicle and from producing the energy (e.g. fuel) to operate it.

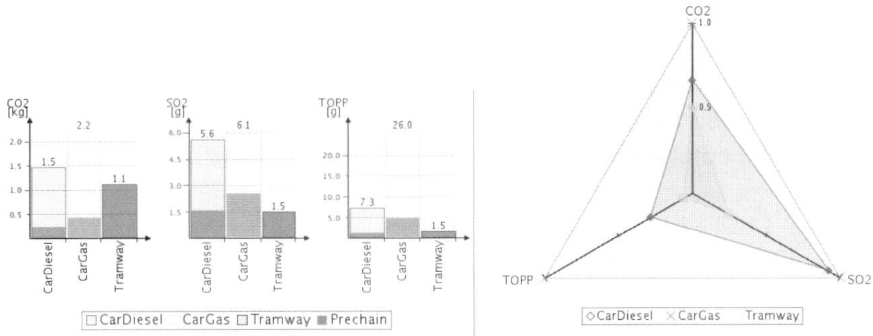

Fig. 4. Bar chart (left) and spider diagram for three sample product systems, representing a diesel-fueled car, a gasoline-fueled car, and a tramway. Shaded boxes within the bars of the bar chart indicate emission from the production of vehicle and fuel/energy.

3.4 Visualization of Results

The calculated results are displayed in two graphical representations (Fig. 4): a bar chart to compare absolute emission values, and a spider diagram to compare relative emissions between product systems. In the bar chart, emissions are grouped by emissions categories, so that differences between product systems are displayed in absolute terms. Additionally, the amount of "gray emissions", which arise from the production of vehicle and fuel/energy, respectively, is shown in order to visualize the share of emissions that can be influenced by users' behavior. In contrast to that, the spider diagram shows three axes representing the three emissions categories. The maximum emission value in a given category is assigned the value 1, while the emissions from other product systems are displayed relative to that. This allows for easy comparison of product systems.

3.5. Comparison with other LCA tools

A number of other LCA tools also claim to be user-friendly, for example SimaPro[2] or GaBi[3]. A closer look reveals some limitations of their usefulness for interactive use with lay participants. Often, users have to leave the graphical process chain editor to input data via large spreadsheets. Simple parameterizations of behavior are not available – instead, quantities like energy use have to be entered (and known) in absolute values. As this en-

[2] http://www.pre.nl/simapro/
[3] http://www.gabi-software.de

hances precision it is preferable when users have some minimum specialized knowledge of the product, like staff of the producing company. Despite of the reduced precision that makes EcoClass hardly suitable for applications like expert opinions, it gives access to LCA technologies to lay users with only common knowledge, which is hardly sufficient for usual LCA tools.

4 Workshop Experiences

EcoClass was used by lay people during the second workshop in the INNOCOPE process. After listening to a plenary presentation and introduction to the software, participants split into three smaller working groups of about 5 persons to use the model by themselves. Each group was equipped with one PC with two monitors, so that the actions of each user could be seen and possibly commented on by the others. The default task was to reflect on personal traffic behavior by assessing the emissions of day-to-day trips and to compare them to trips where a pedelec is used on suitable sections. Users were, however, free to use the software for other investigations as well. If required, users could get assistance in operating the software by the group facilitators.

A lively discussion developed only in two of the three groups. Here, participants started to discuss what settings to modify and assessed the outcomes, e.g. the influence of persons in a tramway on the emissions per person-kilometer. In the third group, adverse group dynamic processes led most participants to refuse working with the model at all.

No major problems were reported with the GUI or the result visualization, although most participants let the facilitators operate the software instead of doing it themselves. The main reason for this was a general time constraint, which did not leave enough time to learn using the GUI for users unfamiliar with computers. However, those users who ran the model themselves were most enthusiastic about the model and its opportunities.

The results caused diverse reactions. For many users, the observed difference of emissions between tramways and mopeds was so surprising that they questioned the credibility of the model, although a straightforward explanation (increased energy use due to vehicle weight) had been provided. Most of the results, however, appeared to be trivial or well known (cars emit relatively more than buses). This raised general questions on the added value from using the model. Model-specific shortcomings (missing suburban train) also led to a disqualification of the model by some users.

Interestingly, the model did rarely cause participants to reflect on their own mobility behavior. This is probably due to an improper timing within the INNOCOPE workshop series, which put EcoClass into a phase that dealt with design issues in detail. Here, a full-featured LCA tool would have been needed, while EcoClass might have been more appropriate to appear in the first workshop to raise awareness on mobility issues.

5 Conclusions

The EcoClass model provides environmental impact data for participatory product development based on the GEMIS life cycle assessment database. EcoClass introduces utilization processes as a new process category. These allow for easy online variation of the product utilization phase even by inexperienced users, in contrast to standard LCA tools where different utilization schemes require the definition of different processes. Since this increased flexibility comes at the cost of slightly reduced accuracy, EcoClass is intended to give an introductory overview on environmental impacts from product utilization. Currently, EcoClass is restricted to urban personnel transport systems and still lacks much of the functionality of fully featured LCA tools; these restrictions shall be dropped in future versions.

The first use of EcoClass during a pilot study showed that much of the design goals have been reached: Most users approached the GUI with minor efforts and understood the graphical results intuitively. The results stimulated discussions among the participants that focused on data credibility and the usefulness of models in general. However, participants were rarely observed to reflect on their personal mobility behavior.

Future uses of EcoClass in its current version should concentrate on raising awareness. This might include introductory sessions in workshop series, but could also be extended to demonstrations at fairs, public information desks, or in schools. Additional fields of use will open up for future model versions that offer full LCA functionality.

Acknowledgements

We thank our colleagues in the GELENA project for their feedback and suggestions during the development of EcoClass. Generous funding of the GELENA project by the German Federal Ministry for Education and Research (BMBF) is gratefully acknowledged.

References

Fritsche U, Rausch L, Simon K-H (1989) Umweltwirkungsanalyse von Energiesystemen: Gesamt-Emissions-Modell Integrierter Systeme (GEMIS), Endbericht, Öko-Institut Darmstadt /WZ III Gesamthochschule Kassel

Fritsche UR, Leuchtner J, Matthes FC, Rausch L, Simon K-H (1994) Gesamt-Emissions-Modell Integrierter Systeme (GEMIS) Version 2.1: Aktualisierter und erweiterter Endbericht. Darmstadt, Freiburg, Berlin, Kassel

Hansen U, Raabe T (1991) Konsumentenbeteiligung an der Produktentwicklung von Konsumgütern. Ergebnisse einer empirischen Untersuchung, Zeitschrift für Betriebswirtschaft, vol 61, edn 2, pp 171-194

Hoffmann E (2006) Consumer-Integration in Sustainable Product Development, paper presented at the 13th International Conference of The Greening of Industry Network, 2006-07-02 – 2006-07-05. Cardiff, UK

Kasemir B, Jäger J, Jaeger CC, Gardner MT (eds) (2003): Public Participation in Sustainability Science. Cambridge Univ. Press, Cambridge MA

Lester RK, Piore MJ, Malek KM (1998) Interpretative Management: What General Managers Can Learn From Design, Harvard Business Review (March-April 1998), pp 86-96

van den Ven A (1986) Central Problems in the Management of Innovation, Management Science, vol 32, edn 5, pp 569-607

Combining LCA with Thermodynamics

Stefan Gößling-Reisemann

University of Bremen, Germany

Abstract: Life Cycle Assessment (LCA) is the most promising methodology to assess environmental impacts of products, services and processes. Its scope of application is constantly evolving, including e.g. application to regional scales and assessing societal consumption patterns. Apart from considering environmental impacts, extensions to the methodology including social and economic impacts are currently being discussed. One of those impacts is resource consumption. It has been argued that the methods for assessing resource consumption in LCA must come from thermodynamics, and must take account of the second law of thermodynamics (entropy law). The challenge arising from this, especially in respect to its applications and software implementation, is the increase in data requirements. While already being a data intensive methodology, including a thermodynamic measure for resource consumption in LCA will increase the data that needs to be handled significantly. This can only be managed by employing thermodynamic data bases and combining these with dedicated LCA software. I will present an approach that makes use of the scriptability of commercial LCA software (Umberto®) and combines LCA data with thermodynamic data where values are stored in a parameterized form. The script then calculates the thermodynamically defined resource consumption and makes it available to the visualization and analysis tools in the LCA software. Processes from the metallurgical sector serve as an illustrative case study.

1 Resource consumption in LCA

Life Cycle Assessment (LCA) has been proven to be one of the most powerful tools to assess environmental burdens associated with industrial products and processes. Its methodology has been stabilized during recent years, which is apparent from an ISO norm (ISO 14040ff, as explained in Guinée and Lindeijer [2002]), specifying standards to which every LCA should adhere. However, there are still many unresolved issues in the LCA methodology, where consensus among scientists and practitioners has not yet been reached. One of these issues is the assessment of resource consumption [see Stewart and Weidema 2005; Finnveden 2005]. Strictly speaking, resource consumption is not really an environmental issue, it is rather an economic issue, adding a complicating factor to the discussion about how it should be implemented in LCA. When talking about resources, we have to distinguish between use, consumption, and depletion. While "use" is too general a term as to attach economic or ecological relevance to, consumption and depletion both have some significance in the debate. We can at least say that both come with environmental impacts attached. When resources are consumed they are transformed, and often wastes and emissions are produced. The depletion of resources, on the other hand, usually is accompanied by large additional material and energy flows, leading to the consumption of more resources and the production of more wastes and emissions. Thus, it is rather the secondary effects or resource depletion and consumption which make them relevant for LCA. In conclusion, and the majority of LCA experts agrees, resources are an important factor for a sustainable development of our society, and should thus be included in LCA.

The approaches towards including resource consumption (or depletion) in LCA are manifold. They range from simple mass or energy oriented measures (as "input of fossil energy carriers" or "cumulative energy demand" [see Guinée and Lindeijer 2002]) to more sophisticated thermodynamic based measures (cumulative exergy consumption [see Cornelissen and Hirs Gerard 2002]; entropy production [see Gößling-Reisemann 2006]) or to methods estimating future impacts of today's resource depletion [Müller-Wenk 1998]. The method chosen here for inclusion in LCA software is based on the second law of thermodynamics, i.e. the entropy law.

2 Entropy as a measure for resource consumption

One of the qualitative aspects of matter and energy transformations is what I call the potential utility. The potential utility P can be thought of as the size of the set of all possible uses of a given amount of resources. When a physical resource is called to be consumed, its potential utility is decreased:

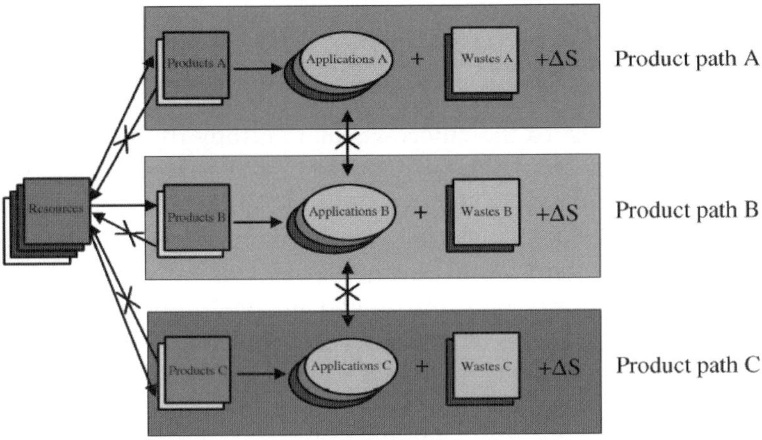

Fig. 1. When resources are consumed in a process, the actual set of possible uses or applications decreases. The size of this set can be viewed as the potential utility of these resources

It should be obvious that by choosing one production path, a certain sub-set of potential uses will no longer be available. It would certainly be desirable to measure the actual loss of potential utility and use this as a measure for resource consumption. Unfortunately, this seems to be impossible, for the same reasons that limit our knowledge of the actual size of the set of potential uses. However, there is something in the transformations which reflects the loss of potential utility on a physical level: the irreversibility of the process. Irreversibility in simple terms means "the impossibility of a process to run in a complete cycle without changing its environment" [Kondepudi and Prigogine 1999]. This definition can be quantified by introducing entropy: an open system undergoes an irreversible transformation if the combined entropy of the system and the environment increases. The second law states that the combined entropy can only remain constant or increase. Thus, the reversible transformation corresponds to constant combined entropy. We have to bear in mind that these statements are only valid for the combination of the system plus its environment. The open system itself can have increasing, constant, or decreasing entropy. If we denote the entropy of the system by S_{sys}, we can express the system's entropy change as

$$\Delta S_{sys} = \Delta_e S_{sys} + \Delta_i S_{sys}, \tag{1}$$

where $\Delta_e S_{sys}$ is the entropy exchange between system and the environment and $\Delta_i S_{sys}$ is the entropy produced within the system. The entropy exchanged with the environment does not increase the overall entropy (of system and environment), since the decrease of the system's entropy (by export) is the increase of the environment's entropy (by import) and vice versa. Thus, the irreversibility of the transformations taking place inside the system are measured by the internally produced entropy $\Delta_i S_{sys}$. The second law simply states that $\Delta_i S_{sys} >= 0$, where the equal sign only applies to the (unrealistic) case of reversibility. Coming back to the discussion of consumption, potential utility and irreversibility, I argue that the more irreversible a process, the more potential utility is lost. The basis for this argument is the fact that irreversibility describes for thermodynamic systems, what the loss of potential utility describes for resources: the loss of potential pathways for development. Since entropy change measures irreversibility, I propose that the change in potential utility ΔP (i.e. the consumption) that occurs inside a system, caused by an internal process or processes, is a negative and monotonically decreasing function of the entropy change ΔS caused by the process(es). Since the entropy change caused by these processes is equal to the internal entropy production $\Delta_i S_{sys}$, we found a physical approximation to the loss of potential utility: entropy production.

3 Implementing the use of thermodynamic data in an LCA tool

In order to calculate the entropy production along the life-cycle of a product or service, one has to balance the entropy of the input and output material flows (assuming that the processing equipment and other infrastructure remains in a steady state). A prerequisite for such a calculation is the completeness of the flow model, meaning that all flows must be specified in terms of mass/volume and be consistent with each other. This is a typical task for any LCA tool. The one used here (Umberto®) achieves this by applying a Petri Net derived methodology. The result of such a calculation is a network of processes (called *transitions*), storages (called *places*) and connecting *arrows* between them. The arrows consist of the material flows (one or many) flowing between the processes and storages. The calculation of the entropy content of flows further requires detailed knowledge of the composition and temperature of the flows, and of the thermodynamic properties of the components. In addition, heat losses from convection and radiation have to be accounted for. These can very often be calculated from an energy balance. In formula, the entropy production within the production-consumption system reads:

$$\Delta_i S = \frac{e_q^{out}}{T_0} - \frac{e_q^{in}}{T_0} + \sum_n \frac{4}{3}\frac{e_{s,n}^{out}}{T_{s,n}^{out}} - \frac{4}{3}\frac{e_s^{in}}{T_{in}} - \sum_j m_j s_j, \tag{2}$$

where e_q are heat flows, $e_{s,n}$ are radiation flows (each having a different temperature), m_j are the material flows, and s_j being their specific entropy. The specific entropy of the material flows can be calculated from the specific entropy of the components $\hat{s}_k$, the molar composition of the flows, given by the molar fraction y_k^j of each component, the total molar amount of the flow n_j, and its mass m_j. The (temperature dependent) specific molar entropy of the components, $\hat{s}_k$, can be calculated from the specific heat capacity:

$$\begin{gathered} s_j = \frac{1}{m_j}\left\{\sum_k n_j y_k^j \hat{s}_k - R\sum_k n_j y_k^j \ln y_k^j\right\}, with \\ \hat{s}_k(T) = s_k(T_0) + \int_{T_0}^{T} \frac{c_p^k(T')}{T'} dT'. \end{gathered} \tag{3}$$

As is evident from the above formulae, the calculation of the entropy balance requires access to thermodynamic properties of (possibly many) chemical components. This can most easily be achieved by linking the material flow model, as provided by a LCA software, to a thermodynamic database. The thermodynamic data can then be stored (temporally of persistently) in the material database of the LCA software, for example in the form of material properties. If a material is a mixture of components, it also has to be noted in the material properties. The actual calculation of the entropy production is then carried out inside the LCA software, using user defined scripts that access the thermodynamic data. For mixtures, the script has to look up the properties of the individual components and calculate the entropy according to equation above. After all material (and energy) flow entropies have been calculated, the overall balance and thus the entropy production can be computed.

As an example of how the above can be implemented, I have set-up a model (in Umberto®) of a process important for copper production: the flash smelter. The main inputs to this process are copper ore concentrate, oxygen-enriched air, flux material (mainly sand) and energy carriers (fuel oil, natural gas, electrical energy). The outputs are copper matte (the actual product), slag, offgas, and waste heat.

The composition of the material flow mixtures (concentrate, air, fuel oil, natural gas) are taken from literature sources [cf. Gößling 2001] and then defined via an array definition in the Python script. The mixtures and pure components that make up the flow model must further be specified regarding their temperature and phase state. This is also done for the whole model in the script which calculates the entropy balance:

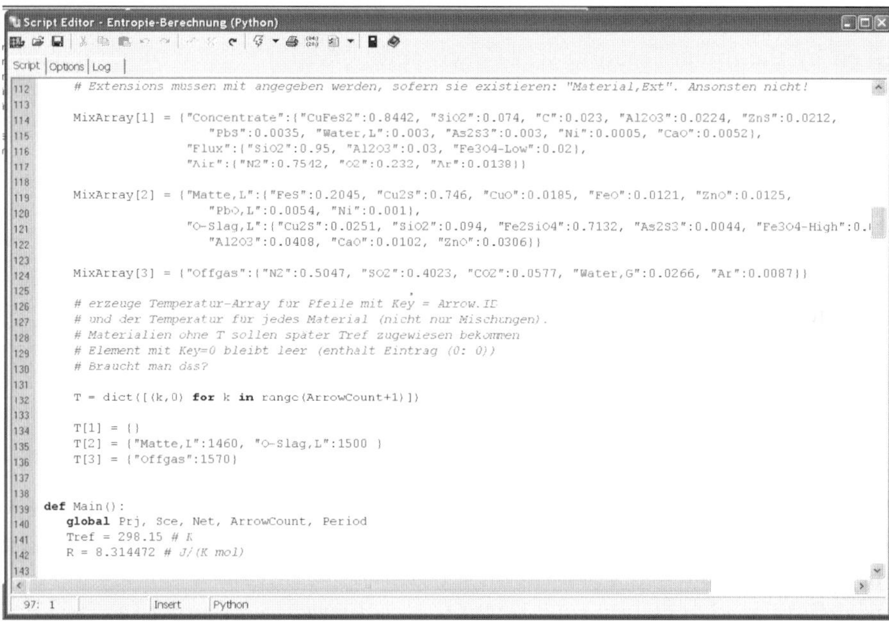

Fig. 2. Excerpt from the script for calculating the entropy balance for the flash smelter example. The array MixArraystores the composition of each flow, while the array T holds the temperatures

The mixture and the components are stored as materials in the Umberto® materials database. In addition, mixtures get the property "IsMixture" and components get their thermodynamic properties added:

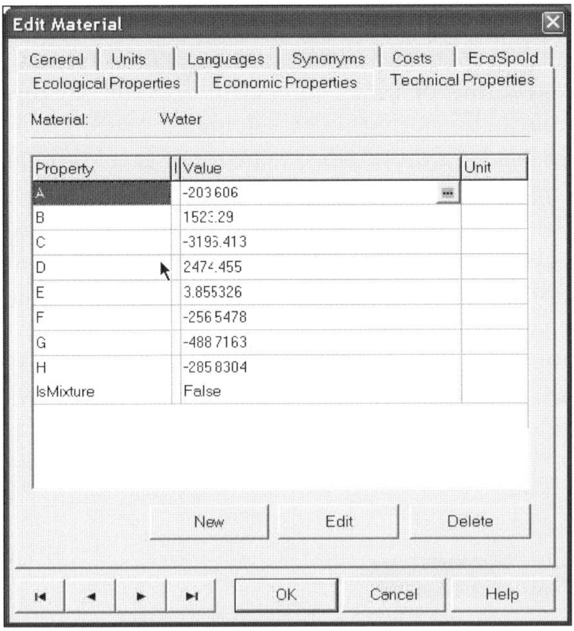

Fig. 3. Material properties for water. The properties A-H are parameters for the Shomate equation, a parameterized form of the (temperature dependent) molar heat capacity. The value "False" for "IsMixture" indicates that water is treated as a pure component

The molar heat capacity and thus the specific entropy of the components is temperature dependent. This requires the thermodynamic properties in the material database also to be given for as wide a temperature range as possible. Whenever available, this has been implemented by using the Shomate equation, as used by the NIST Chemistry Webbook [Linstrom and Mallard 2005]. This equation parameterizes the heat capacity (and by integration the entropy and enthalpy) for a wide range of temperatures:

$$C_{p^\circ} = A + B \cdot t + C \cdot t^2 + D \cdot t^3 + E / t^2, \quad \text{with } t = T/1000$$

$$H^\circ - H^\circ_{298.15} = A \cdot t + B \cdot t^2 / 2 + C \cdot t^3 / 3 + D \cdot t^4 / 4 - E / t + F - H \tag{4}$$

$$S^\circ = A \cdot \ln(t) + B \cdot t + C \cdot t^2 / 2 + D \cdot t^3 / 3 - E / (2 \cdot t^2) + G$$

When the entropy balancing script is executed, it checks for all relevant material properties (compositions, thermodynamic data, temperature) and, when missing, either stops executing or sets appropriate defaults (in this case warnings are issued to allow debugging). The calculated entropy values for each flow are written back to the model, appearing as additional material flows (a "material" in Umberto® can virtually be anything from energy to computer parts). In this way, entropy flows can be visualized with the internal graphics tools. It is then easy to recognize the processes in the network with the largest contribution to entropy production (i.e. resource consumption). This enables a quick judgment on the process steps where resource efficiency might be improved. The flash smelter process can serve as an example. With the process specifications the material flows in the network can be calculated and visualized via a Sankey style diagram:

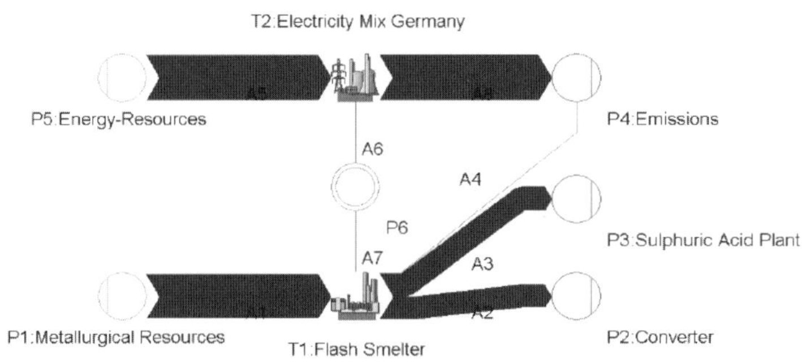

Fig. 4. Material flows of the flash smelting process plus one background process (energy provision). The width of the arrows is proportional to the mass of material flows

After calculation of the entropy contents of the flows, a similar diagram can be drawn for the entropy flows:

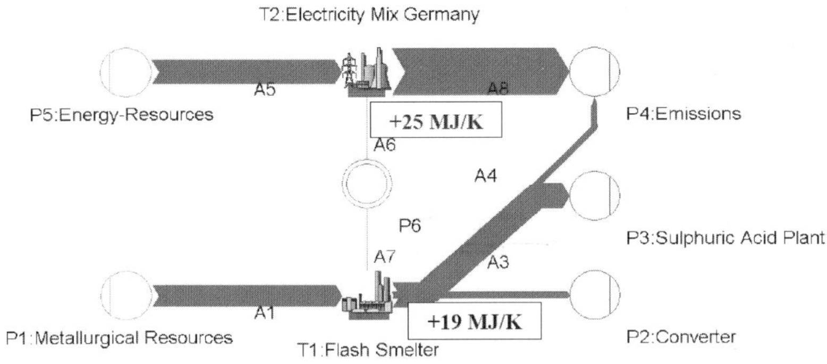

Fig. 5. Entropy flows for the flash smelter process and one background process

There are a few conclusions about the processes, which can be drawn from the diagrams. In the mass flow diagram we can see the conservation of mass quite clearly. The same is not true for the entropy diagram: entropy is not conserved, it is produced whenever material and energy flows are consumed. The amount of entropy production can be interpreted as a measure for resource consumption. In this respect it is interesting to note that the electricity production has a higher resource consumption than the actual smelter process. This is mainly due to the low efficiency of the power plants (typical German mix with high contributions from coal).

4 Conclusion and Outlook

It was shown that common LCA software can be extended towards thermodynamic assessment of resource consumption. While there are specialized software tools which are more apt to calculate thermodynamic properties and their changes in industrial processes, their use in the field of LCA would be quite inconvenient. It is much more straight forward and efficient to make use of the flow modeling capabilities of LCA software and implement the thermodynamics on top of that, using standard databases. The results for resource consumption can then be displayed alongside the environmental impacts, allowing for a side-by-side comparison of both impact types.

In the above example the thermodynamic properties have been added to the LCA materials database by hand. In the future this should be done automatically, with the help of commercial thermodynamic databases like DIPPR or DDB. The scripting functionality of Umberto® will be very helpful in this.

References

Cornelissen RL, Hirs Gerard G (2002) The value of exergetic life cycle assessment besides LCA. In: Energy conversion and management, vol 43, pp 1417-1424

Finnveden G (2005) The Resource Debate Needs to Continue [Stewart M, Weidema B (2005): A Consistent Framework for Assessing the Impacts from Resource Use. Int J LCA , vol 10, edn 4, pp 240-247]. In: The international journal of life cycle assessment, vol 10, p 372

Guinée JB, Lindeijer E (2002) Handbook on life cycle assessment: operational guide to the ISO standards. Kluwer, Dordrecht

Gößling S (2001) Entropy production as a measure for resource use. Staats- und Universitätsbibliothek Carl von Ossietzky, Hamburg

Gößling-Reisemann S (2006) Entropy as a measure for resource use - Application to primary and secondary copper production. In: von Gleich A, Ayres RU, Gößling-Reisemann S (eds) Sustainable Metals Management. Springer, Dordrecht, pp 195-236

Kondepudi DK, Prigogine I (1999) Modern thermodynamics: from heat engines to dissipative structures. Wiley, Chichester

Linstrom P, Mallard W (2005) NIST Chemistry WebBook. National Institute of Standards and Technology, Gaithersburg MD, 20899. http://webbook.nist.gov, access: 2007-02-14

Müller-Wenk R (1998) Depletion of abiotic resources weighted on the base of "virtual" impacts of lower grade deposits used in future. Inst. für Wirtschaft und Ökologie, Univ. St. Gallen, St. Gallen

Stewart M, Weidema BP (2005) A Consistent Framework for Assessing the Impacts from Resource Use – A focus on resource functionality. In: The international journal of life cycle assessment, vol 10, pp 240-247

RFID-based Information Management in the Automotive Plastic Recycling Industry

Zheng Lu[1], Hui Cao[1], Paul Folan[1], David Potter[2], Jim Browne[1]

[1] CIMRU, National University of Ireland, Galway

[2] INDYON GmbH, Germany

Abstract: While metallic fractions in the automotive industry have been the subject of significant and successful research, the recycling of plastics in vehicles is less well understood and is now vital to the success of effective automotive recycling, according to legislative requirements. Moreover, it is a critical challenge given the difficulty of plastic identification, the low technical level of logistics processes, and the lack of availability of historical information on plastics. This paper introduces RFID system, which helps to automatically identify, record, transmit and search product information throughout the product lifecycle. This research produces a feasible solution for decision support in a case study of automotive plastic recycling, using an RFID-based information management framework in the product lifecycle.
Keywords: product lifecycle; RFID; plastic recycling; decision support

1 Background

The number of end-of-life vehicles in the world is increasing as the number of cars increases. At present the automotive industry generate about 5% of the world's industrial waste, whether from the vehicles themselves, or from the plants which produce them [Bellmann and Khare 1999]. In the European market, car manufacturers face the critical challenge of anticipating and responding to recycling legislation. In EU accession countries, a total increase of 124 % in the number of scrapped cars is projected between 2000 and 2015. Reasons include the ageing and growing car stock [EEA 2003]. Every year end-of life vehicles in the Community generate between 8 and 9 million tones of waste, which must be managed correctly [Directive 2000/53/EC]. The EU directive on end-of-life vehicles [Directive 2000/53/EC] has a strong focus on reuse and recycling and it has set a target of 85 % of an end-of-life vehicle is to be reused or recycled after 2015, by average weight per vehicle and year.

Plastics are used in automotive vehicles to meet customer requirements for light weight, inexpensive, and durable components, which results in greater fuel economy and preservation of resources and the environment. In general, about 25-30 types of plastics typically are found in today's automobile. Plastics account for nearly 33% of the non-metallic components, which is 21.1% of an automobile [Duval and Maclean 2006]. While metallic fractions are already a well understood area in recycling [Ferguson and Browne 2001], the recycling of plastics, however, is a complex issue owing to technological, political and economic barriers [Duval and Maclean 2006]. This research focuses on the problems of a lack of plastic identification and the low technical level of its associated logistics processes in the plastics recycling industry. Mixed quality/quantities of plastic material, and a high distribution level of vehicles, has resulted in difficulties in the recycling of automotive plastic, since some of the necessary historical information on the plastic (in terms of material composition, user behavior and changes of the material or product components during the product life of the vehicle) are not available for decision-making. Information management in recycling remains at a relatively unsophisticated state and is based mainly on paper documents and manual data input [Hribernik et al. 2005].

The EU project PROMISE [IST F6: 507100] involves a case study in the plastic recycling industry using the PROMISE architecture, which deals with developing appropriate technology, including product lifecycle models, RFID technology, and associated firmware and software components for decision-making based on data gathered through the product lifecycle for plastic recyclables from the automotive industry.

2 A Comprehensive View on Product Lifecycle

A product system's life cycle is characterized by the three phases (see Fig. 1): Beginning of Life (BOL), including Design and Production, Middle-of-Life (MOL), including Use, Service and Maintenance and End-of-Life (EOL), characterized by various scenarios such as [DIRECTIVE 75/442/EEC; 2000/53/EC]:

- Reuse: any operation by which components of end-of life products are used for the same purpose for which they were conceived;
- Recycling: the reprocessing in a production process of the waste materials for the original purpose or for other purposes;

- Remanufacturing: the process of disassembly of products during which time parts are cleaned, repaired or replaced then reassembled to sound working condition
- Disposal: the process of getting rid of end-of-life products, such as landfill, incineration etc.

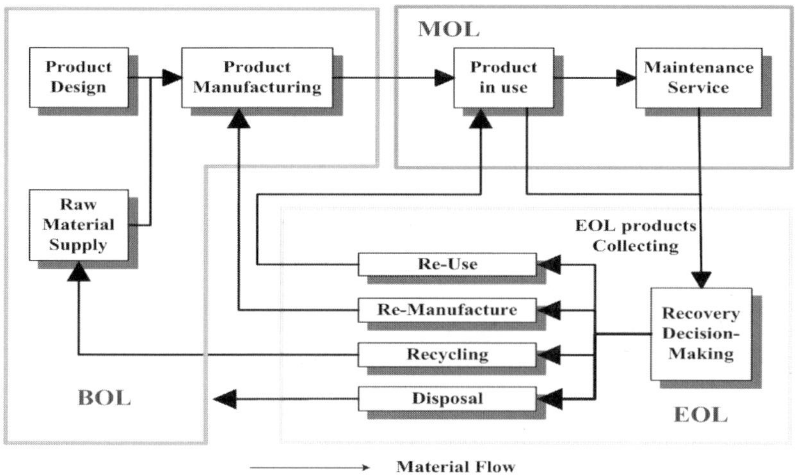

Fig. 1. The Model of Product Life Cycle

Between the BOL and MOL phases of the product lifecycle, the information flow is quite complete and supported by a number of intelligent systems such as CAD/CAM, Product Data Management (PDM) and various Knowledge Management systems that are effectively used throughout industry and, through their influence, by their value chain partners. The information flow becomes less and less complete, however, as we move from the MOL to the EOL phase, and frequently it breaks down after the delivery of the product to the customer. The feedback of data, information and knowledge from MOL maintenance and EOL recovery processes to the BOL designers and producers of the product is interrupted after the product sale.

Incomplete information on the product makes decision-making processes in EOL complicated. Usually decisions are only based on a user's partial but incomplete information sources from BOL, or rule-of-thumb experiences, built-up from an in-depth knowledge of previous transactions; alternatively, experimentation has been used to generate the necessary missing information which increases, however, the operational costs in EOL significantly. Take plastics as an example. Because of the incompatibility of various plastics, the plastic parts dissembled from end-of-life products must be identified and sorted out, in order that the recycling of pure plastics becomes possible. The current practice of plastics marking is of little help, because no indication is available on additives, fillers, flame retardants, etc. Hazardous materials are another remarkable example. There is no sufficient information on the specifications and locations of the problematic materials in the recycling processes. [Scheidt and Zong 1994]

Applying an RFID system may therefore be used to potentially fill information gaps, and make product information accurate, cheaper, real-time, complete and available whenever it is required during its whole lifecycle.

3 RFID Technology and RFID System

RFID, short for Radio Frequency Identification, is a generic technology that uses radio waves to identify objects [Chow et al. 2006]. In 2003, Wal-Mart, the world's largest retailer, announced that it expected that its top 100 suppliers would be RFID-compliant by January 1, 2005, which is commonly regarded as the industrial seal of approval for RFID technology usage. The benefits of this RFID implementation were reported as reducing out-of-stock items and inventory holding, anti-theft, and the reduction of labour costs. According to research from the University of Arkansas, by using RFID technology, Wal-Mart reduced 16% of out-of-stocks and was 63% more effective in replenishing out-of-stocks than their peers [Chain Store Age 2005]. RFID increases efficiency and enables new fields of application by introducing the following abilities:

- To read multiple product identities simultaneously.
- To read identities from distance without having a line-of-sight.
- To interact with the product information without a human intervention.
- To provide the real-time information service.
- Tags with rewritable memory of which the size is extendable.

An RFID system consists of three basic components: (1) an RFID tag, which contains at least an unique identification number to identify the item to which it is attached; (2) a reader, which emits and receives radio waves to read the information stored in the tag; and (3) An RFID middleware application, which promises a seamless collaboration with backend enterprise systems [Røstad and Myklebust 2005; Wu et al. 2006]. Applying RFID systems has opened new possibilities: The ability to store and manipulate data on the RFID tag has enabled the identity and integrity of a product. Therefore, RFID tags are used to store historical data about the products where the tags are attached. Identifying products automatically has been found feasible by a number of companies such as Wal-Mart. Furthermore, because RFID tags can be read automatically from a distance, the technology can be used to acquire positional information of an object that has a tag attached.

Current RFID adaptors are mainly applying RFID in closed environments. That means the information gathered and used by RFID systems is within the same company. However, it seems that the more potential abilities of RFID technology will be emerged in open environments, where information is acquired through the whole life of the product. In many cases, a single product will cross borders of relative companies several times during its whole lifecycle, which strongly implies the need of sharing product information beyond company borders. Thus, a generally accepted solution is needed to achieve this goal.

4 Innovation in Information Management using RFID Technology throughout Product Lifecycle

Implementing RFID systems in the entire product lifecycle model (see Fig. 2) enables us to track products in production processes, automatically identify, record, transmit and search product information throughout its lifecycle, and close the product's information loop.

As described in Figure 2, the BOL material flow begins with the Raw Material Supplier to the Manufacturer; this is paralleled by the provision of design information to manufacturing; and the two combined are subsequently distributed to different Customers in *Product in Use* where the product reaches its MOL phase. Relevant product information is also processed and transmitted to an information management network which named *Product Lifecycle Information Systems*; this contains *BOL Information Systems*, *MOL Information Systems*, and *EOL Information Systems* that enables the immediate, automatic identification and sharing of item-level information in product lifecycle management. The information consist all crucial product information during the whole product lifecycle:

- Design information
- Production information
- Usage information (e.g. data collected from sensors and recorded in RFID tags)
- Maintenance record
- EOL instructions and processing record

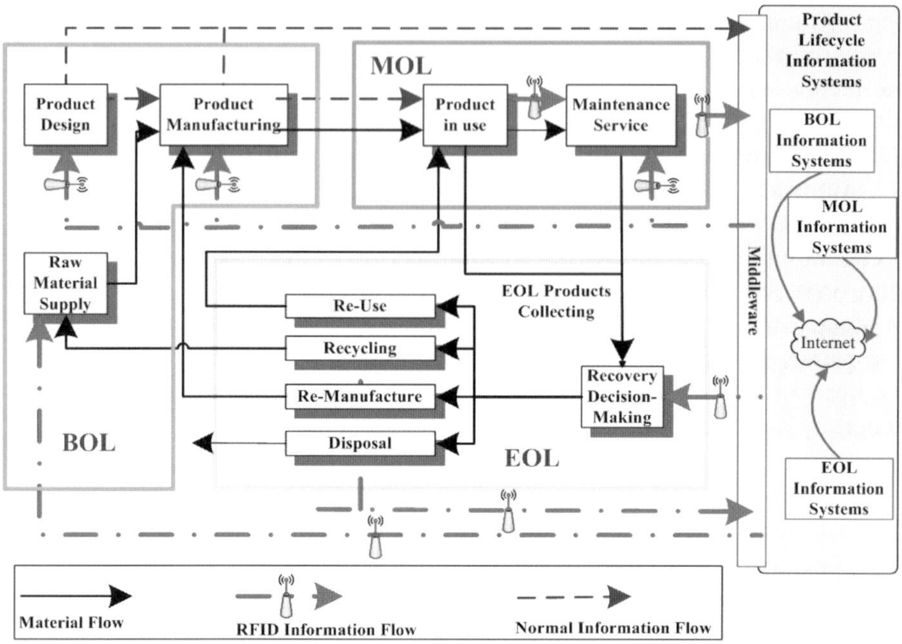

Fig. 2. RFID-based Information Management Framework in product lifecycle

The information provided from BOL might be, for example, from a structured application such as a CAX system (like CAD/CAM, etc.), whereas the various information generated from MOL and EOL shop floor, may be captured and transmitted by using RFID technology. In EOL, the information needed to make *Recovery Route Decisions* can be retrieved from the *Product Lifecycle Information Systems* according to the unique ID stored in the RFID tag. This is also an important issue in PROMISE, but not further discussed in this paper.

Implementing RFID systems in the entire product lifecycle model leads to the innovation of decision support, since it makes accurate, real-time, and complete product information available throughout the whole product lifecycle. A case study that utilizes this model and focuses particularly on its EOL aspects as *applied to recycling* is discussed in the following section.

5 Case Study: Information Management in Plastic Recycling

The automotive plastic recycling operation that is examined here is derived from studies in PROMISE.

5.1 Scope

This case introduces how an automotive plastic recycling plant operates with the support of EOL Information System, based on the *RFID-based Information Management Framework in product lifecycle* (see Fig. 2). It only represents the *Recycling* section of the EOL phase in the model of Fig. 2. The scenario of this case could be:

- Unloading of goods from the truck
- Identification of the delivered products
- Storing of goods identified for recycling in the warehouse system
- Schedule of recycling processes according to customer requirements and available materials in the warehouse
- Storing of recycled goods in the warehouse system
- Loading of goods on the truck

It does not only consider one specific kind of plastics, but for all forms of automotive plastics, such as car bumpers, engine components, dashboard inserts, inner compartment plastic supports etc.

5.2 Objectives

EOL Information System is implemented for the following objectives [Hribernik et al. 2005]:

- To use *RFID-based Information Management Framework in product lifecycle* in order to enhance the processing of plastics identification for recycling.
- To reducing cost and manual mistakes by reducing human intervention of using RFID technology.
- To track and trace materials and manage the availability, security, accuracy and integrity of all relevant product data at every stage of the recycling process.
- To use all available information of the chosen product to optimise decision making at every stage of the recycling process and to meet customer and market demand.

5.3 Technical Infrastructure

The technical infrastructure of the EOL Information System applied in the plastic recycling plant (see Fig. 3), is an example of the *EOL Information System* of the *Product Lifecycle Information Systems* in Fig. 2.

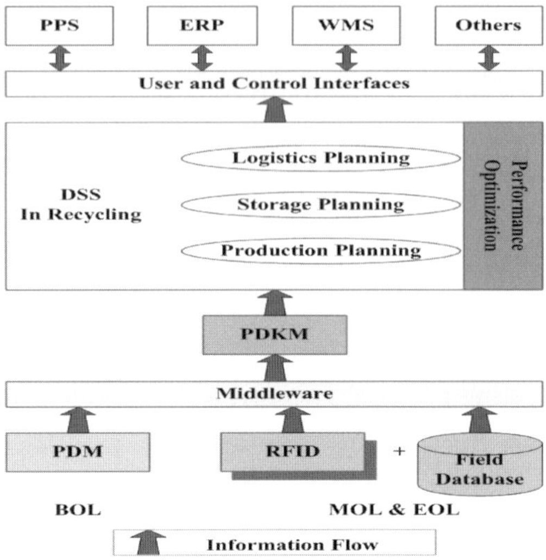

Fig. 3. Technical Infrastructure of the EOL Information System

Data resources: (1) data from the PDM, containing product design and manufacturing information; (2) data from RFID tags attached to the product; and (3) data from Field Databases, supporting the daily operations from the shop floor of recycling. The data will be transformed / aggregated / unified into a consistent representation through the Middleware, and passed to PDKM.

PDKM: PDKM integrates and manages data from all lifecycle phases of products (no matter whether it comes from databases or RFID tags), in particular, from design, development, production, through use and maintenance, to recycling, and finally, to support comprehensive data analysis.

PDKM integrates product-related data, including data from real-time operations which is dynamic and voluminous, into a comprehensive uniform data management platform, on which various kinds of data querying and analysis methods are performed. [Anke et al. 2005]

Decision Support System (DSS): DSS is the core of this EOL Information System, which aims at performance optimization of the plastic recycling processes. It integrates a number of algorithms in order to make better decisions, with the functions of: logistics and storage planning and production planning. See the next section for details.

User and Control Interfaces: Implements the interface of the DSS to users and to other applications, e.g. PPS (Production Planning System), the ERP (Enterprise Resource Planning) and the WMS (Warehouse Management System).

Applications: The ERP system is the source of order data for both incoming materials and customer orders for products. The WMS manages the execution of goods movements within the recycling plant and will communicate information about stock locations to DSS and receive movement requests from it. The PPS manages the planning and scheduling of recycling production. The PPS will communicate status information to the DSS and receive product data from it.

5.4 The Role of the DSS

Decision-making is an important aspect of the management activity, which is a principle function of organizations. High-quality decisions will lead to more productive actions, quicker problem solving, and better organizational performance [Eierman et al.1995]. The concept of the Decision Support System (DSS) has evolved from the original simple text-interface programs supporting individual decision-makers, to friendly graphic-interface systems helping workgroups and teams. Web-based DSS makes it possible that decision makers from geographically distributed locations can capture decision-relevant knowledge and gain decision support easily [Cao et al. 2006].

In the scenario of plastic recycling, the role of the DSS should be to use appropriate rule sets to support, as examples, the following kinds of decisions:

- What to do with incoming materials (identification of material (PDKM))
- Where to store it (stock locations requested from WMS)
- Where to get them from (stock locations requested from WMS)
- What is the requirement of customers (orders from ERP)
- What plastic product to make next (schedule from PPS)
- What ingredients are required (knowledge of stock, formulae in PDKM)
- etc.

Figure 4 details the decision support that occurs in the automotive plastic recycling industry diagrammatically. It gives an IDEF0 model of the DSS, which is structurally described by inputs (left), outputs (right), controls (top), and mechanisms (bottom).

Fig. 4. IDEF0 model of DSS in Recycling

The inputs of the DSS consists of plastics information acquired from the PDKM and shop-floor processes, decision maker's knowledge or expertise, customer orders from the ERP system, and production or transportation invoices. Instructions will be given to recycling processes, transportation systems, and other applications, after the DSS has analyzed this input data with predefined algorithms. The specifications of the DSS inputs and outputs are summarized in Table 1.

Table 1. Specifications of the DSS inputs and outputs

IDEF0 Functions	Description	Specification
Inputs	Materials Historical Data	From PDKM. The product-related data collected throughout product lifecycle.
	Field Data	The real-time feedback from the EOL shop floor.
	Customer Order Data	From the ERP system
	Decision Maker / Expert Knowledge	Human–computer interaction. Manually Input by the decision makers.
	Production Invoices & Delivery Requirements	Production schedules from PPS; and stock locations status / transportation requests from WMS.
Outputs	Recycling Process Instruction	Plastics identification and Recycling production planning. e.g. what is the information of this

		plastic product, what is the next recycling step, which work station, which operator.
	Instructions to other Applications	Triggers to other applications. e.g. what recyclables has arrived, which operation has finished.
	Logistic Support	Invoices of transportation and warehouse management. e.g. which forklift is available, which location in warehouse is available.

The plastic recycling processes will be supervised under the legislative requirements and recycling process policies, e.g. hazardous (or potentially hazardous) material will be identified and be carefully stored and processed. Mastery of market demand and the result of cost-benefit analysis are also the key factors to determine recycling strategies.

The mechanisms that support the execution of the DSS include RFID and PDKM systems, EOL Information Systems, and recycling personnel.

6 Conclusion

Effective automotive plastic recycling is complex in application and execution; however it is urgently required to meet environmental requirements and government legislations. An *RFID-based information management framework in the product lifecycle* is discussed in this paper, which gives a feasible solution to offer complete information on plastic recyclables throughout the whole product lifecycle. The technical infrastructure and functional analysis of the *EOL Information System* in an automotive plastic recycling process is depicted. Using RFID technology in the whole product lifecycle is becoming increasingly important; however the combination of RFID and traditional systems, such as existing barcode systems, is also an option, and may offer a higher flexibility in the short term, while we await the widespread adoption of RFID technologies and the implementation of specific RFID-oriented standards. This represents future work for this research area.

References

Bellmann K, Khare A (1999) European response to issue in recycling car plastics. In: Technovation, vol 19, edn 12, pp 721-734

European Environmental Agency (EEA 2003) Waste Generation and Management. In: Europe's Environment: The Third Assessment, vol 10, pp 151-164. http://reports.eea.eu.int/environmental_assessment_report_2003_10/en/kiev_chapt_07.pdf, access: 2006-02-12

DIRECTIVE 2000/53/EC (2000) "THE EUROPEAN PARLIAMENT AND OF THE COUNCIL of 18 September2000 on end-of life vehicles". In: Official Journal of the European Communities, L 269/34-EN

Duval D, Maclean HL (2006) The role of product information in automotive plastics recycling: a financial and life cycle assessment. In: Journal of Cleaner Production

Ferguson N, Browne J (2001) Issues in end-of-life product recovery and reverse logistics. In: Production Planning & Control, vol 12, edn 5. Taylor and Francis, pp 534-547

Hribernik K, Plettner A, Potter D, Schnatmeyer M, Folan P (2005) Scope, definition and description of the current state vs. critical issues. In: PROMISE Consortium

DIRECTIVE 75/442/EEC, "THE EUROPEAN PARLIAMENT AND OF THE COUNCIL of 15 July 1975 on waste". Office for Official Publications of the European Communities, L0442-EN

Scheidt L-G, Zong S (1994) An Approach to Achieve Reusability of Electronic Modules. In: Proceedings of IEEE International Symposium on Electronics & the Environment (1994), pp 331-336

Chow HKH, Choy KL, Lee WB, Lau KC (2006) Design of an RFID case-based resource management system for warehouse operations. In: Expert Systems with Applications, vol 30, edn 4, pp 561-576

Chain Store Age (2005) Research Confirms Wal-Mart's RFID Benefit. In: Chain Store Age, vol 81, edn 12, p 80

Røstad CC, Myklebust O (2005) PROMISE Demonstrators and State-of-the-art. PROMISE Consortium

Wu NC, Nystrom MA, Lin TR, Yu HC (2006) Challenges to global RFID adoption In: Technovation

Anke J, Do H-H, Marquard M, Metin A, Røstad CC, Smith B (2005) Design of PROMISE Information Management System (PDKM). PROMISE Consortium

Eierman MA, Niederman F, Adams C (1995) DSS theory: A model of constructs and relationships. In: Decision Support Systems, vol 14, pp 1-26

Cao H, Folan P, Lu Z, Mascolo J, Francone N, Browne J (2006) Design of an End-of-Life Decision Support System using PEID Technology, ICE2006

Material Flow Management in a Motor Fabrication: Identifying Saving Potential in the Mechanical Production by Modeling Energy- and Material Flows as Well as Machine Time Using Umberto

Jürgen Reinhard, Stefan Motsch

FHTW-Berlin, Ostendstr. 25, TGS Haus 4, 12459 Berlin, Germany

Abstract: In April 2006 a co-operation between a motor fabrication and the FHTW Berlin Department of Engineering 4, Industrial Environmental Informatics course was initiated. The aim of the project was to meet the following goals: 1. Activity based allocation of energy flows; 2. Identify saving potential; 3. Monetary valuation of saving potentials.
This paper gives an overview on the proceeding in this co-operation project. In detail it shows how the relevant data was raised, which role the knowledge of co-workers played, respectively the communication with them and how the project could be successfully completed.

1 Introduction

In the mechanical production of the motor plant analyzed in this project, 40 partly different CNC-machines are producing crankcases, cylinder heads, crank shafts and connecting rods in variable allocation. On 14,000 m^2 manufacturing area approx. 20% of the energy costs of the entire production plant are caused. A collection of the material and energy flows were only performed for all machines within the production building. Thus, a statement about the internal distribution of central media (electricity, water) could only be given by estimation.

2 Objective

The main goal in this project was to create a guideline for resource saving based on an energy and material flow analysis. In detail after the Kick-off Meeting the following steps were defined:

1. Activity based allocation of the central media (electricity, water) in the mechanical fabrication;
2. Identification of saving potential;
3. Monetary valuation of the saving potential.

It was not possible to conduct a detailed energy and material flow analysis within the available project time. Thus, the investigated part of the system was confined to the production area of the crankcases. The material and energy flow analyses focused the production period from August 2005 to July 2006.

3 Approach

As mentioned in earlier publications [Skrzypek and Wohlgemuth 2000, pp 75-94; Fraunhofer Institut für Arbeitswissenschaft und Organisation 2000; Wietschel 2002, pp 40ff] it is often difficult to raise the data variety which is needed for modeling energy and material flows. Therefore this part of the paper gives a detailed overview about the proceeding in general and the data upheaval especially in this co-operation project. In Figure 1 an abstract view to the project proceeding is given.

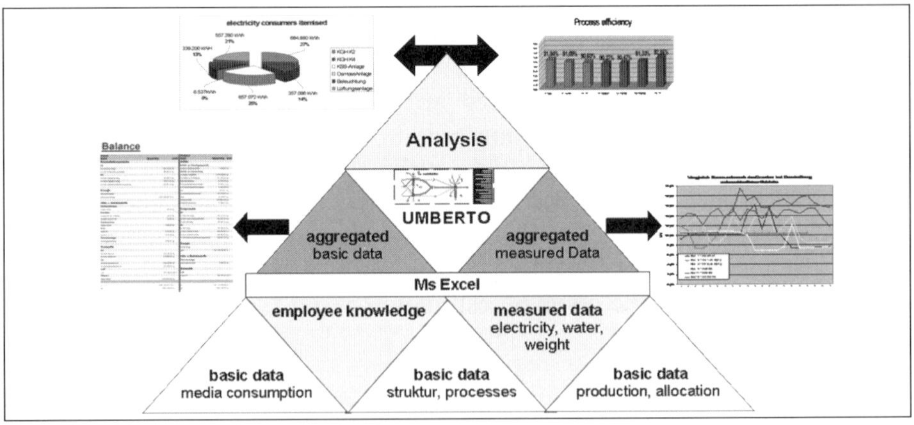

Fig. 1. Proceeding Overview

4 Data Acquisition

The first step was the identification and acquisition of available data for the motor production, meaning the resource consumptions by month of the entire mechanical production as well as process information, the included machines and information about the produced amount of crankcases.

Based on this analysis, electricity measurements could be planed and conducted. In addition the loss of material occurring in every production step was acquired by weight quantifications. Table 1 represents the several data types, their source and quality as well as the data usage in terms of modeling.

Table 1. Several data types, its source, quality and usage

Data type	Data source Department/ Document source	Data Quality	Data usage
Central Media Consumption	Central Media Supply / Ms Excel sheets	High	Model validation
Process and structure	Mechanical Production / Operation chart	High	Model structure
Machine time	Mechanical Production / Operation chart	High	Machine specifications
Production output	Mechanical Production / Ms Excel sheets	High	Simulation terms
Machine Allocation	Mechanical Production / Ms Excel sheets	High	Machine specification
Electricity consumption	Central Machines[a]: detailed measurements	High	Machine specifications
	Others: Manufacturers instructions	Medium	Machine specifications
Water consumption	Mechanical Production / Manufacturers instructions	Medium	Machine specifications
Material weight	Mechanical Production / Weighing after each Machining	High	Machine specifications
Operating supplies	Mechanical Production / Questionnaire	Low	Machine specifications

[a] This refers to all machines which had a connected load more than 5 kWh.

4.1 Employee's knowledge

Initially a meeting for all project related people was held in order to establish a common communication basis and to overcome prejudices related to the project. Using the employee's knowledge obtained by personal discussions (guided by a prior developed questionnaire) the previously assembled information was brought into a logical context. Furthermore it helped on closing knowledge gaps and discovering additional information sources.

4.2 Data Aggregation

The electricity measurements of the central machines yielded more than one million data records. Based on this data set, the average electricity consumptions were determined. All machine specific data was assembled in a data sheet for each machine. Table 2 shows an excerpt of the data sheet of a CNC machine, which is used for the production of crankcases of type K2.

Table 2. Data specification CNC-Machine

Term	Crankcases K2
Input weight	9.6 kg
Output weight	9.1 kg
Power consumption	18 kWh
Time per piece	19 min.

4.3 Modeling

Based on the previously aggregated data, the material and energy flows of every machine of the crankcase production area could be modeled. The model was created as a material flow network in high hierarchically way. Therefore the software Umberto[1] was used which was provided by the Umberto Competence Centre of the FHTW on behalf of Prof. Dr. Volker Wohlgemuth.

The machining times were explicitly considered in the model [Geldermann 2003, pp 75ff], this means they were not only used in the machine specification. Figure 2 shows a part of the material and energy flow model combined with the depiction of machine time and their transition specification.

[1] Umberto is a material flow analyses tool based on the Petri-net theory and is used to model material and energy flows in multi-stage production systems, including its environmental aspects [Wohlgemuth et al. 2006].

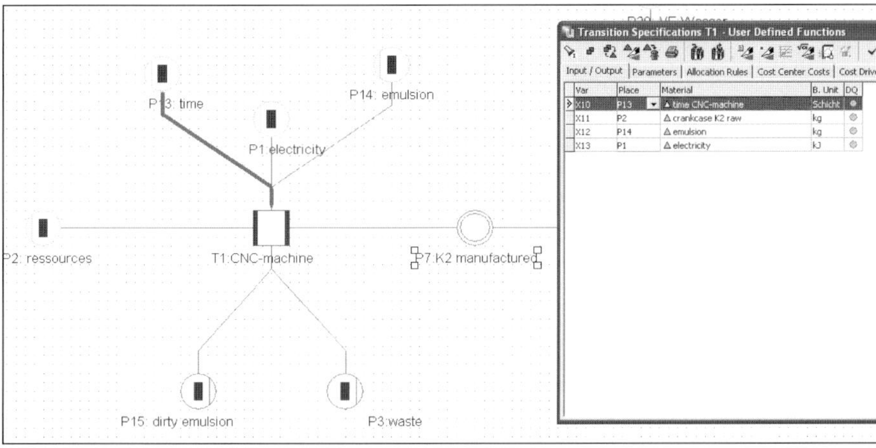

Fig. 2. Depiction of machine time in the material and energy flow model

4.4 Validation

The validation of the model structure and specifications was conducted with the help of the employees. The electricity and water consumptions calculated by the model were compared with the actual electricity and water consumptions. The calculated consumptions deviated approx. 20% from the actual consumptions. The reason for this deviation was that the machine times in the model – adopted from the actual operation charts – did not consider set-up time and production downtimes. Based on these insights an adjustment of the machine times was carried out.

However, the actual electricity usage of the CNC-machines was still higher than calculated. The cause lead to the identification of saving potential and is explained in the following chapter.

5 Results

By modeling each machine in the crank case area, the activity based allocation of all considered media, material and time consumptions and the monetary valuation of the focused media were reached. Thus, the material flow analyses allowed the identification of saving potential.

As an example due to the performed material flow analysis it could be discovered, that the CNC-machines are responsible for approx. 38% of the electricity usage per year.

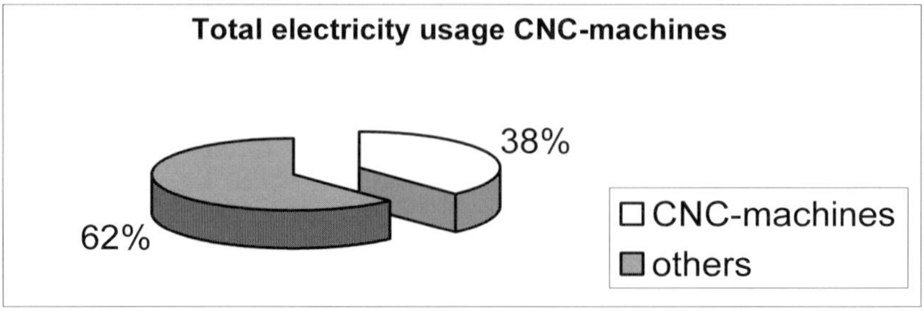

Fig. 1. Part of CNC-machines at the total electricity consumption

The calculated process time of CNC-machines compared with their actual time usage, particularly their production output in the actual time usage, proved a statement of the production manager: the CNC-machines were mostly operated in warm-up phase and not switched of due to production purposes. This problem could be solved by introducing the standby phase, whose existence was discovered during the systematic electricity measurements.

The standby phase needs less than approximately 8 kilowatt hours in comparison to the warm-up phase, and permits a fast production start. Using the difference between the calculated machine times and the time where no production took place – assuming that the machines were running in the warm-up phase – leads to the identification of electricity saving potential.

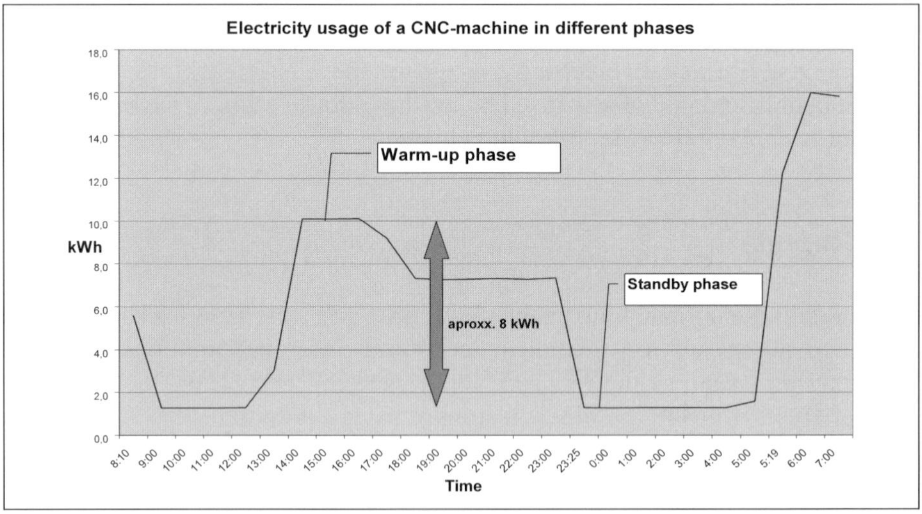

Fig. 2. Electricity usage of a CNC-machine in different phases

In co-operation with the production manager a saving potential of 25% of the total CNC-machine electricity usage per year could be identified, proved and monetarily valuated.

6 Conclusion

The modeling of machine time in combination with energy and material flows proved to be useful in many ways. On the one hand this approach leads to a deeper understanding of relationships between electricity and material usage. On the other hand it was leading to the identification of energy saving potential as mentioned above.

Another important result of this project was the establishment of efficient communication between all relevant project participants. Thus, the employee knowledge could be used to gain better insides of the production processes and played an important role during the modeling process.

In total this project raised the awareness of all participants to further focus on energy and material flows in the future.

References

Fraunhofer Institut für Arbeitswissenschaft und Organisation (2000) Stoffstrommanagement: Erfolgsfaktor für den betrieblichen Umweltschutz. Tagungsband zum 3. Management-Symposium Produktion und Umwelt, 29. November 2000, Institutszentrum der Fraunhofer Gesellschaft, Stuttgart. Fraunhofer IRB Verlag, Stuttgart

Fraunhofer Institut für Arbeitswirtschaft und Organisation (2003) Stoffstrommanagement: Entscheidungsunterstützung durch Umwelt-informationen in der betrieblichen IT. Tagungsband zum 5. Management-Symposium Produktion und Umwelt. Stuttgart

Geldermann J (2003) Die Verknüpfung von Stoffstromanalysen und Zeitstudien. In: (Fraunhofer Institut für Arbeitswirtschaft und Organisation 2003), pp 73-87

Skrzypek V, Wohlgemuth V (2000) Anwendung und Intergration von Umberto in die betriebliche IT-Struktur am Beispiel der Bertelsmann Großdruckerei MOHN Media Mohndruck GmbH. In: Fraunhofer Institut für Arbeitswissenschaft und Organisation 2000, pp 75-94

Wietschel M (2002) Stoffstrommanagement. Europäischer Verlag der Wissenschaften, Frankfurt am Main

Wohlgemuth V, Page B, Kreutzer W (2006) Combining discrete event simulation and material flow analyses in a component-based approach to industrial environmental protection. Elsevier

Semantic Modeling in Farming Systems Research - The Case of the Agricultural Management Definition Module

Ioannis N. Athanasiadis[1], Sander Janssen[2], David Huber[3], Andrea Emilio Rizzoli[1], Martin van Ittersum[2]

[1] Dalle Molle Institute for Artificial Intelligence (IDSIA), USI/SUPSI, Galleria 2, 6928 Manno, Lugano, Switzerland, e-mail: {ioannis,andrea}@idsia.ch, http://www.idsia.ch

[2] Plant Production Systems, Wageningen University, Haarweg 333, 6700 AK Wageningen, The Netherlands, e-mail: {sander.janssen, martin.vanittersum}@wur.nl, http://www.pps.wur.nl

[3] AntOptima SA, Fusoni 4, 6900 Lugano, Switzerland, e-mail: david.huber@antiptima.ch, http://www.antoptima.ch

Abstract: Farming Systems Research studies agricultural systems and their interaction with the natural environment and ecosystems. Agro-ecosystems are highly complex due to the many feedbacks between natural processes, high geographical diversity and human factors involved both as the farmer's decisions at farm household level and as the policy implementations at regional, national or European levels. This paper presents a novel approach for developing an Agricultural Management Definition Module (AMDM), by exploiting ontologies and semantic modeling. Specifically, a declarative approach has been utilized for conceptualizing farming systems and the management alternatives of a farm household. This conceptual model has been implemented as an ontology that ultimately has been used as the basis for software development and integration. This paper presents in detail the methodology used for developing AMDM and a real-world installation, part of the SEAMLESS integrated project.

1 Introduction

1.1 Managing the complexity of farming systems

Farming Systems Research studies the agro-ecosystems, that is agriculture and its interaction with other ecosystems and society, at the farm level. The agro-ecosystems are highly complex [Kropff et al. 2001] due to the many feedbacks between natural processes, human factors, high geographical diversity in agro-ecosystems and the limited knowledge on some of the processes. Frequently, models are utilized to simulate the agro-ecosystems and further the understanding of agro-ecosystems. This has resulted in an abundant number of field, cropping system or farm level models, each developed for specific purposes. These models are realized as software components that are hardly re-usable and it is difficult if not impossible to integrate them with other models, in order to perform integrated analyses [Rizzoli et al. 2005; Athanasiadis et al. 2006]. One of the reasons for this is the poor semantics that usually characterize farm model implementations. In this paper, we demonstrate how ontologies can help to formalize the knowledge captured in these models; in order to subsequently facilitate model knowledge re-usability and exchangeability.

1.2 The SEAMLESS integrated project

European agriculture and rural areas continuously change as a result of an enlarging EU, WTO agreements, introduction of novel agro-technologies, changing societal demands and climate change. Efficient and effective agricultural and environmental policies are needed to support sustainability of European agriculture and its contribution to sustainable development of society at large. Assessing the strengths and weaknesses of new policies and innovations *prior to* their introduction, i.e., "ex-ante integrated assessment", is vital to target policy development for sustainable development. The SEAMLESS integrated project (http://seamless-ip.org) aims at developing a computerized, integrated and working framework (SEAMLESS-IF) to assess and compare, ex-ante, alternative agricultural and environmental policy options, allowing analysis across different scales (from field, farm to region and EU), dimensions of sustainability (economic, social, environmental and institutional) and for a broad range of issues and agents of change.

SEAMLESS-IF will be an open and modular framework and will offer the flexibility to analyze a wide range of issues [van Ittersum et al. 2006]. For specific questions to be analyzed, a subset of models and tools out of the broad range available within SEAMLESS-IF can be used. It will require scientific and technical breakthroughs to enable model integration across scales, disciplines and issues. The *SEAMLESS integrated modeling framework* is aimed to be a platform for the development of integrated applications and used by researchers and scientists to produce applications and outputs for policy makers. Research groups will be able to develop agricultural, environmental, economic and social models, at different scales. These models will be seamlessly integrated, while maintaining the logical independence of data, models and simulation/optimization algorithms. This will be achievable through the development of *independent components* and a *declarative approach to modeling*. Such an approach will permit the clear delegation of tasks within a multidisciplinary research team, strengthening collaboration and improving integration.

1.3 Ontologies, Knowledge Bases and Semantic Web

The last few years, ontologies, knowledge bases and the semantic web attract the interest of the research community. An ontology in computer science is considered as a specification of a conceptualization [Gruber 1993]. It's a formalization that could be expressed in a machine readable format, i.e. as the Web Ontology Language [McGuinness and van Harmelen 2004]. This provisionally allows a software system to "comprehend" a conceptual schema and makes it possible to reason on it. A knowledge base is the result of expressing the information related to a domain in line with a given domain ontology. Typically this activity involves instantiating data with respect to an ontological definition. Ontologies can be seen as a medium for open software environments [Willmott et al. 2002], where software agents provide semantic web services under strict contracts, as for example in the AgentLink project.

In this background, we envision the development of SEAMLESS-IF as an open modular simulation environment for agricultural, science and policy. In SEAMLESS-IF, modelers are considered as communities of "knowledge workers" in the fashion of Warren [2006]. Knowledge captured in environmental models is exposed using ontologies and ultimately using the semantic web can be reused, and combined properly for integrated studies.

The objective of this manuscript is to demonstrate the potential usefulness of semantic modeling in capturing the complexity of the Farming Systems by making knowledge relationships explicit. The development of an ontology formalizing the conceptualization of farming systems considered in SEAMLESS and the use of a Knowledge Manager for linking models and applications together is explained. In this way a Knowledge Manager mediates between various data sources and model algorithms, ensuring data integration based on explicit semantics.

2 The Agricultural Management Definition Module

2.1 Problem definition

A pivot element of the SEAMLESS modeling framework is farm level modeling. Farm level modeling aims to assess the impact of policy changes and technological innovations on farmer behavior now and in the future. To achieve these goals it is required to specify the options available to a farmer for using his/her resources and satisfying his/her objectives. This set of options is the agricultural activities that a farmer can apply. On a purely arable farm, agricultural activities constitute of options related to growing different crops with a range of alternative management practices. In this paper, we define an agricultural activity as a coherent set of crops (a crop rotation) with associated crop management and corresponding inputs, e.g. fertilizer, seed, pesticides, and outputs, e.g. marketable products, production of feedstuffs for on-farm use and environmental effects [Ten Berge et al. 2000; van Ittersum and Rabbinge 1997]. A rotation is a succession of crops in time (cropping sequence) and space (cropping pattern), where the last crop is the predecessor of the first crop (creating a loop). Crop management is a complete set of agronomic inputs (e.g. management practices) characterized by type, level, timing and application technique [van Ittersum and Rabbinge 1997].

A distinction can be made between current and alternative agricultural activities. Alternative agricultural activities are agricultural activities that are not currently used, but are technically feasible alternatives for the future, often technological innovations or newly developed cropping practices, while current activities are agricultural activities that are currently being practiced and can be derived from observed data.

In a farm system model, mathematical programming (usually linear constraints and a non-linear objective function [Janssen and van Ittersum 2006] is used to 'simulate' the allocation of current and alternative activities to the available resources, while satisfying the objective and meeting the policy constraints.

2.2 Abstract architecture and provided services

The main objective of the *Agricultural Management Definition Module* (*AMDM*) is dual. First is to describe, generate and quantify alternative and current agricultural activities that can be evaluated by a dynamic crop simulation model (*Agricultural Production Externalities Simulator: APES*) [cf. van Ittersum and Donatelli 2003] in terms of yields and environmental effects. Second is to generate a set of fully quantified agricultural activities that can serve as inputs to a farm level optimization model in which the possible activities are confronted with farm endowments and farmer's objectives (*Farm System Simulator: FSSIM*). In this respect, AMDM serves both biophysical and optimization models by preparing feasible/possible agricultural activities to be simulated or optimized respectively, as shown in Figure 1. Specifically, in this paper we present the part of the AMDM that formulates alternative agricultural activities for arable farming systems. In principle, a similar procedure could be carried out for livestock farming systems.

The AMDM for alternative agricultural activities consists of three cooperative components:

(i) The Production Enterprise Generator component, that given a set of production orientations and crops generates all feasible crop rotations,

(ii) The Production Technique Generator component, that is responsible for generating the crop management options of the rotations, and

(iii) The Technical Coefficient Generator component, that quantifies, collects and formats the input data for the farm model.

An abstract architecture of the AMDM is presented in Figure 1.

The aim is to develop AMDM components with open interfaces that adhere to a shared ontology. In this respect, all three components were developed on top of a Knowledge Manager shell (*KM*), capable to: (a) register domain ontologies, (b) load data from external sources, (c) realize links between components, and (d) provide interfaces with external applications (in our case APES and FSSIM).

The use of a Knowledge Manager maximizes the substitutability of AMDM components, by expressing all the knowledge related to component interfaces in a declarative way, using an ontology (as discussed for a similar application in Athanasiadis et al. [2006]). Also, based on the AMDM ontology we developed: (a) a database registration facility, that enables different data sources to be used directly as system inputs, and (b) the generation of data types, reflecting the ontology structure, based on which model algorithms can be directly programmed.

In section 3 is shown how the use of ontologies in conjunction with the KM was realized for enabling all AMDM components to access a shared knowledge base for accessing existing databases or sharing data generated the system.

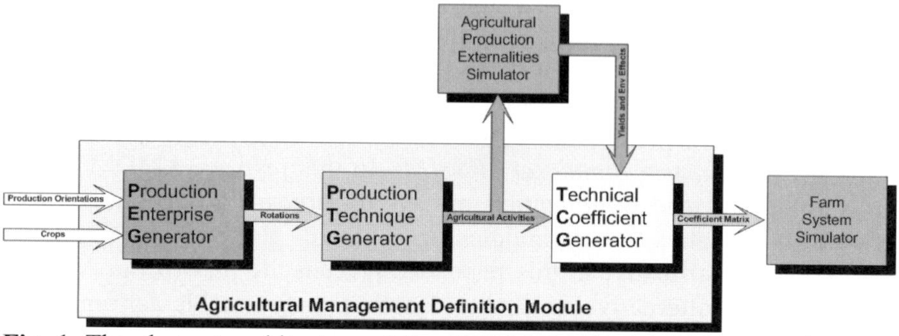

Fig. 1. The abstract architecture of the AMDM. Its three components along with the external applications are shown

2.3 Detailed architecture

The features of each of the AMDM components are detailed below.

The *Production Enterprise Generator* (PEG) is the component that generates a set of feasible rotations of the farm based on suitability filters and rotational filters. In principle, all crops that may be grown in a given biophysical environment can be combined into rotations. However, not all of these combinations are agronomically feasible or desirable. The first part of the PEG contains 10 suitability filters that determine which crops can be grown in a certain biophysical environment, given the biophysical environment the farm is in and given a list of crops. The second part is based on a component called ROTAT, developed by Dogliotti et al. [2003]. From the list of suitable crops, it generates all possible crop sequences and in a subsequent step eliminates all rotations that are not feasible according to 9 rotational filters. Suitability and rotational filters can be switched on or off as desired by the user as suitability filter factory and a rotational filter factory is used to create suitability filter and rotational filter objects at run time. The production orientation limits the length of the rotation and the amount of different crops in a rotation.

The *Production Technique Generator* (PTG) is a component to generate alternative agricultural activities on the basis of the feasible set of rotations by attaching crop management information to each crop in the rotation. A crop management is a complete set of agronomic inputs characterized by type, level, timing and application technique [van Ittersum and Rabbinge 1997]. Crop management exists of five management practices: water management, general management (sowing, harvesting and field inspection), nutrient management, conservation management and weed, pest, disease management. For each of these management practices the PTG has one management generator, which generates a set of events for an aspect of crop management. An event is one operation that takes place during the growing season of the crop, for example sowing, fertilization, irrigation, harvesting, field inspection, etc. The five management generators generate the events based on the implements and inputs a farm uses, the specification of the management practices as part of the production orientation and the rotations that were an output of the PEG. Each of the management generators can be switched on or off independently as a management factory is used to create management generator objects at run time. The output of the PTG is fed into the dynamic crop simulation model APES. Note that APES simulates yields and environmental effects for each crop with associated management in a rotation.

Finally, the *Technical Coefficient Generator* (TCG) links the alternative agricultural activities generated by PEG, PTG and APES to socio-economic inputs and outputs by simple calculations and prepares the inputs for the farm model in an input-output matrix (see Fig. 1). The TCG can produce an input-output matrix for the farm model on different scales: on a daily, yearly or seasonal basis, and on a rotational or individual crop in the rotation basis, as is dependent on the request from the farm model. The simple calculations carried out by the TCG are on variable costs and labor requirements and it has six different variable cost calculators.

2.4 An Ontology as a mediator for the Agricultural Management Definition Module

In order to systematically formalize our knowledge on Farming Systems Research, we developed an ontology for specifying the interfaces of AMDM to external applications (Databases, APES, FSSIM), along with the inter-component communication. In this way, AMDM components behave similarly to software agents providing information services, that are explicitly defined using ontologies. The AMDM ontology links to the core SEAMLESS ontologies (discussed in brief in Rizzoli et al. [2005], and will be available at http://seamless.idsia.ch/ontologies).

A small example of the developed ontology is presented in the following (Fig. 3(a)), where the concept of a Production Orientation is presented in the form of a conceptual map. A Production Orientation (PO; section 2.3) limits the length of the rotation and the amount of different crops in a rotation and directs the management practices associated with the different crops. So, a PO is defined by three data type properties (the minimum and maximum rotation lengths, and the maximum number of crops), and an object property (*hasManagementPractices*) that associates a PO to a set of Management Practices. In Figure 3(a), we present an example instance of a PO, called "Conventional". The Management Practices of a PO on their turn are defined similarly as presented in Figure 3(b), where the characteristics of and the relations between the PO and Management Practices are presented.

Following an iterative development procedure, all intrinsic concepts that AMDM deals with and their properties have been defined in the system ontology. This process involved several iterative reviews of the ontology among the domain experts and the knowledge engineers. The result of this activity was a declarative formalization of concepts that AMDM deals with. In this respect, we do not restrain the components developed by their current implementation languages or internal structures. In the contrary future extension or substitution could be easily and soundly supported if new components are developed with respect to the same (or an equivalent) ontology.

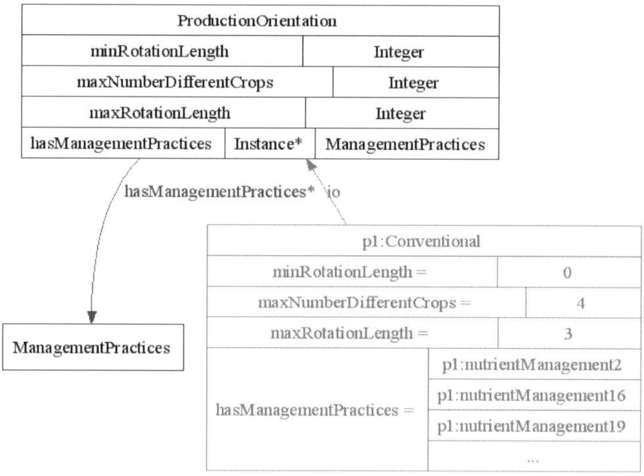

(a)

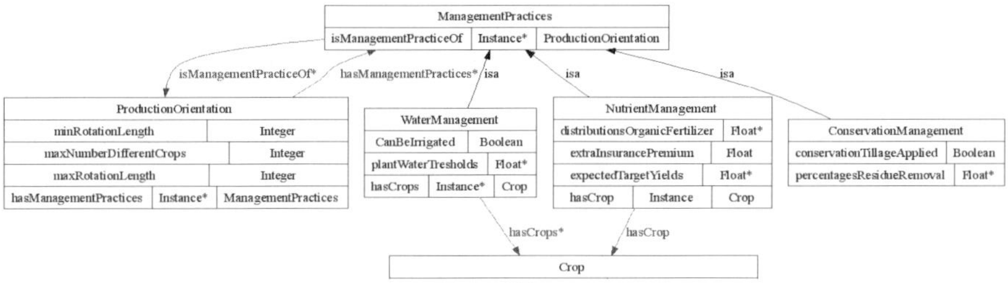

(b)

Fig. 2. Snapshots of the developed ontology: (a) The Production Orientation class and an example instance "Conventional", (b) The Management Practices class and their characteristics.

3. System realization and implementation

3.1 AMDM Implementation

The three components Production Enterprise Generator, Production Technique Generator and the Technical Coefficient Generator of the AMDM for alternative activities are developed in JAVA, while the Knowledge Manager was implemented as an extension of the Java Protégé-OWL API [Horridge et al. 2004].

The AMDM implementation is illustrated in Figure 3. AMDM was implemented on a KM shell that provides interfaces to external sources or applications and facilitates the linking of AMDM components. The added value of such an approach is three-fold:

a completely separates algorithms from data and user interfaces,
b facilitates easy linkage to external database sources and user interfaces, and
c makes algorithms easily extensible and comprehendible.

These objectives were achieved by:

– (i) linking the algorithms to databases through the ontology,
– (ii) developing the user interface at the very end,
– (iii) using design patterns, especially factory and strategy pattern where possible in the algorithms.

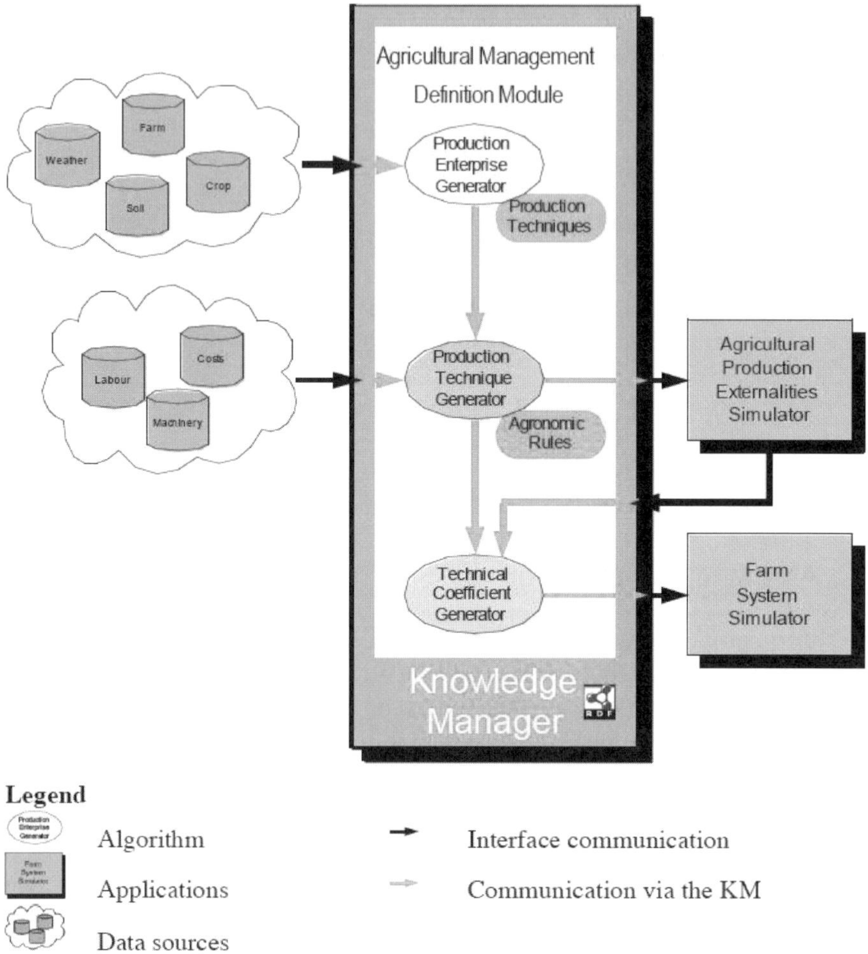

Fig. 3. AMDM for alternative activities and its components: algorithms, databases and connections

Following the workflow of Figure 3, the AMDM process starts with the invocation of PEG. All required inputs related to crops, farm, soils, etc are fetched from external sources via the KM layer. In this respect PEG is not tightened to a specific database(s) structure, rather it relies on the KM to acquire all required information. Then the PEG is executed, creating the rotations that apply to the suitability and rotational filters that the user selected. These are transferred to PTG using the KM as a common repository for sharing results. Then, PTG calculates all possible Agricultural Activities, based on the appropriate management practices for each crop selected by the user, and on data related to machinery, labor and costs. Once again the data sources are decoupled from the algorithm implementation. Next, the Agricultural Activities resulting from the PTG are communicated to APES (an external application), which simulates each one of those, calculating the yields and environmental effects. APES results are captured by the KM interface and fed to the TCG. Finally, TCG is executed for generating the Coefficient Matrix for the FSSIM optimizer. In this respect, the AMDM facilitates the linking of APES with FSSIM, in an open, "loose" coupling based on ontology-specified interfaces.

In the following paragraphs we present how the ontology was utilized for data type code generation, to facilitate on semantically aware development, and how the KM was utilized for connecting to external data sources.

3.2 Use of the ontology for code generation and semantic-rich development

The ontology structure of the AMDM (section 2.3) was used directly for developing the interfaces of the AMDM by generating the source code of the data types exchanged among components and applications. Although Protégé already includes a plug-in for code generation, a new code generator was developed. This was for two reasons. Firstly, because the Protégé code generator is outdated, and takes no advantage of powerful implementation practices, as code annotations and generics. Secondly, Protégé used a class implementation for each interface. In contrast the generator we built uses only interfaces and a common proxy class for accessing the knowledge base. This ensures that the developer will be accessing the Protégé Knowledge Base only via the interfaces, and will either have direct access or duplicate information. In this way, model exchanged information flows via a Knowledge Base, which allows performing semantic checks at runtime for ensuring the soundness of the link among components and applications.

The GUI of the ONTO:Exporter application for generating code interfaces is shown in Figure 4 along with a code segment from the generated interface of the Production Orientation ontology class example, discussed in section 2.4.

3.3 System execution and linking to databases with D2R

The AMDM system developed has been tested on existing data coming from the region of Flevoland, The Netherlands. As mentioned above database sources and schemas are entirely decoupled from the developed system. The latter operates on the Knowledge Manager generated datatypes, which are generic and could "hide" behind various implementations. For software testing purposes and experimentation, we employed the D2R language and library [Bizer 2003] for defining a mapping between any SQL Database (accessed via ODBC) and the AMDM ontology. This functionality is integrated in the Knowledge Manager for accessing external sources that are transformed to instances of the domain ontology.

```
package ch.idsia.domainmanager.generated.fs;

import java.util.Collection;
...

@ClassURI("http://seamless.idsia.ch/ontologies/fssim#ProductionOrientation")
public interface ProductionOrientation extends PEGDatatype {

    // Datatype Functional Property fs:minRotationLength
    @PropertyURI("http://seamless.idsia.ch/ontologies/fssim#minRotationLength")
    public Integer getMinRotationLength();
    @PropertyURI("http://seamless.idsia.ch/ontologies/fssim#minRotationLength")
    public void setMinRotationLength(Integer var);

// (Properties maxNumberDifferentCrops and maxRotationLength omit-
ted for simplicity)

    // Object type non-Functional Property fs:hasManagementPractices
    @PropertyURI("http://seamless.idsia.ch/ontologies/fssim#hasManagementPracti
ces")
    public Collection< ManagementPractices> getHasManagementPractices();
    // Object type non-Functional Propertyfs:hasManagementPractices
    @PropertyURI("http://seamless.idsia.ch/ontologies/fssim#hasManagementPracti
ces")
    public void setHasManagementPractices(Collection<ManagementPractices>
val);
    // Object type Propertyfs:hasManagementPractices
    @PropertyURI("http://seamless.idsia.ch/ontologies/fssim#hasManagementPracti
ces")
    public void addHasManagementPractices(ManagementPractices val);
    // Object type Propertyfs:hasManagementPractices
    @PropertyURI("http://seamless.idsia.ch/ontologies/fssim#hasManagementPracti
ces")
    public void removeHasManagementPractices(ManagementPractices val);
}
```

(a) **(b)**

Fig. 2. (a) Code segment of the generated java interface 'ProductionOrientation' that is exported from the ontology. (b) The Onto:Exporter GUI through which the user may access more than one ontologies

Finally, we demonstrated the use of the system with data from Flevoland. Specifically, PEG was run for 10 crops, a conventional production orientation with a maximum rotation length of 3 years and a farm type in Flevoland. The crops were Carrot, Onion, Pea, Springbarley, Springwheat, Tulip, Lucerne, Fibrehemp, Grassseed and Sunflower. All Suitability Filters and Rotational Filters in the PEG were used, and this ultimately led to 2 rotations as a result of the PEG execution. These rotations are *'GrassSeed-SpringBarley'* and *'FibreHemp-GrassSeed-SpringBarley.'*

Next, the two rotations were fed into the PTG, which used data obtained from the database and specification of management practices on the production orientation, and ultimately generated 252 alternative agricultural activities. Every agricultural activity contains a set of crops (rotation), each of which is associated with a certain year in the rotation and with a set of management events, which is all the information required by APES. The grass seed crop in agriculturalActivity1 with *'GrassSeed-SpringBarley'* has 5 different events, e.g. 1 sowing event, 1 harvest event, 3 nutrient events, and no irrigation events. The execution of APES associated yields and environmental effects to each agricultural activity, that ultimately was directed to the TCG. TCG execution ultimately resulted 720 production coefficients that were finally forwarded to the FSSIM to select the optimal set of production activities given the farmer objectives. For the creation of production coefficients again the data was retrieved from the database via the KM, as in the TCG information on variable costs and labor requirements were attached to each production coefficient. For example, a grass seed crop in productionCoefficient1 with rotation *'GrassSeed-SpringBarley'* has associated a labor requirement of 20 hours per hectare per year and variable costs of 450 euros per hectare per year.

4 Discussion

4.1 Benefits of the approach

Through the case of the AMDM it was explained how semantic modeling, ontologies and the knowledge management practices can be used in modeling agricultural systems. The ontology structure of AMDM helped to capture the complexity of the AMDM by making explicit the knowledge that the agricultural scientist holds. By the use of ontologies, the agricultural scientist is *forced* to define concepts he/she commonly refers to by specifying their properties (datatype properties) and relationships to other concepts (object properties) in a detailed formalization. Also, the concepts in the ontology could subsequently be made available for modeling by allowing the agricultural scientist to write the algorithms. As farming systems approaches require interdisciplinary studies [Kropff et al. 2001] and therefore require the use of several different models and techniques, the different techniques and models need to be able to communicate with meaningful objects. In the AMDM the developed ontology played a vital role in clarifying exchanged information between AMDM and its external peers (i.e. APES, FSSIM, and the databases), as well as among the components of AMDM (PEG, PTG and TCG). This allowed the AMDM to operate on two different scales, both in space in time. Firstly, AMDM is able to exchange meaningful objects with a point scale model, operating on a daily basis, as APES is a dynamic crop simulation model. Secondly, it is able to exchange meaningful with a farm scale model, using annual data, as FSSIM is a usually a static, mathematical programming farm model.

Using ontologies in farming systems research requires a close cooperation between disciplines, in this case agronomy, agricultural economy and information technology. A close cooperation can be achieved by frequent iterations and discussion on the concepts used in both disciplines. Using ontologies implies an additional layer to the modeling exercise, which enforces the other layers (databases, algorithms and model structure), while at the same time making the modeling exercise more distributed.

In particular, in this paper contributed with a modular architecture and implementation of the Agricultural Management Definition Module, by exploiting ontologies and semantic modeling. Also, we examined the performance of existing knowledge engineering tools, particularly related to linking ontologies with legacy database sources and generating programming interfaces. Although effective the use of D2R language and library for Object-Relational Mapping can be significantly improved, if the process becomes semi-automated, i.e. through a formal data registration process. Therefore our future developments will drive towards this direction.

Future work will concentrate on expanding the current implementation. We plan to exploit further the developed ontology by expressing both production techniques filters and agronomic rules in a declarative fashion. Due to the nature of the rules, it is very hard, if not impossible to express them using description logics in OWL-DL. Thus extended frameworks like RuleML and SWRL will be considered for incorporating reasoning capabilities. This will further advance the benefits of using ontologies and semantic modeling in the agricultural modeling domain. Also, alternative implementations of the same system design using web services or software agents will be investigated for deploying the system across a distributed network. The same ontology could be used for describing model interfaces, which in such a case will be implemented as web services or agent communicative acts. Finally, we consider within the context of SEAMLESS project to promote further the use of semantic modeling, and the development of ontologies for the agricultural sector in order to maximize the reusability and the extensibility of the systems developed. Parallel efforts are adopting declarative approaches for the APES and FSSIM applications. The adoption of a set of shared ontologies within these modularized applications will lead us to a semantic-aware modeling and simulation framework for the agriculture sector.

Acknowledgement

This publication has been partially funded under the SEAMLESS integrated project, EU 6th Framework Programme for Research, Technological Development and Demonstration, (contract no. 010036-2). For further reference, the SEAMLESS project website is: http://www.seamless-ip.org.

References

Athanasiadis IN, Rizzoli AE, Donatelli M, Carlini L (2006) Enriching software model interfaces using ontology-based tools. In: 3rd Biennial meeting of the International Environmental Modelling and Software Society', iEMSs

Bizer C (2003) D2R MAP - A Database to RDF Mapping Language. In: Proceedings of the Twelfth International World Wide Web Conference (WWW2003), Budapest, Hungary

Dogliotti S, Rossing WAH, van Ittersum MK(2003) ROTAT, a tool for systematically generating crop rotations. In: European Journal of Agronomy, vol 19, pp 239-250

Ewert F, van Ittersum M, Bezlepkina I, Oude Lansink A, Brouwer F, Andersen E, Flichman G, Heckelei T, Olsson J, Olsson L, Rizzoli A, van der Wal T, Wery J (2005) Development of a conceptual framework for integrated analysis and assessment of agricultural systems in SEAMLESS-IF. In: Technical report, SEAMLESS-IP, http://www.seamless-ip.org/Reports/Report_01_PD1.2.1.pdf

Gruber TR (1993) A Translation Approach to Portable Ontology Specifications. In: Knowledge Acquisition, vol 5, edn 2, pp 199-220

Horridge M, Knublauch H, Rector A, Stevens R, Wroe C A (2004) Practical Guide To Building OWL Ontologies Using The Protégé-OWL Plugin and CO-ODE Tools. Online Tutorial, The University Of Manchester and Stanford University. http://protege.stanford.edu/plugins/owl/, access: 2007-02-14

McGuinness D, van Harmelen F (2004) OWL Web Ontology Language Overview. In: Technical report, WWW Consortium. http://www.w3.org/TR/owl-features/, access: 2007-02-14

Kropff MJ, Bouma J, Jones JW (2001) Systems approaches for the design of sustainable agro-ecosystems. In: Agricultural Systems, vol 70, edn 2-3, pp 369-393

Rizzoli A, Donatelli M, Athanasiadis I, Villa F, Muetzelfeldt R, Huber D (2005) Semantic links in integrated modelling frameworks. In: Zerger A, Argent RM (eds) 16th International Congress on Modelling and Simulation (MODSIM 2005), MSSANZ, pp 704-710

Ten Berge HFM, van Ittersum MK, Rossing WAH, van de Ven GWJ, Schans J (2000) Farming Options for the Netherlands explored by multi objective modelling. In: European Journal of Agronomy, vol 13, pp 263-277

van Ittersum MK, Donatelli M (eds) (2003) Modelling Cropping Systems: Science, Software and Applications. In: Special Issue of European Journal of Agronomy, vol 18, edn 3-4

van Ittersum MK, Rabbinge (1997) Concepts in production ecology for analysis and quantification of agricultural input-output combinations. In: Field Crops Research, vol 52, pp 197-208

Warren P (2006) Knowledge Management and the Semantic Web: From Scenario to Technology. In: IEEE Intelligent Systems, vol 21, edn 1, pp 53-59

Willmott S, Dale J, Charlton P (2002) Agent Communication Semantics for Open Environments. In: Technical report f-in-00066, FIPA

Simulation and Optimization of Habitat Network Permeability

Jens Finke, Michael Sonnenschein

Oldenburg University, Department of Computing Science,
Environmental Informatics Group, 26111 Oldenburg, Germany

Abstract: Landscape fragmentation is a major threat to a lot of wildlife species with a large spatial activity range. To ease the migration through the landscape, a lot of different measurements can be implemented. Though, due to limited resources, these should be placed where they are most effective.

The simulation framework SimapD presented in this paper is capable of assessing the network permeability and incorporates effects of anthropogenic obstacles, like roads, into the assessment. Through an individual-oriented simulation an overall permeability index for a habitat network is obtained. Based on this simulation approach, an algorithm is presented for optimizing such habitat networks concerning its permeability. It is intended to be applied to find optimal mitigation strategies. The complete procedure is illustrated by an example. *Keywords: landscape fragmentation, individual-oriented simulation*

1 Introduction

Today's landscapes are affected by a lot of requirements of human beings, e.g. people's transportation, mobility, and spare time needs. To meet all these needs the landscape is permanently transformed by building of roads, channels, railroads, recreation areas, residential areas, etc. All these transformations lead to a significant fragmentation of the landscape, where more and more large scaled natural areas become fragmented into smaller parts. One of the major causes of landscape fragmentation is the dense road network [Forman and Sperling 2002; Oggier et al. 2001].

This creeping process of landscape fragmentation is a significant threat to wildlife species which depend on spacious migrations, either daily, seasonal or during their lifetime [Baier et al. 2006; Forman 2003]. Depending on the species, the spatial scales for these migrations can range from a few hundred meters to some hundred kilometers of range. The large scale migration is especially important for gene exchange between different local populations and the establishment of new populations. The ability for local, i.e. small scale migrations is a vital trait of habitat quality which also determines the survival probability of a local population. In fragmented landscapes the migration is often impaired by linear obstacles either due to increased barriers (e.g. fenced areas) or loss of individuals while conquering the obstacle (e.g. road kills) [Forman and Sperling 2002].

Due to the threats on wildlife species there are political attempts to enforce a de-fragmentation of the landscape and to ease the migration of wildlife species again. Measurements include but are not limited to the construction of green bridges or underpasses [Trocmé 2006; Hlaváč 2005]. Since there are generally only limited financial resources for the implementation of mitigation measures, information where they are most effective for the de-fragmentation of the landscape are crucial.

The goal of this work is to support the identification and mitigation of conflict hotspots due to human activity and wildlife requirements, especially with regards to wildlife migration. The main questions are: Where does a potential high risk of conflicts exists, how does these conflicts affect the permeability of the landscape for wildlife, and where are the most effective locations for mitigation measurements leading to a more permeable landscape structure? To answer these questions we propose our simulation based Java programming framework called SimapD, an abbreviation for "simulation of anthropogenic population disturbances".

In section 2 some related approaches are discussed, while section 3 introduces the model language used to define the habitat network within SimapD. The individual-oriented simulation algorithm for assessing the network permeability is presented in section 4. Based on our simulation approach, an algorithm for the optimization of habitat networks will be applied, which is outlined in section 5. After that, some remarks on the quality of the optimization will be stated in section 6. Finally, a conclusion and outlook will be given.

2 Related Approaches

Of course, there are known applications and simulation approaches which also deal with landscape fragmentation problems. Meta-X [Grimm et al. 2003] for example simulates meta-populations and evaluates their mean population viability. A meta-population constitutes several subpopulations living in so called patches which are connected by migration corridors. These corridors can be used by individuals to migrate between populations, so that even previously extinct patches can be repopulated. Each corridor has a distance which can be interpreted as it's resistance for an individual's migration. For example, if there is a large road that crosses a wildlife trail the respective corridor has a large distance.

Meta-X' quite abstract notation of models does not directly match properties of the real landscape. The simulation results are obtained by a Markov-chain algorithm and are difficult to interpret regarding concrete landscape characteristics. The mean population viability used by Meta-X as the main metric is a good measure to compare different meta-population scenarios, but it fails when it comes to linking the results to traits of the meta-population network. The size of the habitat network is always determined by the size of the studied meta-population, due to the underlying Meta-X simulation model that requires each patch to be a suitable habitat for the species. Hence, only modeling relevant parts of a landscape with anthropogenic disturbances is difficult.

FRAGSTATS [McGarigal 2002] is a tool for computing several landscape metrics on raster maps. This includes metrics to determine the degree of landscape fragmentation, like the effective mesh size [Jaeger 2000]. Most of these metrics are based on a statistical analysis of the landscape pattern. The dynamics of (artificial) obstacles in the landscape, e.g. the road traffic volume, are not considered. The metrics used by FRAGSTATS are especially useful for evaluating landscape changes over time. An identification of main conflict zones between wildlife and human activity is not possible in an intuitive manner, since population related attributes like wildlife migration corridors are not considered.

3 Habitat Network Model

Comparable to Meta-X, the habitat network in SimapD is described by an undirected graph where the edges represent possible migration corridors in the landscape, and the nodes link two or more corridors. The location of possible migration corridors can be obtained by GIS-Analysis for example [Strein et al. 2006]. A node may represent a habitat or patch of a population, but it is also allowed to simply use it as a juncture for multiple corridors. This definition applies to all regional scales, especially to small scales, where it is not sensible to assume that every node represents a suitable habitat for a species.

Specific properties of the species, like population size, or mortality and fertility rate are not explicitly modeled in SimapD for two reasons: First, these parameters are rarely known for most wildlife species and second, these factors are supposed to play not an important role for measuring the network permeability. This is of course different for models which try to assess the development of the population size over time, like Meta-X. SimapD basically measures the permeability of a network. For this purpose, it uses an individual-oriented simulation approach where individuals are sent for a walk through the network. This replays what happens in reality when individuals utilize the landscape for migration.

To avoid the problem of assessing crossing or mortality events per time unit, SimapD models corridor attributes statically. Each corridor has two properties: a resistance rate which determines the probability that an individual will use the corridor for a transition, and a mortality rate which determines the probability that an individual gets killed when using the corridor for transition. By these two properties all migration related events generally caused by anthropogenic obstacles can be described: successful corridor transition, death during transition, and holding the individual off from crossing. Depending on the intensity of the anthropogenic disturbances these attributes can be adapted per corridor. SimapD also supports the use of sub-models, which may calculate the corridor attributes on base of disturbance specific parameters [Finke et al. submitted].

4 Permeability evaluation

The permeability of a habitat network is evaluated by an individual-oriented simulation algorithm. During the simulation a large number of individuals are sent through the graph for a random walk (Fig. 1).

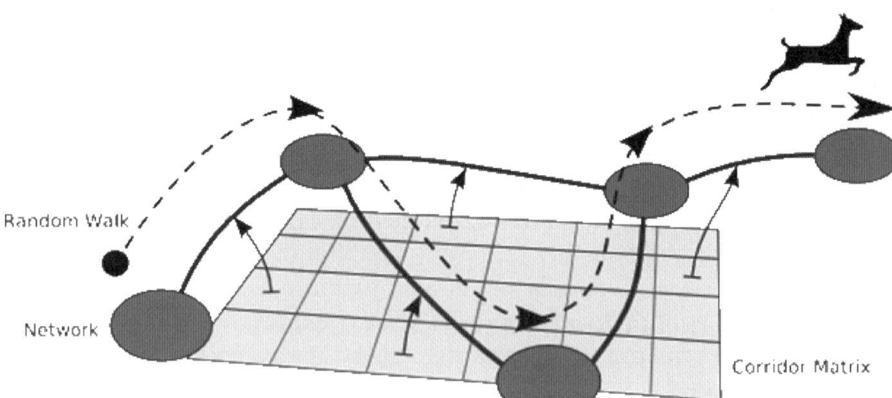

Fig. 1. Composition of a SimapD model

Based on the resistance and mortality rates, provided in the corridor matrix, a Monte-Carlo simulation determines whether the individual gets killed during a corridor transition, reaches the distant patch, or is hold off by the resistance and therefore stays in the same patch. The random walk stops, if either the individual is killed or a maximum path length is reached, specified as a simulation parameter. To treat all nodes of the network in an equal manner, each node releases exactly one individual per simulation run.

A number of statistical parameters are collected during these walks, including the usage and killing rate of each corridor, the visiting rate for each node, etc. The simulation terminates, if the relative frequency distribution of the number of node visits becomes stable for an adequate number of runs, that is, if it does not change more than a small epsilon value between these runs. The SimapD user can control the required number of runs with a stable frequency distribution, and their maximum allowed variability by specifying the used epsilon value. Depending on these values, the simulation time may vary.

A measure for comparing different network configurations is deduced from the performance of each random walk and is denoted by the network permeability index λ_k. For each random walk p the path length s(p) and the number of different graph nodes visited v(p) are used to calculate the path index $\varphi(p)$ by: $\varphi(p) = s(p)*v(p)$. Based on these values a relative frequency distribution of the path indices is calculated. If the network has a low permeability, this path index distribution is shifted to the lower end of the distribution range (low values). The better the permeability of the network is, the more the distribution moves to the right (higher values) and converges to a normal distribution [Finke in press]. The distribution describes nicely the characteristic of the overall network permeability. The expected value of this distribution is used as an aggregated measure and denoted by λ_k. Here, k is the maximum path length which determines the possible range of values for $\varphi(p)$, and hence also for the expected value of the path index frequency distribution.

The measurement λ_k can be used for decision support on different mitigation scenarios. An example can be found in (Finke in press).

5 Optimizing habitat networks

To improve the permeability for a given habitat network, often different measurements can be applied. However, in the majority of cases there are only limited financial resources available, so the investments should be placed where they are most effective. On this account the question for a given network is: Where should a fixed amount of financial resources be invested into the structure to gain best overall permeability increase? In case of SimapD such investments can be represented by a cost distribution that assigns a certain amount of effort (represented by its cost) to each corridor attribute for improving the corridor mortality respectively resistance rate.

5.1 Optimization algorithm

As proposed by Grüning and Sonnenschein [2005] an optimization algorithm based on tabu-search can be applied to obtain such a requested optimal cost distribution. The tabu-search algorithm [Michalewicz and Fogel 2000] requires a lot of iterations for achieving good results, which demands for a very fast evaluation of a specific solution, i.e. a specific cost distribution. To avoid time consuming simulation runs for each scenario to be processed, a meta-model (not to be confused with the Meta-X model) is used which calculates the output based on regression analysis instead of simulation. Grüning and Sonnenschein [2005] used a support vector machine (svm) algorithm for learning such a meta-model that maps the input cost distribution to the simulation output of a specific habitat network. Such a meta-model can be applied without simulation and thus leads to a significant performance gain. However, the quality of the optimization highly depends on the quality of the meta-model (see section 6).

For the optimization of the network, virtual cost units are distributed to the corridor attributes affecting them positively. The influence is expressed by a cost function c(x) that calculates the mortality or resistance rate when a given amount x of costs is applied [Grüning and Sonnenschein 2005]:

$$c(x) = 1 - 2 * \left(\frac{1}{1 + e^{-x}} - \frac{1}{2}\right) \tag{1}$$

The inverse function $c^{-1}(r)$ determines the already invested costs from a certain mortality or resistance rate r. By concatenating these two functions, the improvement r' is calculated when additional costs x are assigned to a corridor attribute: $r' = c(c^{-1}(r)+x)$

The optimization goal is to maximize the network permeability index λ_k, where the maximum path length k is twice the network diameter. This requires a meta-model that maps the cost allocations applied to the corridor matrix to the λ_k values, yield by simulation.

The procedure for obtaining such a model is split into different steps according to Hsu et al. [2006]. First, an adequate number of training data is generated by the simulation model. This data set is used to train the svm model. There are different flavors of svm kernels and learning algorithms [Smola and Schölkopf 1998]. The kernel determines how the training data is moved into a higher dimensional space to find appropriate support vectors, which classify the data properly. By applying a Lagrange interpolation on the resulting support vectors, a function approximation can be obtained instead of a simple classification of the data. The "radial basis function"-kernel (rbf-Kernel) is suitable for such function approximations and is used in the following.

Most svm kernels have to be parameterized with proper values to gain best prediction results. These values are problem dependent and must be determined in advance according to the available training data. Hence, in a second step a grid search combined with cross validation is applied to the training data set to find best parameterization for the rbf-kernel [Hsu et al. 2006]. With the best performing kernel parameters the final meta-model can be trained in a third step and used for λ_k prediction.

5.2 Optimization Example

An example scenario for illustrating the approach sketched so far is depicted in Figure 2. This is an adapted version of the example scenario from Grüning and Sonnenschein [2005] who applied the optimization approach to a Meta-X model. The scenario represents a set of six patches, which are divided by a larger road into north and south compartments. For SimapD each corridor is described by a resistance rate r, which is correlated to the corridor length *l* from the original example. The mapping is done with the function $r(l) = 1 - e^{-l/a}$ where a is a scaling factor, set to a=45. The corridors affected by the road have additional mortality rates associated with them (see Table 1, unoptimized columns).

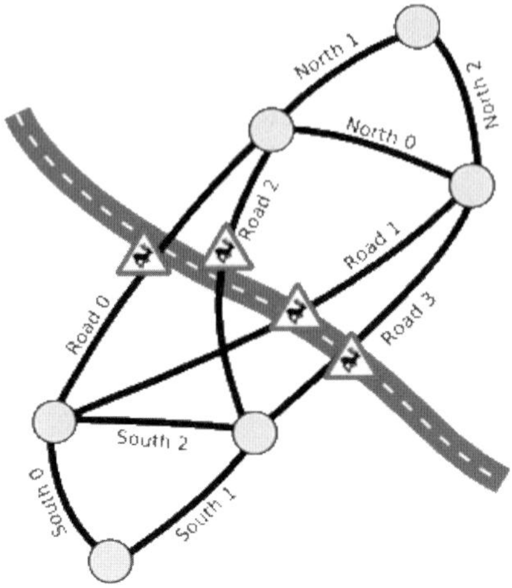

Fig. 2. Layout of the example habitat network used for the optimization

For the actual optimization of the network a total amount of two cost units is assumed, allocatable in steps of 0.1 units. Initially, the costs are equally distributed over all corridors; i.e. each corridor gets 0.1 cost units to reduce the mortality as well as the resistance rate. For the results presented below, the tabu-search runs 10,000 iterations with a tabu-list length of 150 entries. Other variations of these tabu-search parameters did not yield better results.

As a result, the optimization algorithm distributed the costs as noted in Table 1. The main amount of costs is allocated to the SOUTH 1 and ROAD 3 corridors to improve their resistance rate. Minor allocations are applied to the ROAD 2, ROAD 1 and NORTH 1 corridors. The distribution suggested by the optimization algorithm yields a network permeability index gain of 16% regarding the initial unoptimized version; that is from $\lambda_6 = 7.67$ to $\lambda_6 = 9.24$. This is nearly the same improvement achieved with the Meta-X approach (15%). To increase the overall permeability of the network, the optimization algorithm focused on improving a path from the south patch to the north patch along the corridors SOUTH 1, ROAD 2 and NORTH 1. The key patch is the linkage of the SOUTH 1 and ROAD 2 corridors. Except for SOUTH 2 the optimization algorithm applied costs for reducing the resistance rate to all corridors at this linkage point. For this linkage point the improvements lead to a mean probability of 0.51 for successfully reaching a neighboring patch, whereas in the unoptimized case the mean probability is 0.372.

Table 1. Corridor attributes (R = resistance rate, M = mortality rate) before and after applying the optimization procedure; cost distribution allocated by the SimapD optimization and, for comparison, by the Meta-X optimization

	Unoptimized		Optimized		Cost distribution	
Corridor	R	M	R	M	SimapD	MetaX
SOUTH 0	0.4665	0.00	0.4665	0.00		
SOUTH 1	0.6298	0.00	0.3423	0.00	0.8	0.55
SOUTH 2	0.3588	0.00	0.3588	0.00		
NORTH 0	0.3588	0.00	0.3588	0.00		0.12
NORTH 1	0.4665	0.00	0.3987	0.00	0.2	0.39
NORTH 2	0.3588	0.00	0.3588	0.00		0.09
ROAD 0	0.7118	0.27	0.7118	0.27		
ROAD 1	0.7692	0.33	0.6770	0.33	0.2	0.39
ROAD 2	0.6527	0.23	0.5282	0.23	0.3	
ROAD 3	0.7118	0.27	0.5020	0.27	0.5	

For comparison, the corridor cost distribution yield by the Meta-X approach is mentioned in Table 1, too. Since Meta-X assigns a minor amount of costs also for patch improvements, the total sum of the values given in Table 1 is slightly below 2.0. Anyway, it stands out that the main amount of costs is applied to the same corridors SOUTH 1 and NOTH 1 by both models, but by Meta-X only ROAD 1 is optimized, whereas SimapD suggests improving the resistance rate for three road-affected corridors. It can be presumed that this is caused by Meta-X considering only the corridor distance for evaluation. Reducing the longest corridor distance significantly will result in a reasonable improvement. Corridor ROAD 1 is by far the one with the longest distance in Meta-X [Grüning and Sonnenschein 2005]. In comparison SimapD distinguishes resistance and mortality rates. A reduced resistance rate will increase the permeability, but only when the individuals will not be killed by the disturbance too often. Therefore, the optimization algorithm reduced the resistance rate for corridors having a relatively low mortality rate. Altogether the selection of road-affected corridors done by a SimapD based optimization seems to be reasonable.

As a consequence of these results, measurements should be concentrated at the suggested path SOUTH 1, ROAD 2, and NORTH 1. For SOUTH 1 and NORTH 1 possible measurements would be to adapt the environment to the requirements of the examined species (e.g. reforestation). For the road-affected corridors ROAD 2 or ROAD 3, a green bridge may be sensible, that connects the south with the north compartments. However, concrete measurements depend on the species or species family examined.

6 Optimization quality

The results so far were all obtained by using the meta-model to determine the network permeability index λ_k. Therefore, the quality of the optimization process depends on the quality of the meta-model. In turn, the quality of the svm model depends on the appropriateness of the training data.

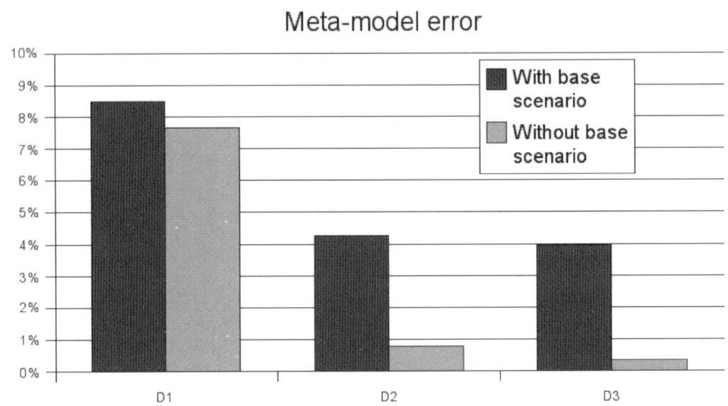

Fig. 3. Mean error of predicted λ_k-values in relation to simulation outputs for three kinds of training data (see text for explanation)

Figure 3 shows the mean prediction error of the meta-model in comparison to the simulation outcomes for three different kinds of cost distributions to train the meta-model. For D1 random costs from the [0, 2.0] interval have been applied to each corridor attribute to create training scenarios. For the D2 distribution 0.1 costs were allocated to a random corridor, until the total amount of 2.0 costs were reached. The D3 distribution allocated a random fraction of the available costs to a random corridor. For D2 and D3 the sum of all applied costs is 2.0. The different training sets were used to train three svm-models and these have been used for optimization. The predicted λ_k values for the base scenario (no cost allocation), the initial solution scenario, and the best solution scenario of the optimization were then compared with the appropriate values achieved by simulation. The comparison of the mean λ_k prediction error regarding these three scenarios is depicted in Figure 3. All three meta-models do not predict the lambda value for the base scenario very well, because they are trained to modified networks with applied costs. For D2 and D3, the support vectors are limited to the search space of the tabu-search algorithm for the optimization, causing mean prediction errors of less than 1% for the initial and best solution scenarios. Since the D3 trained meta-model has been used in the example addressed in section 5, it is assumed that the results are highly accurate compared to the simulation model.

Since Grüning and Sonnenschein [2005] missed a similar comparison, no statement can be made about the quality of the optimization approach for Meta-X.

7 Conclusions

It is an important task to improve the overall permeability of habitat networks to aid the migration requirements for a lot of species. Our proposed individual-oriented simulation method is a promising approach to identify regions of high conflict potential between wildlife needs and human requirements. By explicitly incorporating important anthropogenic disturbances (like roads) into the model, a detailed insight of the effects of these artificial obstacles regarding network permeability can be obtained. The λ_k measure, resulting from the simulation, eases the comparison of different development scenarios.

The optimization procedure applied in this paper helps to place mitigation measures where they are most effective. This is especially important in terms of limited financial resources. The presented example obtained reasonable results, built up on a high accuracy of the meta-model predictions. The use of a meta-model makes the algorithmic optimization feasible in the first place, since the simulation based evaluation of all cost allocations during the optimization process would be much too time consuming.

References

Baier H, Erdmann F, Holz R, Waterstraat A (eds) (2006) Freiraum und Naturschutz: Die Wirkung von Störungen und Zerschneidungen in der Landschaft. Springer, Berlin

Finke J (in press) Ein Simulationsframework zur Bewertung von großräumigen Habitatnetzen. Modellierung und Simulation von Ökosystemen – Workshop Kölpinsee 2006

Finke J, Strein M, Sonnenschein M (submitted) A Simulation Framework for Modelling Anthropogenic Disturbances in Habitat Networks, submitted to Ecological Informatics, Special Issue 3: Bioinspired Computation and Hybrid Modelling of Aquatic and Terrestrial Ecosystems

Forman RTT (2003) Land Mosaics: The ecology of landscape and regions. Cambridge University Press

Forman RTT and Sperling D (2002) Road Ecology: Science and Solutions. Island Press

Grüning F, Sonnenschein M (2005) A Method for Optimizing a Metapopulation. In: Leal Filho W, Marx Gómez J, Rautenstrauch C (eds) ITEE 2005 - Second International ICSC Symposium on Information Technologies in Environmental Engineering. ISBN 3-8322-4362-3, pp 365-380

Hlaváč V (2005) Increasing Permeability of the Czech Road Network for Large Mammals. GAiA Themenschwerpunkt Landschaftszerschneidung, vol 14, edn 2, pp 175-177

Hsu CW, Chang CC, Lin CJ (2006) A practical guide to support vector classification. http://www.csie.ntu.edu.tw/~cjlin/papers/guide/guide.pdf, access: 2006-10-31

Jaeger JAG (2000) Landscape division, splitting index, and effective mesh size: New measures of landscape fragmentation Landscape Ecology, vol 15, edn 2, pp 115-130

McGarigal K, Chushman SA, Neel MC, Ene E (2002) FRAGSTATS: Spatial Pattern Analysis Program for Categorial Maps. http://www.umass.edu/landeco/research/fragstats/fragstats.html, access: 2006-10-31

Michalewicz Z, Fogel DB (2000) How to Solve It: Modern Heuristics. Springer, Berlin

Oggier P, Righetti A, Bonnard L (eds) (2001) Zerschneidung von Lebensräumen durch Verkehrsinfrastrukturen COST 341. Bundesamt für Umwelt, Wald und Landschaft; Bundesamt für Raumentwicklung; Bundesamt für Verkehr; Bundesamt für Strassen, Bern

Smola AJ, Schölkopf B (1998) A Tutorial on Support Vector Regression. NeuroCOLT Technical Report NC-TR-98-030. Royal Holloway College, University of London, UK

Strein M, Müller U, Suchant R (2006) Artunspezifische Modellierung einer Korridor-Potenzial-Karte für Mitteleuropa – Methodik und erste Ergebnisse einer landschaftsökologischen GIS-Analyse. Naturschutz und biologische Vielfalt, vol 17

Trocmé M (2006) Habitat Fragmentation due to Linear Transportation Infrastructure: An overview of mitigation measures in Switzerland. 6th Swiss Transport Research Conference (STRC 2006) http://www.strc.ch/pdf_2006/Trocme_STRC_2006.pdf, access: 2006-10-31

System Architecture for Coverage and Visualization of Underground Pipeline Structures in Detectino

Jana Görmer, Helmut Lessing, Dirk Bettels

Clausthal Institute for Environmental Technologies, 38678 Clausthal-Zellerfeld, Germany, e-mail: {jana.goermer, helmut.lessing}@cutec.de

Abstract: The damages caused by construction machines in underground pipeline structures cause significant economical losses. The project Detectino is combining methods of mechatronics, geophysics and informatics to investigate surface near underground to perform real-time detection and documentation of pipeline structures. Different sensor systems like electromagnetic (EM), ground penetrating radar (GPR) and a new seismic method to generate different data sets of the same objects in the ground. The data fusion and automated interpretation result in high probabilities of detecting and classifying buried objects. Those objects will be visualized three-dimensionally including data from geographical information systems (GIS) and georeferenced data. This paper introduces the basic concepts of Detectino and its system architecture. The system consists of a mobile platform moving over the surface and exploring it using different sensor methods. The control of this platform and its sensory systems is undertaken by an accompanying vehicle (AV). Additionally, there is a communication between the AV and a control center (Trust Center), where work orders are exchanged over Internet Technologies. The results of the investigation are sent back to the Trust Center (TC) and will be saved and updated. The architecture is divided into separate modules representing the different responsibilities and using the expert knowledge of each partner.

1 Introduction

The construction of earth-laid supply and return pipeline systems is widespread business in all industrialized societies. Cities are provided with pipelines of water, gas and electricity supply, as well as communication lines such as fiber cables for internet and, where applicable, caloric supply and waste water disposal. In the last decades, the different pipeline structures developed slowly and in parallel. Therefore in the majority of cases an adequate and precise documentation does not exist.

Excavations are mostly essential for the construction of new pipeline systems, as well as for sanitation of old ones. Thereby construction machines often damage existing pipeline systems, with more then 65 % of accidents caused by digging equipment [Münnich 1995]. This happens because of missing or imprecise documentations of the underground such as cadastral plans. The economical disadvantage for the construction and sanitation of pipeline structures are, only for Germany, about 210 Million Euros per anno [Münnich 1995]. Furthermore the reparation of damaged units – especially electricity and communication lines – usually cannot reestablish their former quality. Thus, there exists a great interest in a contact-free detection system with the ability to visualize the relevant underground section before starting to dig.

The objective of the Detectino project is the development and trial of a locating system which can detect contactlessly the different pipeline systems in the ground and provide their real-time visualization on site. The system is understood as a mobile unit and should detect the linear structures resulting in a virtual reality.

This contribution introduces the concept of Detectino, its work flow and system architecture. A mobile platform scans the underground and explores it with the help of different sensor methods. The control of this platform and its sensory systems is undertaken by an accompanying vehicle (AV). There exists a communication e.g. for work orders between the AV and the Trust Center (TC). The results of the underground investigation are sent back to the TC and will be saved and updated.

The system architecture is divided by responsibilities into modules like mechatronics for the vehicles, sensor systems to scan the ground, computational intelligence to extract the relevant information and visualization to transform the data into a 3D world.

The following sections present this work in further detail. Section 2 contains the related works. The concept and requirements of the Detectino system including the workflow are detailed in section 3. In section 4 the system architecture is described. Finally, conclusions and future work are presented in section 5.

2 Related works

Three projects having the common target of detecting objects in the underground can be remarked: HADU [Wittmann 2006 and Geotechnologien 2006],

Sensoft [Sensors and Software Inc.] and IDSgeoradar [Ingegneria dei sistemi]. HADU combines ground penetrating radar (GPR) and ambient vibration methods to find salt caverns difficult to measure. It provides a three-dimensional visualization and interprets the data on the example of the urban city Hamburg. Besides the concept idea of measuring the underground, visualizing and interpreting, the ambient vibrations method is not precise in the range of 0 to 3 meters revealing not to be appropriate for pipeline structures.

Sensoft and IDSgeoradar are based exclusively on the GPR system but being advanced in using it for several application areas as well as buried utilities and infrastructure assessment. They introduce the idea of having a mobile unit to scan the underground. The two working systems show that information from GPR with optional GPS responds to buried metallic utility pipes, cables and conduits. But it lacks of other geophysical sensors which make data fusion and 3D visualization more precise and efficient.

3 Concept and Work Flow of the Detectino System

The Detectino project combines research and development of a system for detecting different pipeline systems in diverse surface near soils and providing its real-time 3D visualization on site. The Detectino system may be employed at any site having pipeline structures in the underground with the help of mobile units. One unit consists of a platform and an accompanying truck carrying all the needed equipment including the Information Technology (IT) infrastructure for communicating with a Trust Center (TC) for getting orders as shown in Figure 1.

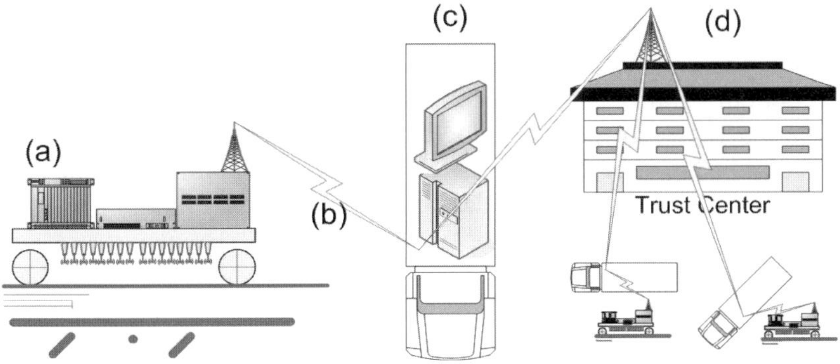

Fig. 1. Detectino System elements: a) platform; b) communication link; c) accompanying truck; d) Trust Center communicating to mobile units

3.1 The platform

The platform will be adapted for a specific job and at place condition before scanning the required ground. The control and data transfer is wireless via broadband communication with the accompanying truck. This truck receives the orders for examining an area from the TC. Those orders contain the precise location and sufficient reference data (cadastral plans, topographic data, etc.) if known and available. Due to this exact context information and the support of GPS signals the necessary localization at place is possible. After finishing a job, the new detected data will be put in context and submitted to the TC which provides this information depending on further orders to the mobile units and customers. The Trust Center stores the data in a memory unit. Because of this technical flexible solution concept all processes can be done at any place without paper. Different orders can be combined in a logistic profitable way before units come back to the central station. The Detectino system can distribute and coordinate both few and many mobile units as a scalable system, which allows a continuous increase depending on the expected demand.

One of the requirements for the platform is a high flexibility in order to manage uneven ground or problematic zones such as higher foot-paths in comparison to the street. So each wheel will have extra control, a robust construction and will be well optimized for the sensors i.e. isolated against electromagnetic waves etc. It is also considerable an autarkic automated steering to be able to drive independently from the accompanying truck to measure wirelessly the subsurface data. Once the functionalities are working, minimization of the platform is to be achievable. Each sensor needs to be optimized to get the best view of the underground in detail with high resolution up to 3 meter depth. Then the sensors should not interfere with each other while the measurement of the subsurface data takes place in parallel. Especially for the ground penetrating radar (GPR) [Sensors and Software Inc.] and seismic it is essential to lie on the ground to avoid the air-to-ground embedding because of a strong signal from the surface. The communication between the platform and truck is wireless for the measured data submission. The problem is to send the information on time and without data loss. The buffering of data streams is done by a high performance computer in the accompanying truck where they are also processed and visualized.

3.2 The accompanying truck

The accompanying truck receives all platform information over the communication line and stores them in a database before they are processed and extracted by the computational intelligence. In order to support the measurements with appropriate equipment, e.g. resistance to humidity and shaking, the AV needs to provide enough storage space. With the established components, we will start with a 3 t truck and intend to minimize once the functionalities work. Inside the truck there will be a High Performance Computer (HPC) which a) receives and stores the measured data, b) provides them for the different modules to process i.e. Computational Intelligence, and c) contains a database and a self-created key of the different possible objects where the measured results will be matched.

Extracting the important Cartesian coordinates the data will be visualized in a 3D reality combining the underground data with information about the house fronts or landmarks by the import of georeference data. In a second stage existing plans and the real measured data can be compared and synchronized. The truck will have different monitors for the visualization as well as for the control, steering, energy supply of the platform. So the person in the truck can take action to support the measurements without being trained to read anomalies of curves which are sent back by normal sensors.

3.3 The Work Flow

After explaining the requirements of the platform and the accompanying truck, the whole process needs to be described. The system workflow mirrors the concept which means the logic development from the raw data to the interpreted visualization. The main issue is the supply chain regarding the successive application of methods on the measured data which come into the system. The workflow steps are as follows:

1. Sending orders received from TC to the mobile unit; initial preparations (with plans of the region and necessary information)
2. Sensors on the platform scan the underground data in the assigned region
3. Industrial computer collects data from the sensors and sends them via wireless communication to the truck
4. In the truck: receiving data and storing them in a database
5. Pre-processing and extraction of relevant data
6. Comparison with reference data
7. Processing, interpretation and classification of data

8. Sending results to the visualization unit
9. Import of GIS data and other information
10. Visualization of GIS data, cadastral plans and measured data (e.g. with different colors)
11. Transmitting the data results to the Trust Center, storing, securing and providing them for customers/new jobs.

4 System Architecture

Detectino is structured in several modules and the complete system will consist of established components deriving from different areas in sensor and information technology, such as ground penetrating radar (GPR), electro magnetism (EM), seismic, geographical information systems (GIS) and methods of pattern recognition including computational intelligence (CI) such as neural networks (NN). The general system architecture is presented in Figure 2. In this section each box of those elements is explained whereas they represent hardware elements as well as software modules. The combination and integration of those established components are innovative and new concerning the requirements and goals of the Detectino project.

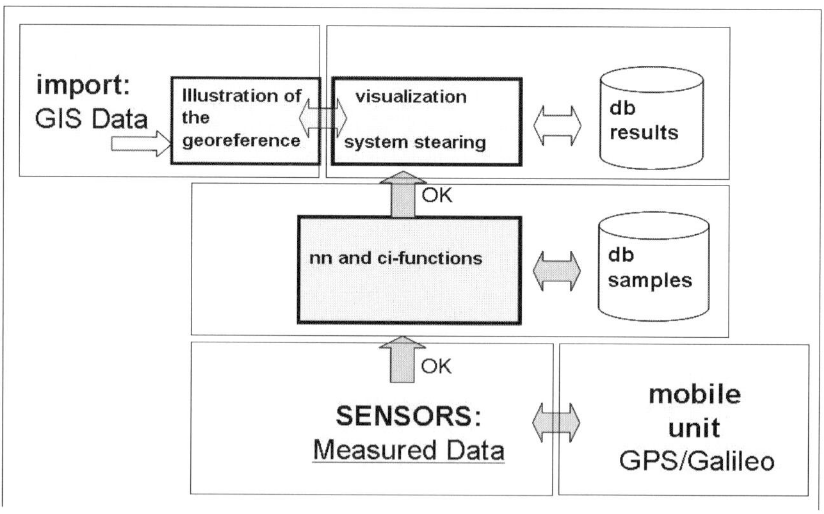

Fig. 2. The system architecture of Detectino

The design of the system architecture is concentrated on the organization of geophysical data. This will include the structuring of data models and 3D visualization of data material. Therefore the IT has, besides processing and providing the measured data, the task to support methods for evaluation and utilization as well as a quality control for the entire workflow to realize the scientific work.

Through former works [Sensors and Software Inc.; Peyerl et al. 2002; Tadjine et al. 2006] and already made examinations measured data exists as well as implemented methods which are used for storing and evaluating the data. Those are specified solutions by the editor and the measuring situation. Analyzing the whole situation, there is already much knowledge of collecting data of different examining regions which are not or only partly adoptable to pipeline detection. All data sets are pre-processed by the investigator which happens interactive but is not documented. To visualize the measured results there exist individual techniques and tools which are proprietary and in general not well documented. Due to those aspects the data exchange and the evaluation between the project groups are restricted. This results in the following tasks for IT:

- Creation of a uniform user interface for all tasks in the project
- Complete georeferencing of the data material
- Complete visualization of the examination region
- Useful supplementation of incomplete data i.e. with NN
- Cross-referencing of data sets

Considering existing proprietary methods seems appropriate for the conception of the entire system to avoid double developments and know-how loss. To demonstrate the division of responsibilities each box of Figure 2 will be explained in the following sections.

4.1 Mobile unit and Sensors

Detectino targets to have a platform with several different sensors for cross data to improve the detection precision since the pipelines of the systems can have either very small (fiber optic cables) or large diameters (sewers). They can consist of ceramic, metal or synthetic construction material or different combinations of these materials. The pipeline systems have variable depths, as a general rule between 0 and 3 meters and are buried into diverse soils. Besides the soil structure and composition the humidity can vary strongly. With all those varying characteristics the only constant may be the man-made linear structure in the underground.

Detectino intends to work with GPR, state-of-the-art EM and a new seismic method. Additionally, it should be possible to add and remove other sensors in a modular way depending on the measuring conditions.

The platform needs the mechatronics of a robot whereas the material has a key role in order to avoid inconsistency for the sensor systems to work in parallel. Flexibility and an expandable modular construction is necessary, as well as a suitable energy supply. Another requirement is the leveling of the platform to ensure that the antennas of the sensors have the same distance to the ground whenever possible to avoid deviations during measurements or to reduce the processing time needed for corrections. Besides the sensor systems, the platform holds an industrial computer to buffer, filter and pre-process the measured data for being fault tolerant and send them over the communication line to the truck.

4.2 Neural Networks and Computational Intelligence

The Neural Networks and Computational Intelligence target to focus different depth horizons between 0 and 3 meters where the same area needs to be scanned with various sensors, especially to identify unused horizons for new pipeline systems. Disruptions in the signal structure are caused by layers and coverage like stones whereas normally there are continuous, linear pipeline structures in the underground. The CI can detect and interpolate those interruptions with mathematical methods [Tadjine et al. 2006; Harneit et al. 2006].

In order to handle fuzzy and inconsistent information, there exists the possibility to combine the information with the use of pattern recognition to achieve new values and receive essentially improved, sometimes complete results of earth-laid pipeline structure.

From nowadays view, it seems to be elementary for the project target to detect and filter disturbances and background noise out of the specific ground signals in order to differentiate and select the information carrying important contents. This requires a calibration beforehand or during the process, since the ground conditions (structure, moistness, etc.) can vary with depth and within a small zone. Once it is possible to calculate the noise from the signals, the important information (cables, canals, etc.) is clearly receivable through a functional combination of the information received from the different sources (pattern recognition).

Whenever data examples about the pipeline structures are available, the information should be stored and used for georeferencing and to enhance pattern recognition. This data can help to calibrate the system and to support the classificators in the conditional phase. In the end, the conditioning needs to be done with real measured data. In a final verification and classification phase, all available information about pipelines are compared with its classificators, to avoid mistakes in the classification as well as to detect errors in the data sets.

4.3 Visualization

The project Detectino demands a solution for a 3D real-time virtual reality. It should illustrate underground systems of pipes and cables. Additionally, as a reference point for the surface, it also demonstrates surroundings: houses, streets, path walks, trees and fields. The possibilities of the arrangements and presentations play an important role and are very complex.

Virtual Reality is developed over years and there exist various ways and solutions of how to create images smooth and in real-time. Modern graphic cards have abilities for projection, transformation and the calculation of textures and lights. So called Application Program Interfaces, such as Microsoft DirectX or OpenGL, are packages with convenient functions. In addition, there are existing frameworks like Irrlicht and Ogre3D which create a dynamic vector based 3D world. The fundamental data model administrates the objects as vector data in the form of dots, lines and fields or as raster data in form of pixels.

Secured information regarding the real pipeline system and their exact matching with signal examples should be stored in a database. Therefore the system can evolve to make comparable analysis because of similar examples.

4.4 GIS Data Import

A geographic information system (GIS) is a computational system to collect, store, process, control, analyze, use, expand, dispose, transfer and demonstrate information like digitally collected and graphically represented georeferenced data.

For Detectino the database unites data of the information systems from states, communes, environment, ground and networks of special applications in order to process and demonstrate the data with useful methods. Through precise data received by GPS, an exact georeferencing of the pipeline objects can be made including information of form, size, depth and orientation. In many cases it will need transformations, conversions and interpolations to generate the exact position. For those mathematical methods also steps of the elimination of geometric distortion and the mutual integration of two data layers need to be implemented.

Georeferencing of images is often based on the interpolation with a reference point of landmarks and a finalizing re-sampling, which is a recombination of the data/objects. The data quality due to the formal information content is important to describe the reality, meaning to simulate a model of the real situation.

5 Conclusions and Future Works

The lack of precise documentation of underground pipeline systems causes economical losses through damages made by subsurface constructions. To reduce this economical impact, the Detectino project proposes a system for coverage and real-time visualization of underground pipeline structures. A mobile unit can be used on site to detect linear structures generating a real-time 3D visualization that supports the construction workers to avoid damages caused by inaccurate existent structural plans.

Detectino is designed to solve location problems, protects valuable assets, reduces collateral damage and works safely. Furthermore it will provide subsurface imaging solutions, locates non-metallic utilities, unravels complex sites and updates the model while processing.

This paper described the concept of Detectino, its work flow and system architecture. This paper shows that three elements are essential in order to reach the real-time in site visualization goal: a modular mobile unit, an accompanying truck and a Trust Center.

The modular mobile unit is necessary to provide flexibility to the scanning of different types of scenarios with the ability to have different types of sensors, specialized for the work that will be done. An accompanying truck provided with computation power is required for processing data in real-time generating the underground images that can be used by the workers for decision process. The Trust Center provides the units with data, receives and stores the scanned and processed underground information in a geographic oriented database for further use.

The system architecture is divided into three main parts: the mobile unit with the sensors measuring the underground, the Computational Intelligence to process the data and extract the important information, so that in the next part it can be visualized in 3D with the help of GIS data import.

As further works it is possible to remark the specific research in each of these main elements as well as with the designed modules of the system architecture. The Detectino project requires further and innovative research in the fields of the seismic sensor, mixing the sensor data for data fusion [Llinas and Hall 1997] and the geographic data integration.

References

Münnich K-H (1995) Volkswirtschaftliche Bedeutung einer Schadensprophylaxe durch richtige Trassenangaben. In: Lenz, J. (ed) Ortung, Schriftenreihe aus dem Institut für Rohrleitungsbau an der Fachhochschule Oldenburg (Band 9). Vulkan-Verlag, Essen

Wittmann J (2006) HADU: Hamburgs dynamischer Untergrund, Anforderungen und Lösungskonzepte des Teilprojektes Informationstechnologie. In: Wittmann J, Müller M (eds) Simulation in Umwelt- und Geowissenschaften, Workshop Leipzig, Berichte aus der Umweltinformatik. Shaker, Aachen

Geotechnologien (2006) Science Report No. 6. BMBF 2005, pp 124-140

Sensors and Software Inc., www.sensoft.ca

IDS Ingegneria dei sistemi, www.idsgeoradar.com

Peyerl P, Sachs J, Richter T, Arnold J (2002) ORDOSYS – Erkundung des unterirdischen Bauraumes mittels Radar. Tagungsband des Kongresses Rohrbau. Weimar, pp 121-129

Tadjine HH, Reuter M, Harneit S (2006) Localisation and Detection of Buried Objects like Mines in Diverse Soils with Soft-Computing Methods. Proc. of 9th European Conference on Non-Destructive Testing. ECNDT, Berlin

Harneit S, Tadjine HH, Reuter M (2006) Multi-Stage Neural Supporting Systems for Time Domain Metal Detectors. Proc. of World Automation Congress WAC 2006, 6th International Symposium on Soft Computing for Industry (ISSCI). TSI Press, ISBN: 1-889335-26-6, IEEE Catalogue N°: 06EX1486, Budapest, Hungary

Llinas J, Hall DL (1997) An Introduction To Multi-Sensor Data Fusion. In: Proc. of the IEEE, vol 85, edn 1, pp 6-23

Finding Optimal Directions for Forest Trees Harvesting Using GIS-based Techniques

Wassim Jaziri

MIRACL laboratory, Institut Supérieur d'Informatique et de Multimédia de Sfax, Route Mharza km 1.5; BP 130, 3018 Sfax, Tunisia,
e-mail: wassim.jaziri@insa-rouen.fr

Abstract: Forest trees harvesting is an important component of the forest resources management process. However, it causes several ecological damage on the forest ecosystem. Thus, approaches for automating or aiding the trees harvesting activities are essential to reduce damage and to ensure a sustainable management of the forest ecosystem. Moreover, an efficient and low-cost trees harvesting is financially important, as it is likely to impact the profitability of a particular management plan. This paper describes a spatial optimization approach to include ecological constraints in scheduling directions for harvesting trees using GIS based techniques. We present an experiment on real data in the forest of "Montagne Tortue" in French Guiana.
Keywords: Geographic Information System, Spatial decision aid support, Ecological optimization, Forest ecosystem management, Reduced impact logging.

1 Introduction

Forest harvest modeling is a particularly difficult problem when approached from the traditional direction of optimization. There are several reasons. One is that in a real-world situation, it is often very expensive and difficult to collect geographic data required to make decision. Consequently there is only a limited variety of data (wood volume, species, etc.) available in the detailed data base. Another is that in practice, forest managers must take into account a wide variety of imprecise data and rules that are difficult to handle and to analyze.

In fact, inaccessibility of required geographic data and difficulties in synthesizing various recommendations are primary obstacles to spatial problems solving. In an environmental management context, the quality of decisions and the ability to provide efficient and realistic solutions can be improved if these obstacles are lessened or removed through an interactive process of spatial decision analysis [Carver 1991; Jankowski 1995].

The forest management planning process is a complex task and typically involves numerous harvesting operations. Furthermore, the need in forest management to include spatial configuration of forest conditions, as well as their aspatial composition, increases the difficulty [Baskent and Jordan 1995; Grumbine 1994]. The trees' cutting is an important and ecologically very expensive operation in this process. Actually, the choice of direction to cut trees is made arbitrary based on the viewpoint of the forest logger. Generally, before cutting a tree, a forest logger analyzes the neighboring area and chooses the easiest direction of cutting which allows transporting wood through the road towards a logging park (a place to store wood). This process is made independently of the potential ecological damage which can be caused on forest trees and on the soil. This working method is inefficient because it causes the destruction of neighboring trees, the degradation of the soil quality and thus a bad forest resources management. In addition, the destruction of future trees[1] (which constitute a future source of profit) represents a financial loss. For this reason, it is necessary to better organize the operation of trees cutting by using new methods able to optimize the direction of cutting. More preciously, to determine the least expensive (financial and ecological cost) cutting direction for each exploitable tree while satisfying spatial constraints (slope, river, etc.).

This important activity in the forest resource management process requires expertise and integration of multi-disciplinary fields. In this paper, we outline a real-life problem from a project between Cirad[2] (France), Embrapa[3] (Bresil) and ONF[4] (French Guiana). This collaboration gathers researchers from various disciplines: computer science, statistics, forestry, ecology and environment sciences.

In the following sections, we present the context of forest management planning problems. Then, we present the developed methodology based on Geographic Information Systems. Finally, we present and discuss an experiment in a French Guyana forest before concluding this work.

[1] Future trees are likely to become exploitable (to cut) at the next rotation cycle.
[2] French centre for agronomic research: http://www.cirad.fr
[3] A multi-disciplinary research centre in Brazil.
[4] National office of forests (France): http://www.onf.fr

2 Forest management planning problems

This work deals with the reduction of forest logging damage by optimizing the direction of trees cutting. This problem can be considered as the minimization of costs to cut neighboring trees dispersed in space. In the literature, there are no yet approaches which led to concrete and satisfactory solutions for this problem.

In the context of forest management planning problems, both simulations, optimization and decision aid techniques have been used by scientists over the last two decades. With the introduction of ecosystems management that focuses on the sustainable production and maintenance of ecological, social and economical values, neither approach provided a credible solution technique to help design the complex structure of forest management activities. This is due to three reasons. First, most scheduling problems involve large number of social, spatial, financial and ecological criteria that are difficult to take into account simultaneously. Second, it is not easy to express and formulate the experience about the terrain, essential to propose adequate and realistic solutions. Finally, some knowledge is based on the uncertainty such as the process of trees regeneration. When all of these factors are considered together in a given environmental problem, it is clear that finding a true optimal solution to an unrestricted problem is virtually impossible. As a result, every spatial planning approach has focused on finding good or near-optimal solutions, but in most cases on simplified problems.

A forest management planning problem might contain more than a thousand trees and include several spatial, ecological and economic restrictions on wood cutting and transport. Problems such as this are difficult combinatorial optimization problems. Consequently, purely mathematical programming techniques are limited by the size of the problem that can feasibly be addressed and the high cost of finding solutions [Weintraub et al. 1995]. Others have combined optimization techniques and heuristics to find good solutions to spatial planning problems [Baskent 1990; Clements et al. 1990; Yoshimoto et al. 1994; Weintraub et al. 1995]. However, most of these approaches have been demonstrated to work for simplified problems.

A number of another various techniques have been employed to solve spatially constrained harvest scheduling problems. These include various mixed-integer programming formulations [Jones et al. 1991; Yoshimoto et al. 1994], heuristics [Anderson and Bare 1994; O'hara et al. 1989], multicriteria methods [Johnson and Scheurman 1977; Roise 1990] and decision aid techniques [Thompson and Weetman 1995].

3 The developed approach

Forest ecosystem management emphasizes the control of the spatio-temporal dynamics of forest landscapes by planning interventions on the forest resources. These interventions and their timings are identified so that spatio-temporal characteristics of the forest landscape, for example, size, shape, distribution, proximity and dispersion of forest trees, can be predicted and measured with respect to objectives. It, therefore, embodies two challenges: first, defining, quantifying and translating diverse financial and ecological values into forest objectives, and second, designing spatially explicit management to achieve those objectives.

Increasingly, forest managers are convinced to include, in addition to the financial objectives, spatial and ecological restrictions in their harvest schedules. These restrictions are imposed to maintain biodiversity, soil quality and forest resources. However, multiple forest restrictions and objectives are difficult to integrate in a forest management model [Baskent and Jordan 1995].

Problems of handling space typically involve a set of geographically-defined alternatives (events) from which a choice of one or more alternatives is made with respect to a given set of evaluation criteria [Jankowski 1995]. In this work, we based our approach on spatial, economic and ecological criteria. Spatial requirements relate to size, shape, slope and adjacency restrictions. In addition, topological data is needed for identifying trees to be accessed as well as for estimating the cost of cutting and transporting them. Spatial data may be obtained from a raster-based geographical information system (GIS) containing appropriate information attributes (such as slope, soil type) as well as a suitable spatial resolution [Goodchild et al. 1993].

3.1 The use of Geographic Information System

Geographic Information Systems (GIS), with their capabilities for spatial analysis and modeling of diverse data sets, enhance the ability to address several natural resource and environmental issues that have a spatial component (biodiversity, land use planning, etc.). GIS facilitates the organization, manipulation and analysis of spatial and aspatial data. Its data structures and analytical techniques are being incorporated into a wide range of management and decision-making operations that pertain specifically to natural resources [Goodchild et al. 1993].

Today, GIS is thought of as an enabling technology for science because of the breadth of potential uses it has as a research tool. Developed and applied correctly, a GIS can provide forest resource managers with powerful information that will aid them in the management process. GIS technology can be extremely useful in resolving conflicts in spatial decision aid systems and provide a potential user with the means to analyze and compare various alternatives based on a spatial analysis.

3.2 Ecological, spatial and financial parameters

In a spatial decision aid problem, the characteristics of space are often expressed through multiple and complex spatial constraints. To simplify the space analysis, we divide it into small units (cells) easier to analyze. In addition, we take into account the neighborhood relations and dependency between elementary units.

Given a grid surface (raster data representation), an appropriate information is defined for each cell. From this, forest trees may be identified as well as the values of the spatial attributes. The task to be accomplished is to choose among the possible directions, the least expensive one. A direction is a succession of neighboring grid cells. Three parameters are taken into account to choose the least expensive direction (Fig. 1, 2 and 3 allow illustrating the problem):

- The direction of the existing road: a tree must be cut in a direction allowing its transport through the existing road towards the logging park (Fig. 1). A tree is drawn in bottom and not by the tree top. Consequently, after cutting, the tree foot must be, in preference, in the direction of the road. In a contrary case (the tree top is in the direction of the road), the skidder must turn the tree in the opposite direction and thus causes damage on the neighboring trees. To favor cutting trees in the direction opposed to the road (the privileged direction), we affect a cost for each direction according to its adaptation according to the transport of trees towards the logging park (Table 1).

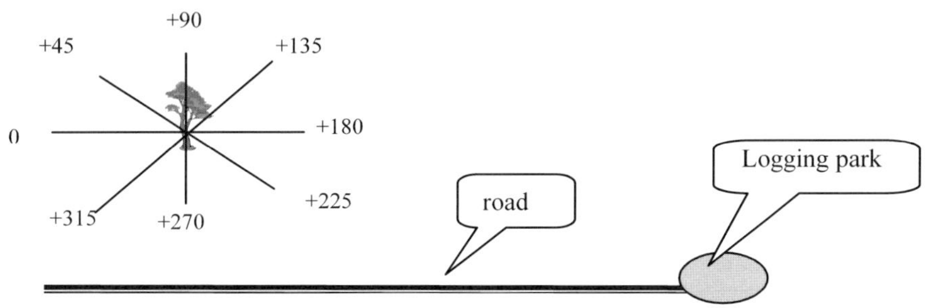

Fig. 1. Directions of a tree cutting according to an existing road.

- The neighboring area for each exploitable tree (in a circle of radius equal to the height of the tree). To reduce damage on the neighboring trees, a cost is affected in a given direction according to the type of trees that can be damaged (future tree, seed-bearer, etc.) and their financial and ecological impact (Table 1). Because a long term look is required to address sustainability (usually several rotations), we take into account future trees in the cost affectation process.

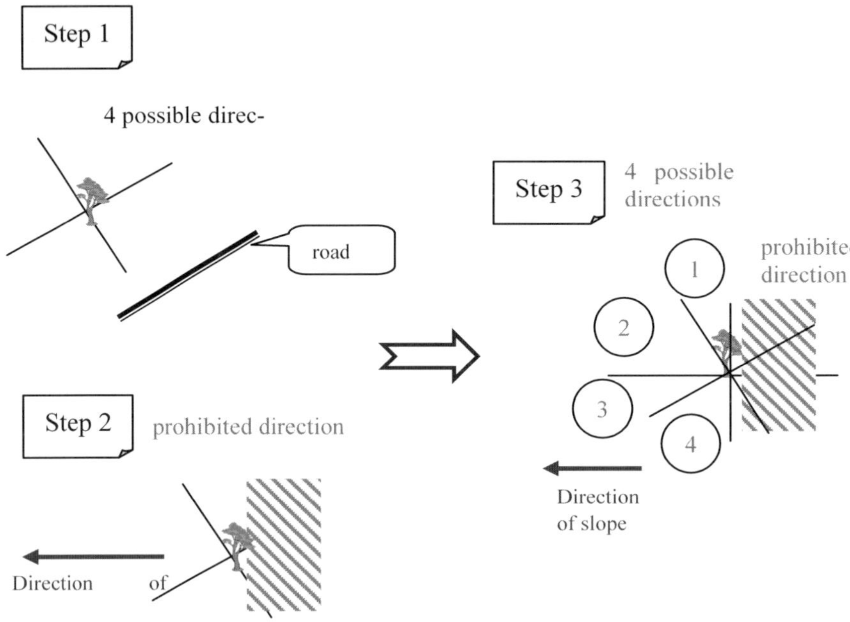

Fig. 2. Combination of the slope (if slope > 15°) and the road to determine the directions of tree cutting.

- Topographic and spatial conditions: the tree should not fall in the direction of a hydrological (a river) or a spatial obstacle (a high slope) to not obstruct its transport. The constraint of slope intervenes only if the slope > 15° (Fig. 2 and 3).

Fig. 3. For each exploitable tree, we consider 4 potential cutting directions.

3.3 Ecological, spatial and financial parameters

With the grid cell unit specification and the definition of possible directions of trees cutting, the problem may be structured as the choice of the least expensive direction to cut all exploitable trees.

Table 1. Spatial, financial and ecological costs. Financial costs are allotted according to the point of view of the forest loggers. Ecological costs are allotted according to the point of view of ecologists.

	Tree to be cut	Future tree	seed-bearer	Tree not to be cut	
				rare species	Another trees
Financial cost	20	4	0	0	0
Ecological cost	0	2	2	2	0
Total	20	6	2	2	0

Obstacle	Slope >= 15°	river	talweg	split
Spatial cost	100	20	10	4

	Good direction [45, 135]	Non adapted direction [45, 135], [135, 225]	Bad direction [225, 315]
Ecological cost	0	5	10
Financial cost	0	5	10
Total	0	10	20

To integrate spatial constraints, we affect a spatial cost to each cell, according to the type of obstacle (for example: the existence of a high slope). To minimize ecological damage, we integrate a value of ecological costs in each cell to express the existence of forest resources. The objective function is thus based on 3 criteria: spatial, financial and ecological costs. In this paper, we use the sum function to ponder the two types of costs.

The developed approach consists in treating all exploitable[5] trees independently. The treatment of a tree consists in choosing the least expensive cutting direction according to the objective function. Example, if in a given direction, there are two future trees and a river then the cost of cutting the tree in this direction is the sum of costs of destruction of the two future trees raised with that of cutting the tree on the river. This cost represents the value of the objective function if this cutting direction is chosen.

[5] Trees intended to be cut and marketed.

4 Experiments

4.1 Study area

Located in French Guiana (Latin America), the study area contains 53 exploitable trees (to cut), 211 future trees and 2,336 others trees (of different species and diameters) dispersed in 6 ha (200m x 300m) (Fig. 4).

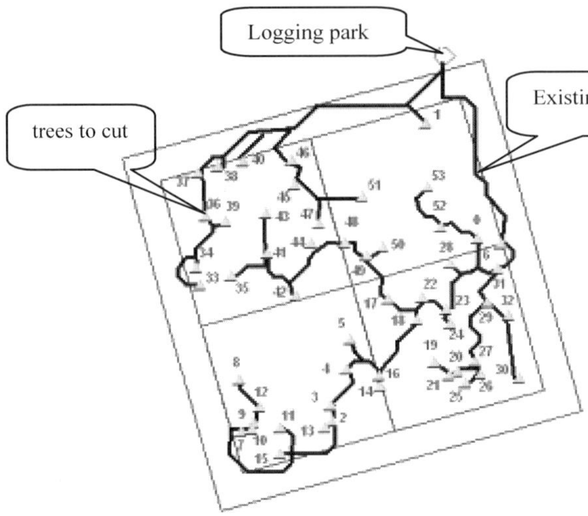

Fig. 4. The study area

In this study, a geographic information system (ARC/INFO) is used to store, treat and analyze the data and the characteristics of the study forest area.

The space of the study area is cut out in cells of square form (10m × 10m). Since the size of a cell is largely larger than the tree diameter, only the cell affiliation of each tree is used. The layers used to choose directions of trees cutting are:

- The existing road;
- The exploitable resources (trees to cut);
- The future trees (to avoid their destruction);
- The spatial, financial and ecological costs (that determine the total cost of a direction).

4.2 An example of trees treatment

We present in this section an example of treatment of the tree 52 (see Fig. 4). The value of slope is 16.15°. The value of the slope direction is 54.17 (Fig. 5 and 6).

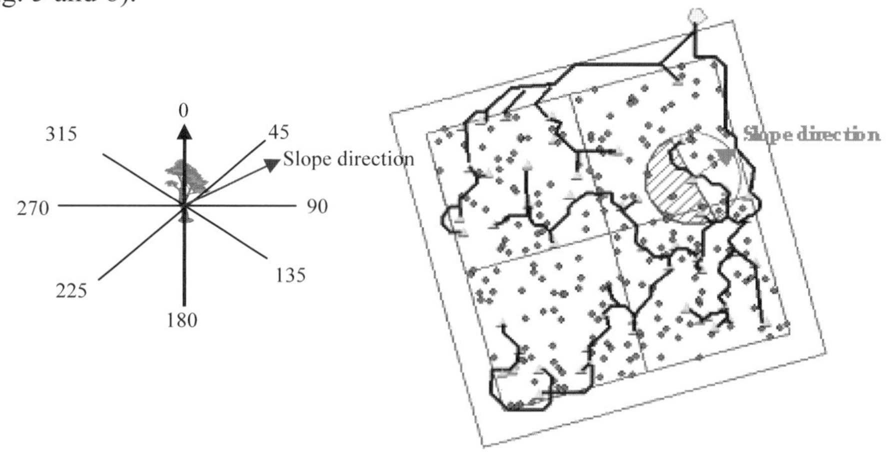

Fig. 5. Analysis according to the slope direction.

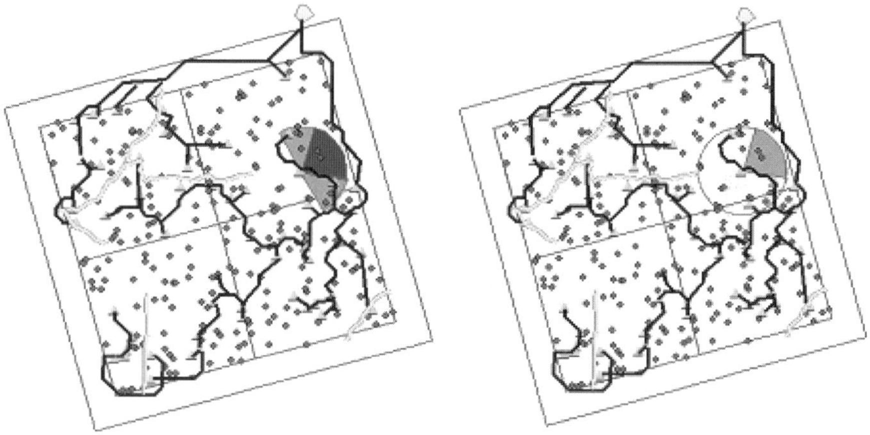

Fig. 6. After combination of the road and the slope direction, we obtain 3 potential directions of tree cutting. Direction 4 is the least expensive to cut the tree 52.

The costs in each direction are respectively:

- *Direction 1* (a good direction): this direction is not taken into account due to the slope direction (Fig. 5).
- *Direction 2* (not adapted): 4 future trees + 1 exploitable tree + cost of direction (non adapted direction) = 4 * 6 + 20 + 10. Total cost = 54 units.

- *Direction 3* (not adapted): 2 future trees + 1 exploitable tree + cost of direction (non adapted direction) = 2 * 6 + 20 + 10. Total cost = 42 units.
- *Direction 4* (a bad direction): 2 future trees + cost of direction (bad direction) = 2 * 6 + 20. Total cost = 32 units.

The direction 4 is retained because it has the lower cost (Fig. 6) although it represents a bad direction according to the transport of tree towards the logging park.

4.3 Improvements of the model

The directions of wood cutting retained for all exploitable trees are shown in Figure 7. Although the developed method was applied to real data, the generated logging plan has not yet been executed.

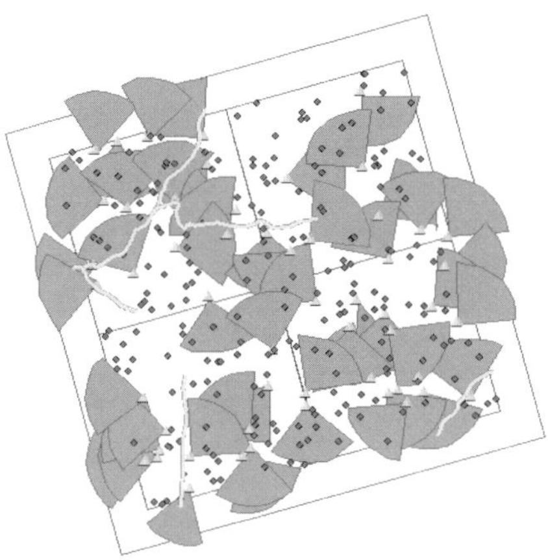

Fig. 7. The least expensive trees cutting directions in the study area.

The judgments of these results by forest managers and loggers are satisfactory. However, the information of slope was necessary but certainly not sufficient to propose a realistic cut direction. Thus, it would be advantageous to take into account the natural direction of fall for the exploitable trees (this information is not currently available) in the same way which we take into account the direction of slope. The natural direction of fall can replace the information about the slope since it expresses the high slope and also when it is flat.

In future work, two improvements can be brought: (1) to raise, at the time of the inventory, the natural direction of fall for each tree to use it as input datum instead of the slope information; (2) to trace new elementary ways to join the trees from an existing road. In the second case, an additional cost will be generated to express damage caused by the extension of the new road portions. The first solution allows solving the problem but it is expensive. The observation of the tree being long if we want to obtain a reliable diagnosis. The prospectors must in addition be trained with this use because they haven't the required competence. Moreover, an additional constraint must be included in a future work: we must not cut neighboring trees in the same direction in order to avoid large openings in the forest.

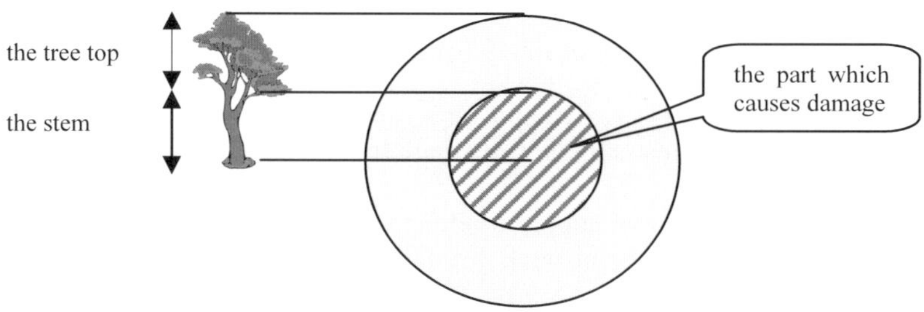

Fig. 8. Damage is essentially caused by the stem and not by the tree top.

In our approach, we consider only 4 potential directions of trees cutting and it would be interesting to refine this approach. In addition, we noted on the terrain that damage caused by a tree cutting on the neighboring trees relate generally to the stem and not to the tree top of a tree. For this reason, we want to compute the costs in each direction with radius equal to the height of the tree stem (without the tree top) instead of the whole height of the tree (Fig. 8).

5 Discussion and conclusion

The problem of reduced impact logging becomes more and more important. This includes the problem of determining the optimal cutting direction for harvesting trees. This paper addressed this problem on a technological sound basis using GIS for determining an optimal cutting direction. We presented a spatio-ecological optimization approach to reduce damage caused by the forest trees cutting process. This paper contributes to research in this area and demonstrates the larger context in which this particular forest planning problem is based.

The optimization of trees cutting directions activity is an important step to ensure a sustainable management of forest ecosystems. Several spatial, ecological and financial factors must be taken into account and increase the complexity of the problem solving. The trees cutting process includes a great deal of information about the terrain, which is based on experience rather than formal rules. This means that the results of such process should sometimes be modified and readjusted to provide efficient and realistic solutions. For this reason, we direct our work towards the design of a spatial decision aid system to assist forest loggers in the choice of trees cutting direction instead of deciding for them.

References

Anderson DJ, Bare BB (1994) A dynamic programming algorithm for optimization of uneven-aged forest stands. Canadian Journal of Forest Research. vol 24, pp 1758-1765

Baskent EZ (1990) Spatial wood supply modeling: concept and practice. M.Sc.F. Thesis, University of New Brunswick. Fredericton, NB

Baskent EZ, Jordan JA (1995) Designing forest management to control spatial structure of landscapes. Landscape and Urban Planning, vol 34, pp 55-74

Carver SJ (1991) Integrating multi-criteria evaluation with geographical information systems, Geographic Information System. vol 5, pp 321-339

Clements SE, Dallain PL, Jamnick MS (1990) An operational spatially constrained harvest scheduling model. Canadian Journal of Forest Research. vol 20, pp 1438-1447

Goodchild, MF, Parks BO, Steyaert LT (1993) Environmental modelling with GIS. Oxford University Press, New York

Grumbine RE (1994) What is ecosystem management? Conserv. Biol. 8. pp 27-38

Jankowski P (1995) Integrating geographical information systems and multiple criteria decision making methods, Geo. Inform. Sys. vol 9, pp 251-273

Johnson KN, Scheurman HL (1977) Techniques for prescribing optimal timber harvest and investment under different objectives, discussion and synthesis. Forest Science. vol 23, pp 1-31

Jones JG, Meneghin BJ, Kirby MW (1991) Formulating adjacency constraints in linear optimization models for scheduling projects in tactical planning. Forest Science. vol 37, pp 1283–1297

O'hara AJ, Faaland BH, Hare BB (1989) Spatially constrained timber harvest scheduling. Canadian Journal of Forest Research 19, pp 715-724

Roise JP (1990) Multicriteria Nonlinear Programming for Optimal Spatial Allocation of Stands. Forest Science, vol 36, pp 487-501

Thompson WA, Weetman GF (1995) Decision support systems for silviculture planning in Canada. The Forestry Chronicle, vol 71, pp 291-298

Weintraub A, Jones G, Meacham M, Magendzo A, Magendzo A, Malchuk D (1995) Heuristic procedures for solving mixed integer harvest scheduling transportation planning models. Canadian Journal of Forest Research, vol 25, pp 1618–1626

Yoshimoto A, Brodie JD, Sessions J (1994) A new heuristic to solve spatially constrained long-term harvest scheduling problems. Forest Science, vol 40, pp 365-396

Evaluating Water Quality Standards Through Time Series Modeling by Wavelets

Jean Duclos Alegue, Albrecht Gnauck

Department of Ecosystems and Environmental Informatics at Brandenburg University of Technology Cottbus, 03013 Cottbus, Germany,
e-mail: jeduclo@yahoo.com, albrecht.gnauck@tu-cottbus.de

Abstract: Setting realistic water quality standards should not only take into consideration the hydrologic, ecological conditions and land use of the region but also the time varying behavior of the indicators. Investigating the time varying structure of water quality indicator using wavelets multiresolution analysis reveal a cyclic behavior of certain indicators. Using data from the Havel River sampled at 10 minutes interval from 1998 to 2002 at the Potsdam measuring station in the state of Brandenburg, Germany, LAWA [1998] water quality standards for certain indicators were investigated. It was found that most indicators portray a cyclic behavior due to internal and external driving forces. A comparism of this behavior and the standards show that some standards seam inappropriate. An upper and or lower limit that takes into consideration the time varying structure of the indicators is more appropriate.
Keywords: water quality standards, cycling indicators, time series analysis, multiresolution analysis, wavelet modeling.

1 Introduction

The necessity to efficiently manage and conserve freshwater resources is becoming more and more evident as water quantity and quality are becoming the main limiting factors to sustainable development [Uhlmann 1991]. For a sustainable management of freshwater ecosystems, an understanding of the physical, chemical and biological components, their functioning and interrelationship is imperative. As a result, water quality indicators (variables) are recorded across time in order to assess the state of the water body, track changes, forecast future conditions using models as well as study the interrelationship between the different indicators.

Based on the desired use of the freshwater resource, water quality standards, which provide the concentration of substances considered safe for human and ecosystem health, are established. These standards are desirable because they serve as a reference point for assessing the present water quality and its suitability for different purposes. The establishing of water quality standards should be realistic, taking into consideration the climatic and geologic characteristics of the watershed, the time varying structure of the indicator as well as the latest scientific knowledge relevant to the indicator [Shear et al. 2005]. This is not always the case in most developing countries whereby standards from developed nations are simply imported without taking into consideration the local realities. Investigating the internal structure of water quality indicators so as to extract information on the inherent changes or fluctuations at different time scale is of great importance in establishing standards. Classical signal analysis methods like the autocorrelation function and the periodogram are not able to reveal the inherent changes and their intensity at different time scale. Wavelet analysis offers such a possibility for decomposing and inspecting the changes occurring and their intensity at different time scale.

2 Water quality standards

Both the ambient water quality of a water body and the release of pollutants to the water body (effluents) which influence the ambient water quality need to be controlled. There exist different water quality standards for different freshwater uses. For example, drinking, fishing and bathing water quality standard. They generally comprise two elements, namely, the ambient standard (maximum allowable concentration) and the effluent standards (maximum permissible discharge).

The ambient water quality standard provides the concentration of a given substance above which the water resource is unsuitable for a particular use. Above this concentration, human and ecosystem health are affected [Viesmann and Hammer 1998]. The ambient water quality standard serves as a basis or reference point for assessing the current status of water quality and its suitability for various uses. The rationale for setting water quality standards is to ensure that the concentration of certain substances in freshwater ecosystems should have no direct or indirect harmful effect on human and ecosystem health. The standard assumes that concentrations below or above a certain threshold posses no threat or risk to human and ecosystem health under the designated use of the freshwater ecosystem [Koren 1991]. Hence, the standard is established based on the designated use for the water resource. The European Union for example has different standards depending on the use assigned to a freshwater ecosystem (for example fishing, recreation, drinking water abstraction).

In setting a realistic water quality standard, the climatic, geologic and land use characteristics of the watershed must be taken into consideration. In Germany for example, each state establishes the standard for substances based on the characteristics of the different watersheds of the state. Figure 1 illustrates the existence of great differences between the standards of the same substances from one state to another. Given that the watersheds of different eco-regions in the different states possess different characteristics, it is not possible to have one standard for all water bodies in the country.

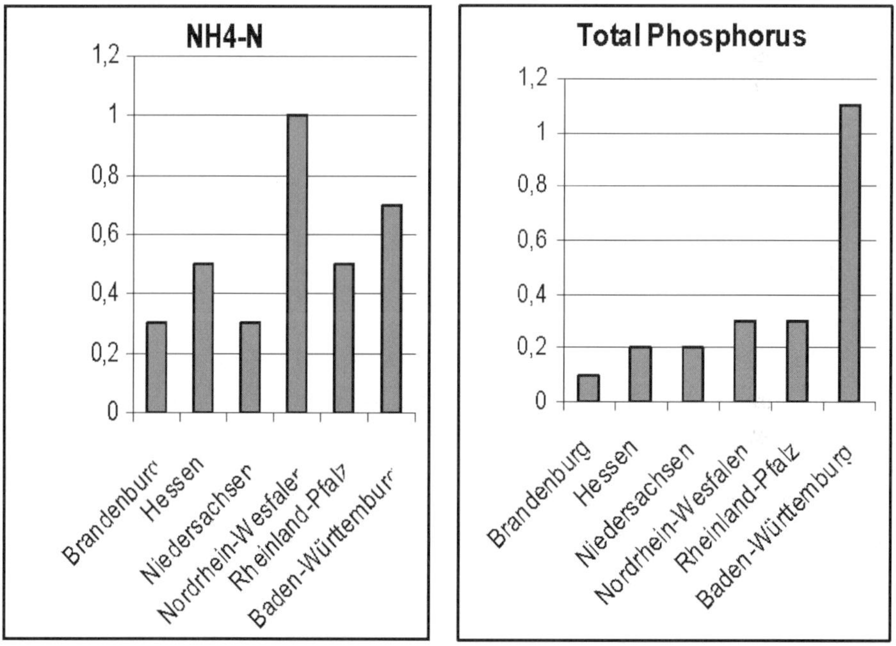

Fig. 1. LAWA class II water quality standard for some states in Germany

Effluent limitations refers to the maximum amount of a pollutant that a polluter may discharge into a water body taking into consideration the effluent from all other source and is generally derived by the help of mathematical models [Viessmann and Hammer 1998]. The appropriate effluent standard helps in the attainment of the ambient water quality standard for a given water body. The effluent standard may allow some discharge or no discharge of pollutants based on the type of pollutant [Koren 1991]. Generally, the discharge of certain pollutants like high level radioactive wastes, biological warfare materials and any radiological wastes into freshwater bodies are strictly prohibited. Such pollutants are able to cause death, disease, behavioral abnormality, cancer, genetic mutations, physiological malfunctions, etc, to humans and other organisms [Koren 1991]. In setting effluent standards, an adequate margin of safety must exist. Industries generally have to meet effluent limits that require or reflect the current best practical technology. Due to the present increasing challenges facing freshwater ecosystems as a result of anthropogenic pollution, zero discharge of pollutants should be the standard when possible.

A number of freshwater quality indicators portray a cycling time structure (cf. Figure 3) as a result of parallel acting seasonal and periodic internal and external driving forces. Establishing an appropriate water quality standard without taking this cycling behavior into consideration will render the standard inappropriate as a result of the water body complying and not complying to the standard in the lower and upper regions of the cycle respectively. In order to assess the compliance of a freshwater ecosystem to the established standard for a designated water use, monitored indicator data is compared to the standard. It is quite interesting to closely examine cycling water quality indicators and standards.

3 Wavelet evaluation of standards

Wavelet analysis is a mathematical tool that serves as a lens for inspecting the time varying structure of signals and relationships between signals. It decomposes a signal on a scale by scale basis by projecting it onto a basis function called a mother wavelet. By projecting a signal X(t) on to a particular mother wavelet Ψ, a translated and dilated version of the signal W(u,s) is obtained and is given by

$$W(u,s) = \int_{-\infty}^{\infty} X(t)\Psi_{u,s}(t)dt, \tag{1}$$

Where

$$\Psi_{u,s}(t) = \frac{1}{\sqrt{s}}\Psi\left(\frac{t-u}{s}\right) \tag{2}$$

gives a translated and dilated version of the original wavelet function. "*s*" is the scale parameter, "*u*" is the location parameter. Changing "*s*" produces dilating effects ($s > 1$) or contracting effects ($s < 1$) of the function $\psi(t)$. Changing "*u*" analysis the signal X(t) around different points of "*s*". The coefficients that are obtained are a function of the location and scale parameters. Applying shifted and scaled versions of a wavelet function decomposes the signal into simpler components.

In multiresolution analysis, the wavelet transform is applied in the form of a filter bank, comprising two filters. The scaling filter known as the father wavelet which is a low pass filter and the wavelet filter known as the mother wavelet which is a high pass filter. Given a signal X(t) of length $N=2^j$, the filtering procedure can be performed a maximum of j time, giving rise to j different wavelet scales. The wavelet coefficients or detail coefficients are produced by the wavelet filter while the scaling filter gives rise to the smooth version of the signal used at the next scale. Given the respective father and mother wavelets,

$$\Phi_{J,k} = 2^{\frac{-J}{2}} \Phi\left(\frac{t - 2^J k}{2^J}\right) \qquad (3)$$

$$\int \Phi(t)dt = 1$$

and

$$\Psi_{j,k} = 2^{\frac{-j}{2}} \Psi\left(\frac{t - 2^j k}{2^j}\right) \qquad (4)$$

$$\int \Psi(t)dt = 0$$

where $\Phi_{J,k}$ is the father wavelet and $\Psi_{j,k}$ is the mother wavelet with the scale parameter "*s*" being restricted to the dyadic scale 2^j. When a signal is projected onto a given basis function, we obtain

$$S_{J,k} = \int X(t)\Phi_{J,k} \qquad (5)$$

and

$$d_{j,k} = \int X(t)\Psi_{j,k} \qquad (6)$$

with $S_{J,k}$ being the coefficients for the father wavelet at a maximum scale of 2^J (the smooth coefficients) and $d_{j,k}$ being the detail coefficients from the mother wavelet at all scales from 1 to J, to the maximal scale. Based on these coefficients, the function X(t) can be represented by

$$X(t) = \sum_k S_{J,k}\Phi_{J,k}(t) + \sum_k d_{j,k}\Psi_{J,k}(t) + ... + \sum_k d_{1,k}\Psi_{1,k}(t) \tag{7}$$

and can be equally represented by

$$X(t)=S_J + D_J + D_{J-1} + \ldots + D_j + \ldots D_1 \tag{8}$$

where

$$S_J = \sum_k S_{J,k}\Phi_{J,k}(t)$$
$$D_j = \sum_k d_{j,k}\Psi_{j,k}(t) \tag{9}$$

The detail represented by D_j gives the changes or fluctuations occurring in the signal at different time scale. The wavelet coefficient at a particular time scale measures the amount of fluctuations or changes that occur at that particular time scale. Given a signal sampled at daily intervals, table 1 shows the frequencies or time scale associated with the different details at different levels of decomposition.

Table 1. Frequencies associated with level of decomposition

Level	Detail	Associated daily frequency
1	D1	1
2	D2	2
3	D3	4
4	D4	8
5	D5	16
6	D6	32 (1month)
7	D7	64 (2 months)

The Wavelet multi-resolution analysis was used to decompose the fluctuations or changes of chlorophyll-a from the Havel River in the state of Brandenburg, Germany at different time scales as shown in Figure 2. The indicator was sampled at daily interval and the Daubechies 4 mother wavelet was used for the analysis.

The results of the multiresolution decomposition shown in Figure 2 reveal the micro structure of the indicator at different time scale. It can be observed for level 1, 2, 3 and 4 that there exist an alternation of period of high fluctuation and periods of very low fluctuation from one season to the other, thereby producing a cycling behavior. Levels 5 and 6 exhibit cyclic fluctuations of greater intensity irrespective of the season. A combination of these effects produces an overall cyclic pattern in most water quality indicators. This is quite interesting because most standards are fixed values which do not take into consideration these effects. If the measurements are done during periods of low fluctuations, the freshwater body will seam to comply with the established standards. On the other hand, if they are carried out during the periods of high fluctuation, the water body will seam not to comply, creating a situation of alternating compliance and none compliance within a very short interval of time. This phenomenon raises questions on the appropriateness of a fixed value being a standard or cycling water quality indicators.

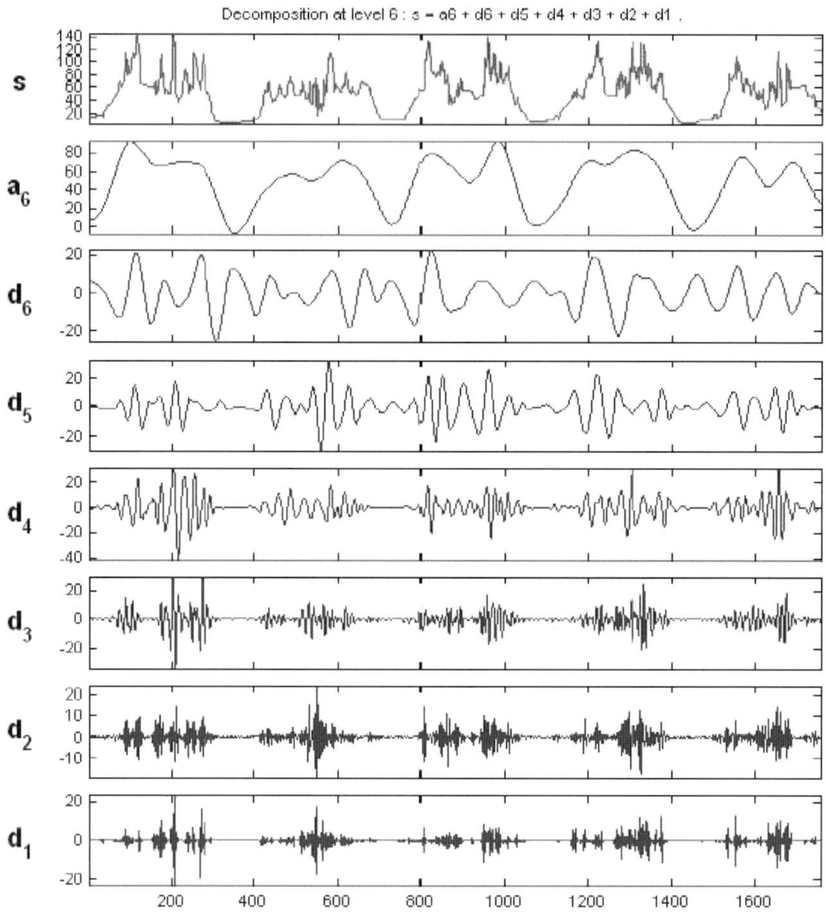

Fig. 2. Multiresolution analysis of chlorophyll-a using Daubechies 8 mother wavelet

To further investigate indicators having a cycling internal structure as a result of internal and external driving forces, some water quality indicators, sampled at a daily interval, from the Havel River in the state of Brandenburg in Germany were compared against the [LAWA 1998] water quality standards for the state of Brandenburg.

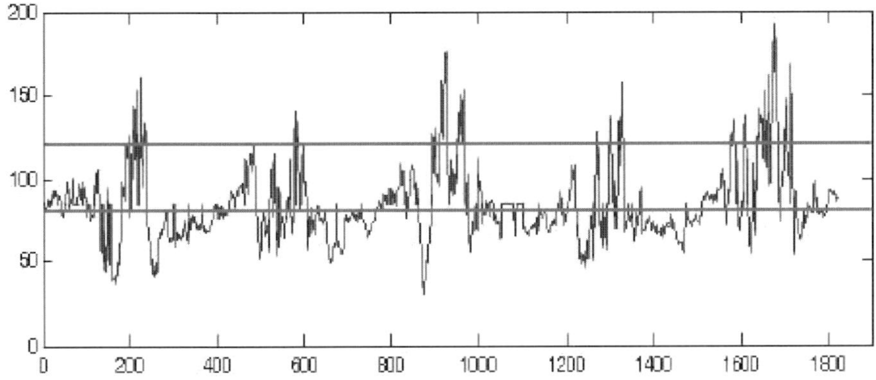

Fig. 3. Water quality standard for oxygen saturation with an upper limit of 120 % and a lower limit of 80 % for the state of Brandenburg, Germany

It can be observed from Figure 3 hat the lower limit for oxygen saturation is more or less the mean value (83.3%) for this indicator of respiration and productivity. Due to the fluctuations, there exist an abnormal situation of compliance and non compliance within an extremely short period of time, putting in question the appropriateness of the standard for this indicator.

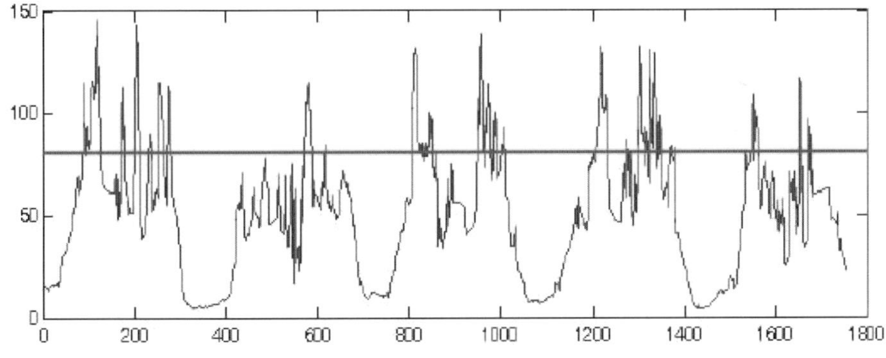

Fig. 4. Water quality standard for chlorophyll-a with an upper limit of 80 μg/l

In Figure 4, the cyclic behavior exhibited by chlorophyll-a, an indicator of biomass or productivity also causes some difficulties in assessing the compliance of this freshwater ecosystem to the required standard. It is observed that within a relatively short period, the water body complies and does not comply with the standard. It may be necessary for such an indicator to have different standards for different relevant seasons taking into consideration the cyclic behavior.

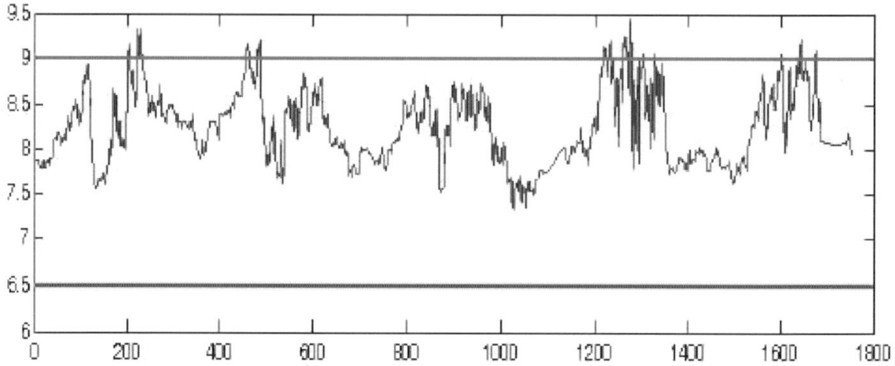

Fig. 5. Water quality standard for pH with an upper limit of 9 standard units and a lower limit of 6.5 standard units

Figure 5 clearly shows that the pH, an indicator of acidity of the water body, complies with the lower limit of the standard. With respect to the upper limit, there exist periods when the standard is met and when it is violated.

4 Conclusion

Water quality standards are established in order to ensure that the concentration of certain substances in freshwater ecosystems should have no direct or indirect harmful effect on human and ecosystem health. They serve as a basis or reference against which water quality is assessed. In establishing water quality standards, the climatic, geologic, land use characteristics of the watershed as well as the time varying behavior of the indicator have to be taken into consideration. Most water quality indicators as revealed by wavelet multiresolution analysis portray an inherent cyclic behavior as a result of internal and external driving forces. This inherent cyclic behavior of certain indicators has to be taken into consideration in order to establish water quality standards. Failure of integrating this cyclic behavior will result in standards which are quite inappropriate for cycling water quality indicators. Other signal analysis methods like time and frequency domain methods reveal changes at a particular time scale and the periodic components and frequency components in a signal respectively, but are not able to decompose the signals to reveal their inherent behavior at different time scale.

References

Koren H (1991) Handbook of Environmental Health and Safety: Principles and Practices. Lewis Publishers, Florida

LAWA (1998) Beurteilung der Wasserbeschaffenheit von Fließgewässern in der Bundesrepublik Deutschland – Chemische Gewässergüteklassifikation. Kulturbuchverlag, Berlin

Shear H, Bertram P, Forst C, Horvatin P (2005) Development and Application of Ecosystem Health Indicators in the North American Great Lakes Basin. In: Jørgensen ES, Costanza R, Xu F (eds) Handbook of Ecological Indicators for Assessment of Ecosystem Health. Taylor and Francis, New York

Uhlmann D (1991) Anthropogenic Perturbation of Ecological Systems. A Need for Transfer from Principles to Applications. In: Ravera, O. (ed) Terrestrial and Aquatic Ecosystem Perturbation and Recovery. Ellis Horwood, New York, pp 47- 61

Viessman W Jr, Hammer JM (1998) Water Supply and Pollution Control. Addison Wesley Longman, Menlo Park

Simulation of Shallow Lake Eutrophication

Albrecht Gnauck

Brandenburg University of Technology at Cottbus

Abstract: Natural and man induced water pollution affects the functioning of freshwater ecosystems, restricts various water uses and risks human health. Especially, internal pollution by nutrient remobilization from sediment plays an important role in shallow water bodies. In this paper results of a water quality simulation framework for shallow lakes of the River Havel are presented. The framework consists of a simulation model combined with a multicriteria optimization procedure. Model state variables are phytoplankton, zooplankton, orthophosphate phosphorus, ammonia nitrogen and nitrate nitrogen. External driving forces are photoperiod, solar radiation and water temperature. For water quality management two control strategies are discussed
Keywords: eutrophication, shallow lakes, water quality modeling, multicriteria optimization

1 Introduction

One of the most important water quality processes is the eutrophication of freshwater bodies [Uhlmann 1988; Uhlmann and Horn 2001]. Water pollution by organic and inorganic as well as toxic substances affects the functioning of freshwater ecosystems, and risks human health. The eutrophication process is stimulated by nutrient remobilization from sediment where water temperature is one of the most important control variable [Goltermann 2004]. Additionally, the eutrophication process of freshwater ecosystems is supported by intensive man-made activities in river basins. They cause drastic changes in the biocoenotic structure and in the functioning of freshwater ecosystems. Sediments have been accumulated nutrients over several decades to such an extent that they now function as internal nutrient sources [DiToro 2000]. Compared with the amount of nutrients in the pelagic zone of eutrophic lakes, the nutrient content of sediment of shallow water bodies is considerable higher. Managing eutrophication problems water quality models are widespread used [Biswas 1981]. To control the water quality of shallow lakes of a lowland river an eutrophication simulation framework was developed [Gnauck et al. 2003a]. To get optimal management scenarios the simulator was coupled with an optimization processor [Krug 2002; Gnauck et al. 2003b]. In this paper, selected simulation results are presented for important water quality indicators like phytoplankton biomass and phosphate phosphorus, ammonia nitrogen and nitrate nitrogen.

2 The eutrophication simulator

The eutrophication process is stimulated by external nutrient input from the catchment due to man-made activities, but also by internal pollution due to nutrient remobilization from sediment supported by meteorological and hydrochemical conditions. Dead organic matter as algal biomass is mineralized by micro-organisms. This process needs electron acceptors which are supplied from the water column. They enter the sediment by molecular diffusion, convection, or bioturbation. The presence of electron acceptors in the water body, as dissolved oxygen, nitrate, iron (III), manganese (IV), sulphate, is a prerequisite for the remobilization of nutrients from sediment. Nitrate and dissolved oxygen are the major electron acceptors before iron is consumed. The order of consumption is determined by the Gibbs free energy gained in the reaction. In the case of aerobic conditions phosphate phosphorus will be fixed in the sediment. The dissolved inorganic phosphorus is bounded by $Fe(OH)_3$ or $Fe_3(PO_4)_2$. Anaerobic conditions and formation of hydrogen sulphide cause a destruction of the iron(III)phosphate and iron(III)hydroxide layer of the mud-water interface which prevents phosphate phosphorus remobilization under aerobic conditions [Uhlmann 1988; Uhlmann and Horn 2001]. The species iron(III)phosphate will be reduced to iron(II)phosphate under formation of iron sulphide and phosphate phosphorus is released into pore water. This process leads to a diffusion of phosphate phosphorus from pore water to the water column. Because of internal pollution of a water body is one of the main components which affect the water quality Hoffmann [Hoffmann 1999; Hoffmann et al. 2003] developed a process model describing the phosphate phosphorus release from sediment. For modeling he used the AQUASIM developers software [Reichert 1994]. Figure 1 shows the conceptual model.

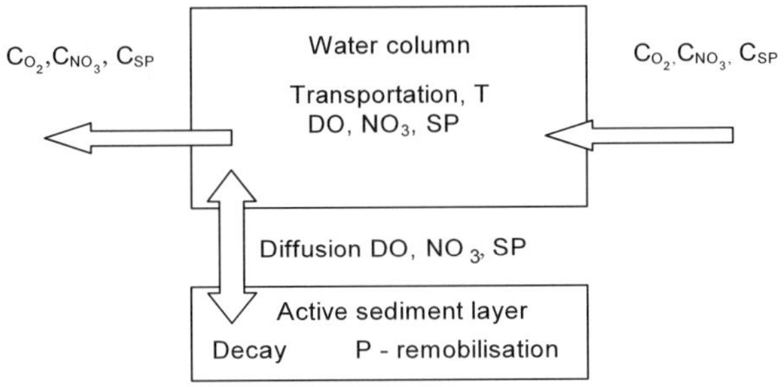

Fig. 1. Model concept of P-remobilization from sediment (SP-soluble phosphorus, DO-dissolved oxygen, NO3-nitrate nitrogen), (according to Hoffmann [1999], modified)

To simulate the eutrophication process in shallow water bodies a stationary 1D-model was developed within a MATLAB environment. It includes the sub-model of phosphorus remobilization from sediment. Figure 2 shows the model concept.

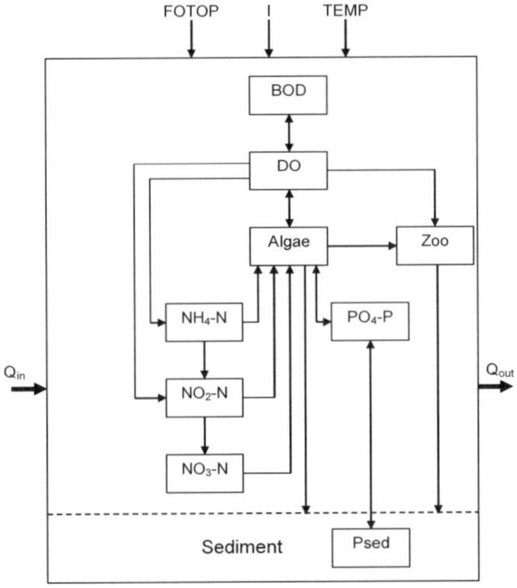

Fig. 2. Conceptual model of the eutrophication simulator

Model state variables are given by the water quality indicators phytoplankton (Algae), zooplankton (Zoo), orthophosphate phosphorus (PO_4-P), ammonia nitrogen (NH_4-N), nitrite nitrogen (NO_2-N), nitrate nitrogen (NO_3-N) as well as by dissolved oxygen (DO) and biochemical oxygen demand (BOD). Detailed descriptions of model equations, parameters, site constants and system specific parameters are presented by [Gnauck et al.

2003a; Gnauck et al. 2003b]. Q_{in} and Q_{out} mean the water flow into and out of the lake under consideration. External driving forces are photoperiod (FOTOP), solar radiation (I) and water temperature (TEMP). The following equations are used:

Phytoplankton biomass A (mg CHA/l)
dA/dt = Q/V(AIN-AOUT)+GROW-UA·f·A-FRZ·CR·Z·A-RESP·TEMP·A

Phosphate phosphorus P (mg P/l)
dP/dt = Q/V(PIN-POUT)+FRZ·CR·Z·A·((1-AZP)KSA/(KSA+A))
+RESP·TEMP·A-GROW+1/4·PSED

with $dPSED/dt = (-1)^{\Theta}\cdot phi\cdot hs(-Dsp/(1-\ln(phi^2))(P-(P_{SED}/(hs\cdot phi)))/hs/2 - \Theta\,(KFe\ Cp+qp))$

and $\Theta = 1$, if $CpEA \le cpcrit$, or $\Theta = 0$, if $CpEA > cpcrit$.

(phi - sediment porosity, hs - thickness of active sediment layer (m), Dsp - diffusion coefficient of dissolved phosphorus, P - dissolved phosphorus, CpFe - iron concentration in pore water, qp - ratio P/Fe of reducible iron, Kfe - iron concentration in pore water with KFe = K1(TW)/36, cpcrit - critical value of CpEA, K1(TW) - temperature dependent decay rate of organic material in pore water with $K1(TW) = K1(20)^{(0.1\cdot lg(2)/lg(K120)\cdot(TEMP-20)+1)}$, K1(20) - standardized decay rate of organic material in pore water at 20°C, CpO2 - dissolved oxygen concentration in pore water with CpO2 = O2/31,998, CpNO3 - nitrate concentration in pore water with CpNO3 = NO3/14.007, CpEA - electron acceptor concentration in pore water with CpEA = 2CpO2+5 CpNO3).

Ammonia nitrogen NH4-N (mg N/l)
$dNH4/dt = Q/V\cdot(NH4IN-NH4OUT)+B_3\cdot NORG_{in}-B_1\cdot NH4-FA1\cdot FUP\cdot G$

Nitrite nitrogen NO2-N (mg N/l)
$dNO2/dt = Q/V\cdot(NO2IN-NO2OUT)+B_1\cdot NH4-B_2\cdot NO2$

Nitrate nitrogen NO3-N (mg N/l)
$dNO3/dt = Q/V\cdot(NO3IN-NO3OUT)-(1-FUP)\cdot FA1\cdot G+B_2\cdot NO2$

Filtrating zooplankton Z (mg C/l)
dZ/dt = Q/V·(ZIN-ZOUT)+FRZ·A·Z·CR·AZP·C·KSA/(KSA+A)
-MORT·TEMP·Z

Dissolved oxygen DO (mg/l)
$dDO/dt = Q/V\cdot(DOIN-DOOUT)+K_2\cdot(DO_{sat}-DO)+(a_3\cdot G/A$
$-a_4\cdot RESP\cdot TEMP)\cdot A-K_1\cdot BOD-K_4/zmix-a_5\cdot B_1\cdot NH4-a_6\cdot B_2\cdot NO2$

$-a_7 \cdot MORT \cdot TEMP \cdot Z$

Biochemical oxygen demand BOD (mg/l)
$dBOD/dt = Q/V \cdot (BODIN\text{-}BODOUT) + K_1 \cdot BOD - K_3 \cdot BOD$

According to [Straškraba and Gnauck 1985] temperature dependencies of physical water quality variables are modeled by sinusoidal functions. Saturation concentration of DO is expressed by a third order polynomial [Thomann 1972].

3 The optimization processor ISSOP

To combine simulation and optimization procedures the processor ISSOP was used. It contains an open interface of MATLAB models [Krug 1997; Krug 2002; Wiedemann and Krug 2001]. The dialogue between external and internal models and optimization methods is realized by an universal parameter interface (Fig. 3).

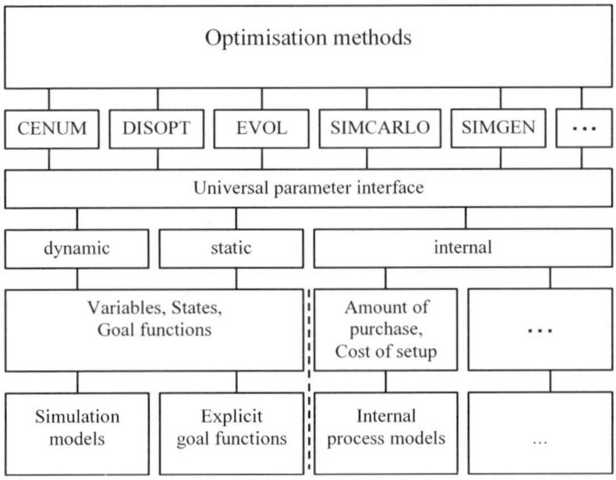

Fig. 3. The architecture of optimization processor ISSOP

Where CENUM - component wise enumeration, DISOPT - a quasi-gradient method, EVOL - an evolutionary optimization strategy, SIMCARLO - optimization by Monte Carlo method, SIMGEN - optimization by a genetic algorithm. Before starting an optimization run each simulation problem is automatically transformed into the standard problem of optimization [Krug 2002]. On the lowest level of this architecture simulation models, goal functions and internal process models are given explicitly. External static and dynamic simulation models can be implemented without any restriction. Figure 4 shows the general structure of coupling. Input variables of the simulation system are denoted by $\alpha_1 x_1,\ldots, \alpha_k x_k$, while outputs are symbolized by $y_1,\ldots,y_m$. Goal functions are represented by $f_1,\ldots,f_n$ with $f_i(M(\alpha_1 x_1,\ldots, \alpha_k x_k)) = f_i(y_1,\ldots,y_m)$ for $i = 1, \ldots$ n where arbitrary continuous functions f_i can be used. They will be optimized simultaneously. There are no restrictions for goal functions (e. g. convexity). If $n > 1$, then goal functions $f_1,\ldots,f_n$ are aggregated to a weighted sum $S = \Sigma w_i f_i$. For weighting factors w_i the condition $\Sigma|w_i| = 1$ is valid.

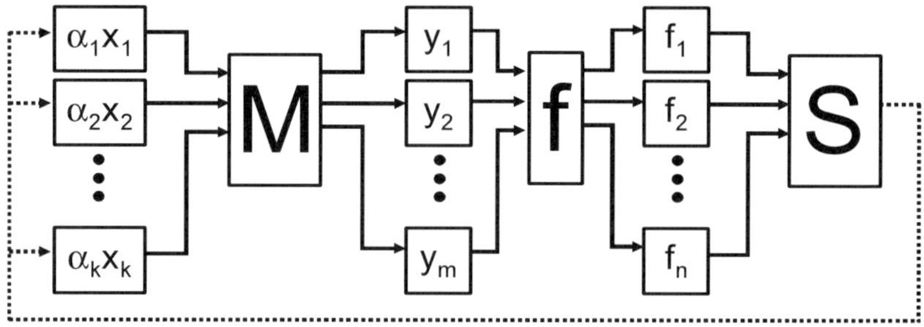

Fig. 4. Coupling of ISSOP with a simulation model

The data transfer between both software tools is shown in Figure 5. The processor ISSOP uses the model variables and target values as input data and gives optimized state variables back to the eutrophication simulator. If the goals are not reached, then the input parameters can be changed.

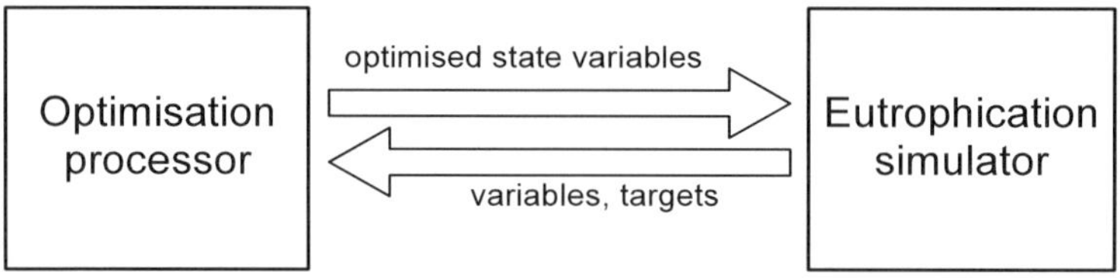

Fig. 5. Data transfer between optimization processor and eutrophication simulator

4 Experimental Area and Data

Experimental investigations are carried out in the river basin of Lower Havel River. It belongs to the greatest tributaries on the right hand site of the Elbe River and is strongly influenced by the River Spree. The catchment is characterized by shallow lakes, wetlands and marshy country. Figure 6 shows a generalized graph of the environmental area.

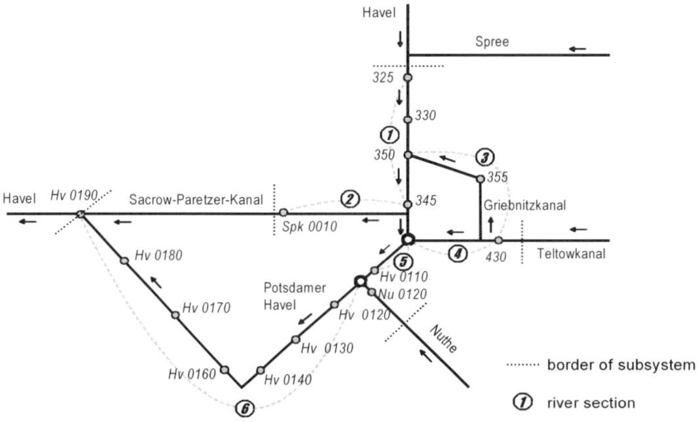

Fig. 6. Generalized graph of the experimental area

The River Havel is a typical lowland river with a very small elevation difference between source and the mouth into the Elbe River (about 41 m). For low-flow conditions a slope of the water level of 2 cm/km is observed. Hydraulic works, banked-up water levels and high evaporation rates (only 25% of precipitation contributes to discharge) influence flow rates and retention times of the water as well as the intensity and kinetics of chemical and biological water quality processes. The active sediment layer is given by 2 cm to 6 cm. For water quality simulation the river basin of the Lower Havel River was divided into several segments of different length. Time series of water quality data from 1997 are taken into consideration as references, while time series from 1998 to 2002 with daily, weekly and two-weekly sampling intervals are used for modeling, parameterization and further simulation runs. After calibration and validation the eutrophication simulator was used to simulate water quality changes.

5 Results

Figure 7 shows basic simulation runs for phytoplankton biomass, and nutrient dynamics (Fig. 8 and 9) of the Lower Havel River at different observation points (see Fig. 6) for the reference year 1997.

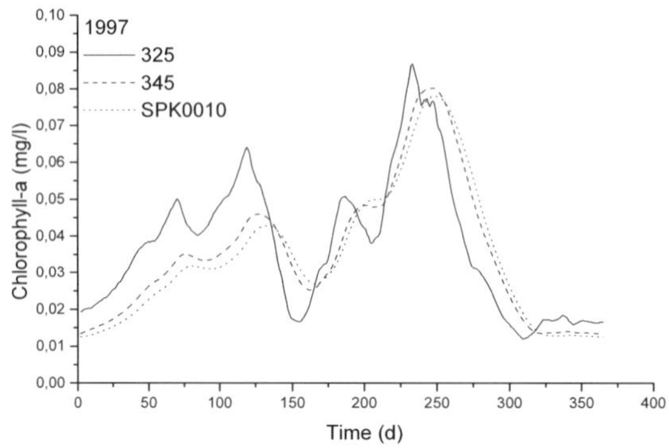

Fig. 7. Phytoplankton dynamics of the Lower Havel River

Dominant algal species are diatoms in spring and cyanobacteria in late summer and fall. Because of nutrient rich water body the bioproduction is high in spring and late summer (upper part). During the first four months (at low temperature levels) the growth of diatoms can be seen while in late summer (high temperature levels) cyanobacteria dominate. According to decreasing temperature levels in fall algal blooms collapse.

Phosphorus concentration dynamics (Fig. 8) is determined by two different processes. In spring, a decrease of phosphate phosphorus due to phytoplankton uptake by diatoms is observed. In opposite of that, during fall an extremely increase of phosphorus can be seen. An intensive decay of dead organic matter due to temperature dependent high decay rates accompanied by high rates of oxygen consumption leads to anoxic conditions at the sediment-water interface. These effects cause internal pollution by remobilization of phosphate phosphorus from sediment.

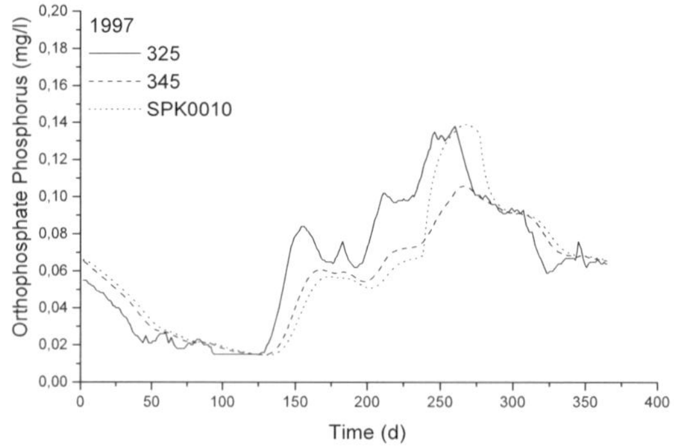

Fig. 8. Dynamics of phosphate phosphorus

The dynamics of nitrate nitrogen is presented in Figure 10. In correspondence with Figures 8 and 9 the nitrogen consumption due to uptake by cyanobacteria is well demonstrated. After algal bloom the nitrogen pool is filled up again by nutrient rich water inflow.

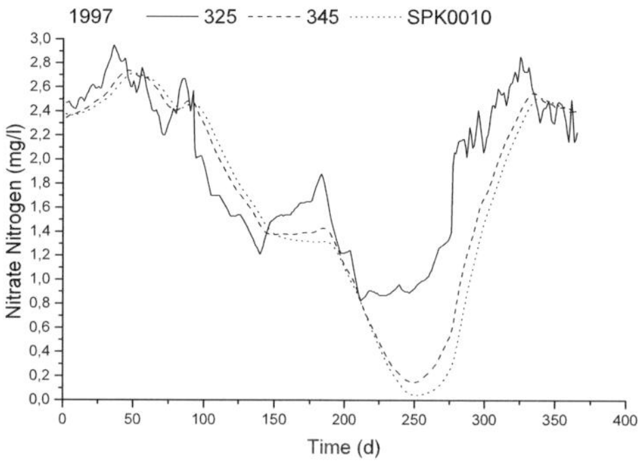

Fig. 9. Dynamic changes of nitrate nitrogen

In Figure 10 simulation results of nitrate nitrogen are presented for another river stretch (cf. Fig. 6). The upper line shows high variations of the nitrate input of a polluted tributary (430) with an average of approximately 5mg/l over the year. The other lines indicate nitrate nitrogen changes at different observation points within a chain of shallow lakes. As can be seen, the dynamics of nitrate nitrogen concentrations within this river stretch follow the phytoplankton dynamics. Small influences of the tributary can be slightly seen for the observation point Hv0110 only. The water mass of lakes equalizes high input concentrations.

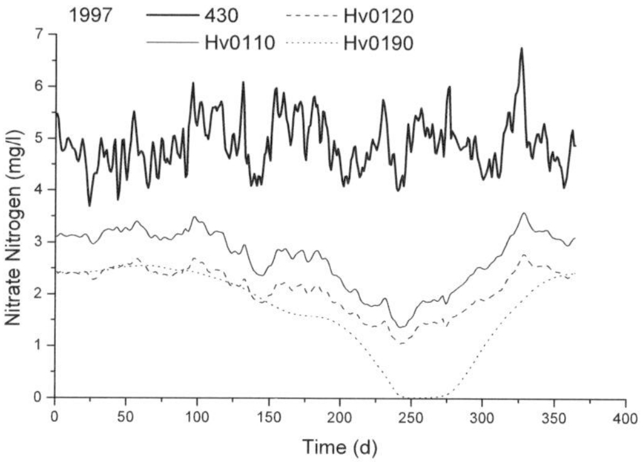

Fig. 11. Simulation of nitrate nitrogen

For water quality management two management strategies are considered. The first one is based on the limiting nutrient concept of algal biomass [Uhlmann 1988]. The second one refers to target values of German Working Group LAWA regulations [LAWA 1998]. The goal functions phytoplankton biomass $f_1(t) = \Sigma_x \Sigma_t\, y_1(x, t) \rightarrow \min$, orthophosphate phosphorus $f_2(t) = \Sigma_x \Sigma_t\, y_2(x, t) \rightarrow \max$, and nitrate nitrogen $f_3(t) = \Sigma_x \Sigma_t\, y_3(x, t) \rightarrow \max$ are used for both strategies

The weights according to the limiting nutrient concept are $w_1 = 90.5$, $w_2 = -1.1$ and $w_3 = -8.4$. It can be stated that an eutrophication control according to the limiting nutrient concept leads to a diminished phytoplankton maximum in late summer due to optimized nitrate concentrations (Fig. 10). No effect of optimized orthophosphate phosphorus concentration can be stated. Optimal averages of goal functions are $f_1 = 44.991$ µg/l, $f_2 = 1.472$ µg/l, and $f_3 = 1.54$ mg/l.

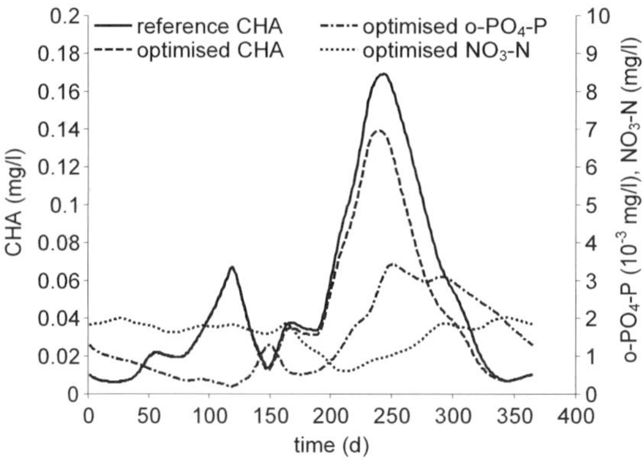

Fig. 10. Management options according to limiting nutrient concept

The weights according to LAWA are $w_1 = 42\%$, $w_2 = -57\%$ and $w_3 = -1\%$. Eutrophication control according to LAWA target values leads to nearly the same behavior of phytoplankton biomass in spring but to smaller differences of phytoplankton maxima and to low nutrient concentrations in late summer. Optimal averages of goal functions are $f_1 = 48.762$ µg/l, $f_2 = 0.166$ µg/l and $f_3 = 0.08$ mg/l.

On the other hand, management options according to LAWA regulations lead to nearly the same behavior of phytoplankton biomass in spring but to smaller differences of phytoplankton maxima and to low nutrient concentrations in late summer (Fig. 11).

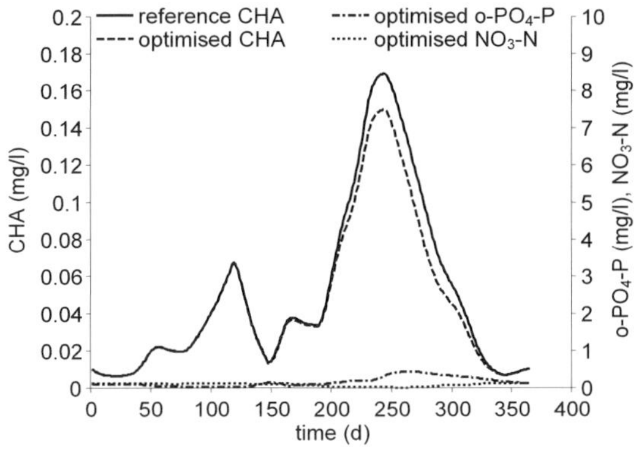

Fig. 9. Management options according to target values [Lawa 1998]

As a result the LAWA concept leads to significant lower nutrient concentrations but to a slight increase of phytoplankton biomass. In opposite of that an eutrophication control according to the limiting nutrient concept results in lower phytoplankton concentrations but higher admissible nutrient inputs.

6 Conclusions

A sustainable management to control freshwater ecosystems can only be achieved by using powerful informatic tools. The use of combined simulation-optimization procedures to manage the water quality of rivers, lakes and reservoirs is an approach promising more theoretical understanding of complicated natural processes and software engineering methods. Direct interrelations exist between sediment and water quality influenced by physical, chemical, biological, hydraulically, hydrological and geomorphologic driving forces. In dependence of these conditions the sediment layer works either as a nutrient trap or as a nutrient source. Phosphate remobilization from sediment can be considered as a result of some contradictory processes of matter changes. Especially, phosphate remobilization depends on dissolved oxygen conditions, on the load of organic substances, on water temperature, on pH-value, on bacterial activity, on the total phosphorus content of the sediment and interstitial waters, on methane convection, on wind stress and hydraulic conditions, and also on zoobenthic organisms and fishes. The different model approaches are constrained by the amount and the type of available data. Perspectives of developments of simulation frameworks for water quality management on a river basin scale may be seen in combinations of water quality simulation models, multi-objective optimization procedures and visualization tools. The use of such a framework promises more theoretical understanding of complicated natural and management processes.

Acknowledgements

The author is very thankful to Prof. M. Freude from Environmental Agency of the State of Brandenburg for supporting this work and providing the data. Thanks are also given to Bernhard Luther for fruitful discussions and computational help and to Shafi Noor Islam for technical assistance.

References

Biswas AK (1981) Models for Water Quality Management. McGraw-Hill, New York

DiToro DM (2000) Sediment Flux Modelling. Wiley, Chichester

Goltermann HL (2004) The Chemistry of Phosphate and Nitrogen Compounds in Sediments. Kluwer, Dordrecht

Gnauck A, Gnauck A, Luther B, Wiedemann T, Krug W (2003a) Coupling of Simulators for Optimal Control of Water Quality. In: Gnauck A, Heinrich R (eds) The Information Society and Enlargement of the European Union. Metropolis, Marburg, pp 373-380

Gnauck A, Heinrich R, Luther B (2003b) Modellierung und Simulation von Seen in der Unteren Havel. Forum der Forschung, vol 8, edn 15, pp 43-48

Hoffmann A (1999) Mathematical modelling of phosphorus dynamics in rivers with special regard to phosphate remobilization from sediment. Diploma thesis, Dept. of Ecosystems and Environmental Informatics, BTU Cottbus

Hoffmann A, Gnauck A, Tesche T (2003) Modellierung der Phosphorrücklösung aus dem Sediment eutropher Gewässer. In: Gnauck A (ed) Theorie und Modellierung von Ökosystemen. Shaker, Aachen, pp 251-262

Krug W (1997) Intelligentes Simulations- und Optimierungssystem für Prozesse der Fertigung, Organisation und Logistik – ARENA/ISSOP. SCS International Publishing House, San Diego

Krug W (2002) Modelling, Simulation and Optimisation for Manufacturing, Organisational and Logistical Processes. SCS Europe Publishing House, Delft

LAWA (1998) Beurteilung der Wasserbeschaffenheit von Fließgewässern in der Bundesrepublik Deutschland - Chemische Gewässergüteklassifikation. Kulturbuchverlag, Berlin

Reichert P (1994) Concepts Underlying a Computer Program for the Identification and Simulation of Aquatic Systems. Dübendorf: Report ETH Zürich/EAWAG

Straškraba M, Gnauck A (1985) Freshwater Ecosystems. Elsevier, Amsterdam
Thomann RV (1972) Systems Analysis and Water Quality Management. McGraw-Hill, New York
Uhlmann D (1988) Hydrobiologie. Fischer, Jena
Uhlmann D, Horn W (2001) Hydrobiologie der Binnengewässer. Ulmer, Stuttgart
Wiedemann T, Krug W (2001) Multikriterielle Optimierung von MATLAB-Modellen mit ISSOP. In: Panreck K, Dörrscheidt F (eds) Frontiers in Simulation. Erlangen, 2001, pp 199-204

GIS Use in the Aral Sea Area: Water Quality and Population's Health

E. Kamilova[1], D. Fayzieva[2], B. Nurtaev[3], L. Orlovsky[4], D. Reible[5]

[1] Institute of Genetics and Plant Experimental Biology, Academy of Sciences, Republic of Uzbekistan, Tashkent, Uzbekistan

[2] Institute of Water Problems, Academy of Sciences, Republic of Uzbekistan, Tashkent, Uzbekistan

[3] Institute of Geology and Geophysics, Academy of Sciences, Republic of Uzbekistan, Tashkent, Uzbekistan

[4] Ben-Gurion University of the Negev, Sede Boker Campus, Israel

[5] The University of Texas at Austin, USA

Abstract: This research represents the integrated study of ecological and sanitary status of water in selected areas of the Zeravshan River Basin, and interrelations between the level of their contamination and the population's health status. For the first time in this region, the rate and structure of morbidity were used as an indicator of the environment status. The main results of this research are: identification of the most vulnerable population groups; recommendations on assessment and control of the environmental conditions necessary for the stable management of water resources; and introduction of a Geographic Information System to monitor, control and forecast the water sources and population's health.
Keywords: Geographic information systems, Environmental health risk, Environmental pollution and control, Environment, Ecotoxicology

1 Introduction: Formation of water quality in water-bodies in the basin of the river Zarafshan

The basin of the river Zarafshan is one of the most populated areas in Central Asia. The river Zarafshan starts in the territory of Tadjikistan where its water is actively used and polluted. The river is the main source of potable water supply to the local population. At the same time, the river accumulates industrial, house-hold and irrigation waste water. In the basin of the Zarafshan, a tense situation with water supply and environment is developing because the social and economic needs of the region require a significant increase of water supply to developing industry, house-hold services and irrigation of fields. Such situation, no doubt, manifests itself in the sanitary and epidemiological changes as most of inhabitants of the rural area are using water for drinking and house-hold needs from surface water-bodies and canals.

In small rivers, the qualitative and quantitative shifts in microbe cenosis have been registered resulting from poor flowage caused by huge water withdrawal for irrigation and intensive pollution of water with organic and biogenic substances. In Samarkand region, the phenol concentration, i.e. the main indicator of pollution of water-bodies with organic substances, reached 0.02 mg/l (20 MPC) while the maximum permissible concentration is 0.001 mg/l.

In the period from 1998 to 2005 there was an increase of COD in the lower reaches of the Zarafshan compared to the middle reaches. The highest values of COD were registered in the samples of water taken below the waste water discharge of "Navoiazot" near the town of Navoi.

Intensive biogenic and organic pollution of water results in misbalance events in the ratio of microorganism species in water-body causing an increase of the level of pathogenic bacteria (Salmonella) in the range of 16 – 33 mpv per one liter of water even when the content of the indicator microflora is decreased. This demonstrates the potential epidemic danger of the water-body and permits to consider the water-body to be a source of spread of acute intestinal diseases.

2 The Development of GIS Concept

The studies of physical-geographic and other nature factors influencing the condition of water-bodies in the Zarafshan basin and evaluation of anthropogenic factors that might impact on water quality in water-bodies being used for potable water supply were based on the documents of some agencies and institutions dealing with nature protection, e.g. the Goscompriroda of the RUz, Glavhydromet of the RUz and NIGMI under the Glavhydromet of the RUz (the received data are stored in digital form).

The environmental situation has been analyzed in Samarkand, Bukhara and Navoi region, a detailed plan of taking samples of water, soil, bottom sediment was designed. Soil samples were taken under the major agricultural plants, in places close to big industrial enterprises; the control samples were taken from virgin land. Samples of potable water in the first year of the Project implementation were taken in the town of Samarkand and district centers of Samarkand region from the following water-bodies: water-pipe systems, river, water from aryks (irrigation ditches).

The data on population morbidity were taken in the department of the State Sanitary and Epidemiology Surveillance, Statistic department of the Health Ministry, regional health administrations and regional sanitary and epidemiology surveillance administrations of Samarkand region of Uzbekistan.

The territory of Samarkand region was chosen for the research using a GIS because the epidemiological situation with enteric fever and bacterial dysentery in the last decade was poor there. Figure 1 shows the living settlements, roads, relief and water-bodies in the territory of Samarkand region. The rivers and canals were divided into four types by their water volume. Figure 1 shows the main water sources here to be the river Zarafshan and the canals offtaking water from the river. The spatial analysis was made using the GIS MapView that enabled to request the geographical database and to reveal the most affected areas in view of the acute intestinal diseases rate among the population. The map (of 1:500,000 scale) was one of the bases for the GIS.

Such major indicators of water quality as water discharge, pH, water temperature, levels of chlorides, sulfates, dry residue, oxygen dissolved in water, nitrogen, ammonium, nitrites/nitrates, specific pollutants, BOD and COD, total microbe number, Coli-index, level of pathogenic bacteria were analyzed by the statistic methods.

The information for five years (2000-2005) obtained for each indicator to be predicted was summarized. In addition, the data for 1996-2005 on morbidity of enteric fever, bacterial dysentery and other acute intestinal infections, as well as viral hepatitis A among the population in the area under study were analyzed.

A territory of a district was taken as a unit of taxonomic characteristic that is convenient in terms of administrative relations and statistic data. However, these areas do not correspond to the areas of contamination with ecotones and actual landscape and geographic borders. This required creation of specific maps based on topographic and landscape maps. The obtained monitoring data and field study findings were statistically treated, after that they were averaged both in space and time dynamics.

Draft outlines of areas with unsafe environment were made using the Mapinfo Geographical Information System through superimposing different maps with areas of contamination exceeding the MPCs or background values.

The analysis of multiple or pair correlation between components of environment pollution and certain diseases and general morbidity of the population as well as the relation to social-economic, sanitary-hygienic factors of life, traditions and customs of local population made the ground for integrating the ecological, medical-hygienic and social-economic modules.

Each of the suggested modules includes 10-15 characteristic criteria. The absolute value of a criterion for each administrative area has been calculated. The number of scores has been assigned by experts on the basis of the obtained correlation between the impact of certain components, worsening the habitat, on human health and the share of each criterion. The shares of natural-climatic, environmental, medical-hygienic, social-economic criteria are differentiated in points into four classes: mild (background) – medium – high (tense) – very high (crisis).

The ecological module includes the following criteria (in the order of their priority): the disparity between the quality of potable water, food raw materials and the State Standard requirements (%), the index of air pollution. These criteria indicate the direct effect of toxic substances on human body and receive a maximum number of points. They are followed by indicators showing the area of soil contaminated with pesticides, ME, HM in concentrations exceeding the MPCs; the same areas with agricultural plants contaminated with the same substances; soil salinity; underflooding of areas; potentially unsafe, in terms of environment, enterprises; indices of underground water contamination (IUWC); those of contamination of surface water (ICW); areas of pastures degradation; the yield class of soil; the index of climatic seasonable air pollution (ICSA).

The medical-hygienic module includes statistical indicators of the most common and typical diseases related to changed environment in the area: rates of total and children morbidity, viral hepatitis, acute intestinal infections, acute respiratory diseases, neoplasia, endocrinologic diseases, TB, infant mortality. Deviations of specific for the area indicators of population morbidity from those averaged for the country and region reflect indirectly the environment situation in the area. The structure and diagnostics of diseases, divided in 14 classes according to the ICD in the areas under study, enable to analyze the endemic pathology, cause and effect links between certain diseases and the environment impact, in children diseases in particular.

The social-economic module characterizes the living conditions of population, the condition of balanced nutrition and other factors related to environmental and medical-hygienic changes. The relation between poverty and environment degradation is evident as they both have social roots. The "Methodology of division into environment areas" takes into consideration the following parameters: population density, life-span, education level, provision of population with centralized water-pipe and sewage systems, children institutions, population income.

The priority criteria used to evaluate areas with unsafe environment. When using the "Methodology of division into environment areas" 20 priority criteria were selected:

- *CPC* – the climatic potential of contamination characterizes meteorological events: mild wind, temperature inversions, stagnant state of air, repeated fog, area orography, etc. The absolute values of the CPV alter from moderate 2.0, 2.4 to very high 3.6, 3.9.
- *ICC* – the index of climate condition is evaluated by the degree of recurrence of extreme values of climatic parameters: temperature in winter (from – 10oC to – 25oC) and in summer (from +30oC to +45oC), relative humidity (<30% and >80%), strong wind (from 8 to 20 m/sec), intensity of precipitates (>20 mm), dust storms, etc. The ICC values are measured within the range 1.8-2.8 according to the data of the State Meteorological Committee.
- *Yield class of soil* reflects physical-chemical, mechanical characteristics of soil, climatic conditions, the relief of the area, salinity and erosion ability of soil, the presence of humus, etc. The yield class of soil expressed within 80 – 100 shows the best soil, while the value <20 shows the worst one.
- *Soil salinity* (in % of the irrigated area) is evaluated by the method of the Uzdaverloyikha. The sum of lands that have medium- and strong-salinity is taken into account. The values 0-10% of salinity are considered to be mild, while 40-50% show very strong salinity.
- *Area of soil contamination with pesticides* (in %), exceeding the MPC, is evaluated by the residual amount of COP. The areas where up to 10% of land is occupied with contaminated plants are defined as that one with mild contamination, if the value is 41-50%, the land is badly contaminated.
- *Area of contamination of plants with pesticides* (in %), exceeding the MPC, is evaluated by the residual amount of COP. The areas where up to 10% of land is occupied with contaminated plants are defined as that one with mild contamination, if the value is over 40% the area is badly contaminated.
- *Contamination of soil with ME and HM* (in %). The areas with soil contamination under 5% of the MPCs are defined as mildly contaminated; those with 25% of the MPCs were defined as badly contaminated.

- *Areas with plants contaminated with ME and HM* (in %) over 10% of the MPCs are defined as mildly contaminated, 25% contamination referred the area to badly contaminated one.
- *Discrepancy between the State Standards requirements and the quality of food stuff* (in %) is evaluated by the data of the SESS, MoH, separately by chemical and microbiological indicators. The value of 50% and more is considered to be a wide discrepancy, that one under 10% - as a mild discrepancy.
- *Discrepancy between the State Standards requirements and the quality of potable water.* The level of water pollution under 15% is mild one, while 60-70% of pollution means badly contaminated water. In some areas of the country, 50-100% of water does not meet the standards that influence significantly the population health (39).
- *ICW, the index of contamination of surface water*, takes into account, according to the proposal of the Institute for Water Problems (the Academy of Sciences of Uzbekistan), all ingredients of pollution exceeding the MPCs. The amount of contaminating ingredients and the class of their danger are corrected by the correction factor. The ICW values under 1.0 are considered good, while those of 5-10 are dangerous.
- *ICUW, the index of contamination of underground water*, is determined by the method of the VNIIVO that takes into account three mandatory ingredients (dense sediment, general hardness, nitrates) as well as other components exceeding the MPCs. The same way as for the ICW, the correction factors are used. The ICUW values are ranging from mild (0.3-0.1) to high (over 6).
- *IAP, the generalized index of air pollution*, is calculated by the method suggested by the State Meteorological committee taking into account the class of danger:
-

$$\mathrm{IAP} = \frac{(C_1)^{q_1}}{MPC_1} + \frac{(C_2)^{q_2}}{MPC_2} + \frac{(C_n)^{q_1 n}}{MPC_n}, \tag{2.1}$$

Where C is the averaged daily value of admixture concentration in air; MPC is the averaged daily value of emissions concentration; q is the class of danger of emissions into air. The value of IAP under 5 is interpreted as mild, that of over 20 – as very high.

- *Unsafe, in terms of environment, enterprises* include big industrial enterprises, chemical plants, electric power stations, thermoelectric power stations, tailing dumps, areas of contaminated soil, animal-breeding farms, poison burial grounds, etc. If the area has 5 such enterprises, the load is considered to be mild. If there are over 20 of them, the load is very heavy. The number of the most big enterprises is taken into account by an additional coefficient.

- *Land underflooding* (in % of total area). The autumn level of ground water after vegetation watering is taken into account if the value is under 1.5-2.0 m. Underflooding of 5-10% of the total area is defined as mild, that of 40% and more – as very bad.
- *Morbidity rate* (per 1,000 population) includes adults and children under 14 years. The data of the MoH for last 5 years were used, the data being averaged by morbidity rate. The morbidity rate under 650 per 1,000 population (the minimum average country level) was interpreted as low, that one over 750 per 1,000 population was interpreted as high.
- *Population density* (thousand/km2). The density of 1.0-1.5 thous./km2 refers the area to the ones with a mild load, those with the density over 6.5 have very heavy load. An aggregate of priority criteria, ranges of changes in their absolute values determine the quality of environment, habitat and the extend of environmental trouble in each area.

3 Population's health

Comparison of data on population health status in different regions of Uzbekistan in 2004 and 2005 showed that Navoi region leads in total morbidity of its population. In 2004 the highest level of total morbidity was recorded in Navoi region (the highest one on the country); in 2005 the region is the forth by the rates of total morbidity (after the Republic of Karakalpakstan, Bukhara and Khorezm regions). In Navoi region, high morbidity rates were recorded for the following diseases: diseases of blood and hemopoietic organs (with anemia being the most common), respiratory diseases, diseases of alimentary and nervous systems. Children under 14 years showed sharply increased rates of total mortality from 1992-2005. In 1997 a decrease of the rates almost to the level of total mortality recorded in 1992. This probably may be explained by bias in diseases registration due to less often seeking medical care services in health care facilities associated with these diseases.

The rates of infections in 2005 increased dramatically in comparison with those in 2004, with the prevalence of the rates in Navoi region is lower than in Tashkent-city and Tashkent region. The structure of infection morbidity by some infectious diseases does not comply with the data of the National Department of Statistics.

The prevalence of neoplasms in Navoi region is also lower than in Tashkent-city and Tashkent region. In 2005 oncological morbidity in Navoi region slightly reduced. The trend to increase was observed for prevalence of diseases of the endocrine system, impairment of metabolism and immunity deficiency – the region heads the list according to the data for 2005. Diseases of blood and hemopoietic organs became more often in 2005 in comparison to 2004; and by their rates Navoi region is the third after the Republic of Karakalpakstan and Bukhara region. The prevalence of cardio-vascular diseases has also increased. However, in the past year the rate is lower than that ones in Khorezm, Tashkent, Bulkara regions.

4 GIS-analysis of the rates of water-borne acute intestinal diseases among the population of Samarkand region

The figures given below present the information on the rates of enteric fever, dysentery, acute intestinal diseases and viral hepatitis A. Figure 1 illustrates their average values obtained during a long-term observation (1996-2004) conducted in different districts of Samarkand region. These data were obtained during a comprehensive surveillance of the sanitary and epidemiology status of the region.

Recently the unfavorable dynamics of bacteriological indicators of water quality had been registered in the places of water intake for potable water supply in Samarkand region. In 1996 the sanitary standard was not met by 14.5 per cent of water samples taken in surface water-bodies, while in 1998 the figure increased to 24.7 per cent. This can indicate an increased microbe pollution of water-bodies and a higher risk of intestinal diseases morbidity for the local population. In 1996 the water from water-pipes did not meet the sanitary standard in 4.5 per cent of samples, in 1998 their number increased to 4.8 per cent.

Taking into account the information on availability of centralized water supply in the urban (83.5%) and rural areas (71.5%) of the region, one can assume how much the rural people are liable to getting infected with intestinal infections. This assumption can be confirmed by the fact that in 1996 and 1997 most of patients with enteric fever (97.9% and 93.5%, respectively) in Samarkand region were people living in the rural area. This situation is probably caused by the character of their job, too, as most of them are working in agriculture and, hence, have often contact with water and soil.

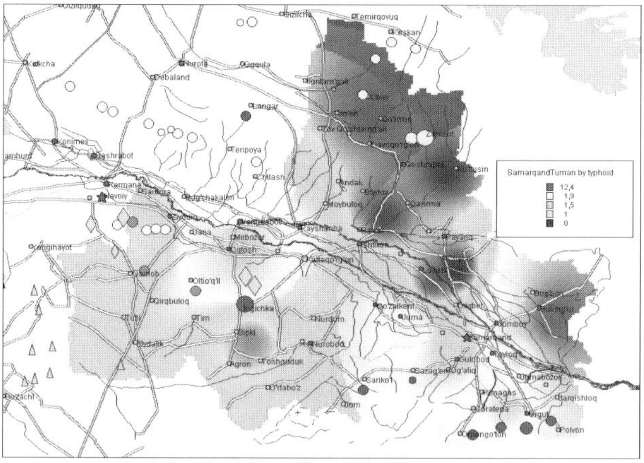

Fig. 1. Enteric fever morbidity among the population (the number of patients per 100,000 population)

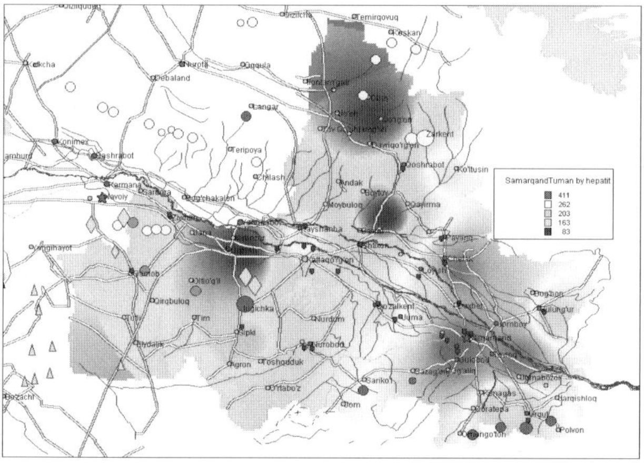

Fig. 2. Distribution of hepatitis A morbidity among the population (the number of patients per 100,000 population)

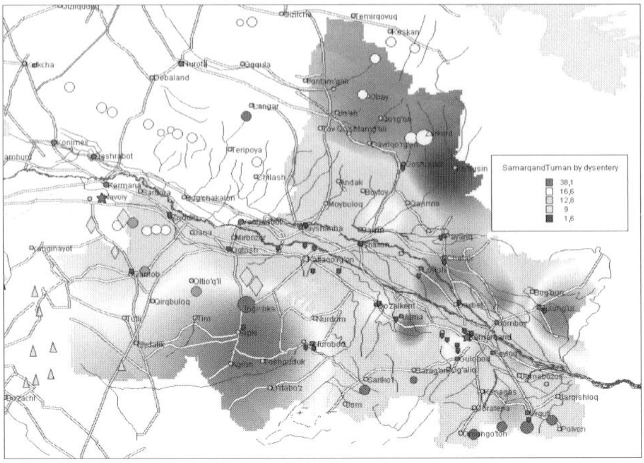

Fig. 3. Distribution of dysentery morbidity among the population (the number of patients per 100,000 population)

Local contamination with heavy metals recorded in some areas of Samarkand (surface water, soil) has resulted in higher incidence of musculoskeletal diseases. High mineralization of water (including the water on water-supply systems) and a high concentration of strontium in Bukhara region cause a sharp rise of urolithiasis rates, formation of stones in the kidneys, bladder, etc. Unfavorable environment status (contamination with radioactive substances, PCB, pesticides, nitrites, nitrates, and organic substances) in Navoi region leads to a dramatic increase of rates of oncological, allergic diseases, diseases of the nervous system, antepartum, intrapartum and posrpartum complications, as well as congenital anomalies (malformation).

5 Conclusion

The quality of potable water in the basin of the river Zarafshan tends to deteriorate every year judging by the sanitary and epidemiological indicators.

Today a serious danger for human health in the region is constituted by poor provision of sewerage system services and lack of water-treatment plants. Only towns and few district centers have these plants. The major part of the population lives in the rural area where waterpipe-systems are poorly developed.

For the last decade the total morbidity of the population has increased. The rate of water-borne infections is higher in the rural area in comparison to the urban one.

The research findings show a marked discrepancy of spatial distribution of enteric fever cases that depends on hyper-endemic character of some areas that once more necessitates a differential approach to evaluation of the situation and further revealing the conditions and agents of the epidemic process activation. The epidemic surveillance needs to be improved to address the epidemiological situation at definite time and place.

Revealing the cause-effect relation of these diseases development is rather difficult process as the impact of other environment factors on disease development is complex and requires further in-depth ecological and epidemiological research before development of sound preventive actions in the region.

The approved methods of the epidemiological analysis using the geographic information system helps improve evaluation of environment risk for human health. The GIS technology ensures joining up the materials of different agencies, organizations and institutions of industrial, nature protection and medical profiles and enables to form a holistic conception on the risk of getting infected with water-borne diseases in a definite area.

This technology needs to be used for monitoring and prediction of water-bodies status and evaluation of the impact on human health that are necessary both for the practice of sanitary-epidemiological service and departments of environment expert examination in nature protection agencies.

References

Aral MM, Maslia ML (1996) Evaluation of human exposure to contaminated water supplies using GIS and modeling. In: Kovar K and Nachtnebel HP (eds) Application of Geographic Information Systems in Hydrology and Water Resources Management. IAHS Publications No 235. IAHS Press, Wallingford, pp 243-252

Barr R (1996) Data, information and knowledge in GIS. GIS Europe. vol 5, edn 3

Frisch JD, Shaw GM, Harris JA (1990) Epidemiologic Research Using Existing Databases of Environmental Measures. Archives of Environmental Health, vol 45, edn 5, pp 303-307

Elliott EP, Cuzick J, English D, Stern R (1996) Geographical & Environmental Epidemiology. Methods for Small Area Studies. Oxford Medical Publications, p 382

Malone JB et al. (1992) Use of LANDSTAT MSS imagery and soil type in a geographic information system to assess site-specific risk of fascioliasis on Red

River farms in Louisiana. Annals of the New York Academy of Sciences, p 653, pp 389-397
de Lepper MJC, Scholten HJ, Stern RM (1995) The Added Value of Geographical Information Systems in Public and Environmental Health. Kluwer Academic Publishers, The Netherlands, p 355
Martynenko AI, Bugaevsky UL, Shibalov SN (1995) GIS fundamentals: theory and practice. Moscow, p 75
Mirtazaev OM, Norov G, Hasanov AH, Toshpulatov E, Tursunov PB (1999) Epidemiological characteristics of enteric fever outbreaks. Infections, Immunity and Pharmacology edn 1, pp 78-80
Mirtazaev OM, Niyazmatov BI, Norov G, Kantemirov MP (1999) Epidemiological characteristics of enteric fever in Uzbekistan. Epidemiology and Infectious Diseases. edn 4, pp 16-18
Reshetov VV, Gorky AV (1994) Comprehensive evaluation of radiation, geochemical, medical and sanitary situation in towns of Russia. Scientific report on implementation of the 2nd stage of the Federal Program "Ecological Safety of Russia"
Utkin VV (1996) Problems of formation of the territory information space. Information Technologies. edn 3, pp 16-19

Evidence of Heavy Metal Pollution in French Jura Lakes: Observed Impacts and Countermeasures

Rachid Nedjai

Institut de Géographie Alpine – Université Joseph Fourier Grenoble 14 bis Avenue Marie Reynoard 38100 Grenoble, France
e-mail: nedjai@ujf-grenoble.fr

Abstract: Like other lakes in France and Europe, French Jura lakes are currently the subject of fierce controversy following a large number of research studies carried out in a variety of disciplines. The presence of abnormal quantities of certain trace metals has created a climate of apprehension among lake users and caused administrators sit up and take notice. As water reservoirs for medium- and high-altitude mountain towns and villages as well as leisure and recreation areas for large numbers of tourists, the lakes are highly-valued heritage sites. They also generate considerable income for local authorities and to a certain extent for the state.

Long-term hydrological monitoring and analysis of the sediments of nine lakes in the Franche-Comté region of eastern France has enabled the physical-chemical processes regulating their functions over the last 300 years to be understood and questions concerning the origin and extent of the lakes' pollution to be answered. The analyses confirmed that the lakes' water is of local origin, from within their respective catchment areas, and that it stays in the lakes for between 10 and 12 months. Isotopic analyses showed that surface water is young (H3 around -10 UT), while a graph of H2 as a function of O18 gave a surface precipitation altitude corresponding to the mean altitude of each lake's catchment area.

The relatively high rates of heavy metals, mainly Pb, V, Zn, Cu, Ni, Co and Cr, originated from outside the basin. The isotopic ratios of Pb (204Pb, 206Pb, 207Pb and 208Pb) allowed three sources of deposits to be identified: the first is lithogenic at deep lake levels and corresponds to the composition of the continental crust; human for upper levels at depths of 0-17 cm, mainly due to industrial discharges (leaded petrol, the coal industry, steelworks, etc.); while an intermediate origin mingles both these sources. Metals are first recorded from the 1880s, the era of the Industrial Revolution. These trace metal deposits peaked in the 1960s and 1970s and have

mostly begun declining. Enrichment factors were calculated by using three lithogenic elements (Th, Zr, La) and by taking the core sample from Lake Saint-Point as a base sample. This confirmed the human origin and rate of succeeding deposits since the middle of the Industrial Revolution. There are frequent exchanges between metal and water, and recent increases in the content of certain elements have caused concern.
The aim of this study was to understand the exchanges taking place at the deepest levels of the water column following changes in the physical-chemical conditions of the lakes' environment.
Keywords: Lake, heavy metal, Jura, lead isotope, water, sediment

1 Introduction

Like other European lakes, French medium- and high-altitude mountain lakes are suffering from the fact that environmental policies are mainly turned towards large lakes in low-lying areas, among them Geneva (Léman), Annecy, Aiguebelette and Le Bourget, where the political and economic issues at stake are much more pressing. The latter have tended to overshadow small-scale mountain lake systems, perceived as being remote from the concerns of decision-makers and beyond the reach of anthropogenic activities likely to damage their hydrological quality and quantity. As mountain lake districts have steadily opened up to recreational and leisure pursuits, especially fishing, managers have become increasingly aware of the prospect of a hitherto-unsuspected economic resource. Along with these new orientations, the growth of mountain towns and the need for further zones for building purposes have touched off public debate on the future of so-called "fragile" wilderness areas, among which lakes and ponds are prominent. Increasing pressure on such systems has forced managers to adopt quality approaches in order to control their quality and evolution, and to draw up good-behavior charters to preserve the balance between human actions and the environment. Nevertheless, and despite the sometimes drastic regulations enforced in lake areas, there has been growing evidence of deteriorating water quality, while studies carried out on several lakes in the Alps and Jura have confirmed research results from other European lakes [Shotyk et al. 1996, 1998, 2001, 2003, 2005; Arnaud 2003; Arnaud et al. 2004; Boyle et al. 1999, 2001; Ethan et al. 2001; Norton et al., Nedjai et al. 2005; Givelet et al. 2003; Davis et al. 1983; Perry et al 2005]. Such disturbances result from the combination of several factors classifiable under two headings:

1. Natural ones linked to the normal evolution of lake areas after their initial formation, which for French lakes was generally during the Quaternary glaciation era. The steady passage of a lake from a young state to a peaty one is accompanied by the development of a belt of vegetation around it and thus by the release of mainly organic matter. The peaty stage indicates a lake's ultimate degree of eutrophication.
2. Anthropogenic factors, some of them direct, e.g. the pumping of water and fishing, and others indirect: with mountain lakes mainly farming and to a lesser extent industry. These two types of interference often go hand in hand, and it is by no means rare to be confronted with conflicting situations of tangled interests between farmers and industrialists. Confronted with this type of situation and given the state of medium-altitude lakes, deciders have taken watch-dog measures to enhance protection by regulating activity on the scale of catchment areas and by reducing sources of disturbance blamed for the problems, such as waste discharge, agriculture and industry.

Development work carried out on the Jura lakes' catchment areas has been in line with water management policies drawn up in the early 1990s, and more recently with the European Framework Directive, and has aimed at restoring satisfactory water quality to the area's lakes and rivers. Like other French lakes, those in the Jura region have reaped the benefits of these water policies, to the extent that two of them, Grand-Maclu and Etival, are currently used as models for managers in the Rhône River drainage basin. While the measures have undoubtedly reduced local pressure on the lakes, they have done nothing to eliminate noxious inputs, especially of heavy metals. Recent studies of Jura lake sediments have revealed abnormally high levels of methyl mercury, lead, zinc and other metals. This casts doubt on the measures' efficacy and raises a number of questions about the sources of the inputs and the geochemical processes involved (dumping, adsorption, etc.) The present study, begun in 1998, has undertaken to examine the state of the lakes in order to discover the sources of pollution and the processes at work. It has attempted to be both wide in its conception and precise in its approach by investigating the entire set of factors involved the lakes' functioning (hydrology, morphology, geology, geochemistry and so on).

2 The location, morphology and geology of the lakes and their drainage basins

In all, nine lakes were studied, and their water quality and general physical and chemical states analyzed. The French Jura region is in eastern France, bordering Switzerland. The Jura lakes lie in an area covering around a hundred square kilometers, bounded to the east by the French-Swiss border and to the west by the town of Lons-le-Saunier (Fig. 1). The area's southern limit is the town of Saint-Claude and the northern one the town of Pontarlier. The latter's water supply comes mainly from Lake Saint-Point, classified as a heritage reserve.

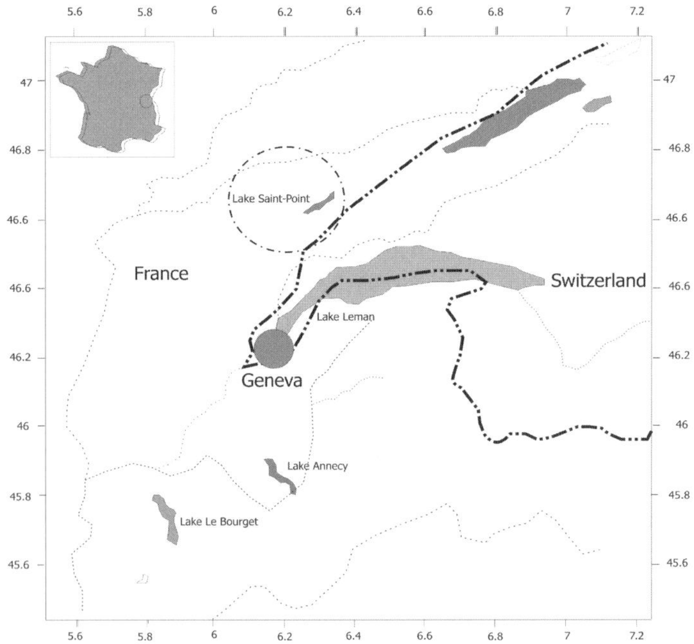

Fig. 1. The study area

Altitudes are from around 400 meters asl at Lons-le-Saunier to 1203 meters at the summit of Crêt de la Neige, located in the eastern part of Jura Department. The mean altitude is around 750 meters, overlooking the lake district. The 9 lakes are unevenly distributed around the area; six are in the region between the towns of Champagnole to the north, Saint-Laurent to the south-east and Clairvaux-les-Lacs to the west. Lakes Le Val and Chalain lie towards the north, Etival and Clairvaux towards the west, and Lakes Abbaye and Saint-Point towards the east. The lakes' spatial distribution is the determining factor when evaluating and measuring their similarities and differences. Several studies have in fact highlighted the geographical aspect as a factor in explaining and analyzing the physical processes involved in the way the lakes function [Touchart 2001; Dearing et al. 2000].

All the lakes are elongated in shape and lie in two main directions, 10°N and 90°N. These were the directions often taken by recent glaciers, dating from the Quaternary (Würm), which originally formed the various areas of standing water. Several of the lakes lie in steep-sided basins surrounded by a network of fault lines several kilometers long belonging to the Syam fault network. The bottoms of most of the basins are covered in a thick morainal layer which is clearly visible on the lake shores. It is made up of calcareous debris enveloped in clay, while the lakes' outlets are obstructed by morainal deposits; the latter act as dams and prevent the water from flowing out. The fault lines of the Syam network run through the middle of the Jura massif and divide it into three tablelands. These are, from north to south: the Lower Jura plateau, where most of the water courses converge towards the Ain River; the intermediate plateau, where two thirds of the Jura lakes lie; and the high plateaus with Lake Abbaye. The predominant geological formations are mostly carbonated limestone, with domolitic limestone and marls here and there. Tectonic contact zones are made up of relatively thin conglomerate and breach layers. With air masses moving in from the ocean, this north-south arrangement causes climatic disparities between higher-altitude areas, where there is considerably more precipitation (2,000 mm on average) and areas lower down, where atmospheric inputs have much less effect. The climate is oceanic, with spring and autumn rainstorms and snow in winter.

Temperatures recorded at Le Frasnois village show a mean of around 7.63°C, with a maximum of 35°C recorded in 2003 (a dry year) and -15°C in 2002 (a rainy year). Such temperatures directly affect the lakes, resulting in a double stratification, direct in spring, summer and autumn and inversed in winter. In Dussart's classification [1966], the lakes come into the dimictic category. Lake Narlay is the only one of the nine whose slow stratification and mixing rates would put it into the meromictic category. All the lakes go through two homothermic phases, one at the beginning of spring and the other in autumn, connected with stratification and destratification respectively.

3 Instruments and measurements: building a data base for geographical limnology

The geographical limnology studies of the Franche-Comté lakes were undertaken on the basis of an in-depth bibliographical analysis of previous research data from the DIREN, University of Besançon, fishing associations and so on. The first analyses were used to draw up an instrument protocol for hydroclimatic monitoring with a view to assessing data and understanding the influence of climate factors on the physical and chemical processes at work in the water and sediments. They were also used to draw up a protocol for sampling water and sediments. The second stage focused on laboratory work and on techniques of analysis.

So far, all the work undertaken has concerned only the top 5 centimeters of the sedimentary layers, thus minimizing temporal aspects in favor of spatial ones. This precludes the possibility of going back in time and tracing the history of the events which underlie the paleo-environmental evolution of entire lake systems.

Climate measurements were made at Le Frasnois village by a Campbell CR10X station fitted with three sensors for temperature, humidity and precipitation respectively. The rain gauge has a reception area of 40 cm. The station's data were supplemented by data from other stations of the French Meteorological Agency *(Météo France)* at Lons-le-Saunier. In spite of its distance, the latter was taken as the benchmark station because of its data length and also because of its measurements of wind velocity and direction. The first stage enabled the relationship between the lake and its drainage basin to be established, and both drainage basin and lake dynamics to be accurately determined.

The second stage of our protocol examined the actual bodies of water and their physical and chemical characteristics, e.g., water temperature, pH, redox, conductivity, oxygen level. An Xtroll Professional multi-parameter probe with five high-precision (5% error margin) sensors was used to make selective measurements, mainly in the centers of the lakes. It was also used to spatialize the set of parameters so that possible spatial disparities, and thus limnological regions, could be identified. Monitoring got under way in 2004 and enabled a physical and chemical data base to be drawn up for the whole of the study area. This was used to compare the major physical and chemical parameters of Lake Saint-Point with other Jura lakes in order to identify any points of similarity or difference which could be ascribed to human activity. Acidity levels have risen in several of the lakes; this could be linked to acid rain or to an increase in organic matter inputs and the consequent formation of H_2S. This hypothesis led to the third stage of our research study focusing on solid matter, i.e. sediments. There are numerous exchanges with lakebed solids; such movements are often the result of changes in the physical and chemical characteristics of open water.

The chemical and mineralogical state of sediments were analyzed in two ways. A series of sedimentary layers about 30 cm thick was analyzed first. Core samples were taken with a piston sampler two meters in length and fitted with airtight stoppers. The sampling margin of error does not exceed 2 or 3 centimeters. The cores were then sampled over their entire length at 1 cm intervals. Samples were stocked at low temperature before being freeze-dried at low pressure and sifted so as keep particles of under 63 μ only. Each sample was then analyzed for its heavy metal content as well as for a number of lithogenic elements in order to assess the weathering processes of the drainage basin and determine enrichment factors and major pollutants such as Pb, Zn, Sb, As and so on.

Pore water was analyzed simultaneously with the study's second stage. Pore water is where metals are mainly stored, and are the main source of exchanges with both sediments and open water. The protocol used was the Piper technique which allows around 50 ml of water to be transferred through a porous membrane. The whole device was first prepared in the laboratory by placing Piper plates in a low potential redox solution or one identical to actual conditions prevailing in the lake. This enabled any oxidization of the water to be prevented before sampling. As with the sediments, analyses focused mainly on heavy metals to determine type and proportion. Heavy metal analysis protocols are shown in Figure 2 (B).

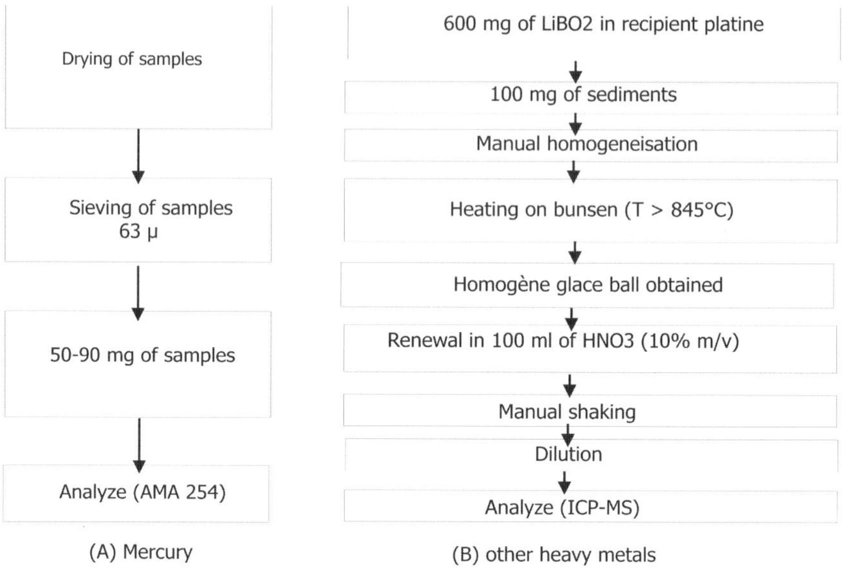

Fig. 2. Analysis protocol of heavy metals in sediments

Table 1. Characteristics of the French Jura Lakes.

Lac	Alt	S(ha)	L mx	lmx	Zmx	Z my	Zmy/ Zmx	Ic	V(Mm3)
Clairvaux	525	56.5	1250	650	20	8.84	0.44	2.66	4.99
P. Maclu	779	4.5	500	120	11	7.6	0.69	5.19	0.34
G. Maclu	779	21	1120	300	24	12.7	0.53	5.24	2.65
Ilay	774	72	1900	400	32	11	0.34	3.77	7.7
Etival	793	15	1000	150	9.5	8	0.84	2.45	-
Val	520	51	1375	425	25	15.4	0.61	3.5	7.85
Narlay	748	41	950	625	40	20	0.5	6.25	8.2
Chalain	486	230	2700	1100	32	16.2	0.6	2.11	44
Abbaye	879	91	2125	280	19.5	11	0.51	2.04	5.8
Bonlieu	791	17.4	725	300	15	8	0.53	3.12	1.5
St. Point	850	419	7200	1000	43	20	0.49	-	-

The analyses of heavy metals were partly carried out by ICP-MS in the CNRS-SARM Laboratory in Nancy and at the LGIT (Laboratoire de Géophysique Interne et de Tectonophysique) in Grenoble, France. Mercury only was analyzed in the latter laboratory with an AMA 254, following the protocol in Fig. (A).

A final series of analyses focused on lead isotopes with the aim of distinguishing between sources of contamination and identifying the origins of the lead and other elements associated with it. Isotopic dosing was carried out by ICP-SMS to determine rates of ^{206}Pb, ^{207}Pb and ^{208}Pb after digestion by HNO3. The ICP-SMS method is more accurate than TIMS; it varies between 0.03 and 0.05% [Krachler et al. 2004]. Throughout this stage the top 20 centimeters were dated by the Appelby et al. [1996] method using lead 210. The characteristics of the lakes and their catchment areas are shown in Table 1.

3 The lakes' hydrological and refilling characteristic

The only way to assess sedimentary and chemical inputs into a lake is to determine the characteristics of flows and exchanges between the lake and its catchment area on the one hand and the lake and potential aquiferous formations on the other. It is also the way to evaluate how mineral content is distributed among the different lake sectors. Broadly speaking, part of the organic and mineral matter flowing into these lakes is deposited on the lakebed among the sediment and adsorbed by it, while the other part reaches the outlet and flows downstream towards the Ain River.

The Jura lakes, half of which lie at the heads of drainage basins and the other half at the bottom, get their water either directly from rain or snow, with mean annual total precipitation amounting to around 1,800 – 2,000 mm, or via diffuse forms of trickling, streaming and hypodermic flows from within their respective basins, half of which are small-scale.

The basins have elongated shapes ($1.2 < Kc < 1.56$) and cover relatively small areas, except for the Lower Jura lakes, which are somewhat larger. Concentration time is between three days and a week for most of the lakes in periods of moderate rainfall (8-10 mm/h). State-owned forests - mainly deciduous and conifers - cover 80% of the catchment areas, with scattered grassy meadows in the areas of the larger lakes. Seventy percent of the lakes are surrounded by a belt of hygrophilous vegetation, some of them also having the beginnings of tree growth, including pines and deciduous species. On the whole, flow rates are between 100 l/s for the lakes on the intermediate and high plateaus (Ilay, Bonlieu, Etival, Clairvaux, etc.), reaching 500 l/s during periods of thaw and heavy rainfall, and 1 m^3/s for the Lower Jura lakes (Chalain and Val). Lake Saint-Point, further east, drains a sizeable basin, and flow rates can be in excess of 2 m^3/s. The extent of the catchment area and the abundance of the water supply influence and regulate the flow of solid matter and contribute significantly to cleaning and purifying lake waters and sediments. Research has shown a drop in the mercury content of certain lakes which drain extended surface areas.

Geographically, there are clearly observable east-west and uphill-downhill distinctions. The lakes at higher altitudes are well exposed to air masses moving in from the ocean and receive relatively abundant rainfall.

Lower down they are surrounded by limestone cliffs which collect rain water, and get much of their water input from intermediate plateaus.

The region's prevailing winds are west-east, blowing locally from 10°N to 30°N at velocities of around 5 m/s. The mostly carbonated karsts favor infiltration and underground water courses. Speleological investigations carried out in the area by teams from Saint-Claude and Champagnole have confirmed the presence of a number of natural drains several kilometers in length running beneath the Jura plateaus, some of them used by water from Lakes Vernois and Narlay. Water flow combined with the hydrogeological complexity of the region has resulted in two types of lake systems.

4 Lake water and sediments: the current and past states of lakes in the Franche-Comté area

Two separate lakewater sampling expeditions were undertaken and chemical analysis of major and trace elements carried out. Generally speaking, the chemical content of the water is characterized by calcium bicarbonate, except for Lake Abbaye, which has slightly higher sodium. Concentrations are around 50 mg/l and reflect the preponderance of contact and circulation of water in limestone and dolomitic limestone formations of the Secondary Era, mainly the Jurassic and Cretaceous periods. Differences in concentration between the lakes reflect differences in water circulation patterns and in the degree of fracturing and karsification of the limestone landforms. They also show how and where the water flows. Lakes Narlay, Le Vernois and Chalain have the highest calcium content. Lake Clairvaux, with a highly developed underground water system, can also be added to this group. It gets its water from the limestone landforms of Meussia plateau, which is drained by large underground channels, giving rise to substantial flow rates (m^3/s). The pH of surface water is around 8 in summer, when most of the carbonates precipitate and are added to the sediments on the lakebeds. There is a close connection between free-flowing water and sediment, so that exchanges between them are plentiful, often regulated by complex physical-chemical processes, e.g., reduction, adsorption, absorption and so on.

The oxygen quality of some of the nine Jura lakes has steadily worsened, and the lakes are showing signs that water quality has deteriorated. Oxygen levels have fallen appreciably, while the deepest layers (the hypolimnion) have become more acidic (pH ~5). Metals in the sediments are thus more easily released into the free water and some of them transformed; the formation of methyl mercury is the main example. One of the major causes of the changes is the abundance of organic matter in most of the lakes. It comes from two main sources, mostly the surrounding forests (C/N ratio > 20), but also from within the lakes themselves and in the immediate vicinity of the shoreline. The most affected lakes are surrounded by wide peaty areas which produce organic matter, act as buffer zones and retain water flowing from uphill. As it penetrates slowly through the layers, the water picks up metals and other chemicals which end up in the lakes and their sediments. C/N ratios were over 40 in Lake Clairvaux, which has relatively moderate plant growth around its shore, and under 20 in Lake Bonlieu, considered as eutrophic and partly filled with peat. The contamination of Jura lakes by noxious elements, essentially heavy metals, has been caused essentially by industries outside the region. This calls into question the work undertaken by the lakes' managers over the last few decades in an effort to reduce the pressure of local economic activities on the lakes' catchment areas. It has been hypothesized that the contamination may be regional, and the studies carried out on the lakes tend to confirm this. Analyzing the geochemistry of the sedimentary metals was another means of identifying the sources of pollution. Several core samples were taken from the lakes' sediments in order to assess both spatial disparities and the evolution of the quality of the lakes over a period of centuries. Lake Saint-Point was taken as the benchmark; it was sampled to a depth of 70 cm, representing 200 years. Core samples were also taken from the other lakes to depths of between 10 and 30 cm.

5 Results and discussion

The study mainly focused on two metals: lead and mercury. The content of both was high, and in several of the lakes was well over the limit for drinking water. There are several possible hypotheses which could explain the origin of the two metals, including from within the catchment areas themselves, thus implicating the few local industries that exist. Analysis and dating of the samples [Appelby et al. 1978, 1986, 1992] of lake sediments gave the following results:

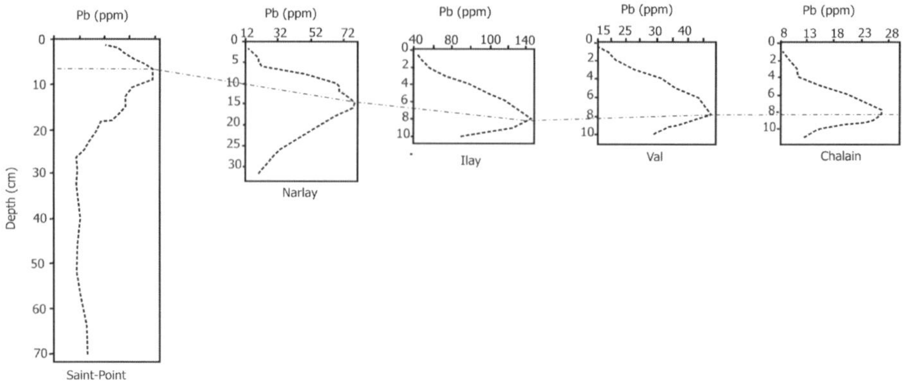

Fig. 3. Lead profiles in five typical Jura lakes

All the lakes showed an increase in lead content from the late 19th century onwards, a period which coincided with the Northern Hemisphere's Industrial Revolution after the mid 19th century (Fig. 3). Content increased steadily in the early 20th century after lead started being added to petrol for motor vehicles, first in the United States in 1923 and then in Europe in 1930. The release of lead into the atmosphere peaked in 1973. This was when lead additives in petrol began being phased out.

Locally, disparities between lakes are most noticeable in those where there is a high organic matter content. The latter differ from the other lakes by having significant plant growth around their shores. Thus, Lakes Ilay, Narlay and Le Val are all surrounded by marshy areas covering over 10% of the lake area.

The evaluation of enrichment factors for lead and mercury as measured against three lithogenic elements (titanium, zirconium and thorium) confirmed the intensity of the inputs of both from the late 19th century onwards and right throughout the 20th.

Isotopic analyses carried out on the five lakes showed $^{206}Pb/^{207}Pb$ ratios varying from 1.22 at the base of the core sample to 1.14 at the top. The latter value is considered indicative of lead from hydrocarbons.

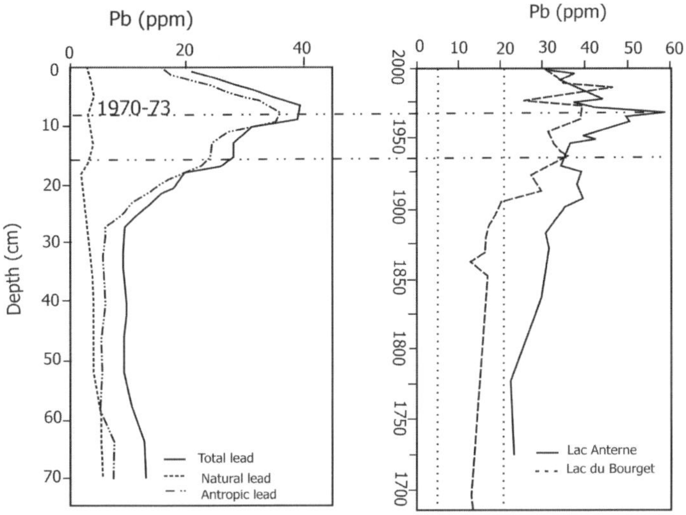

Fig. 4. Comparative analysis of lead profiles with lakes in Savoie and Haute-Savoie departments

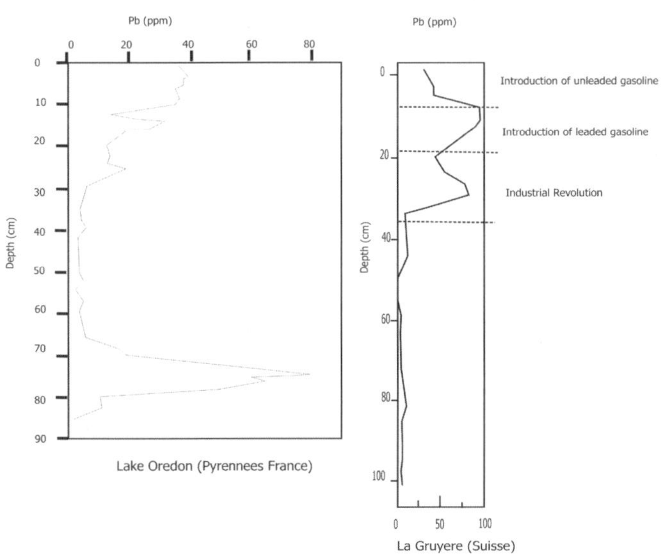

Fig. 5. Comparative lead profiles of Jura lakes, Lake Orédon (the Pyrennees) and pond in Gruyère canton (Switzerland).

A comparative analysis of the lead profiles of Jura lakes with lakes in Lake Anterne in Haute-Savoie and Le Bourget in Savoie (France) [Arnaud

2003; Arnaud et al. 2004] (Fig. 4), Gruyère canton in Switzerland [Shotyk et al. 1998, 2001, 2003, 2005], Lake Orédon [Ariès 2001] in the French Pyrenees (Fig. 5), confirmed the element's regional origin and the amplitude of industrial emissions in the Northern Hemisphere.

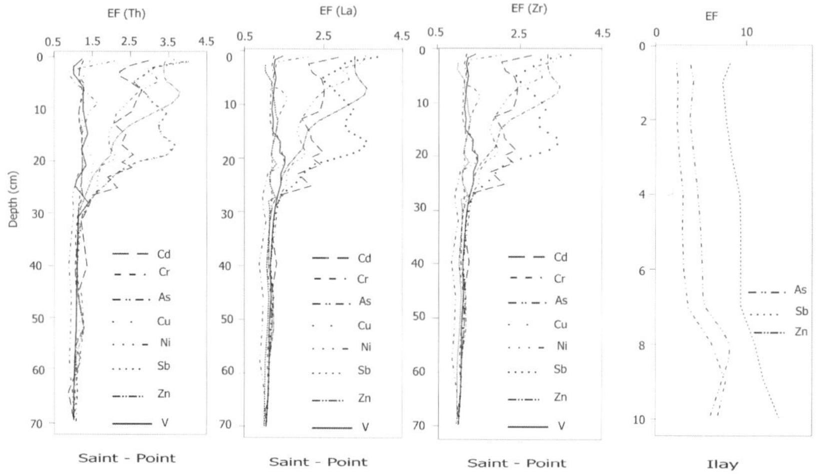

Fig. 6. Factor enrichment of heavy metals estimated with Th, Zr, La and comparison with Ilay lake

Similar tendencies were observed for all the other metals found (Fig. 6), while patterns for copper, zinc, arsenic and antimony were almost identical. The latter are considered as able to be taken up by aerosols and thus potentially volatile. Zinc, which fixes itself to dust, follows the same trajectories as lead and mercury and can travel long distances to contaminate remote areas. The benchmark Lake Saint-Point confirms these hypotheses, and displays increases in a number of heavy metals along with that of lead, which had been observed previously. The evolution of zinc content has been similar, but with occasional peaks which may correspond to internal inputs, probably due to the spreading of an animal wastes on fields, a common local farming practice.

6 Conclusion

Recent evidence of pollution by heavy metals, particularly volatile ones, has raised questions about certain methods of protection. The narrow focus of some scientific studies of the region's lake systems did not take into account wider areas of interaction, namely catchment areas. Lakes, like other environmental systems, are collectors of chemical and other inputs from both within their catchment areas and from more distant areas. Several signs of disturbance, initially physical and chemical with falls in deep-water oxygen levels, have been accompanied by a rise in acidity and the development of conditions favorable to the transformation of certain metals into noxious compounds such as methyl mercury. The presence of significant amounts of lead and other metals in deep water and at the water-sediment interface has resulted from a slow but steady transformation of physical and chemical conditions under the influence of organic matter, the main factor in the mobilization of such metals. The lakes which have been affected the most are those with a zone of peat and a highly developed lakeside vegetation belt. There is also a second source of the same importance, i.e. the relatively dense vegetation covering the lakes' catchment areas. Not all the effects of reafforestation efforts in mountain areas at the beginning of the last century have been positive, and have actually created long-term threats to the lakes because of the large amounts of organic matter they produce. All this shows how important it is for lake managers to take into account spatial aspects and scales of action from smaller – lake surface – to larger – catchment area and regional.

References

Aries S (2001) Mise en évidence de contaminations métalliques historiques à partir de l'étude d'enregistrements sédimentaires de lacs de haute montagne. Thèse de Doctorat de l'Université de Toulouse III, p 243

Arnaud F (2003) PhD thesis. Université de Lille, France, p 195

Arnaud F, Revel-Rolland M, Bosch D, Winiarski T, Desmet M, Tribovillard N, Givelet N (2004) J. Environ. Monit., vol 6, pp 448-456

Appleby PG, Oldfield F (1978) The calculation of ^{210}Pb dates assuming a constant rate of supply of unsupported ^{210}Pb to the sediment. Catena. vol 5, pp 1-8

Appleby PG, Nolan PJ, Gifford DW, Godfrey MJ, Oldfield F, Anderson NJ, Battarbee RW (1986) ^{210}Pb dating by low background gamma counting. Hydrobiologia, vol 141, pp 21-27

Appleby PG, Richardson N, Nolan PJ (1992) Self-absorption corrections for well-type germanium detectors. Nucl. Inst.& Methods B, vol 71, pp 228-233

Campy M (1982) Le quaternaire franc-comtois: essai chronologique et paléoclimatique. Thèse de Docteur Es Sciences naturelles. Faculté des Sciences et Techniques de l'Université de Franche-Comté, p 573

Davis R B, Norton SA, Hess CT, Brakke DF (1983) Hydrobiologia. vol 103, pp 113-123

Der Knaap, O, Norton SA, Lohse C (2005) Accumulation rates and prédominant atmospheric sources of natural and anthropogenic Hg and Pb on the Faroe Islands. Geochimica et Cosmochimica Acta. vol 69, edn 1, pp 1-17 Givelet N, Roos-Barraclough F, Shotyk W (2003) J. Environ. Monit. vol 5, pp 935-949

Boyle JF, Mackay AW, Rose NL, Flower RJ, Appleby PG (1998) Sediment heavy metal record in Lake Bailkal: natuel and anthropogenic sources. Journal of Paleolimnolgy, vol 20, pp 135-150

Krachler M, Le Roux G, Kober B, Shotyk W (2004) J. Anal. At. Spectrom. vol 19, pp 354-361

Norton SA (1986) Water, Air and soil pollution, vol 30, pp 331-345

Perry E, Norton SA, Kamman NC, Lorey PM, Driscoll CT (2005) Ecotoxicology. vol 14, pp 85-99

Shotyk W, Cheburkin AK, Appelby PG, Franckhauser A, Kramers JD (1996) Two thousand years of atmospheric arsenic, antimony, and lead deposition recorded in an ombrothrophic peat bog profile, Jura Mountains, Switzerland. Earth and Planetary Science Letters, vol 145, E1-E7

Shotyk W, Weiss D, Heisterkamp M, Cheburkin AK, Gloor M, Reese S (2001) Geochemistry of the peat bog at Etang de la Gruère, Jura Mountains, Switzerland, and its record of atmospheric Pb and lithogenic trace metals (Sc, Ti, Y, Zr and REE) since 12,370 14C BP. Geochimica et Cosmochimica Acta, vol 65, pp 2337-2360

Shotyk W, Goodsite ME, Roos-Barraclough F, Frei R, Heinmeier J, Asmund G, Lohse C, Hansen TS (2003) Anthropogenic contributions to atmospheric Hg, Pb and As accumulation recorded by peat cores from southern Greenland and Danemark dated using the ^{14}C “bomb pulse curve”. Geochimica Cosmochimica Acta, vol 67, edn 21, pp 3991-4011

Shotyk W, Goodsite ME, Roos-Barraclough F, Givelet N, Le Roux G, Weiss D, Cheburkin AK, Knudsen K, Heinmeier J, Van Wedepohl KH (1995) The composition of the continental crust. Geochim. Cosmochim. Acta 59, pp 1217-1232 (à mettre dans le calcul du plomb lithogénic)

Touchart L De la température de l'eau à la géographie des lacs. - Thèse d'HDR, Univ. de Limoges, p 480

Water Supply and Water Demand of Beijing – A Game Theoretic Approach for Modeling

Shouke Wei, Albrecht Gnauck

Department of Ecosystems and Environmental Informatics at Brandenburg University of Technology Cottbus, 03013 Cottbus, Germany, e-mail: sigmundwei@yahoo.com, albrecht.gnauck@tu-cottbus.de

Abstract: Water allocation is an important and complicated issue in natural resource management. It usually involves water demand and water supply conflicts. However, most of the existing water allocation approaches usually do not function well in solving those conflicts. On the one hand, market-based water allocation is efficient for different users. On the other hand, however, water market is difficult to establish in most countries. One common feature of water allocation is that it is involved in such conflicts with multi-users of contradictory interests, goals and strategies. In this paper, a game-theory based concept for modeling of water allocation problems is presented. The demand-supply principle (DSP) is used as a platform to formulate non-cooperative games for decision support of water management plans. The results of such games allow not only better comparisons of the different groups of water users (including environment), but also give benefits to the administration and water supply companies. Examples are given from Beijing municipality.
Keywords: water allocation, water demand, water supply, game theory, decision support

1 Introduction

Water is essential for agriculture, industry, and even human existence. However, due to temporally and spatially uneven distributed precipitation and exponentially increasing water consumption [Wetzel 1983; Al Radif 1999; Wang et al. 2003], with increased population growth rates, degradation of water quality [UN-CSD 1994] and the loss of potential sources of freshwater resulting from unsustainable water management practices [Wang 2005], water supply has been increasing scarcity in many developing countries. It is estimated that by 2025, 4 billion people will be living in areas with insufficient water supply [AusAID 2003]. Conflicts are usually increasing due to the growing competitions among multi-actors over the scarce water, such as the disputes between Arabs and Israelis, Indians and Bangladeshes, Americans and Mexicans, and among all 10 Nile basin co-riparians [Wolf 1999]. The multi-users usually have contradictory or conflicting interests [Fang et al. 2002; Hipel et al. 1997; van der Veeren 2003], goals and strategies. One approach frequently cited in the literature is the use of water markets [Burness and Quirk 1979; Colby 1990; Green and O'Connor 2001]. On the one hand water market methods can provide water users incentives to use water efficiently. However, it requires defining the original water rights, creating institutional and legal mechanisms, and establishing basic infrastructures for water trade [Holden and Thobani 1996; Yang and Zeng 2004] before water market can operate well. Water market is only a good theory, while it is hard to establish a real water market in most countries. However, game theoretic approaches are appropriate to model and simulate such conflicting situations. Game theory is usually divided broadly into two big approaches, the non-cooperative and the cooperative approach. Non-cooperative game theory is strategy oriented and studies what players expect to do and how they do it; while cooperative game theory studies what the players can achieve and how they can achieve it.

Many studies have been made about the application of game theory on water allocation [Tisddell and Harrison 1992; Yang and Zeng 2004; Wang et al. 2005]. Wang [2005] has developed the Cooperative Water Allocation Model (CWAM) for modeling equitable and efficient water allocation among competing users at the basin scale, but water rights in this model are not so easy to define clearly in reality. In this paper, a concept for modeling of water allocation problems as a non-cooperative game is presented.

2 Methodology and Data

The existing and potential conflicts on water supply and demand involved in Bejing are modeled as a non-cooperative game. This model aims at analyzing and forecasting water demands of agriculture, industry, domestics and ecology in Bejing. The game modeling process consists of defining the conflicts, formulating these conflicts as a game, solving the game and interpreting the results. For forecasting water demand of the Municipality of Beijing during the next 10 years, statistical and econometric methods are used. In detail, regression models (linear regression, semilog regression, double-log regression, polynomial regression), Autoregressive (AR1) and/or moving average (MA1) are used to establish models of added values of industry and agriculture. AR and/or MA terms included in the equation of the models are one of the common methods of accounting for serial correlation. Water demands in different sectors of the years up to 2015 are predicted through those models. To estimate the parameters of these functions ordinary least squares (OLS) are used.

Mainly, data collected are taken from yearly books in related fields at different administration levels, official reports and planning documents, previous studies as well as other local online information systems and data resources. The main type of data includes socio-economic data (population, industrial added value, agriculture added value, and per capita net income), water quantity data (water supply and water consumption data in different sectors, such as agriculture, industry and domestic life) and hydrological data (surface water, ground water, precipitation and evaporation).

3 Conflicts of Water Demand in Beijing

Situated in the north-eastern part of China, Beijing Municipality covers an area of 16,808 km^2, stretching 160 km from east to west and over 180 km from north to south. There is a population of 15.38 million people [BSBC 2006]. It is facing serious problems of water scarcity due to uneven distribution of precipitation in the southern and northern China. Besides of the lack of precipitation, water scarcity in Beijing municipality is caused by water pollution. Figure 1 shows the total amount of water resource (water supply) and total water consumption during 1986 to 2005. Water consumptions per capita is 225 m3/person in 2005 [CSB 2006], and water consumption per unit value of GDP is 50.10 m3 per 10,000 yuan in 2005 [BSBC 2006]. According to the internationally recognized standards of extreme water deficit ($\leq$ 500 m3/person), Beijing belongs to these areas. Water consumption per unit value of GDP is one of the indicators which reflect the efficiency of water use in Beijing.

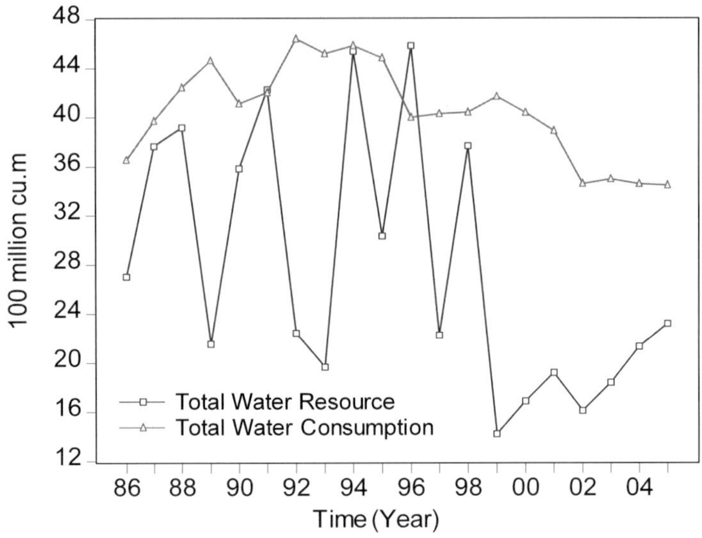

Fig. 1. Total amount of water resource and water consumption (1986-2005)

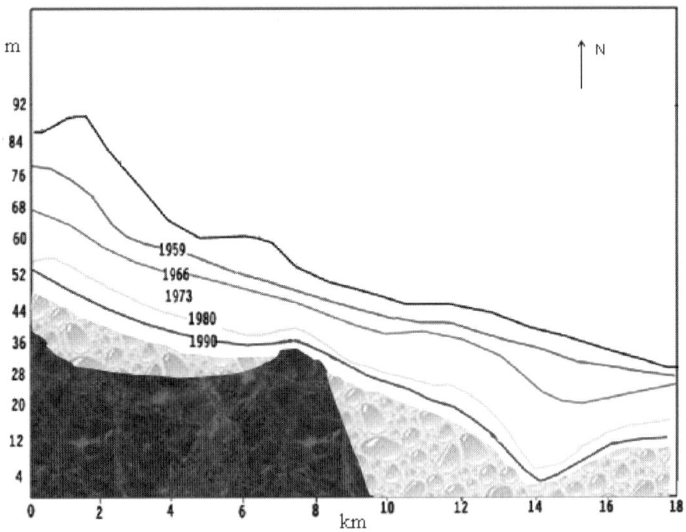

Fig. 2. Ground water table of Beijing

To ensure water supply for cities, the municipality has to overuse ground water and overtake water from agriculture and urban ecology. It is estimated that the total overexploited groundwater is about 4 billion m^3 during 1961-1995. Figure 2 shows the changes of groundwater table of

Beijing. Guanting Reservoir and Miyun Reservoir supplied 920 million m³ of water for agriculture in 1980, while it has dropped to less than 200 million m³ during the recent years. The contradiction between water supply and demand sharpens water conflicts between different water users. Fighting for water among downstream and upstream users, between the right bank and left bank of the rivers, are not rare [IWRPYZ 2005] (Fig. 3).

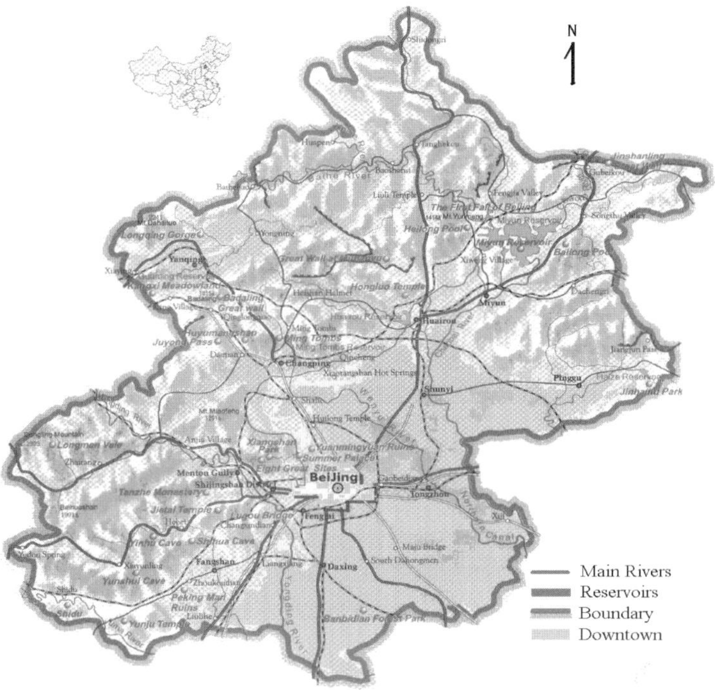

Fig. 3. Water resource of Beijing

4 Formulating the game

4.1 Assumptions

The game is finite, dynamic and with completed information. All the players are rational, and their aims are to maximize their welfares through using water efficiently. There is no intervention of administration during game processing; but the game processing is influenced by the current policy. Technology on water saving will be developed in future. All data are authentic.

4.2 A Non-cooperative Game Model for Water Demand

Action (Moves or Strategies), P - Payoff (or Utility), I - Information, O - Outcome and E - Equilibrium (NAPI-OE). NAPI are collectively known as the rules of a game. OE are the game results. The formulating or defining of a game can be referred to [Fundenberg and Tirole 1996; Gibbons 1992].A non-cooperative game for water demand of Beijing is defined as follows:

$$G = \langle N, S_i, U_i \rangle \tag{1}$$

where
- N is the set of players,
- Si is the strategy of every player i,
- Ui is the payoff function of every player i.

In this game, agriculture (1), industry (2), domestic life (3) and ecology (4) are defined as the players:

$$N = \{1,2,3,4\} \tag{2}$$

The strategies of every player i are the measures or plans to obtain enough water for his or her use in different periods of time t (year) based on their abilities (innate potentials for growth, development, or accomplishment), W_i^t i.e.

$$s_i = W_i^t \in S_i = [0, \infty), \quad i = 1, 2, 3, 4 \tag{3}$$

Every player wants to get sufficient water. In this game, therefore, the payoff function is the function of water demand which is given by:

$$w_i^t = f(T, P, L, I, V \ldots.) + u \tag{4}$$

Where
- w is water demand of every player (dependent variable);
- T, P, L, I, V are independent variables,
- time (T), water price (P), labor (L), income (I), value of productions (V),

– u is the usual residual that contains the non-explainable variation.

The Nash equilibrium (NE) is $(w_1^{t*}, w_i^{t*}, ..., w_n^{t*})$, which means player i will maximize w_i^t (water demand) under the condition of efficient water uses where the others players choose $(w_1^{t*}, w_{i-1}^{t\ *}, w_{i+1}^{t\ *}, ..., w_n^{t*})$. NE is a best strategy combination for all n players. In the equilibrium, each of the players can maximize his or her payoff, thus no one has incentive to choose other strategy and leave automatically this equilibrium.

4.2.1 Payoffs Forecasting

In this game model, the payoff functions of different players are defined by water demand models. The model of water demand per unit of agriculture added value (W/AAV) (Fig. 4a) is given by:

$$LOG(W/AAV) = 177.3120501 - 0.08483514172*T + [AR(1)] \tag{5}$$

where AR(1) = 0.4868903787 with $R^2 = 0.969$ and adjusted $R^2 = 0.965$.

The model of agricultural added value (AVV) (Fig. 4b) is given by:

$$LOG(AAV) = 0.03931995424*T - 74.26815683 + [AR(1)] \tag{6}$$

where AR(1) = 0.6720168331 with $R^2 = 0.968$ and adjusted $R^2 = 0.964$.

The model of industry water demand per unit of industry added value (W/IAV) (Fig. 5 a) is given by:

$$LOG(W/IAV) = -0.1793124776*T + 363.3022987 + [AR(1), MA(1), BACKCAST] \tag{7}$$

where AR(1) = 0.6720168331, MA(1) = 0.4105110315, BACKCAST = 1987 with R^2=0.993 and adjusted $R^2 = 0.992$.

The model of industry added value (IAV) (**Fig**. 5 b) is given by:

$$LOG(IAV) = -1900.352764 + 250.9060073*LOG(T) + [AR(1), MA(1), BACKCAST] \quad (8)$$

where AR(1) = 0.5405098917, MA(1) = 0.5313474127, BACKCAST = 1979 with R^2=0.997 and adjusted R^2 = 0.997.

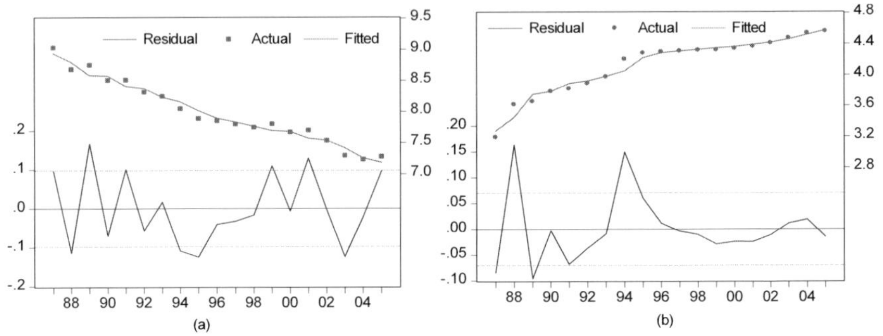

Fig. 4. (a) Water demand per unit of agriculture added value (1,000m^3/10,000) (1985-2005), (b) Agriculture added value (1000 million yuan at current prices) (1986-2005)

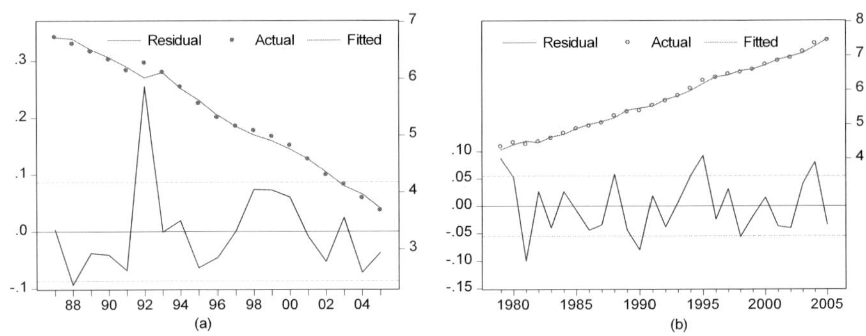

Fig. 5. (a) Water demand per industry added value (100m^3/10,000 yuan) (1986-2005); (b) Industry added value (1million yuan) (1978-2005)

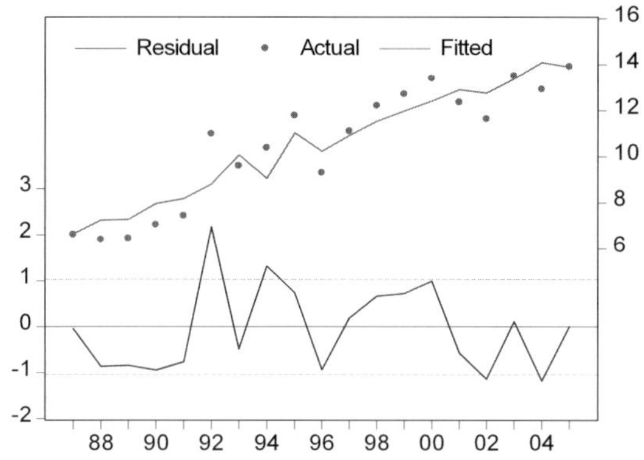

Fig. 6. Domestic water demand (100 million m^3) (1986-2005)

The model of domestic water demand (DW) (Fig. 6) is given by:

$$DW = -832.5563317 + 0.4224079361*T + [AR(1), MA(1), BACKCAST] \quad (9)$$

where AR(1) = -0.6656456604, MA(1) = 0.9972668958, BACKCAST = 1987 with $R^2 = 0.869$ and adjusted $R^2 = 0.842$.

In Beijing, the multi-annual mean precipitation is 542 mm and the multi-annual mean evaporation is 1,100 mm. The area of water surface is about 598 km^2, water quota for public green area is 1 m^3/km^2 per year, and water quota for one tree is $3m^3$ per year [BJWB 2002]. The public green area and trees in future years are predicted based on the growth rates in the past 11 years (1995-2005). The ecological water is calculated based on the water demand of water surface, public green areas and trees in the city. The model is as followed:

$$w_e{}^t = Q_s^t + Q_g^t + Q_{tr}^t \quad (10)$$

$$Qst = (Et\text{-}Pt)*Sw*1,000 \quad (11)$$

$$Qgt = Sgt*Dgt*1.06 \quad (12)$$

$$Qtrt = Ntrt*Dtrt*1,000 \quad (13)$$

Where

- t : a period of time (year);
- $w_e{}^t$: ecological water demand;
- $Q_s{}^t$: water demand of water surface;
- $Q_g{}^t$: water demand of public green area;
- $Q_{tr}{}^t$: water demand of trees;
- E^t: evaporation of water surface;
- P^t: precipitation;
- S_w: area of water surface;
- $S_g{}^t$: public green area;
- $D_g{}^t$: water quota for one unit of public green area;
- $N_{tr}{}^t$: the number of trees in a year; $D_{tr}{}^t$: water quota for a tree.

4 Results

Through the game theoretical approach and the water demand models of four players established above, the following equilibria results up to the years of 2015 are obtained (Table 1). Water demand models of player 1, 2, 3 are established according to their actual water consumptions in the past 20 years. For the ecological water demand, this paper only calculates the minimum water to maintain the ecological balance in Beijing. Therefore, no player would like to leave these equilibria. If one player breaks these balances, there will be conflicts. These results are determined by the socio-economic parameters used in models. If these parameters were changed due to irresistible forces, such as wars, natural disasters, social changes and so on, the results would change.

Table 1. Water demands 2006-2015 (100 million m³)

Year	Player 1	Player 2	Player 3	Player 4	Total
2006	12.55	6.70	14.79	7.05	41.09
2007	11.99	6.35	15.22	7.49	41.05
2008	11.46	6.01	15.64	7.97	41.08
2009	10.95	5.69	16.06	8.50	41.20
2010	10.46	5.39	16.48	9.10	41.43
2011	10.00	5.1	16.91	9.77	41.78
2012	9.55	4.83	17.33	10.51	42.22
2013	9.13	4.58	17.75	11.34	42.80

2014	8.72	4.33	18.17	12.26	43.48
2015	8.33	4.10	18.60	13.30	44.33

5 Conclusions

Water allocation is one of the important and complicated issues in natural resource management, and it usually involves water demand and water supply conflicts. Game theory is an appropriate modeling approach to solve such kinds of conflicts. Due to the uneven distribution of precipitation, Beijing is facing serious problem of water scarcity. The total water resource amount cannot meet the water demands of agriculture, industry, domestic and ecology. The ground water is overused, and water for maintaining urban ecology and irrigating agriculture are overtaken for industry and domestic life. Therefore, water conflicts and disputes among different water users are common phenomena. In order to find the water demands of agriculture, industry, domestic and ecology, a non-cooperative game theoretic approach is used. For analyzing and forecasting the water demands of the four player (agriculture, industry, domestic and ecology), statistical and econometric methods are used. The results of the work are determined by socio-economic parameters, such as industry added value, agriculture added value, population, etc. These results are not only giving better comparisons of the different groups of water users (including environment). Also the administration and the water supply companies might benefit from these results.

The non-cooperative game results show what the real need of each player is. In the case of Beijing, water supply cannot meet the water demand of each player. The conflicts of competing use of water between players are not avoidable. Therefore, the equilibrium of non-cooperative game cannot be reached in the reality. If all the players cooperate and distribute water among them based on its non-cooperative results, all of them will be better off.

References

Al Radif A (1999) Integrated water resources management (IWRM): an approach to face the challenges of next century and to avert future crises. In: Desalination,vol124, edn 1-3, pp 145-153

AusAID (2003) Making Every Drop Count. Commonwealth of Australia, Canberra

BJWB (2002) Water Use Quota for the Main Sectors in Beijing. Beijing Water Bureau, Beijing
BSBC (2006) Beijing Yearly Book 2006. China Statatistic Press, Beijing
Burness HS, Quirk JP (1979) Appropriative water rights and the efficient allocation of resources. In: American Economic Review, vol 69, edn 1, pp 25-37
Colby BG (1990) Transaction costs and efficiency in western water allocation. In: American Journal of Agricultural Economics, vol 72, edn 5, pp 1182-92
CSB (2006) China Yearly Book 2006. China Statatistic Press, Beijing
Fang L, Hipel KW, Wang LZ (2002) Gisborne water export conflict study. In: Schmitz GH (ed) Proceedings of the Third International conference on Water Resources and Environmental Research, Dresden, Germany, vol 1, pp 432-436
Fundenberg D and Tirole J (1996) Game Theory. MIT Press, Cambridge
Green GP, O' Connor JP (2001) Water banking and restoration of endangered species habitat: An application to the Snake River. In: Contemporary Economic Policy, vol 19, edn 2, pp 225-237
Gibbons R (1992) Game Theory for Applied Economists. Princeton University Press, Princeton
Hipel KW, Kilgour DM, Fang L, Peng X (1997) The decision support system GMCR II in environmental conflict management. In: Appl. Math. Comput., vol 83, pp 117-152
Holden P, Thobani M (1996) Tradable water rights: A property rights approach to resolving water shortages and promoting investment. World Bank Policy Research Working Paper No. 1627
IWRPYZ (2005) Report on the Environmental Impacts of Middle Route of South-to-North Water Transfer Project. Wuhan, China
Tisdell JG, Harrison SR (1992) Estimating an optimal distribution of water entitlements. In: Water Resources Research, vol 28, edn 12, pp 3111-3117
UN-CSD (1994) Review of sectoral clusters, first phase: human settlement and fresh water, fresh water resource. In: Report of the Secretary General of the UN, General Overview, Paragraphs 3-10
Van der Veeren RJHM, Tol, RSJ (2003) Game theoretic analyses of nitrate emission reduction strategies in the Rhine river basin. In: International Journal Global Environmental Issues, vol 3, edn 1, pp 74-103
Wang LZ (2005) Cooperative Water Resources Allocation among Competing Users. Diss. of the University of Waterloo, Waterloo, Ontario, Canada
Wang LZ, Fang L, Hipel KW (2003) Water resource allocation: A cooperative game approach. In: Journal of Environmental Informatics, vol 2, edn 1, pp 11-22
Wetzel GR (1983) Limnology (2nd Ed.). Saunders College Publishing, Fort Worth, Philadelphia, San Diego, New York, etc.
Wolf AT (1999) Criteria for equitable allocations: The heart of international water conflict. In: Natural Resources Forum, vol 23, pp 3-30
Yang ZF, Zeng Y (2004) Mathematical model for water conflict and coordination in transboundary regions. In: ACTA SCIENTIAE CIRCUMSTANTIAE, vol 24, edn 1, pp 71-76

Workshop: Corporate Environmental Management Information Systems

The workshop "Corporate Environmental Management Information Systems" (CEMIS) managed by the University of Applied Sciences (FHTW) Berlin is organized and conducted under the supervision of Prof. Dr. Horst Junker.

Modeling a Material Flow Network of a Casting House

Alexander Bock

FHTW Berlin, Germany, e-mail: Alexander.bock1@googlemail.com

Abstract: The present contribution shows the process-modeling and analyzing of a foundry with the software Umberto. Besides, the special difficulties which appear at this enterprise should be indicated. In addition to that, a suggestion should be discussed to solve the problem of the data capture.

1 Introduction

The modeling and analysis of operational material flow systems has become increasingly important. It is particularly suitable for putting the synthesis by economics and ecology on the operational level [Stappen 2006]. And it is especially suited to save the resources and to increase the resources efficiency, which is the cause of a lasting development in the enterprise. The article shall point out the goal to illustrate practical experiences of the production of a material flow model in the case of the BAE batteries GmbH. This should be given a contribution for the energy conservation of the enterprise. An internal analysis shows that the foundry is an important power consumer. Therefore they decide to build up the complex system connections foundry as a model. That should include all necessary raw materials, supplies, auxiliaries and fuels, wastes and sources of energy.

2 Method

It is tried to lean on the procedural model of Bullinger. Bullinger and Beuker [2000] divide a material flow management cycle into a multiple-stage model e.g. analysis, control and so on. They include into the stage analysis a material flow analysis cycle. This article looks at the stage of the analysis. For the solution of the concept the software Umberto is used. It is based on the concept of the material flow nets, which are developed according to the principles of the Petri - network theory [Möller et al. 1997]. This represents a formal method to model systems and/or transformation processes [Schmidt and Keil 2002]. At first an investigation period was specified (03.10.2005 - 31.12.2005). A second stage used activities, which are necessary to get basic information for energy conservation. In addition to that the concept (structuring of the nets) and the data acquisition of the data is important. Measurements of the casting machines were accomplished, because a lot of data were not present. Primary measuring points were thereby the lead boilers because these produce the largest current consumptions. It is also interesting to know, if there are energetically beneficial to drive down the lead boilers for a special time period. It was badly need to organize the data measurements and discussions and to evaluate existing data bases. With the getting knowledge of the data the model could be realized and balanced.

3 Realization

The data is divided into raw materials, supplies, auxiliaries and fuels, wastes and sources of energy and products. The different lead alloys are assigned to certain machines. The experience shows that the production of the grids is very efficient. The input quantities of the lead could be determined over the incoming goods for the period of the fourth quarter 2005. A problem is that the single quantity is not directly assigned to the order. Therefore some total quantities had to be scaled on the machines. The consumption of tin could be determined indirectly, because for 100kg lead calcium 700g tin are needed. Similarly it behaved with the fleece bags. One is assigned exactly to one grid. This consumption based on the calculation of the output grid types. It is much more difficult to calculate the consumption of the auxiliary and fuels. All auxiliary and fuels are led only in SAP. Cork flour e.g. that e.g. cork flour (parting agent) is ordered only once in a year, but is used by five machines. Therefore all values consumed in Umberto are based on statements of the master of the foundry. The sources of energy electric power and natural gas could be well determined. A special current tool (electric measuring instrument) acquired data, which was written by a displaying measuring instrument of the enterprise Jumo. This was adjusted in such a way that every two minutes a value of the temperature and the used up ampere-hours are written into his memory. With emissions above all the lead emissions are meant. These were proven and based on a gravity casting machine. But a personal dosimeter was used, which was screwed on a mounting plate. The mounting plate is hung into the boiler departure of the machine. Due to the same design and the relatively similar extent of utilization of the machines a comparable emission is proceeded. All producible grid types are designated as products.

3.1 Concept

During the investigation of complex system connections human beings push soon at limits. Without additional aids it is hardly possible, to realize, to understand and to reconstruct various dependence and dynamic procedures in a larger system. This applies equally to the systems of the natural environment (e.g. ecological systems), to the technical created by humans (in the available case a foundry) and economic systems. Therefore we use the characteristics of the system, which are relevant in the context of the respective problem definition and illustrate them in a model of the system. The system in the model is substantially simplified represented by aggregation, abstraction and idealization. The investigation takes place at the model of the system, because complexity of the systems becomes manageable. So the model shows the substantial system properties. Therefore it can be transferred the results, which were obtained with the model, to the system. Which system properties are substantial depends in each case on the question. This applies also to the definition of the system border. You define what has to be adding to the system and which can be regarded as system environment. Important is that each model describes a certain scenario in the context of a problem definition.

3.2 Structuring of the nets

The Figure 1 shows the backbone network of the foundry. The orange-colored framework represents the system border. A characteristic comes to the modeling of the grid blend. This is in- and output, because the blend from the last production step of a grid again in the first production step.

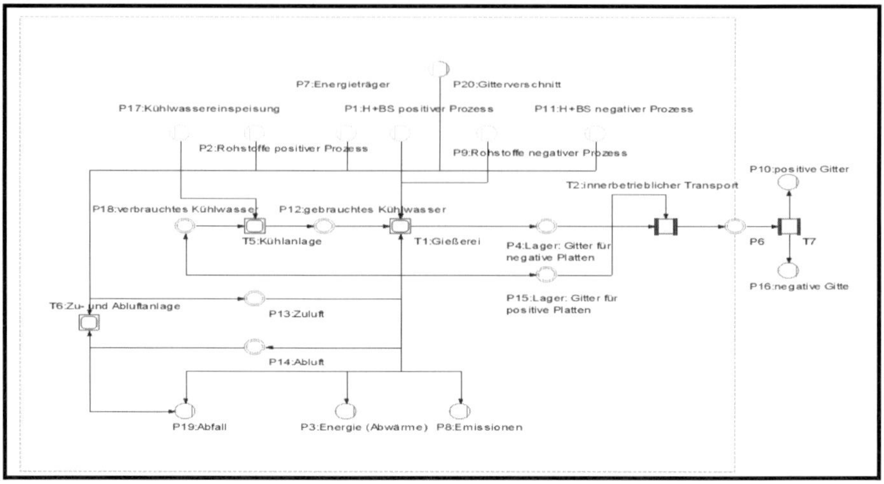

Fig. 1. Main-net of the foundry

The figure point out the three main energy consumers with the Transition, T1 foundry, T5 cooling system and T6 exhaust air system in the backbone network. This organization was made, in order to make clear that the foundry consists not only of the production department, but also of other ranges, which cause the energy consumption. It is attending to that in all nets a separation between in and output is clear, in order to be able to keep the nets well arranged. This formalism was broken into all sub-nets inside. The foundry is subdivided in two main production chains. On the one hand the production of positive, and on the other hand the production of negative grids. For the positive production process four die casting machines are used, which are appropriate for the production of certain grid types. The positive production process meant the production of grid types, which apply to the production step “shake”. In contrast to it the negative production process meant the production of grids, which apply to the coming production step “compound”. The nets differ between gravity casting (negative production process) and die casting machines only in the last production step. As you can see in Figure 2, the positive grids are drawn after cutting into tubing bags. That is necessary, because into the “shake” the red lead normally would not remain sticking to the grid. Red lead is a basic module for the building of a battery. It is needed for the construction of the batteries.

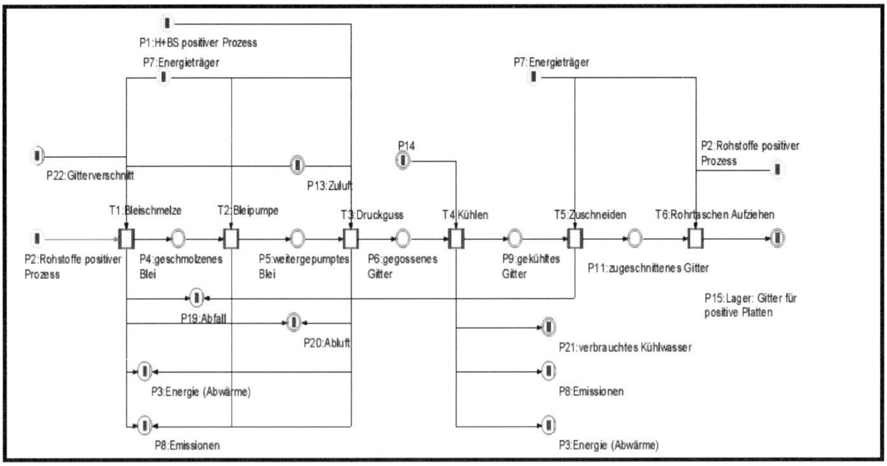

Fig. 2. Positive production process

3.3 Computation of the nets

The computation method works completely locally, continuously related to an individual transition or an individual place. This beginning reduces the qualifications for the application of the computation method and permits the execution of the computations, even if the specifications are not yet complete in a material flow net. The computation direction is fixed on the direction of the material stream. The computation toward the material stream was made for the computation of the operating statement. The computation starts from the sub-nets, at which the input lead is the starting point of the computation in each case. This becomes clear by a pink colored connection to the first Transition. The third picture exemplary shows the computed backbone network of the foundry in a sankey-diagram. The net shows thereby the energy consumption of the main energy consumers, whereby the consumption of the foundry is the largest.

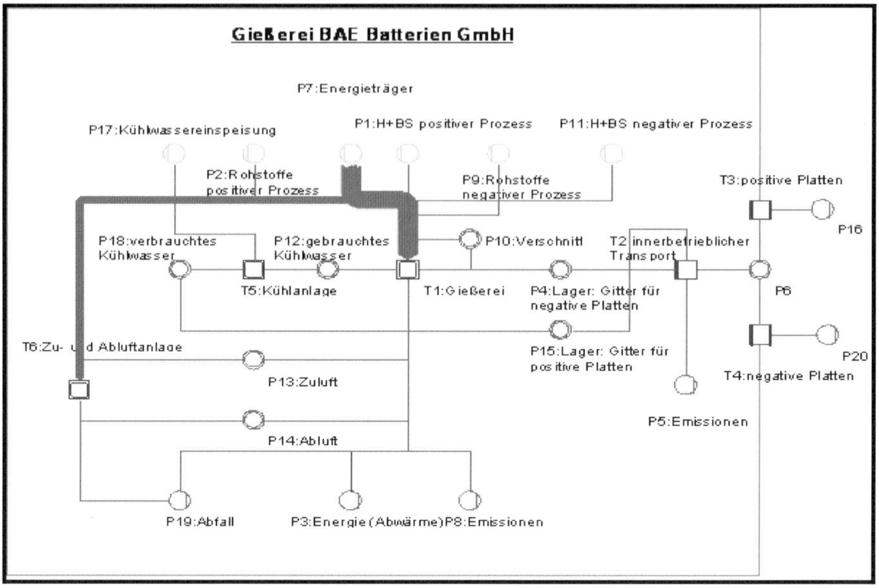

Fig. 3. Sankey-diagram of the foundry

4 Results

The results are based on the measurements and a weak point analysis. First results of the analysis are to be regarded. It is determined, who are the three main consumers of the foundry. The consumption by the lighting resounds is neglected. This should be for instance about 1% of the total consumption of the foundry. Interesting is here the portion of the exhaust air system. Approximately a third of the current consumption of the foundry reveals the fears that the installed plant is nevertheless too largely laid out. Since this the plant can drive only the conditions "enterprise" and "non-enterprise". An optimization is to be seen if you use more casting machines in the enterprise, because is the higher the efficiency of the plant. It is further interesting, as the monetary portion of the product grid behaves. The energy consumption at the grid is the second highest cost factor, but it is not to compare with the material costs, which are caused by the lead alloy. In addition, on the other hand it must be understood, that the enterprise does not have influence on the lead price, because these daily fluctuations are subjected to the conditions at the world market. The lead quantity of a grid is fixed to certain tolerances, which may not to fall below or exceed a limit. So the current consumption as next cost factor is much more important for the enterprise, because an influence on the consumption can be exerted here.

Further results arose e.g. with the measuring of the warming up phases of the machines. So a preheating time could be based by 7h on the gravity casting machine 3, while the other gravity casting machines lay within the allowable times for the heating phase of the boilers. The reason is to be seen in only five functioning heaters. This affects adversely on the current consumption and the assignment of the machine. Further was to be demonstrated, when it is worthwhile to drive down a machine completely.

If we look to the gravity casting with an average preheating time from 4h and 7 heaters a'5.5 KW, we would save energy starting from the 14h, that means that the machine would has approx. two shifts not to be used, when you want to have an energetically advantage. It can re-capitulatory be said that it is meaningful to have more heaters at a boiler. So the energy consumption in the warming up phase can be lowered. With all nine functioning heaters a machine uses least electricity. This has on the other hand the advantage that you could redefine switch-on times for heating the machines in the week planning. By a temporally smaller heating duration you would save not only energy costs, but could be also more flexibly to the situation concerning orders react.

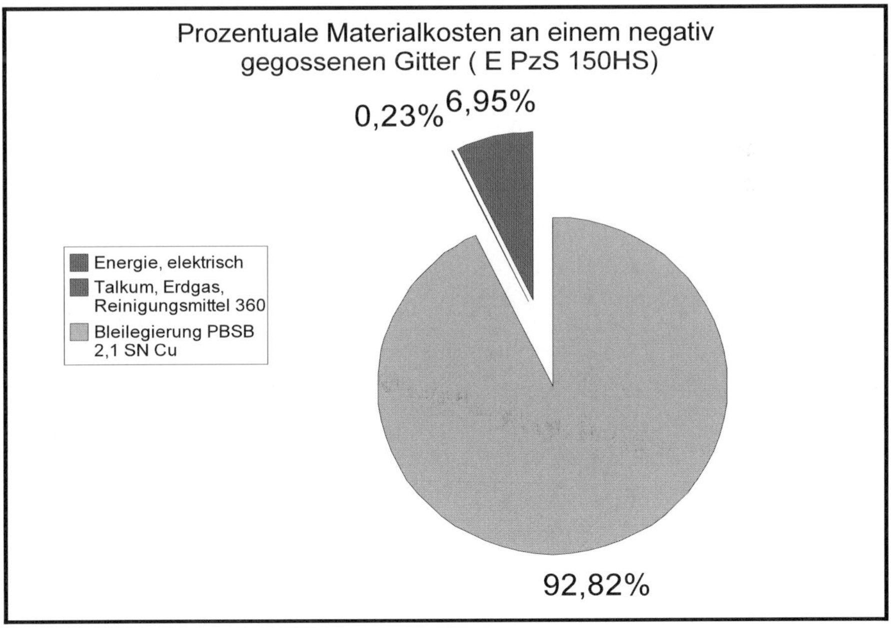

Fig. 4. Material costs of a negative produced grid

Switching a boiler off is worthwhile itself, if you does not have to produce two shifts. From several measurements a characteristic number for the current consumption at the boiler in the operating condition could be determined. This lies with about 42.1 % of the energy consumption, which is needed when warming up. From the view with the work on the model it can be said that it would be meaningful to assign the orders to the machines. Thereby you could determine individual lead consumption more qualitatively. An optimization is still possible, if the energy consumption of the product is taken up. You should think about the point after a suitable measuring period whether you can take up the flow coefficients at a machine. So you can be transfer the consumption of all necessary in- and outputs to a grid.

5 Outlook

I think that there are possibilities to improve the vague statements to the auxiliary and fuels of the foundry. Some auxiliary materials are ordered e.g. once in a year for the sub-range foundry and used up then in roughly one year (e.g. cork flour). The quantitative employment at the individual casting machines is not documented, therefore the backtracking of the employment of cork flour at an individual grid is only possible over a rough approach estimation. This can be faulted. That means that the admission of certain data at the machine should be still made available. In order to avoid no larger expenditure, I would like to submit the following suggestion.

If you would change that the once in a year bought cork flour is divided into all gravity casting machines and in the same course you equip the gravity casting machines with a counter, you could measure after many gone through grids, how much cork flour was used. Thus an average value could be determined for each machine. So you could get more exact statements about the final consumption of the individual operating supplies, if you should use this procedure with every other auxiliary material. After a certain measuring period a relatively good average value would result, which could be in-maintained again as meaningful calculation size in the provided model of Umberto. On the other hand it could be uncovered whether there are machine-specific differences in consumption. The model provided with Umberto could be designed to the individual grid for the flow coefficients and the internal computation algorithm could be made dependent on the flow coefficients. Thus could be more easily to assign the operating supplies to the grids. In my opinion the BDE of the foundry should be considered again.

References

Bullinger HJ, Beucker S (2000) Stoffstrommanagement und Betriebliche Umweltinformationssysteme (BUIS) liefern neue Impulse für das Umweltcontrolling. In: IAO 2000, pp 1-18

Schmidt M, Keil R (2002) Stoffstromnetze und ihre Nutzung für mehr Kostentransparenz sowie die Analyse der Umweltwirkung betrieblicher Stoffströme. In: Hochschulausgabe Fachhochschule Pforzheim, vol 103

Möller A, Häuslein A, Rolf A (1997) Öko – Controlling in Handelsunternehmen. Berlin Heidelberg New York

Stappen RK (2006) A Sustainable World is Possible - Der Wise Consensus. Eichstätt

Conception and Realization of a Web-based User Interface for an Environmental Simulation with Gridfunctionality

Stefan Hinze

FHTW Berlin, Germany, e-mail: Stefan.Hinze.Berlin@t-online.de

1 Introduction

Many available software products are overcrowded with functionality, so that a lot of users are overwhelmed. Often, the result is a failure of the according product. Especially environmental simulations need a lot of data as input and results should be simple to handle.

ERAMAS (Environmental Risk Analysis and Management System) is such an environmental simulation. Due to the fact that ERAMAS is a grid-based software product, it should be published for demonstration purposes on the Instant-Grid CD. Instant-Grid is a project of Fraunhofer FIRST (Fraunhofer Institut - Rechnerarchitektur und Softwaretechnik). Goal of this is a bootable Knoppix-based CD which autonomously builds a grid-network. Advantage of grid-computing is that extensive tasks can be calculated in shorter times. That means that a certain amount of personal computers or supercomputers are parallely processing the same task and thus, minimize the machining time. The problem here is the fragmentation of the current object of investigation and the according distribution across the connected computers in the grid. The fragmented results have to be assembled thereafter again.

Because the Instant-Grid CD will be added to a magazine, users with no background knowledge about the functionality of grid-computing and environmental simulations could be a general user among specialized users.

The result of the project is a web-based user interface which allows an intuitive and fast handling for users with no background knowledge. The interface comprises of a simple input mask for some basic meteorological data like wind direction, wind strength and special scenario data like toxic substances and volume. The place of accident can be chosen by a simple click on a geographical map. After the calculation, the results are displayed with a simple graphic in the centre of a web browser for fast interpretation and to create countermeasure plans.

2 ERAMAS

ERAMAS is a highly efficient, computer-based environmental risks analysis and management system. It allows the forecast of dispersion of toxic substances in the atmosphere, the ground as well as in the groundwater (Fig. 1).

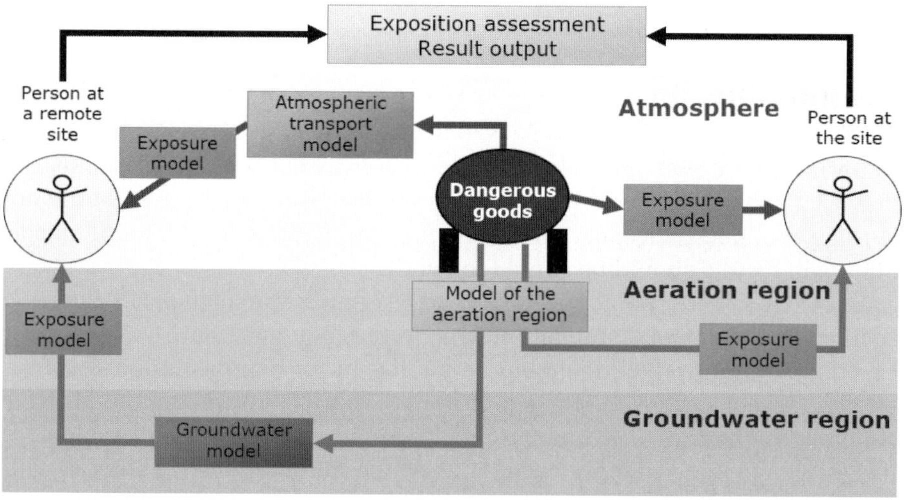

Fig. 1. ERAMAS – models and operational area (source: http://www.eramas.de, access: 2007-02-13)

After an environmental contamination, with ERAMAS it is possible to perform a fast and safe analysis and evaluation of the danger potential for humans and the ecological environment. Because of the multitude of influencing factors, environmental simulations are time-consuming, so that ERAMAS exploits the grid-computing technology. Under real time terms it can be used to analyze and evaluate accidents at industrial plants, hazardous materials transportation or terror impacts. Furthermore it can be used for authorization procedures and the creation of countermeasure plans.

The version of ERAMAS on the Instant-Grid CD comprised the atmospheric transport model [cp. ERAMAS 2002].

3 Conception of a GUI development

The conception, development and realization of the interface was done according to DIN EN ISO 9142 and 13407 which support the design of ergonomic software products.

Especially part 10 and 11 of DIN EN ISO 9142 include basic terms concerning the usability of software products - principles of dialog creation and usability requirements.

The DIN EN ISO 13047 is an important standard regarding high quality software with a focus on usability and should be exploited for the development of software products.

To ensure a uniform control interface, the product is embedded in a web portal named gridsphere. The portal supports the Java Specification Request (JSR) 168. This is an application programming interface (API) for the development of portlets. By using the standard, a high reusability of the developed source code is possible.

3.1 DIN EN ISO 9142 "Ergonomics of human-system interaction"

DIN EN ISO 9241 gives a lot of design recommendation for a graphical user interface. DIN EN ISO 9241-110 "Dialogue principles" defines important basic principles of dialogues [cp. Grunenberg 2001]:

- Suitability for the task:
  The dialog should support the user to do his working task actually and efficiently.
- Self-descriptiveness:
  For actually and efficiently working selfexplicatory concepts should be used.
- Conformity with user expectations:
  The dialog should be consistent and correspond to the expectations of user.
- Controllability:
  The user should be able to begin the dialog expiry as well as to influence direction and speed.
- Error tolerance:
  Possible input errors from the user are intercepted and visualized in form of an understandable error messages.
- Suitability for individualization:

The dialog system must admit adaptations to the requirements of the working task as well as to the individual abilities and predilections of the user.

- Suitability for learning:
  A dialog should support the user while learning the dialog system and train.

These principles lead to part 11 in which the usability is defined. Adequacy characterizes the suitability of a user interface in a specific context. DIN EN ISO 9241-11 "Guidance on usability", is part 11 and includes three criteria for the usability of a software product [cp. Grunenberg (2001)]:

- Effectiveness
- Efficiency
- Satisfaction

With the help of these characteristics a measurement of this adequacy is possible.

3.2 DIN EN ISO 13407 "Human-centered design processes for interactive systems"

The DIN EN ISO 13407 defines the human-centered design process for interactive systems (Fig. 2).

The development process of a graphical user interface must be adapted to user to user requirements in multistage process. This kind of development process is important for the usability.

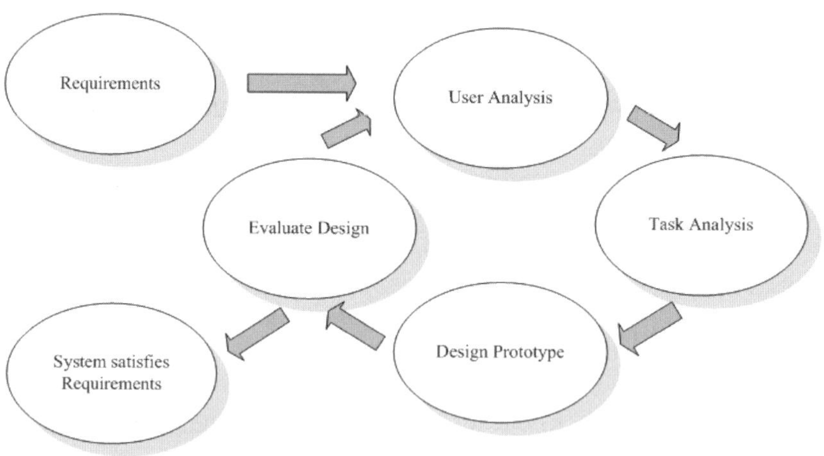

Fig. 2 User-centered development process according to ISO 13407 (source: http://blog.nutzbar.ch/200510, access: 2007-02-13)

3.3 Theoretical procedure for interface development

In consideration of the described standards follows the theoretical procedure for the development of a GUI [cp. Hogenkamp 2006].

- Analysis of user groups
- Analysis of requirements
- Analysis of surrounding conditions
- Development of an ergonomic product in consideration of DIN EN ISO 9142

Therefore functions and requirements must be determined corporately to user groups. Subsequently, the surrounding conditions will be defined and a user friendly GUI developed.

4 Realization of the GUI

4.1 Analysis of ERAMAS and Instant-Grid user groups

The Instant-Grid CD will be added to a magazine. So there will be two user groups for ERAMAS. It exist user without background knowledge about the functionality of grid-computing and environmental simulations and user with special knowledge about simulation.

4.2 Analysis of requirements

The analysis of user groups has devoted that there will be two user groups. For this reason the development concentrated on the first user group so that not all functions of an environmental simulation are available. It will be limited to the following input fields and parameter:

- wind direction
- wind speed
- location
- temperature
- harmful substance
- CAS-number
- volume

For seasoned user input fields for more input parameter can be added easily at a later date.

4.3 Analysis of surrounding conditions

An extensive analysis of surrounding conditions was not needed in this project (Fig. 3). Surrounding conditions were defined by this project.

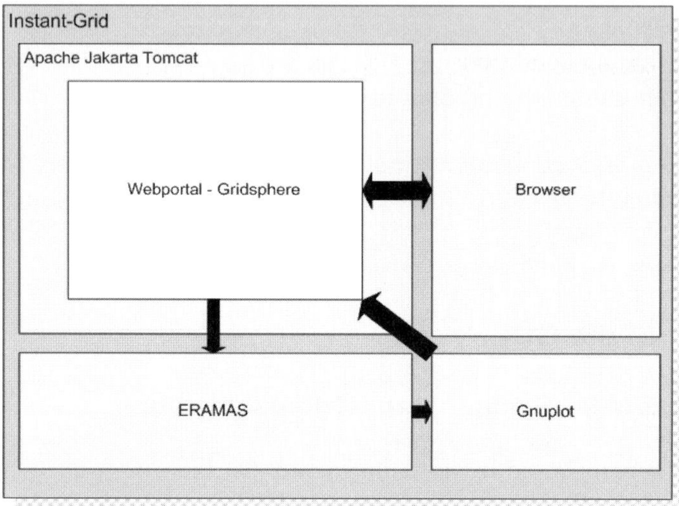

Fig. 3. Surrounding conditions of ERAMAS and GUI

Because ERAMAS should be also used independent of workplace, a web-based user interface will be developed. For the user friendly individualization, the ERAMAS GUI will be implemented directly into the web-portal of Gridsphere.

4.4 Development of an ergonomic product in consideration of DIN EN ISO 9142

After the functionality of a user interface, the aesthetic aspect is an important component for an ergonomic software product. If design and expressiveness are well combined, an aesthetic GUI should be the result of the development.

Following points results from the DIN EN ISO 9142 and should be taken into consideration:

- Use of a uniform structure of window
- Each window is divided in up to 3 fields for gradually work
- No high-contrast color design of the sides
- A short assistance is always at the right side edge
- Using of meaningful and short inscriptions

For the realization of the interface the languages Html (Hypertext Markup Language), Java, Javascript and JSP (JavaServer Pages) were used.

```
...
    String action = req.getParameter( "action" );

    if (action != null) {
    createInput(wind,speed,kelvin,stoffe,CAS,menge,x,y);
        doSim();

    }
}
...

...
public void doSim()
{
    logger.error("####### doSIM methode #######");

    Process p = null;
    String start = "Pfad - ERAMAS";
    try{
        p = Runtime.getRuntime().exec(start);
    }
    catch(IOException e){
        logger.error("####### Fehler #######");
                logger.error(e);
    }
}
...
```

Fig. 4. Catching a predefined parameter and start external application

ERAMAS is a separate application on instant-grid CD. The problem of HTML, JSP and Javascript is, that they don't have the rights to start external applications like *.exe-Files. So how do you start an external application from a web-based user interface?

It was required to use the following way to resolve this problem. To execute external applications a java class was developed which catch a predefined parameter of a clicked button on a HTML-site. If the parameter is sent by the web interface the java class starts the application ERAMAS (Fig. 4).

On the same way was realized the visualization of the outputs of the simulation. The visualization will be done with gnuplot.

Gnuplot is a small but extensive, scientific plot program to visualize data like measured data. To start this separate application a java class has to catch the special parameter like the parameter of ERAMAS [cp. Völker 2002].

5 Results

Goal of this project was the conception and the realization of a web-based user interface for an environmental simulation within the scope of the disasters management system ERAMAS and Instant-Grid (Fig. 5). The developed user interface is today on the status of a prototype and does not contain all functions for a simulation of pollutant dispersal in atmosphere.

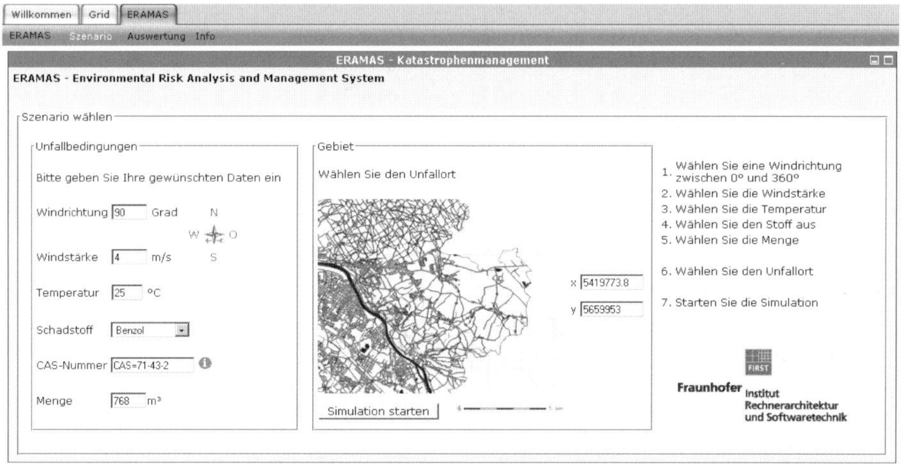

Fig. 5. Construction of a scenario

Also, unversed users can create a scenario in a fast and easy way. The attitudes were limited for the sake of simplicity to a minimum. For a fast and uncomplicated entrance into this user interface, a help area is generally at the right edge.

The users have to provide the wind direction, wind speed, temperature, harmful substances (chlorine, benzene and trichloroethane) and volume. Possible input errors from the user are intercepted and visualized in form of an understandable error message.

The location can be selected with an easy click on the map in the centre. The results of simulations are visualized on a new site with the same side construction (Fig. 6).

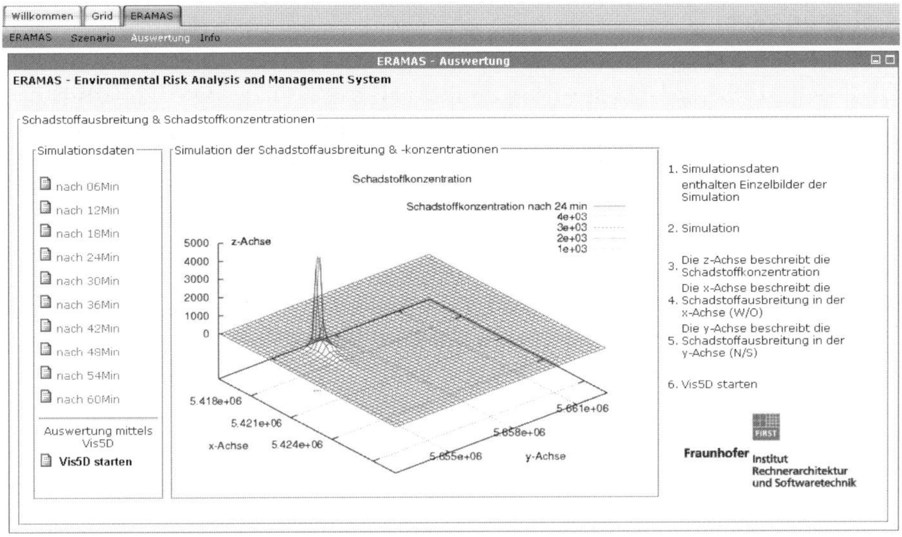

Fig. 6. Evaluation of the simulation results

For a fast overview of pollutant dispersal in the atmosphere an animated picture displays the dispersion in center. On the left side the user can choose between 10 timesteps and the special pictures of dispersion. Additionally, evaluation of the simulation results can be made with Vis5D+. The standard visualization is done with Gnuplot.

6 Conclusion and Outlook

The development of Instant-Grid and the user interface is still continuing. At the end of the year 2007 a first Instant-Grid version with ERAMAS and the graphical user interface will be released.

In distant future the use of this interface is also conceivable for the full version of ERAMAS. For this reason the developer team set a high value on extensibility during the development.

In a later Version of Instant-Grid and ERAMAS seasoned users can use all the input parameters like:

- Simulation date
- Weather temperature
- Weather temperature error rate
- Weather stability class
- Substance volume error rate
- etc.

References

ERAMAS (2002) ERAMAS - Environmental Risk Analysis and Management System. http://www.eramas.de, access: 2007-02-13

Grunenberg B (2001) Bildschirmarbeitsplätze, vol January 2001. München

Instant-Grid (2007) Ein Grid-Demonstrations-Toolkit. http://www.instant-grid.de, access: 2007-02-13

Völker D (2002) Gnuplot-Grundkurs. FU Berlin. http://userpage.fu-berlin.de/~voelker/gnuplotkurs/gnuplotkurs.html, access: 2007-02-13

Hogenkamp P (2006) Von Usability-Testing zu User Centered Development, Zeix AG, Zürich. http://blog.nutzbar.ch/200510, access: 2007-02-13

Concept to Migrate an Access Based Environmental Information System into a Web-based Front-end Solution for the Environmental Management of a Waste Disposal Company

Thomas Kästner

FHTW Berlin, Germany

Abstract: Many Enterprises do still use MS-Access based Applications for the assistance of the tasks of environmental management. It is increasingly postulated to bring web based applications in use. Therefore many solutions are possible. This development will be shown on an example of the migration of an environmental information system for a waste disposal company. In this case the enterprise given conditions lead to three possible solutions. The usage of the Java Server Pages technology turned out as most suitable.
Keywords: environmental information system, MS Access, JSP

1 Introduction

A waste disposal company uses for the support of the environmental organization an environmental information system. The old system was realized as a MS Access application with front-end and backend. It was intended to reengineer the old system, to take advantage of a web-based architecture. The task is to develop a strategy for the migration of the environmental information system.

This paper discusses three solutions. Further on, development regarding maintenance and servicing is to be considered and in preparation for a decision, an evaluation catalogue has to be provided. A comparison will lead to an optimal solution for the conversion to a web application.

If the method is chosen, a demonstration prototype will be realized. Therefore a suitable module of the environmental information system is selected. The prototype will be tested in the active system. For this prototype emphasis was laid on gathering information and becoming more experienced with the development and maintenance of web-applications.

2 The Situation of the example Project

First of all the old system should be introduced. A short impression of the tasks and uses of the Environmental Information System will be given. The aim is to figure out the importance of the system for the company. On the other hand the range of possible solutions is determined by the actual conditions of the enterprise. These conditions will also be managed in this chapter.

2.1 The Environmental Information System

The Environmental Information System is a self-development of the company. It is used as a source of information for the fulfillment of environmental protection. As a tool of the controlling it illustrates all relevant ranges for the fulfillment of environmental protection. The work of the responsible persons within their ranges is supported by the system. In this case the System leads to an increased security in law.

The work of the environmental assignments during the monitoring of the many decentralized locations of the company is substantially simplified by the system. The system is used to represent the historical process of the environmental relevant development and serves as a long-term proof demanded by the legislator. A normal record keeping in environmental protection is enabled. The wide range of the existing information allows reporting and also prognostic views. Data can be seized standardized and made accessible for the decentralized ranges. Information of the technology and organization can coherently be used in one system.

The system is realized as a MS Access application. It underlies a constantly development process. 2003 a material data base module was added. It is used as a base for a dangerous material register. The modular structure makes simple maintenance and servicing as well as their extension possible. The data is stored in a MS Access backend data base, which is accessed over the company network. The Clients work with a local front-end, which connects to the backend, so that in fact, a simple Client server architecture is copied. But MS Access is not an active data base management system, but a desktop base management system. That means all work is done by the Clients. This variant is sufficient for the existing data sets and user numbers. Actual the Environmental Information System contains over 2000 environmental relevant plants and over 1000 permissions from the environmental and building laws. The individual modules illustrate all relevant ranges for the fulfillment of environmental protection in the enterprise.

The modular structure of the System and the simple integrated development environment under MS Access enables the company to handle the continuous extension and maintenance by itself. Suggestions for improvement can be handled very fast and efficiently. These are the feedback of the approximately 250 users of the system and changed legislation.

2.2 Conditions of the Enterprise

The Environmental Information System cannot be changed by the project. A way must be found to not impair the active system. Therefore the same backend data base should be used by both systems.

The server administration makes conditions for the web server and the surrounding field of server side application. The server world of the enterprise is dominated by Linux. There are considerations for using Linux also for the clients. This makes a new development of the system necessary. So far there are still no concrete plans concerning a change to a free operating system. From these considerations the task is to develop a possible way of accessing the "mdb" file format of MS Access under Linux. These restrictions led to the different alternatives.

Apart from the technical access to the MS Access data base there are defaults, which concern the layout and the integration into the existing Intranet. The organization of web applications differs from classical applications. The organization of the elements should be adapting the uniform occurrence of the Intranet. The recognizing value should be as high as possible. Users, who know the classical system, will get along fast with the web application. In addition, first time users should have a simple entrance.

An important aspect with the development is the conversion on it own as well as maintenance and servicing. Further maintenance and servicing of the system take place in the providing division. The used programming languages should remain limited. A development environment which can be served simply is also of great importance.

3 Advantages and Disadvantages for the Enterprise

Web applications are characterized by the fact that they only need a web browser on the client. So no installations on the clients were necessary. After the web concept the main load of the treatment is no more carried by the client. While web browsers are very common, a certain independence of platform is reached. Unfortunately an absolutely uniform appearance in all browsers cannot be achieved because web browsers do only interpret the HTML data. A disadvantage with the migration of an MS Access application into a web application is the necessary reengineering of the application. Depending on the selected web technology, a suitable development environment is to be found.

The migration of existing applications usually leads to technical problems. The file format tiff is for example an industry standard, but no web-compatible file format. A possible solution would be a transformation into the PDF file format at run-time. The organization of web applications, e.g. the menu guidance and the presentation of contents is different from conventional applications. A compromise between the design of the existing System and the design guidelines used in the intranet has to be found. A web application is a lasting and economical alternative to a conventional application.

4 The managed solutions

In this paper three solutions were discussed. They differ in the use of the server side programming language and the database system. These solutions will be presented more detailed.

4.1 Solution: PHP and MySQL

The first alternative is based on the use of an active data base management system. The data for the web application will be stored in an own data base. In regular intervals a data alignment with the backend data base of the system will take place. For the interaction between data base and web page the programming language PHP is used. It is a script language for the production of dynamic web pages.

PHP is open source software. The company already uses PHP for different applications in the Intranet. An advantage is the simple connection to numerous SQL data bases. Common is a connection to the data base management system MySQL. In connection with the web server, MySQL is also already in use. It has to be paid attention to characteristics and differences between MySQL and MS Access. The repeating adjustment of data may cause anomalies or inconsistencies. For the automation of the adjustment a command line tool for Linux is necessary. But after intensive search no usable tool for this task was found. A change of the data base is connected with various changes in the PHP source code. This is caused by the native functions of PHP. For PHP there are a number of suitable development environments available.

4.2 Solution: direct PHP

The second alternative is to get direct access to the MS Access backend under Linux. A number of small tools are necessary to accomplish that. For the dynamic production of web contents PHP is used as well. However PHP does not support this technology automatically. There is a tool for reading Access files under Linux. In connection with an extension for PHP this can be used for generating web pages. This solution is characterized by the use of many small Tools. The problem is there is no support for these tools. Particularly the PHP extension is to be regarded as experimental. It only provides read-only access. It can only represent a temporary solution.

4.3 Solution: Java Server Pages

The programming language Java offers another approach. For the connection to a data source the Java Database Connectivity is used. This is special universal data base interface for relational data bases. By using an abstraction layer the data base can be changed by only loading the correct driver. For the use in this case a type four driver is needed. A type four driver is completely platform independent implemented in Java. Even a type four driver for MS Access is available. For the production of a prototype an evaluation version of the driver can be used. To generate dynamic web pages, Java provides the Java Servlets and the Java Server Pages (JSP). In order to simplify the development of web applications Java server Pages offer similar conditions to work like with PHP or ASP. With the implementation of XML tags is a separation of the presentation, the internal processing and data storage possible. A JSP has the task of the presentation. The processing of the client inquiries and the delivery of the data from the data base is provided by Servlets. For the execution of Servlets and JSPs is a special JSP/Servlet container necessary. There is for example the Apache Tomcat. The used Linux distribution already includes Tomcat and the linking module for the used Apache web server. The modular structure of Java makes it easy to use further libraries. For example a library can be loaded with PDF functions. So it is possible to create PDF documents from the TIFF files.

5 Evaluation

The process of the evaluation can be divided into three phases. In the first phase the requirements are to be specified. The definition of the quality criteria takes place following the evaluation characteristics suggested in the ISO 9126. The used quality criteria were functionality, efficiency, maintainability and portability. The criterion of the usability is not used for the evaluation, because the organization of the user interface and software ergonomics is in all three solutions the identical. Some sub criterions were defined after the purposes of the enterprise.

In order to be able to use the criterions, it is necessary to define suitable units and classification levels. A defined scale makes it possible to evaluate the criteria quantitatively. A classification level assigns a meaning to the numerical values.

The evaluation of the individual criteria takes place on the basis of a value scale. For the distinction of the suitability of a criterion, a simple scale is sufficient. The selected scale possesses the values one to four. If a criterion obtains the value one, the method fulfils the requirements badly or not. A criterion obtains the value four, if the requirements are fully fulfilled.

When the individual criteria were evaluated, a problem arose. The criterions have to be weighted to gain a value of quality from the criterions. The weighting was conducted with the department of environmental protection of the enterprise and gives information on the priorities.

Functionalities of the web application and their maintainability are regarded as very important and count in each case with 30%. Efficiency represents 25% of the quality value. The priority of the portability is set to 15% due to the well-known, stable server environment.

The result of the evaluation is a clear order of the three solutions. The variant with the direct access achieves the last place. While this way turned out as unsuitable, the use of an active data base in connection with PHP offers potentials under a certain trend. The use of the JSP seems to be most suitable. As proof for the practicability a module of the environmental information system will be converted into a prototype using Java Server Pages.

6 The Prototype

A suitable module of the environmental information system is selected for the conversion. The material module was chosen. The information becomes available for all users of the Intranet. The information of the working materials and the depending documents become visible in the browser. The following functions were implemented by the prototype:

- The complex material search functions of the environmental information system.
- The environmental relevant information to a working material is to be arranged and presented according to special categories.
- The existing documents of the working materials can be viewed in the Browser.

A first step during the conversion is to establish the needed environment for the Java Server Pages. The installation of the JSP/Servlet container and the linking module for the web server was finished with no problems. For the development the development environment Eclipse is used. Especially for the development of web applications with Java an additional web module for Eclipse is used.

The evolutionary growth of the prototype includes the changes by the enterprise during the development. The design of the prototype adapted the appearance of the Intranet. So the web front-end can directly be integrated into the structure of the existing Intranet. The design of the web application was developed in close cooperation with the IT department. For further development of a web front-end with the full functionality of the environmental information system a new design has to be developed.

The web application was transferred into the active system of the Intranet without problems. The user is able to select different criteria to search for a working material. According to the search criteria a list of the existing working materials is presented. On the basis of the results, the user is informed immediately about the release status and the existing documents. Afterwards the user is enabled to view detailed information of a working material or viewing the existing documents.

If the user selected a working material, he gets access to the material data. The information is divided into different categories. The range leads from industrial safety to environmental relevant information.

Especially older documents were stored using the TIFF format. With the developed prototype it is possible to convert a set of TIFF files on request into a PDF document.

7 Conclusion

The prototype made clear, that the software ergonomics of a web application is different in comparison to conventional applications. A new corporate design for the conversion of the forms of the MS Access application has to be developed. An immediately conversion of the application into a web-based application is not necessary. Nevertheless there can be situations, which require an alternative for the MS Access application. A big problem for a conversion is the high amount of VBA code in the forms. VBA is a programming language, which is not wide spread outside of the Microsoft world. The code must be manually translated into languages. Tools for automation do not exist.

With the described JDBC driver it is possible to get reading access to the MS Access data base. A parallel employment of Java and MS Access should be avoided. The JDBC driver uses its own mechanisms ignoring MS Access safety restrictions. This can lead to errors during the access. For this reason it is recommended to convert the backend data base into an active data base management system. The MS Access front end can easily be operated with such a backend. The web application can be switched over without any problems to another data base.

References

Balzert H (2000) Lehrbuch der Software-Technik: Software-Entwicklung. vol 2, Heidelberg, Berlin

Balzert H (1998) Lehrbuch der Software-Technik: Software-Management, Software-Qualitätssicherung, Unternehmensmodellierung, Heidelberg

Berliner Stadtreinigungsbetriebe (2005) GO Grün Orange Umweltbericht 2004/2005. Berlin

DIN Deutsches Institut für Normung e.V. (1994) DIN 66272 Bewertung von Softwareprodukten, Identisch mit ISO/IEC 9126:1991. Berlin

Kijewski K, Spitznagel M (2005) Umweltinformationssystem des BSR. In: Das aktuelle Handbuch für den Entsorgungsfachbetrieb. Merching

Sun Microsystems Inc.: Java Server Pages Homepage, http://java.sun.com/products/jsp/

Sun Microsystems Inc.: Java Servlet Technology Homepage, http://java.sun.com/products/servlet/

The PHP Group: PHP Homepage, http://www.php.net/, access: 2007-02-16

Development of a Software-based Environmental Reporting According to the "Bundes-Immissionsschutzgesetz"

Jan Goscimski

FHTW Berlin, Germany

Abstract: Subject of this work is a software-based reporting according to paragraph 19 of 13th "BlmSchV". Paragraph 19 calls for an annual reporting. The realization took place at a mineral oil company. On the enterprise's behalf, an automated reporting was to be introduced in order to support the person responsible for emission protection. The reporting is based on ecological data, located at databases within the company's intranet. The result of the implementation is a "Microsoft Excel" file. Various tasks have been realized through the usage of "Visual Basic for Applications".

1 Introduction

In the year 2004 the 13th regulation for the execution of the federal immission control law ("13.BImSchV") was adapted in accordance with European Union right amends. The "13.BImSchV" is the so called "large firing plant regulation". Therefore an annual report must be handed over to the supervising authority for all heating systems of an enterprise, which exhibit an authorized thermal output of more than (or exact) 50 megawatts. In this report the entire energy employment for the reporting period is reported. Additionally every three years an accumulated report is given to the authority. For the reporting, the regulation differentiates between plants of a mineral oil refinery and other heating systems. Report is submitted as a table with the fuel inventory for the reporting period and supplementing data of the individual plants. Essentially, this data contains the assigned fuels (in "ton per year" and "Tera Joule"), as well as the yearly emissions of the plants in SO_2, NO_x, and total dust.

For the reporting year 2005, the report above was still provided manually. Therefore, data was gathered from different sources. It became obvious that the majority of the data was already available in electronic form on the enterprise's intranet. However, it was stored at various places. Only little data, e.g. the authorized firing thermal outputs is available in written form only.

During the creation of this report, the question arose whether and how this report could be automated, particularly since this would be a substantial support for the person assigned to immission control of the enterprise in his tasks. Further research within the enterprise showed, that all necessary environmental data is deposited in the company's intranet and available in data bases. In accordance with legal requirements environmental data must at least be kept for five years.

2 Legislation in Germany

Superordinate in Germany is the Federal Immission Control Law (BImSchG). For the conversion of the Federal Immission Control Law regulations are issued. Such regulations serve in order to concretize the demands of the Federal Immission Control Law. In these regulations, the emission limit values attainable after the state of the art are specified among other things.

The 13th BImSchV "regulation on large firing and gas turbine systems" affects all plants with a firing thermal output of 50 megawatts (MW) or more, unless a plant is expressively excluded (this concerns in particular facilities which burn exclusively or partially waste – here the 17th BImSchV applies).

Reporting obligations are defined in the regulation, too. This paragraph is the basis of the written work at hand. The development of a software-supported reporting takes place because of the obligations specified here [Feldhaus 2006]

3 The environmental management system of the enterprise

For the accomplishment of the tasks of environmental protection and constant control of the emission data of all plants of the enterprise, the environmental management system "D-EMS 2000" of the company DURAG was installed. The abbreviation "DEMS" stands for "DURAG - emission management system".

With this system, all data of emission measurements on any computer workstation can be made visible within the enterprise. The data of the individual emission measurements are stored in the "Submaster Umwelt". With "D-EMS 2000", a multiplicity of evaluations of the data is possible. For example, the minimum or maximum values for any period can be determined for the emissions of any plant. Moreover, one has created an interface to "Microsoft Excel" in "D-EMS 2000". Over an ActiveX interface - out of "Microsoft Excel" - access to various functions of the "D-EMS 2000" is possible [D-EMS 2000].

4 Requirement analysis

In cooperation with the immission control-assignee of the enterprise, the necessary information was gathered. In first discussions the desires of the client were documented. The result covered less than a DIN A4 sheet. Based on that, first searches on the Intranet of the company and first sample inquiries of the environmental data were accomplished. In the following discussions and interviews, functionalities which could be developed were more exactly defined and a first layout was provided. This could then be "filled" with data for test purposes. In cooperation with the immission control-assignee, the final functionalities could be defined using the simple prototype in several iterations.

5 Layout and contents of the message

For the report's content, there are defaults defined by the legislation. For the layout of it, there are not. Legal demands to the content could partially be taken from the legal text. They were later concretized by the national environmental office. Upon the company's request, additional data is added to the layout; this may be data about CO and CO_2.

With the preceding search no suggestions were found to the layout – probably, because this report was to be provided for the first time in the year 2005. The layout was then set up in cooperation with the immission control-assignee of the enterprise.

The report for 2005 was sent to the responsible national environmental office in the finished layout draft. It served at the same time as a suggestion of the enterprise, in which form the message was to take place in the future. This was accepted by the national environmental office and since then the layout of the message is mandatory.

Berichterstattung nach §19(1) der 13.BImSchV
für den Zeitraum 01.01.0 - 31.12.0

Anlage	Nr.			FWL [MW] genehmigt	FWL [MW] IST	Brennstoffeinsatz Flüssig [t]	Flüssig [TJ]	Heizgas [t]	Heizgas [TJ]	Summe Energie [TJ]		Jahresemissionen SO2 [t]	NOx [t]	Staub [t]	CO [t]	CO2 [M]
IKS	0001	HES	B[1]	-	-			-	-	-	K[2]	-	-	-	-	-
		Visbr.R		-	-			-	-	-		-	-	-	-	-
		HSC-R		-	-			-	-	-		-	-	-	-	-
		CLO		-	-			-	-	-		-	-	-	-	-
	Σ IKS															
Raffinerie	0004	Rohöl 1	B[1]								K[2]					
	0006	Rohöl 3	B[1]								K[2]					
	0002	Mitteldruck	B[1]			-	-				E[2]					
	0013	Aromizer	B[1]			-	-				E[2]					
	0015	VT3/Desust	B[1]			-	-				E[2]					
	0023	Visbreaker	B[1]			-	-				E[2]					
	0023	HSC	B[1]			-	-				E[2]					
	0035	H2B-Anlage HG	B[1]			-	-				E[2]					
		DWA				-	-									
		Σ H_2B-Anlage				-	-									
	Σ Raffinerie															
Σ PCK																

[1] Anlagenstatus gemäß EU: N= Neuanlage
B= bestehende Anlage

[2] Art der Ermittlung: K= kontinuierliche Erfassung
E= Einzelmessung

Fig. 1. Layout of the message

5.1 The first section of the report

The first four columns describe the single facilities. In the enterprise, 9 plants are subject to the 13th BImSchV. The regulation differentiates for the reporting between plants of a mineral oil refinery and other heating systems. Therefore, the report is divided into the power station and the plants of the refinery. Finally, indicators of the entire enterprise are shown. For the power station the possible (liquid) fuels are listed, while the individual plants are represented for the refinery. The plant status is proven in accordance with the European Union guideline 2001/80/EG. A distinction takes place between new and existing plants. New installations in the sense of this regulation were approved after 1 July 1987, existing plants before 1 July 1987.

5.2 The middle section of the report

The firing thermal outputs "FWL", individual fuel consumption, and the heat capacity belonging to it are provided here. The firing thermal output is additionally divided into authorized and actual firing thermal output during the reporting period. The authorized FWL is fixed in the note of approval of the respective plant. The actual FWL can be computed in a good approximation from the flue gas volume of the plant. For the fuel inventory between liquid and solid fuels one differentiates: those are indicated in "tons" and are obtainable from the "DURAG" system". The heat capacity belonging to it can be computed with the respective heat values of the fuels.

5.3 The third section of the report

In the last part of the message the yearly emissions of the single facilities are provided. The listing contains data on SO_2, NO_x, dust, CO and CO_2. It further indicates whether the values are determined by continuous measurements or on the basis individual measuring. The computation of the emissions with individual measuring takes place within the "DURAG" system. All emission data can be obtained from the "DURAG" system.

6 The conversion of the concept

6.1 Application layer

All necessary data can be accessed out of the "DURAG" system. Based on the layout some necessary computations and entries result. The authorized firing thermal outputs are firmly stored in the report. The actual firing thermal outputs are computed from the flue gas volume. In addition, via an inquiry, the average value of the flue gas volume is determined and handed over to "Microsoft Excel". Subsequently, this value is divided by "1,000". The result indicates the firing thermal output in "megawatt" (MW). The fuel inventories (liquid and gaseous) can be queried easily from the data bases of the "DURAG" system. For some plants the fuels are only individually provided for the respective furnaces, the summation is then done in their own table within the "Microsoft Excel" file.

The amounts of heat must be computed, but the heat values of the assigned fuels as well as the amounts of fuel are necessary, too. The heat values cannot be enquired; they are obtained by laboratory analyses in different time intervals. They are only available on paper. They were therefore stored in their own table within the "Microsoft Excel" file in order to be able to maintain them manually. For the computations, the application goes back to this table.

The yearly emissions within the third section of the report are present in the "DURAG" system and are selected from there. The emission values of the entire plant are crucial. If a plant consists of several furnaces, then these are supervised individually and the plant values are determined in the emission computer via simple addition. For the determination of the pollutant concentrations a weighted average of the individual furnaces is used.

The kind of the determination (continuous or individual measuring) is stored in the layout and can be changed manually if necessary. The locking accumulation for the assigned energy of the reporting period can take place within "Microsoft Excel". A further table sheet contains the necessary computations. The sums determined there are proven in the finished report.

6.2 Data layer

The local emission computers locally in the plant hand the raw data to the Submaster "Umwelt". To prevent possible interruptions of the data linkage, the raw data is stored up to 7 days in the emission computer.

In the Submaster, all data for the past 5 years is stored on non-removable disks (in accordance with § 15 of the 13. BImSchV). From here, the connection to the network (Intranet) of the company is realized. From the Submaster, the data will be transferred to the server, where they are stored a second time on non-removable disks. The non-removable disks in the server are then copied again. Thus, three-way storage of all environmental data takes place.

6.3 Presentation layer

After activation of the "Microsoft Excel" file, the opening side automatically appears. It contains short explanations to the program as well as two buttons, where one selects whether the report is to be done for one year and the other for three years.

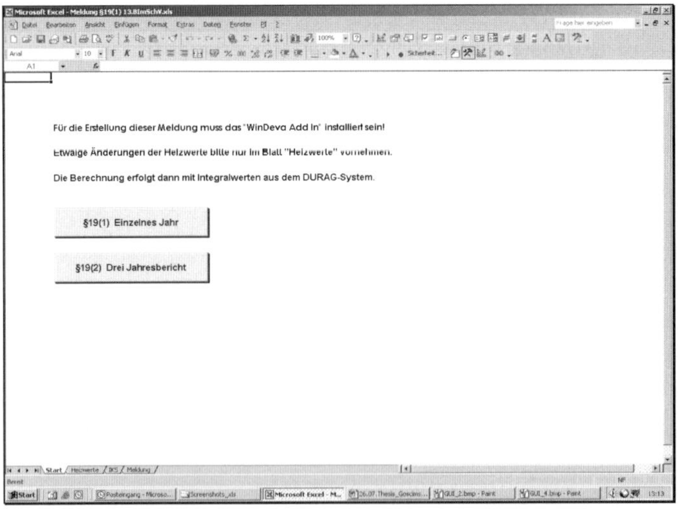

Fig. 2. The opening site

After this selection, a dialog field appears, into which the year of report (for the annual bulletin), and/or which is to be registered first year of the report (for the three-annual report).

If a report year was selected, the exact selection of the period takes place. This takes place again within a dialog field, in which the period "01.01.20xx - 31.12.20xx" is already given, yet can be changed individually if desired. Here arbitrary periods are selectable, which includes individual days, too. It is checked internally whether the period for the annual report is smaller than 366 days (and/or 1,096 days with the three-annual report).

Afterwards, automatically the selection and calculation of the data in the background take place and the announcement of the results within the finished report on the screen. If the finished report is indicated, the user can select the kind of printing. In addition, again two switching surfaces exist, one for "total minutes" and one for "authority minutes".

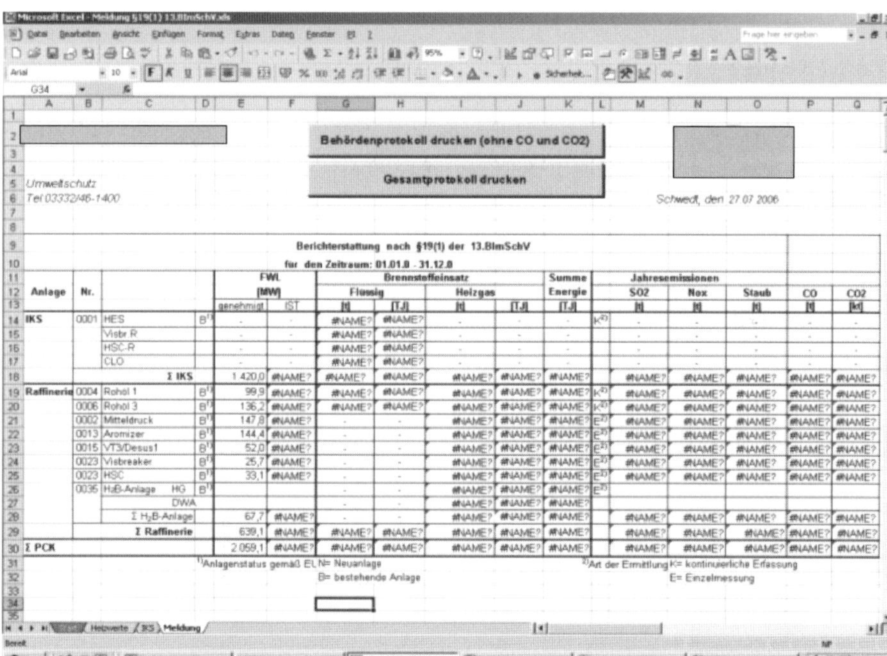

Fig. 3. The finished report

In "authority minutes", the columns for CO and CO_2 are not given. On the other hand, in "total minutes" the whole content of the report is printed.

7 Conclusion

By using the existing "Microsoft Windows" - and "Microsoft Office" environment, all desired functions could be implemented. During further research in the intranet of the enterprise, applications in "Microsoft Excel" were found, that could partly be reused. The integrated development environment "Visual Basic for Applications" (VBA) made all necessary methods available for desired functions. On the environmental server, the computation of the values takes place, as far as possible.

One purposely neglected the development of an independent user surface. The inputs of the user are limited to the input of the reporting year (respectively the first year of a three-year-report) and the selection of the desired print. Programming took place in individual modules. Thus individual parts and whole modules, too, of the software can be used later for other projects again.

References

„D-EMS 2000“ Owners Manual. DURAG Company
Feldhaus (2006) BImSchR. C. F. Müller, Heidelberg

Environmental and Safety Management Support

Thomas Wetzel, Nabil Allam

FHTW Berlin, Germany

Abstract: The legal conditions of industrial facilities that protect environment and staff have been improved in the last years. Independently of the type of enterprise, the appropriate regulations, conditions and documentations can hardly be managed. Under practical conditions these tasks are part of safety, health, and environment management. In 2005 a software prototype was developed at the University of Applied Sciences (FHTW). This prototype used simple technologies (e.g. Access as DBMS and a Microsoft Visual Basic 6 front-end) also contains the substantially functional requirements such as handling hazardous material. A research project was started in 2006 at FHTW that endeavors to realize encompassing environmental information system on the basis of the prototype. The usage of modern web-technologies is a special focus of data handling such as java server faces with hibernate. However such technologies are not only associated with advantages, especially the access to database via hibernate has to be used with caution to avoid data defects.

1 Motivation

The appropriate environmental laws, regulations, and conditions those are relevant to the enterprise constantly change. Further more these laws are important tasks and conditions which are required for certification or by other institutions such as insurance. Not only the administration of such tasks is problematic, but also after few years the growing number of documents becoming increasingly difficult to handle. Accordantly environmental management in enterprises is not separated field. The management delegates are confronted with tasks of safety, health, and environmental management. For an optimal solution of the tasks mentioned before enterprises have to centralize the trans-sectoral data supported by software solution. One possible software system could be an operational environmental system. An operational environment system is a system for storing, documentation, planning, and controlling environmental impacts [Rey 2005, p 3].

The three topics (environment, health and safety) are important for the entire enterprise structure. They cannot be related to a certain operating level or group of staff, but all areas of an enterprise are affected by them.

Those tasks could be divided into following sections:

- Licenses and conditions are a collection of legal regulations for facilities and machines. Besides the regular laws exist different environmental laws which are important for the enterprise. The Technical facilities such as waste utilizing and sewage disposal are especially affected.
- From a certain size of an enterprise, the legislature requires that staff has to take part on training regularly. This includes environmental training. Other institutions like insurance are interested in training, too while the management of such training is difficult to handle. The Staff has to be trained in regular intervals without interferes with everyday business activity.
- All laws and regulations that are relevant for the enterprise and have to be accessible to the staff are collected in a register of laws. If the enterprise is certificated by EMAS II, all laws and conditions have to be accessible. Further all laws and conditions have to be regularly updated [EMASII 2001].
- If hazardous materials are used by an enterprise, a register of hazardous materials is necessary by law. Important data are place of storage, place of utilization, quantities and classification of danger for human and environment. The legislature requires operating instruction for the used hazardous material, which have to be accessible to the staff.
- Parallel to register of hazardous materials, operating instructions are also required for machines and dangerous facilities.

The points mentioned above are partly required by legislature or necessary for different certifications. Such tasks have been solved by 90%, with Office applications (Germany and Austria). This causes different problems:

- Data redundancies cause data defects and needs more time to be managed.
- Centralization might be hardly achieved.
- Topicality of data is not able to achieve.
- Data security and data safety cannot be ensured.
- Delegation of tasks is hardly possible because of difficult structures.
- Substantially higher resources consumption.

The idea was to develop an operational environmental information system, which supports different business structures and solves the mentioned tasks entirely. The system should be a client/server system. A user that has appropriate rights can get all information important to him by a client. The software was built modular to achieve the extendibility.

There are additional important aspects to be considered during the design process of an integrated software system. Among these aspects is the fact that different IT information systems already exist in enterprises. Those systems contain part of the data that are necessary solving the problems mentioned above. The software connects to the existing databases by an interface and requests the necessary data.

2 Software Overview

The software was built modularly to achieve extendibility and adjustability to different problems (see Figure 1). Therefore it is possible to extend with further modules or to use only a selection of modules. Based on tasks that are mentioned in chapter 1 (Motivation), the following modules were developed:

- Licenses and conditions
- Register of laws
- Training register
- Register of hazardous material
- Register of facilities

In addition to the five basic modules, an administration module (see Figure 1) for managing the system is exist, e.g. adjusting the structure or setting access authorizations.

The software system models the structure of the enterprise making it transparent and easy to use by all employees. Hence the managing tasks depends no more on a single person.

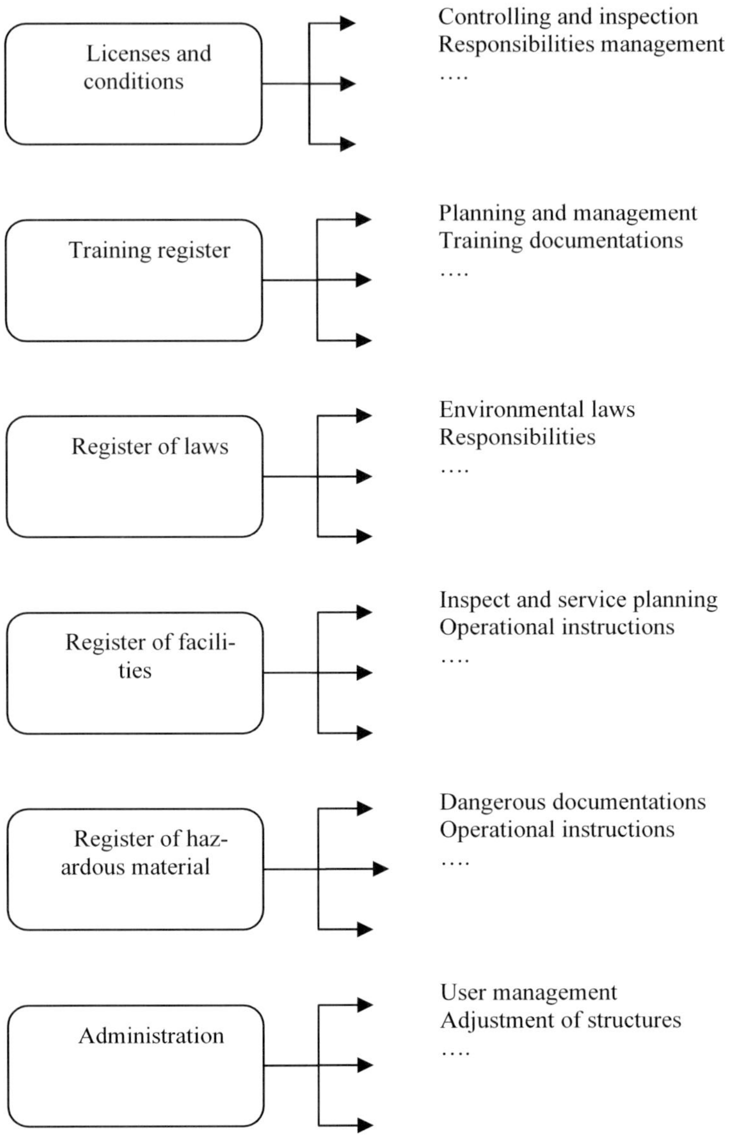

Fig. 1. Software modules

2.1 Licenses and conditions

The module licenses and conditions are arranged according to licenses, conditions, and inspections. Particularly licenses and conditions of the

German emission control law are stored here. Many conditions are stored relative to one license and it is possible to store many inspections relative to one condition. After continuous storage of all licenses, it is possible to observe the appropriate conditions. The early recognition of dates helps to describe the responsibilities, minimize the safety risks, and avoids penalty. It is possible to store permanently the results of the inspections.

2.2 Register of laws

The register of laws is required for different certifications. The frequent use of register of laws is by managements delegates. At the same time, the register of laws enables a relation between environmental management documents and protection of labor and security. The laws that are listed in the environmental management documents are explained in details here. This register contains all environmental laws for the enterprise such as:

- Emission protection
- Water protection
- Soil protection
- Chemical laws

The laws are arranged according to these categories.

2.3 Training register

The legislature and other institutions like insurance companies require their staff to attend on training regularly. a prove of training is required for different certifications. The important data for training register are:

- Staff data such as personal data and a list of training courses attended.
- Training data such as costs, places, trainers, contents, and lengths.
- Date management for the training to avoid the absence of all employees at the same time.

Using a reminder button, the responsible person can access the next important dates. An analyses tool shows the employees that still have to take part on the training.

2.4 Register of hazardous materials

The register of hazardous material is required for different certifications by law. To achieve that, the following items should be stored:

- Hazardous materials that are used in the enterprise like solvents.

- Material data sheets such as operational instructions.

The register of hazardous material is arranged according to section, facility, and buildings of storage.

2.5 Register of facilities

The register of facilities is required for different certifications. It contains an over-view of all facilities and machines of the enterprise with appropriate operational instructions. The register of facilities is arranged according to section, facility, and buildings, too. The responsible person can see the next dates for inspections and services.

3 Implementation

Based on the requirements of operational information systems, it is possible to de-rive requirements for operational environmental information systems. Accordingly the single user or user group should get access to a common database. Therefore the software architecture was implemented as web service architecture (see Figure 2).

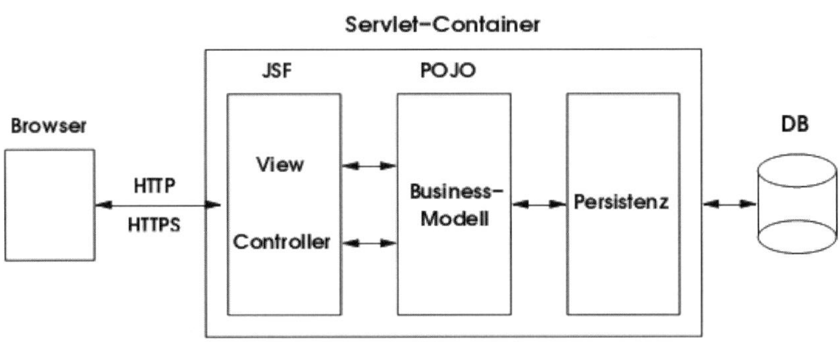

Fig. 2. Software architecture [Müller 2007]

The system is realized with the platform J2EE by the JavaServer Faces (JSF) Framework. JavaServer Faces is a Java-based Web application framework that simplifies the development of user interfaces for Java EE applications. JSF manage the communication with the browser (e.g. Firefox, IExplorer, Opera etc.) by using HTTP.

The business model is realized by java-classes which access the tables of a relational database through persistence service. Since persistence service is using Hibernate. Hibernate is a powerful, high performance object/relational persistence and query service. Hibernate allows you to express queries in its own portable SQL extension (HQL), as well as in native SQL, or with an object-oriented Criteria.

A served container is necessary for using a JSF application. So Apache Tomcat is used which is a web container developed at the Apache Software Foundation (ASF). Tomcat implements the server and the JavaServer Pages (JSP) specifications from Sun Microsystems, providing an environment for Java code to run in cooperation with a web server. It adds tools for configuration and management but can also be configured by editing configuration files that are normally XML-formatted. Tomcat includes its own HTTP server internally.

According to the CMV-model (control, model, view) (see Figure 3), the software was split into three components. Besides the CMV-model other techniques were used to improve the software development qualitatively, e.g. a name convention was fixed for the whole software-project that was partially based on the so-cal led "Hungarian Notations".

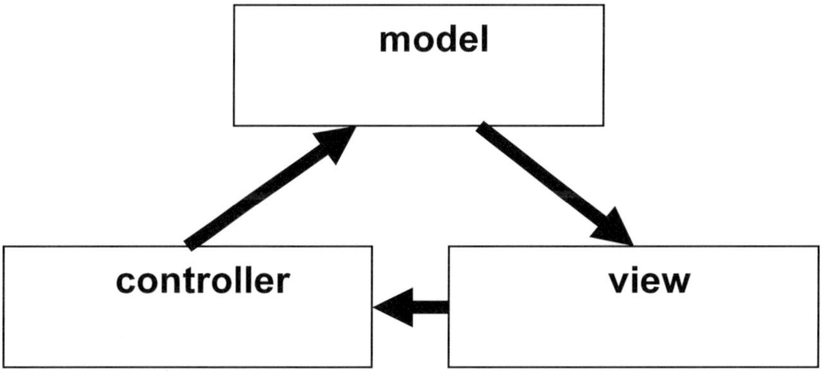

Fig. 3. MVC-Modell [Hansen 2001, p 1]

The database was built according to the rational database model. MySQL was used as database system. However the software provides the option to use any sys-tem of data management. To accomplish this, only the database driver has to be adjusted to the particular system.

4 Costumer's Benefits

As already mentioned, the software solution centralizes and structures several tasks. Redundancies of the data can be prevented by the use of normalized rational databases. The environmentally relevant information is made accessible any time to all staff of the enterprise. The information supply in the enterprise is improved. The Up-to-datedness of information is also provided, which gives a comprehend view of important dates. Data safety and data security are realized by using access authorization. By using a uniform system structure, every employee can solve tasks much faster. Eventually this realizes staff and helps to concentrated on the business. Therefore the resources (human and material) can be conserved.

Worse accessibility of information and unknown responsibilities are main reasons for dates being missed in enterprises. The software helps to eliminate these reasons.

So the system is platform independent when using modern web technologies like Java and JSP.

5 Final Result

Special focus of data handling is the using of modern web-technologies. However such technologies are associated with disadvantages too. Especially the access to a database via hibernates which have to be used with caution to avoid data defects. In this context the Hibernate session handling has to be considered to achieve the up-to-datedness of the persistence objects. The persistence objects will not be al-ways up-to-date during the development process. Changes at the database are sometimes not recognized by the persistence objects. This problem has to be solved in the future.

Furthermore the integration of existing enterprise information systems must be considered and strategies of integration to be developed. Such integration should be detected with small possible changes.

Largely the software gets positive resonance. The software has become problem oriented with the development process held close to the daily problems of enterprises. Also important dates cannot be missed with the help of the reminder functions,. The software built with a uniform structure, which helps the staff to use the software intuitive.

In the software management analyses tools are still missing, these tools help to generate in overview of environmental situation in enterprise e.g. a detailed balance of hazardous material. There are still some common enterprise structures that the software cannot model appropriately, yet. The software should be more flexible for those cases where (The interfaces are enterprise-type specific).

References

Rey U (2005) AEB I Summer semester 2005
Hansen H (2001) OOP Struktur objektorientierter Programme
EMAS II 2001, http://www.emas.de/, access: 2007-02-15
Müller B (2007) http://www.jsfpraxis.de, access: 2007-02-15

Development and Implementation of a Software to Accelerate Data Acquisition as Part of a Corporate Environmental Management Information System for an Energy Company

Mathias Zirkelbach

FHTW Berlin, Germany, e-mail: zirkelbach.mathias@web.de

Abstract: In the present paper I will discuss a way to simplify the acquisition of a fuel data capturing software as part of an industrial environmental in-formation system for an energy producing company.
A lot of data is directly or indirectly linked to financial matters such as gains, costs, tax refunds, fees and necessary CO_2 certificates. Numerous changes in the latest regulations and laws make it necessary for the energy producer to adjust his power stations [cp. Krimmka 2006].
If a system's emission is not up-to-date, sections are rebuilt. If this is impossible, a new power station needs to be build. The available soft-ware often is not able to meet the increasing amount of demands. New possibilities for the fuel data capturing make it necessary to adjust the software in short time. Even though the software is adjusted to the production, there are also several programs to be considered which are subject to the data evaluation. These programs do not have import inter-faces but are manually provided with data. The annual stocktaking, for example, is completely carried out by hand. Its data is saved in excel.

1 Intention

The increasing demand for data and the available time-frame make new software necessary. The software's aim is defined clearly. Useful functions of the FORTRAN predecessor are to be migrated.

Thus, the present basis of the fuel data capturing software is migrated. Furthermore, possible upcoming changes should be easy to implement. This lead is the decision to use a programming language which provides modular adjustment.

2 Realization

The first step was to analyze the technical documentation of the energy producer. It was important to capture fuel-, factory-, environment- and production relevant data. In a second step, the hardware of the technical information system and all affiliated software was analyzed.

The old program (programmed in FORTRAN) which used to capture the fuel data, has been partly migrated and has been extended with new functions in VB.net. Visual BASIC.NET is a modern 100 percent object oriented language [Martin 2002].

Missing functions such as “fuel inventory", "delivered fuel" and "SAP entry" were programmed and integrated in the menu of the daily report.

In the following, the most important parts of the program are described in detail.

The fuel data capturing software starts by opening a window (see Fig. 1) with a menu to choose from for processing the fuel data.

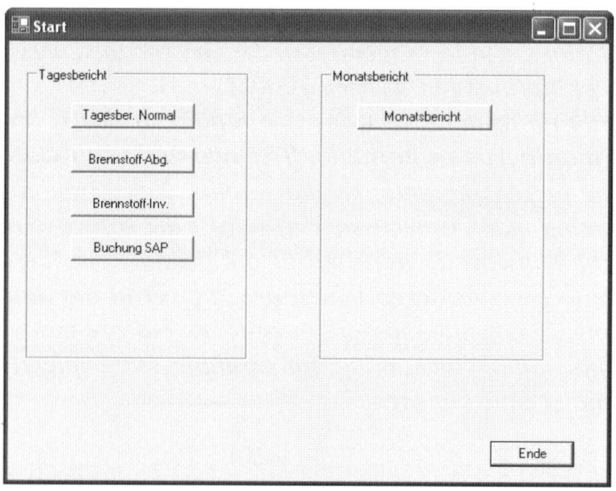

Fig. 1. Menu of the daily fuel report

The daily report offers four important functions: "normal daily report", "delivered fuel", the "fuel inventory" and "SAP entry". They can be called up separately. First, the normal daily report was roughly outlined. It has all necessary functions and modules including the daily changes in the fuel inventory of all power stations. A new fuel delivery therefore results in an increase of the fuel stock. The daily use of fuel, as a countermove, decreases the available stock. All changes in the available fuel stock have an impact on the daily calculation of the day's report. Thus, the actual fuel stock can be determined on a daily basis.

If the daily-report-button pressed the treatment date is to be put down as the first by the user. The user can choose only one date what lies in the past. For the next steps this date is stored temporarily. Therefore the possibility is given the treatment date from every place in the code for a comparison. For the precise calculation of the topical fuel data the processing of the respective day before must have occurred. Only so correct calculations can be carried out. Therefore, there is a module (`mod_Vortag.check`) which checks the completeness of the fuel supplies of the day before. Should data of the day before be absent, a program termination is forced and an error message is given to the user. The topical consumption and accesses of fuels becomes transmitted from the power stations by files ASCII, in the form of figure columns to a server. With the help of another module (`mod_allASCII`) was checked whether all power stations have transmitted her data of the day to be processed. Should a power station not have transmitted his data, an error message with information of which power station the data appears to the user are absent?

These figure columns have a special pattern which they define. After this pattern reading in the single files (see Fig. 2) occurs. The figure columns are disassembled by the parameters standing in clips accordingly and are transferred with the help of several inserts into the suitable tables on Oracle DB.

```
'strings per line must be selected
        Datstzschl = line.Substring(0, 5)
        AnlageTa = line.Substring(6, 3)
        DatumTa = line.Substring(9, 6)

    ' Query whether date in ASCII always correctly
        If DatumTa <> DatumWdl Then
          MsgBox("falsches Datum bei Anlage " & AnlageTa)
           End
        End If
         STD = line.Substring(16, 6)
         EFE = line.Substring(22, 6)
         EFL = line.Substring(28, 6)
         STI = line.Substring(34, 6)
         RBK = line.Substring(52, 6)
         HES = line.Substring(58, 6)
         HEL = line.Substring(64, 7)
      IEG = line.Substring(71, 7)
```

Fig. 2. Cutting of a code to process the figure columns

After the processing of the first two parameters (Datstzschl, AnlageTa) the selections of the date follows. This is important for the capture because the number of the scheme (AnlageTa) and the date together form the primary key by which guarantees the definiteness of the record. For this reason the date is compared explicitly to date of execution. In case of an error, an error message appears to the user. If all figure columns are selected with the correct date, the actual stock is calculated. The cumulativeness as well as the addition and subtraction of the fuels now follow. Afterwards the actual stock is transferred on the DB with the help of an insert.

```
    'DELETE für VERFEUERUNG
      Dim delStrVERF As String = "Delete from
MONFOR.T_PRAKT_TA_VERFEUERUNG
      Where Datum ='" & Datum & "' and Anlage =" & AnlageTa
      Dim delcmdVERF As New OleDbCommand(delStrVERF, OracleConn)

    'INSERT für  VERFEUERUNG
      Dim insertVERFEUER As String = "insert into
MONFOR.T_PRAKT_TA_VERFEUERUNG
      (DATUM, ANLAGE,WERK, STD_BRST, STI_BRST, RBK_BRST, H" & _
       "ES_BRST, HEL_BRST, IEG_BRST, EFE_BRST, EFL_BRST, SORT) VALUES
      (to_date(?,'dd.mm.yyyy'),?, ?,?, ?, ?,?, ?,?, ?, ?,?) "

     'MsgBox(insertVERFEUER)
       Dim insVERF As New OleDb.OleDbCommand(insertVERFEUER, OracleConn)
```

Fig. 3. Code cutting for the proceeding delete and an insert

However, before the insert-order can be executed, it has to be ensured by a Delete that there exists no data on the DB for this record yet (see Fig. 3). Calculating the amount of fuel occurs on a daily basis until the end of the month. If a month' end is reached, the data of the last day of the month will be transferred into another chart (month' chart).

At the end of this rough outline a detailed test of the system is executed in which every day of the last three month is captured by the new software. A comparison between the data captured with the old software and the new data derived by the new software was carried out success-fully.

In the fine draught even other functions have been complemented to the "daily report normally". The function (`fkt_MOselHEIZW.select1`) selects from the Oracle DB heat values of the previous month. If no heat values exist from the previous month, average year values are read in. This can be the case if in the previous month in a power station a certain type of fuel (e.g., fuel oil) has not been burnt. These average year values are firmly integrated in the program code, however, can be changed when required. In the Figure 4 are shown cutting-wise the two power stations with her heat values. Besides, the abbreviations STD (coal german), STI (coal import), HEL (fuel oil extra-easily), HES (fuel oil Hard) mark the fuels [cp. Dern 1960].

```
' Heat values: work-related annual averages of the year before (2005)

    ' Heat values of the first power station
       Dim IHU_STD_MO As Integer = 28437
       Dim IHU_STI_MO As Integer = 28000
       Dim IHU_HEL_MO As Integer = 36279

    ' Heat values of the second power station
       Dim IHU_STD_RW As Integer = 28854
       Dim IHU_STI_RW As Integer = 26916
       Dim IHU_HES_RW As Integer = 41350
       Dim IHU_HEL_RW As Integer = 36189
```

Fig. 4. Code cutting for the average year values of the heat

From the daily burnt amounts of fuel and the heat values of the previous month the calculation of the everyday CO_2 issues occurs. These are calculated at the moment after two different procedures. The first one is the Brand/Baehr procedure which is not a contemporary, but it is still co calculated. The second one is the emission trade procedure which is carried out by the guidelines of issue trade. In the following picture a CO_2 calculation is carried out by the guidelines of issue trade for coal. The task an inter result has to store the variable `STK_U`. If the inter result has occurred greater one a calculation, then it is filled the variable STK_CO_2 with a zero (see Fig. 5).

```
' Calculation of the CO2 amounts by guidelines of issue trade

        If (STK_U > 1.0) Then
          STK_CO2 = ((0.054829 + (STK_U * 0.023736)) / STK_U) _
               * (44 / 12) * (STK_BRW)      ' [t]
        Else
          STK_CO2 = 0
     End If
```

Fig. 5. Code cutting for calculation CO_2

The results consist are likewise written with the help of two inserts (`fkt_TAinsBBCO2.insert` and `fkt_TAinsEHCO2.insert`) into a DB. To be able to check all expiries for mistakes, it is explicitly grasped with the help of a log file (`mod_LOGwriter`) everybody select, delete and insert on the DB and is stored in the log file by name and date. Two forms, 11 modules and 13 functions are necessary for the whole "daily report normally". The preliminary design draught became carried out for the "fuel of delivery", for the "fuel of stocktaking" and for the "reservation of

SAP". The next system test has been cancelled on account of a lack of time. Therefore a development of the fine draught was immediately carried out in the connection. With the "fuel of delivery" additional deliveries are processed by fuels. Additional deliveries are, e.g., cleanings of the oil tanks. On this occasion usable remaining stocks (waste oil) with the mud itself mature in the oil are not pumped off and are de-contaminated. For the additional delivery a form which realizes the "fuel of delivery" with the help of a datagrid and three functions was provided. In the following Figure 6 the fuel delivery with datagrid and the matching fuels is shown. The choice of the power plants was realized about a combo box. About the datagrid the amounts of the fuel which have originated from the delivery are grasped.

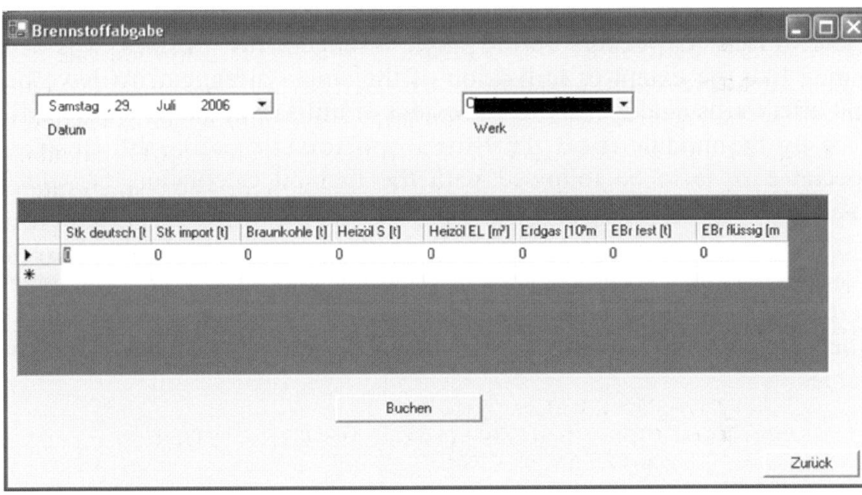

Fig. 6. Form for the fuels

With the registration of the amounts the inputs of the user are to be checked explicitly. If no figures are given in the datagrid which corresponds to the predefined parameters, the user receives a tip. If all inputs are correctly affected, an insert occurs while operating the button "Buchen" into the DB the work about the function (`fkt_insABG.insert`). For the new continuance calculation the topical continuance from the DB with the function (`fkt_TAselBES.select1`) is read in. Then with the help of a subtraction the topical continuance and the amount of the delivery are settled with each other. Afterwards the result from the calculation will again transfer about an insert function (`fkt_TAinsgBES.insert`) into the DB. The inventory likewise belongs to the daily report. Once a year all fuel data are determined by a foreign company precisely and a comparison with the supposedly available continuance data which are on the DB is carried out. The inventory is built up as the delivery and likewise be-comes about a form which a datagrid contain realized. With the inventory some calculations which are accommodated in two modules and three functions are carried out. All fuels supplies ascertained by the inventory are put down on a datagrid of hand and are transferred about the function (`fkt_JAinsBESi.insert`) into the DB. To be able to determine the differences which lies between DB-data continuance and the ascertained data continuance of the inventory, the topical continuance about the function (`fkt_TAselBES.select1`) is read in by the DB. After the difference was calculated and was written in the DB with the function (`fkt_JAselVERFi.select1`). Afterwards a subdivision of the stocktaking differences occurs on the single arrangements. Besides, it is determined like the extent of utilisation of the single arrangements have been and afterwards analogously to the extent of utilisation the inventory difference by the module (`mod_Invent. calculation`) split. Besides, a special case is to be followed with the fuel oil calculation. Should the value of the inventory difference be smaller one, no subdivision on the arrangements occurs more (see Fig. 7). In the first condition (if query) it is checked whether the variable `HEL_verbI` are smaller one and greater zero. If this condition is fulfilled, it comes to a special selection on the DB. Then the selected value is assigned to the arrangement with the highest value.

```
    'Wenn die Verfeuerung kleiner Eins ist, dann wird Diff. der Anlage mit der
höchsten Verfeuerung zugeordnet
    If HEL_verbI < 1 And HEL_verbI > 0 Then

        Dim selectVERFaKlEins As New OleDbCommand("select * from (
select
              anlage, sum(HEL_BRST)" & _
          " from MONFOR.T_PRAKT_BE_VERFEUERUNG" & _
          " where werk =" & WerkInv & "and datum like '%" & yy & "'" & _
          " group by anlage" & _
          " order by 2 desc)" & _
          " where rownum=1", OracleConn)

        dr4 = selectVERFaKlEins.ExecuteReader
        kumulierte Verfeuerungsdaten werden Anlagenweise gelesen
        With dr4
          While .Read()
             AnlageKlEins = .Item("Anlage")
             HEL_verfla_TempSUM = .Item("sum(HEL_BRST)")

          End While
        End With

          OracleConn.Close()

    End If
```

Fig. 7. Code cutting for inventory difference smaller 1

As soon as the subdivision is concluded, the split differences are written with the help of the function (`fkt_JAinsVERFI.insert`) into the DB. For other examinations, which happened mainly in Excel, are transferred in addition still the continuance before the inventory to which continuance is stored away after the stocktaking, the inventory differences and the subdivision of the differences on the single power station arrangements to Excel. The last part of the daily report is the "reservation of SAP" in which two modules are used. A txt file, which contains all relevant re-cords for SAP, is generated. For the transmission of the file the second txt file which gets all necessary address parameters for the file transfer (FTP) cultivated serves (see Fig. 8). The second txt file is provided under the name FTPCOMMANDS.txt.

```
Dim writer As StreamWriter = File.CreateText(PfadSAP &
"FTPCOMMANDS.txt")

        writer.WriteLine("open biztalk1.corp.vattenfall.de" & (vbCrLf) & _
                "EUR\ftpbewageai" & (vbCrLf) & _
                "transferit1$" & (vbCrLf) & _
                "lcd S:\Tagesbericht_SAP" & (vbCrLf) & _
                "cd BTS_FTP\Input\BW_WAERME_FL_Verfeuerung" &
         (vbCrLf) & _
                "put SAP1_20" & yy & mm & ".TXT" & (vbCrLf) & _
                "bye")
        writer.Close()

        Call mod_Batch.SAPsend()
```

Fig. 8. Code cutting of the address parameters

If all necessary parameters are put down in the file FTPCOMMANDS.txt, the delivery occurs to a special server with the call (`Call mod_Batch. SAPsend`). SAP can access this server independently which reads in file and process. Finally all single parts (modules) become the "fuel daily report Normal", the "fuel of delivery", the "fuel of inventory" and the "reservation of SAP" on her function tested. The explicit test scenarios, which were developed only for the fuel data processing, were necessary for this test. Every exceptional case was checked on his precise function. A whole decrease test of the software occurred with the help of original data from the production [cp. Balzert 1998].

3 Results

The new tool, which was programmed in VB.NET, realizes a more convenient and quicker working with combustible data, combustible inventory and also the combustible delivery. The combustible data is used to calculate the burned amounts of fuel and combustibles. The values are provided to the stocktaking of the supplies.

4 Conclusions

As a result of the herein described project, a software tool was provided to the energy producer, which solved a couple of problems. First, the necessity for manual and redundant data input was reduced. Second, several aspects for example stocktaking, have been basically automated.

Besides of the provided value for the energy producer, the project was within time and quality and thus fulfilled the main targets for IT-projects.

References

Balzert H (1998) Lehrbuch der Software-Technik, Band II Software-Management, Software- Qualitätssicherung, Unternehmensmodellierung. Heidelberg, Berlin

Dern E (1960) Zeitschrift der Vereinugung Deutscher Elektrizitätswerke – VDEW Kennzahlen des Kraftwerksbetriebes bei Kraft-Wärme-Kupplung. Berlin

Krimmka T (2006) IV Ablauforganisation/Immissionsschutz. Intranet: http://www.evu/ablauforganisation/immisionsschutz.de, access: 2006-07-06

Martin R (ed) (2002) Einstieg in VB.Net. Bonn

Author Index